Analysis of
Vertebrate
Structure

Analysis of Vertebrate Structure

MILTON HILDEBRAND
Professor of Zoology
University of California, Davis

Illustrated by
Viola and Milton Hildebrand

JOHN WILEY & SONS
New York London Sydney Toronto

Library of Congress Cataloging in Publication Data:

Hildebrand, Milton, 1918–
Analysis of vertebrate structure.

Bibliography: p.
1. Vertebrates—Anatomy.   2. Morphology (Animals)
I. Title. [DNLM: 1. Anatomy, Comparative.
2. Vertebrates—Anatomy & histology. QL805 H642a 1973]

QL805.H64   596′.04   73-11486
ISBN 0-471-39580-3

Printed in the United States of America

10 9 8 7 6 5

Preface

More than a dozen textbooks of comparative anatomy are on the market. Some are very good. I do not presume that I could do better than the authors of those books in presenting the same material in a comparable way. Our subject being a large one, there can be various approaches, and this book is distinctive for its combination of characteristics.

A first overall characteristic is breadth. Description of structure is included, as it always is, because knowledge must precede interpretation. Interpretation of structure in terms of phylogeny is included, as is usual, because organic evolution is one of the greatest stories biology has to tell, and the lineages of vertebrate animals illustrate the story with more continuity and persuasion than do the known lineages of other animals or plants. Also included, however, are interpretations of structure that relate to function, development, and some other factors.

A second characteristic of this book is its placement of emphasis. Phylogeny is stressed, yet is less dominant than in some other texts. More than usual emphasis is given to interpretation of structure on the basis of function. This is currently the emphasis of much research, and thus provides the opportunity to add recent advances to classical knowledge. It brings to attention the variety and perfection of vertebrate structure, and it lends itself to analytical treatment to which, in my experience, students respond with interest.

Moderate attention is given to the evident and engaging relationship between development and adult structure. Developmental biologists now place less emphasis on the broader aspects of evolution than they did several generations ago, yet much excellent work is being done on relative growth and evolutionary morphogenesis. Aspects of development receive special attention in Chapters 1 and 5, but otherwise the subject is integrated rather than isolated in a separate chapter. Interpretations of structure based on body size, age, sex, and individual variation round out the narrative, but are given little emphasis.

v

A third characteristic of the book is its style and coverage. I intend that the presentation be sound and solid, yet not overly technical. The book does offer a lot of material. However, description as an end in itself is minimized — particularly in Part III. Details that do not serve interpretations are omitted. The qualifying words "usually" and "sometimes" are used frequently for the sake of accuracy, but specific exceptions to usual structure are "usually" omitted. A reasonably full vocabulary of the basic terms of morphology is provided without introducing unusual terminology or saying in Latin what can better be said in English. An effort has been made to sort concepts from illustrative material, and free use has been made of parenthetical statements to subordinate the examples and qualifications.

Fourth, this book presents vertebrate morphology as a living discipline. The transformations from fin to limb, from jawbone to ear ossicle, from branchial artery to carotid circulation, and many more, are classic stories which should be retold to new generations of students, yet note is also made of recent studies, unsolved problems, tentative explanations, active areas of research, and current trends in the discipline.

Finally, it is hoped that this book will be found interesting. No field of study as large and complex as this one is likely to interest every student in all its aspects. Application will be needed: The book is not intended as light entertainment. Nevertheless, it is a tragedy that the teaching of comparative anatomy has sometimes been a parade of dull facts, dusty skeletons, and much-preserved specimens. Nothing else in nature has more exquisite structure than the vertebrate body. I will be pleased if the reader occasionally forgets the forthcoming examination and reads on thinking "Wow, that's really something!"

Part I of the book is a survey of the vertebrates. Students must be able to recognize and relate the major taxa in order to follow Parts II and III. Brief descriptions that stress typical features and recognition characters serve as preparation for these parts. Extinct groups are included or not, according to their relation to what will follow.

In Part II, the customary organ system approach is used to present the general structure of the classes and subclasses of vertebrates and to review the structural evidence for their evolutionary relationships. Features that do not characterize major taxa or do not show progressive change between successive categories are deemphasized or omitted. The treatment is less detailed than in several other texts, yet includes, I believe, as much "meat" as can be learned in one course of study. Many students think that teachers of anatomy tend to pack into their courses more detail than will be useful. The proportionately few students who will become professional morphologists can easily find supplementary information elsewhere.

Part III presents knowledge and analysis of the major functional groups of vertebrates. Following two chapters on bone-muscle mechanics, successive chapters consider the major locomotor and feeding adaptations in order. Unrelated, often convergent, groups of animals are taken together to see how evolution has provided for their common requirements. Extreme modifications (which would be distractions in Part II) are included.

Of the nearly 1000 separate drawings comprising the illustrations in this book, about 64% are completely original, about 20% are largely original, and the remainder are all redrawn with some modification. We acknowledge our debt to more than 150 authors whose illustrations have been used to a greater or lesser extent as a guide. Twenty-eight drawings are based on photographs in "Mammals of the World," by E. P. Walker, 18 drawings are taken at least in part from one or another of the books by A. S. Romer, and 15 drawings are patterned on illustrations in *Traité de Zoologie.* No other source was used for more than 7 drawings.

My wife and I worked together closely on the illustrations. It is principally her talent that makes the artwork distinctive; she made all the carbon pencil drawings and some of the pen and ink drawings (about 45% of the total). I selected materials (largely from my teaching collection), photographed, and made most of the pen and ink drawings (about 42% of the total). The publisher's staff artists completed the more diagrammatic illustrations (about 13% of the total) from my sketches and did all labeling. For each subject we sought an appropriate compromise between illustration that is so pictorial as to introduce extraneous detail, and illustration that is so simplistic as to reduce the living body to mechanical analogs.

Selected and partly annotated references are listed at the end of the book to give students a start on assignments and seminars and to indulge the curiosity of anyone having unanswered questions. The meanings of more than 150 word roots are given parenthetically where first used in the text. These are intended as aids to understanding, and hence memory, not as etymology lessons.

A preparatory course in general biology or zoology is assumed; most of the requisite fundamentals and terms are reviewed here, but would come as a big dose if none were already familiar. A prior foundation in embryology, physiology, or evolution is desirable to give the student the benefit of additional familiar ground, but is not assumed. Similarly, recollection of algebra and geometry, and a course in physics would make Part III easier, but more for the security provided than for formulas remembered.

Some students complain that textbooks present more than

can be learned. Well, the gourmet is not able to eat all that is on the menu, but does not therefore desire few selections; others have other tastes and he will himself choose differently another time. I urge instructors not to serve up more of this book than can be assimilated in the time available. In a two-term course, most (though I prefer not all) of the chapters can be presented. In a one-term course, selection is necessary. A balanced diet can be obtained by combining portions of both Parts II and III, thus illustrating more than one of the major approaches to the analysis of structure and abandoning the myth that any one approach would be "covered" merely by assigning all relevant chapters in this, or any other, text. Unassigned chapters can become the bases for special reports. Further, I hope that students and instructors (and their examinations!) will permit the terms and the illustrative and parenthetical material in the book to support but not substitute for the ideas.

Prepublication reviewers are never responsible for the deficiencies of a book, yet they are responsible, if competent and conscientious, for great improvement in the manuscript. I have been fortunate in having such reviewers. Carl Gans and David B. Wake each read the entire text and made hundreds of very valuable suggestions. Five chapters were reviewed by Karel F. Liem, two chapters each by R. Glenn Northcutt and Irving H. Wagman, and one chapter each by Paul F. A. Maderson, Paul S. Moller, Terry A. Vaughan, and Marvalee H. Wake. It is a pleasure to express my gratitude to these people. I credit also the skillful and important contributions made by those who produced the book: editor, production manager, art staff, designer, copy editor, and others.

I shall be grateful to instructors and students who send me corrections, comments, and sources of material for revisions.

Davis, California   *Milton Hildebrand*

Contents

x **Contents**

xii Contents

PART III

STRUCTURAL ADAPTATION:
Evolution in Relation to Habit
and Habitat

CONCLUSION

Analysis of
Vertebrate
Structure

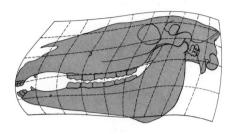

INTRODUCTION

The Nature of
Vertebrate
Morphology

Anatomy is the science of the observation and description of structure. Gross structure, microstructure, and ultrastructure are included, though this book stresses the first and presents little of the last. Morphology is the science of the *interpretation* of observed structure.

Since structure is influenced in many ways, information is needed from many sources if interpretation is to be reasonably complete: (*1*) Much interpretation follows from knowledge of the historical origins, or phylogeny, of form. Thus, morphology relates closely to paleontology, taxonomy, and the principles of evolution, and draws upon other fields such as serum chemistry and analysis of chromosome structure that provide evidence of evolutionary affinity. (2) Also, much interpretation is dependent upon knowledge of functional adaptation. Morphology is therefore related to ecology, ethology, physiology, biophysics, and biochemistry. Some adult structure is interpreted through (3) knowledge of embryology, and some through (4) analysis of the relation between form and size. Other structure is (5) age-dependent or (6) sex-dependent. Finally, (7) individual variation of structure may have genetic, nutritional, pathologic, or other environmental origins.

It is clear that vertebrate morphology relates to many other sciences. It is desirable for the morphologist to supplement his training in the principles of biology at all levels and his grounding in vertebrate structure by gaining familiarity with the concepts and methodology of one, or preferably more, related disciplines.

This book describes the anatomy of the major structural and behavioral groups of vertebrate animals, and interprets their morphological differences primarily in terms of ancestry and function, employing related fields as needed to further this objective.

Why Study Vertebrate Morphology?   In order to generate the interest and application needed to derive full value from any course of study, the student must be convinced that the returns will justify the effort. Different students derive different benefits from studying vertebrate morphology, but the following advantages, in varying proportions, accrue to most students:

(1) Knowledge of anatomy has direct application to many specializations within biology. The surgeon and veterinarian, experimental embryologist and neural physiologist, paleontologist and pathologist, all need to be familiar with the structure of their materials.

(2) Knowledge of animal structure has made important contributions to human health and technology. The selection of experimental animals, the conduct of innumerable studies in basic and applied physiology and medicine, and the design of prosthetic devices are examples. Some engineers are studying animals for clues to improved design of bearings, ships, and aircraft.

(3) Study of morphology increases the biologist's understanding of the structure of his materials. Through interpretation he becomes less of a practitioner and more of a professional, less a technician and more a scholar, less a catalog of facts and more an expert. This benefit is difficult to measure, yet can be of great value to the individual.

(4) Analysis of structure may increase the biologist's inter-

est in, and even fascination with, animal form. This subtle benefit can be very rewarding.

(5) Vertebrate morphology provides particularly favorable evidence for the process and product of organic evolution. It contributes to the answering of questions that have long been important to man: What forces govern the stream of life? How can one gain perspective in time and space? How can one account for the perfection of the animal body?

Study of vertebrate morphology is unlikely to bring the biologist wealth, fame, or influence, but it *is* likely to increase his competence and pleasure in his work.

Analogy and Homology. In order to interpret the structure of any vertebrate, the morphologist must identify and explain points of similarity and dissimilarity with other vertebrates. The concepts of analogy and homology are basic to the process.

Closely related animals tend to be more alike than distantly related animals: Cats and mice (related at the class level) are obviously more nearly alike than cats and sharks (related only at the subphylum level) (see Figure 1-1). Yet in many ways, cats and dolphins (again related at the class level) are *less* alike than dolphins and sharks (related only at the subphylum level). Dolphins and sharks share fins and streamlined form not seen in cats because *any* animal highly adapted for rapid swimming develops these features. The similar structures noted have evolved independently because of *similarity of function* and do not necessarily express close ancestral relationship. Such characters are said to be **analogous.** Analogy is resemblance of structure that results from adaptation to a common function.

On the other hand, dolphins are like cats and different from

Some Principles and Considerations

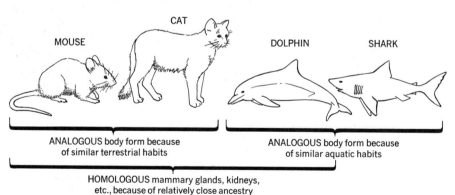

MOUSE · CAT · DOLPHIN · SHARK

ANALOGOUS body form because of similar terrestrial habits

ANALOGOUS body form because of similar aquatic habits

HOMOLOGOUS mammary glands, kidneys, etc., because of relatively close ancestry

FIGURE 1-1
THE DISTINCTION BETWEEN ANALOGOUS AND HOMOLOGOUS STRUCTURE.

sharks in having large brains, three ear ossicles, and mammary glands. These structures correspond because of historical, evolutionary, or *phylogenetic relationship.* Such characters are said to be **homologous.** Homology is equivalence of structure that results from inheritance from a common ancestor.

The importance of this distinction justifies a further example: The pectoral fin of the dolphin and the front leg of the lizard are not analogous because they have different functions (waterfoil as against support). However, in spite of superficial dissimilarity, they are homologous because internal structure reveals a basic correspondence of skeletal parts that can be explained only on the basis of common origin from a remote ancestor (Figure 1-2). Conversely, the fin of the dolphin is analogous, but not homologous, with that of a carp because the correspondence of general function and superficial appearance is accompanied by fundamentally different skeletal structure not shared by a common ancestor. Analogy and homology are not mutually exclusive: Mammary glands of cats and of dol-

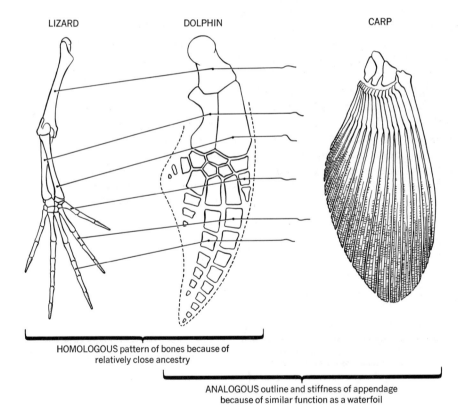

LIZARD   DOLPHIN   CARP

HOMOLOGOUS pattern of bones because of
relatively close ancestry

ANALOGOUS outline and stiffness of appendage
because of similar function as a waterfoil

FIGURE 1-2
THE DISTINCTION BETWEEN ANALOGOUS AND HOMOLOGOUS
STRUCTURE.

phins are both analogous *and* homologous because they are similar in function *and* phylogenetic origin.

(These concepts have been considerably elaborated, and somewhat disputed, by evolutionists. (See, for instance, the article by Bock cited in the References at the back of this book.) Theoretical difficulties may arise over the relative nature of analogy and homology in regard to the level of the taxa involved, over the level of "equivalence" of form or function required in a specific instance, or over acceptable remoteness of a "common" ancestor. Practical difficulties may arise from inability to distinguish analogy from homology among closely related animals. In this book the frame of reference will usually be classes, subclasses, and orders, where the concepts are seldom ambiguous and frequently useful.)

Serial homology is a rather independent concept which is mentioned here because of the similarity of terms. Structures are serially homologous if they occupy different spatial positions in a series of like structures. The separate vertebrae are serially homologous, as are the different teeth of a tooth row, the several gill arches of the series, the successive muscle segments along the back of a fish, and the many tubules in the kidney of a lower vertebrate. It is usual for the structures of such series to form gradients: Vertebrae become larger toward the pelvis, teeth often become more complex toward the back of the mouth, gills may become smaller toward the posterior end of the series. Serial homologs have similar potential for change: Any embryonic vertebra behind the ribs will form an articulation with the pelvis if experimentally placed adjacent to it. Also, change usually affects more than one element of a series: If one tooth becomes larger or more complex, its neighbors tend to do likewise.

Structures have **sexual homology** if they develop from equivalent embryonic primordia, yet are sexually dimorphic. The ovary is the sexual homolog of the testis, and the clitoris is the sexual homolog of the penis.

Considerations from Morphology. It is important for the morphologist to be able to recognize and distinguish analogy and homology in the general sense. The following are some of the considerations that guide him:

(*1*) When structures are studied in combination, it is found that correspondence of many parts bespeaks evolutionary relationship, whereas correspondence of one or several parts may result from other causes. Several examples will illustrate: Numerous cartilaginous and bony fishes have electric organs. This suggests common origin, but the fishes are so different in regard to so many other characteristics that it is virtually certain their electric organs evolved independently. Hedgehogs and porcupines were long classed together because each bears quills, yet these animals have dissimilar teeth, caecum, and reproductive organs. Likewise, horses and cattle each have complicated enamel patterns in their cheek teeth, but their stomachs, dental formulas, and skull structures argue against close affinity.

(2) Noting that homologous structures *may* have similar functions, whereas analogous structures *always* do, the morphologist constantly tries to correlate structure with function. He suspects analogy when similar functions can be demonstrated. Thus, horses and cattle share large size, hoofs, and similar molar teeth because each runs well and eats grass. The common structures evolved in response to common habits, and were not retained from common ancestry. Hedgehogs, porcupines, and one of the egg-laying mammals (echidna) all have quills because quills provide a satisfactory defense for animals that cannot run or fight.

Analogous features are commonly found among those structures that are used in locomotion or feeding. Teeth, alimentary canal, sense organs, and feet are particularly responsive to changes in an animal's way of life. If a morphologist wants to study adaptations, he may compare these organs in unrelated animals with similar habits, or in related animals with different habits. If he wants to establish phylogenetic relationships, however, he then concentrates on characters that may be expected to exhibit similarities only if related by common ancestry—the degree of correspondence suggests the closeness of the relationship. For example, the fact that the trapezius muscle of the shoulder of mammals is innervated by a cranial nerve, rather than by a spinal nerve, is a clue to this muscle's slow backward migration from the pharyngeal position its homolog still occupies at the fish stage of evolution.

It is also important to note the presence of vestigial or degenerate organs having no function at all. It is safe to assume that such structures were functional in the ancestor, and this tells us something about the evolution of the descendant. The blind eyes of burrowing moles and cave fishes indicate that their ancestors could see, and tiny bones in pythons and whales tell us that the remote ancestors of these animals had legs.

(3) The study of complicated structures, although more difficult, may be more rewarding than the study of simple ones. Ribs are just too simple to hold many secrets. The skull, on the other hand, has so many functions and so many bones, foramina, and contours that it reveals much of the habits and his-

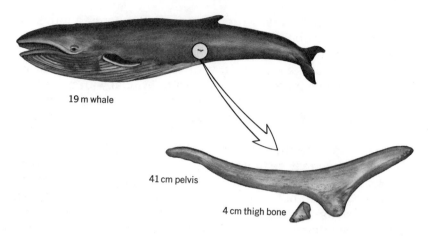

19 m whale

41 cm pelvis

4 cm thigh bone

FIGURE 1-3
VESTIGIAL ORGAN illustrated by the pelvis of a finback whale.

tory of its former owner. Further, unrelated structures that are also complicated are unlikely to be similar merely by chance.

(4) The primitive, or ancestral condition of a structure is more likely to be found in an animal that retains the ancestral condition of *other* parts. One must avoid circular reasoning. One cannot *prove* that a feature is primitive because it is found in an animal that is *assumed* to be primitive. Nor can one assume that a feature is primitive just because it is found in an animal that is *known* to be relatively primitive—even relict species have some highly specialized structures. Nevertheless, an animal that is *known* to be primitive in the expression of several characters is of particular importance. Thus, if one is working on the phylogeny of the types of placentas found among rodents, he would do well to see what sort of placenta the mountain beaver has. It could be quite ordinary, but if it were unusual, it would be the more interesting because this animal (not really a beaver) is known by evidence of its skeleton, muscles, and fossil record to be the most primitive living rodent. Part II of this text will pay special attention to several animals that retain numerous primitive features: *Polypterus* among fishes, *Sphenodon* among reptiles, the platypus among mammals, and some others.

Another, and particularly important, consideration when analyzing structure is variation.

Recognition and Assessment of Variation. The morphologist rarely describes an individual animal *as* an individual; he studies individuals to learn about *kinds* of animals. Individuals of a kind vary so much among themselves that a single specimen is not adequately representative.

Most characters vary independently of one another—a specimen may be average for one yet extreme for another. However, some characters are related in such a way that variation in one can be correlated with variation in another. For instance, among certain mammals, an individual with extra-high tooth crowns is likely also to have extra-large premolars and an extra-long jaw, because these characters are functionally related.

Structures that have relatively great variation among individuals of a kind may be of limited value for interpretation of form. Molar teeth and ankle bones must be just right to function well, so they do not vary much. Breastbones, by contrast, can be flatter, or longer, or more segmented than usual without much impairment of function. Accordingly, there is more variation in breastbones and less can be interpreted from their configurations.

It may be important to study **sexual dimorphism** either so that the gender of a particular vertebrate can be established or so that sexual characteristics will not be mistaken for species characteristics or for adaptive features of another nature. Of course, sexual dimorphism relates most to the gonads, genital ducts, external genitalia, and accessory sex organs. These may be conspicuously different in the two sexes or closely similar. Gender differences of the genitalia may be much more evident at one stage of the breeding cycle than at another because of alterations in size, position, coloration, secretions, or vascular congestion.

Other sexual dimorphism does not relate directly to the genital system, yet correlates with distinctions between the sexes in regard to sex role and behavior. These include differences in the coloration, amount, and distribution of scales, feathers, or hair, in clasping organs and pelvic architecture, and in the presence or development of crests, spurs, antlers, tusks, and scent glands.

Adult body size and rate of growth are frequently sexually dimorphic, and gender differences in body proportions and configuration are common. Thus, males commonly have coarser and relatively heavier skeletons. There may be sexual differences in the muscular and vascular systems, and though not yet identified, structural differences may be expected in the hypothalamus of the brain and in the pituitary gland.

The morphologist also needs to recognize **age variation** and to avoid confusing it with variation of another kind. The accretion of successive rings or layers on hard tissues often accurately reflects alternating periods of slow and rapid growth. Like tree rings, these relate to times of favorable climate or nutrition which usually are seasonal. In favorable circumstances, the scales and otoliths (bones within the inner ear) of bony fishes, the scutes and long bones of turtles, the baleen and ear plugs of whales, and the dentine and cement of the teeth of mammals all show marked growth rings.

Hard parts also correlate with age in other ways: The centers where developing bones first ossify appear in regular sequence, the various teeth of mammals erupt and are replaced at different but regular ages, sutures between bones close on schedule, and epiphyses at the ends of the long bones join the shafts at given times. The ages of young humans and of some other animals (for which the sequences have been worked out) can be determined within fairly narrow limits by x-ray analysis of the skull, wrists, or other joints.

The following may also be age-related, though usually with less precision: size and morphology of the baculum (a bone in the penis of some mammals), structure of the cranial wall (birds), interorbital width (some carnivores) and relative size of orbits, size and configuration of horns and antlers, porosity of bones, development of crests and tuberosities on the skeleton, position of basicranial axis, and relative development of rostrum or facial area.

The structure of soft tissues tends to be less precisely age-related, yet the weight of the lens of the eye (measured after careful removal and dessication) is sometimes a reliable index of age. Changes in the skin, in the progression of molt, in the composition of muscle, and, as all too many men and women learn, in the circulatory system, are also ascribed to advancing age.

Before the twentieth century, it was common for an anatomist to dissect a casualty from a zoo and to publish a description of that kind of animal on the basis of the one specimen. Such animals tend to be immature or senile, pathologic, or suffering from malnutrition and inactivity, and to that extent are atypical of the species. If an animal is extinct, rare, or otherwise difficult to obtain, it may be necessary to work with few specimens, but the morphologist now tries to obtain random samples of adequate size. He may first sort the individuals by sex, age, geographic location, or other criterion, keeping enough specimens in each subsample (about 25 is usually adequate) to assess these variables. The individuals of such subsamples still vary among themselves, however, and this **individual variation** must be analyzed if the true nature of each characteristic is to be identified. The nature of typical structure, the range of variation, and the variability, or tendency to vary, are all population characteristics; they cannot be learned from single specimens.

Statistical analysis of population characteristics is beyond the scope of this book, yet should be studied by aspiring morphologists. Even several relatively simple parameters can be very helpful: The arithmetic mean expresses the "average" condition. The standard deviation shows the degree to which the separate values tend to cluster around the mean. The standard error of the mean enables one to judge if means derived from different populations (or samples from different populations) are significantly different. The coefficient of variation makes it possible to say if animals of different absolute size are comparably variable, i.e., if a shrew varies as much for a shrew as an elephant does for an elephant. Morphologists commonly use these and many other parameters, and also employ various graphical techniques.

Finally, it is important for the morphologist to be able to distinguish among accuracy, reliability, and significance. A hypothetical example will illustrate: Suppose one were to carefully weigh three bald men and three men with hair, and suppose further that each bald man weighed more than any man with hair. Although one could then accurately say that these bald men are heavier, one could not reliably conclude that in general bald men are heavier than men with hair. Suppose now one weighed thousands or even millions of men of each class and again found that the bald men weigh more. The result might be accurate but still unreliable because bald men tend to be older and older men tend to weigh more, or because the bald men were weighed in the evening and the others in the morning, or for other reasons. But suppose that all such factors are controlled and the weighing is carefully done, one might then find accurately that men with hair weigh more, on the average, by the slight weight of their hair. Having a large controlled sample, the result would also be reliable; one would have confidence that it could be repeated. However, one would expect the difference to be so trivial as to have no significance. It would not be useful for health services, insurance agents, or dietitions. Morphologists usually seek results that are accurate, reliable, and also significant.

Contributions from Paleontology. Paleontology is the study of prehistoric life as revealed by fossils. Some vertebrate fossils are mere trackways or body imprints made millions of years ago in mud that was later converted to stone. Rarely an entire animal is preserved: Insects trapped in ancient pitch have retained their delicate structures as the pitch gradually turned to amber, and great mammoths that stumbled into glacial crevasses have been preserved a million years in a natural deep freeze. Usually only teeth and bones are preserved, and even for these the slow process of petrification has replaced most of the original organic tissue by an exact copy in hard minerals.

To be preserved, the skeleton must be protected from the destructive forces of weathering, and this means that it must be covered up soon after it is released by decay from the tissues it supported in life. Landslides, wind-driven sand, and volcanic ash cover some skeletons, but most that become fossils are covered by silt and sand deposited in lakes or streams, or on the flood plains of wide valleys. In time, the mud is converted by pressure to shale and the sand to sandstone. Upheavals of the earth's crust, and subsequent erosion, expose the contained fossils. Sedimentary rocks are thus a gigantic filing cabinet for the fossils they contain: The oldest fossils are in the oldest strata and the recent fossils are in recent strata. Evolutionary changes in the structure of animals progress step by step with the accumulation of earth sediments.

As an interpreter of animal structure, the vertebrate paleontologist is confronted with many difficulties. He usually must work with broken materials and fragments, he seldom finds fossils of animals that lived in arid uplands where skeletons disintegrate quickly in dry air and hot sun, and with rare exceptions, he must work with a single organ system. Nevertheless, paleontologists have described about three times as many extinct vertebrates as there are surviving vertebrates. The morphologist, embryologist, zoogeographer, serologist, and other specialists have provided part of the story of evolution, but the paleontologist has contributed the most, and his part is the most sure.

Evolution and Habitat. Interpretation of structure is made more meaningful by familiarity with the major features of the evolutionary process. Supplementary reading on that subject is highly recommended (see the References at the back of this book). There is space to present only the barest outlines here, and it must be assumed that the reader has general knowledge of the origins of heritable variation, natural selection, and earth change.

A major consideration is the relation between evolutionary change and the stability of the environment. With minor exceptions, evolution results from the interplay between changing environments and adapting organisms. Each kind of animal becomes adapted to, and dependent upon, a particular kind of life (predation, seed-eating, grazing) in a particular kind of habitat (marsh, stream, meadow). If the habitat is large and constant (oceans, tropical forests, coniferous forests), the animal inhabitants have time to become well-adjusted. It is most advantageous for each species to remain about as it is, so natural selection tends to prevent change. Large habitats do move slowly in earth history (e.g., continental ice sheets and rain forests advance and recede), but most of the animals move with them, remain adapted, and change relatively little.

Although the general habitat may shift slowly in space, the restricted habitat of a population of animals may instead slowly change. Thus, a shoreline may gradually become more rocky, or a pine–fir forest that is isolated between mountain ranges may become more alpine in character as timberline shifts southward. The average expression of the characters of a kind of animal is then no longer quite optimum. Natural selection therefore tends to cause the animal to become a somewhat different kind of animal. If the habitat alters as a unit, the evolutionary change is in a more or less straight line and is said to be **linear**. If the old habitat subdivides into different units, different parts of the original animal population become isolated from one another, adapt independently, and evolution is **branching**.

When a habitat alters too fast for a resident species to adapt to its new character, then, so far as that species is concerned, the habitat does not change, but disappears: As the earth slowly tilts, an inland sea drains away leaving only shoreline terraces on dry hills; meadows are invaded year after year by pine seedlings, and finally succumb to the forest. As a habitat becomes less and less satisfactory, the force of natural selection, i.e., **selection pressure**, for the old way of life weakens. Finally, extreme variants in the population may be better suited for life in a new habitat than in the old, provided that one or several new habitats are physically available and not already occupied by effective competitors. Selection pressure then becomes strong and shifts away from the old life style toward the new.

Several factors may contribute to success in this hazardous kind of evolution. **Specialized structures** are those that have become modified to perform restricted functions with great effectiveness, whereas **unspecialized structures** are suited to perform adequately a less restricted function or a variety of functions. Unspecialized structures have more capacity for evolutionary change, and hence favor survival when change is essential. Thus, the ancestral five-toed foot has been converted to a springing support, wing, paddle, grasping organ, and so on. Specialized organs are satisfactory as long as their restricted functions are needed (it is probable that there will long be ants for anteaters and krill for baleen whales), but if a new function is needed, such organs rarely can adapt. The running foot of the horse could in theory be converted once again to a grasping organ, but in reality the process of evolution does not favor such change. The saber-toothed tiger presumably became extinct when its thick-skinned prey died out. Some flamingos now depend on the persistence of brackish inland lakes where they must find the diatoms on which they exclusively feed.

When a species must adapt quickly to altered conditions, it is not granted time to evolve an entirely new complement of structural attributes. It must rely for a time on the intensified or altered use of attributes it already has. Natural selection may "discover" that a structure that was useful in one way before can now be useful for another purpose. Such structures are said to have **preadaptation.** For instance, it was a long and major task for evolution to convert the walking legs of proavian reptiles into the wings of birds. It was relatively quick and simple for several groups of birds to use their wings as waterfoils for swimming instead of airfoils for flying—a shortcut of tens of millions of years in the evolution of effective paddles. Unspecialized form and preadaptations are good cards to hold in the game of find-a-new-way-of-life-or-become-extinct.

Just as habitats can disappear, so they can appear. A new marsh, formed close to existing marshes, will be populated from nearby marshes, and no evolution will occur. A new inland sea, however, if extensive and isolated, is a place of evolutionary opportunity for such animals as can get a start there. Similarly, the habitat may be old, but its availability may be new: Land first became available to vertebrates when reptiles evolved. There were sufficient land plants and arthropods to provide food and shelter. There, inviting colonization, were millions of square miles of diverse new habitats having no competition from established animals. In this infrequent circumstance, evolution rapidly creates various new kinds of animals. It is said to be **radiating.**

Animals that are successful in evolution's stern game of adapt or perish become extinct by progressive change: Ancestors disappear as descendants evolve. Many more kinds of animals leave no daughter species. They become extinct because they cannot meet the competition, because disbalance occurs in predator–prey relations, or because there is failure to adapt to changes in the noncompetitive parts of the environment.

Evolution and Time. Figure 2-5 presents the time scale in relation to vertebrate evolution. The numbers are easy to read but difficult to comprehend. It helps to relate geologic time to something relevant and finite. One example equates all the time since life began on earth with one year: Then, vertebrates appeared about October 20, mammals evolved on December 7, and *Homo sapiens* made his debut at 11:48 PM on December 31.

This example teaches proportion but fails to show the magnitude of the full-scale phenomenon. Perhaps that is not possible, but consider this: Our galaxy is an assemblage of 10 billion stars distributed in the shape of a wheel so large that light requires nearly 100 thousand years to traverse its diameter. Yet, the entire galaxy has revolved once around in space since the coal bogs dried out and the dinosaur radiation began. From time to time, galaxies pass through one another. Such a collision is completed in about one-seventh of the time it took the horse to evolve from one genus to the next. Since vertebrates evolved, the far edges of our expanding universe have receded a distance 40 times the diameter of our galaxy!

The fossil record presents many relatively complete and continuous lineages: Elephants were once numerous and diversified, and camels came in all sizes and were abundant on four continents. Horses, turtles, crocodiles, the kinds of dinosaurs, and many other groups are all assemblages of clearly related genera. Such a lineage is called a **phyletic line,** and is usually represented by a number of genera that are related in time by linear and branching evolution, and through extinction by progressive change. Systematically, a phyletic line is usually a single family but sometimes an order or intermediate taxon. For reasons explained below, nearly all phyletic lines appear suddenly in the fossil record and, if not still surviving, seem to disappear from the record rather than merge into one or more other lines.

What factors influence rates of evolution within phyletic lines? The rate can be high only when the adjustment of the species to its environment is such as to cause selection pressure to be strong. When population size is quite small, evolution tends to be slow, relatively nonadaptive, and not sustained. When population size is very large, it tends to be adaptive and sustained, but slow. Medium population size—perhaps 250–25,000 breeding individuals—favors adaptive, sustained, and relatively fast evolution. Short generations permit the highest rates of evolution for short periods, but do not, in fact, correlate with high rates of sustained evolution. Finally, although mutation must supplement recombination in order for evolution to be sustained, still, mutation rates are usually not limiting.

Different phyletic lines evolve at different rates, each line evolves at different rates at different times, and different characters of one line evolve at different rates at the same time. With these qualifications, however, some examples can be given of slow and fast rates of evolution.

Low rates and long survival are most likely to occur among large populations of medium-to-small-sized animals that have unrestricted food habits and are well-adapted to extensive and uniform habitats. The group characters of rabbits, armadillos, and turtles were established more than 65 million years ago. Opossums have been opossums for 100 million years, and the same is true for crocodiles. Some groups of fishes (coelacanths, dipnoans, sharks) have survived for 300 million years.

High rates of evolution occur when medium-sized populations of animals having sufficient potential for individual variation experience strong selection pressure. Peak rates for vertebrate classes in terms of numbers of new genera evolved per million years have been about as follows: placoderms, 3; cartilaginous fishes, 2; bony fishes, 6; amphibians, 3; reptiles, 12; and mammals, 30. The rate at which genus replaced genus in the horse lineage was 1 per 7.5 million years, which is high for vertebrates generally, but not particularly high for mammals. The highest rates occurred among land vertebrates because they have been the most adaptable and their habitats have come and gone the fastest.

Even though there are gaps in the fossil record between phyletic lines, it is now agreed by most evolutionists that the evolutionary process is the same between phyletic lines as within phyletic lines. New lines appear under the conditions noted above that favor rapid evolutionary change. Two factors contribute to the relatively incomplete nature of the fossil record at the times that new lines are established: First, fewer animals are fossilized because, when evolutionary change is maximum, kinds of animals do not survive long, and there are relatively few individuals in the affected populations. Second, such fossils as are formed are less often found because the animals tend to be small, and the same extreme conditions of earth change that increase selection pressures also tend to erode and destroy the earth's crust, thus destroying the contained fossils.

Evolution and Direction. The direction of evolution is temporarily somewhat random when selection pressure is weak or population size is small, but such change is relatively infrequent and unimportant in the course of evolution. It is a striking fact that *within* phyletic lines, each adaptive change tends to progress in more or less the same direction without stopping, zigzagging, or reversing. Such gradual changes are called **evolutionary trends** and, though not universal, they have been usual for large populations evolving at moderate rates. Evolutionary trends are oriented and prolonged by selection pressure. The characteristic continues to develop because it continues to be advantageous.

A common, though by no means universal, trend has been to large body size. Repeatedly, and in unrelated lineages, there has been gradual increase in size from modest ancestor to

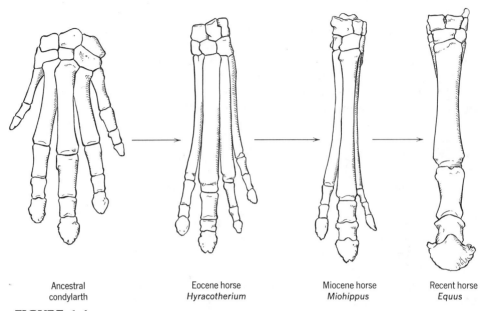

| Ancestral condylarth | Eocene horse *Hyracotherium* | Miocene horse *Miohippus* | Recent horse *Equus* |

FIGURE 1-4
EVOLUTIONARY TRENDS IN PROPORTIONS AND NUMBER OF SERIAL PARTS
of the left forefoot of the horse. (Not drawn to scale.)

gigantic descendant. Examples, the advantages of large size, and adaptations for supporting great weight are included in Chapter 19, and some other structural responses to large body size are noted later in this chapter. A common trend for reduction in number of serial parts is exemplified by teeth of many lineages, lateral digits of hoofed mammals, gill arches of lower fishes, and cranial bones from fishes to mammals. Conversely, teeth and muscles have increased in number in some lineages. Trends leading to change in relative size of parts of the body have been common. The gradual development of tusks, antlers, and unusual beaks are examples. D'Arcy Thompson demonstrated in 1917 that trends in body proportions are admirably described by progressively distorting a grid based on Cartesian coordinates as shown in Figure 1-5. Other trends involve increase in specialization of parts, such as the formation of elaborate enamel patterns in the grinding teeth of horses and elephants. Many examples are found in Part III of this book.

Rarely can one identify a single, isolated trend, because initial or primary changes usually necessitate dependent or secondary trends in response to functional necessity. Thus, increase in body size ultimately requires postural, skeletal, and muscular alterations; increase in the efficiency for grinding of premolar teeth requires lengthening of the face and altered jaw mechanics.

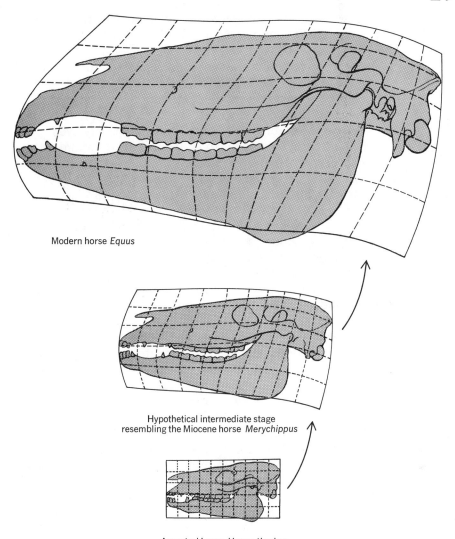

Modern horse *Equus*

Hypothetical intermediate stage
resembling the Miocene horse *Merychippus*

Ancestral horse *Hyracotherium*

FIGURE 1-5
**EVOLUTIONARY TRENDS IN PROPORTIONS AND SIZE of the horse skull
shown by the progressive distortion of a grid.**

A degree of rectilinearity is inherent in the nature of most
trends. Body size, body proportions, and number of parts in a
series can only increase, remain the same, or decrease. Such
trends seldom progress at a constant rate, and often have temporary arrests, yet the direction of change is usually constant
because the direction of advantage to survival is usually constant. Trends in form and degree of complexity are less rectilinear.

Even trends of the stop-or-go variety usually come to an end if, and when, the condition of maximum advantage is reached: The sabers of saber-toothed tigers increased until the Oligocene epoch and then remained about constant for 40 million years. Further, many trends have theoretical end points: Change toward longer-wearing teeth ends with ever-growing roots; change toward loss of lateral digits ends with one toe; change toward loss of vision ends with rudimentary eyes.

Still, some trends do seem to stop short of maximum advantage whereas others go too far, as though evolution, like a flywheel, includes an element of inertia which continues to function after the driving force is removed. The prodigious size of the antlers of the extinct Irish elk is often cited as an example. Some extinct elephants had tusks so large they curved and crossed and could no longer effectively thrust, pry, or dig.

There are several ways in which this sort of evolutionary momentum can come about. First, natural selection makes no provision for nature's senior citizens. It improves the fitness of breeders and potential breeders, but neglects individuals that are no longer productive. If only senile elephants had excessively large tusks, then there was no mechanism for correcting the handicap. Indeed, if the same tusks that would grow to be most burdensome to old animals were also most advantageous to vigorous young breeders, then selection would, without compassion, make the tusks even larger.

If the continuance of a trend is advantageous in one respect but disadvantageous in another, then the trend will stop when the advantages and disadvantages of further change are in balance. This is common, though often difficult to identify in specific instances. Increased dermal armor might give a fish added protection against predation but make it less maneuverable; increased slenderness of limb might give an antelope more speed but make it less able to avoid injury; increased curvature of the beak might make it easier for a bird to secure one kind of food but more difficult to secure another. The result is optimum compromise, not optimum structure in regard to any one function.

Parallelism and Convergence.   Descendant lineages may resemble one another closely in characters that are not present in the common ancestor. **Parallelism** is the independent development of similar characters in lineages that inherit from a common ancestor a potential for such development but do not directly inherit the similar characters. Both lineages change, but in sufficiently similar ways so that they remain about equally alike.

The kangaroo rats of western North America and jerboas of Africa and Asia show striking parallelism: Each has long hind limbs, short forelimbs, loss of lateral toes, lax fur of tan color, long tail with white tip, large eyes, inflated bony ear capsules, and compacted neck vertebrae. It is unlikely that the common ancestor had any of these characters; on the basis of other features, these rodents are placed in different suborders. Similarly, the golden moles of south Africa and the marsupial moles of Australia belong to different orders, yet have in common short robust forelimbs with strong claws, rudimentary eyes, loss of external ears, tough nose pads, and other characters. Parallel lineages are each adapted to nearly identical ways and places of living: kangaroo rats and jerboas to hot, sandy deserts; golden moles and marsupial moles to life in the soil. It is the effectiveness of natural selection that causes basically similar animals having nearly identical functional requirements to evolve similar structural adaptations.

Convergence is the independent development of similar characters in two or more lineages even though they receive from the common ancestor neither the adaptations involved nor characteristics that channel the development of those adaptations. The common ancestor may be (but is not always) more remote than for parallelism. Similarity does not include correspondence of detail because closely similar functional requirements are met in somewhat different ways.

The remarkable similarity of the shark, ichthyosaur (an extinct reptile), and dolphin is a classical example. The common ancestor was a primitive armored fish (or placoderm) unlike any of them. One has no terrestrial ancestor; the others have dissimilar terrestrial ancestors. The tail of each is a waterfoil, but one turns up, one turns down, and one is straight with lateral flukes. Each has numerous simple teeth, but they are rootless for one, rooted in grooves in another, and rooted in sockets in the third. The same sort of general resemblance with variation of detail holds for the spine, eyes, and paired and median appendages. Another example is the similarity of birds and flying reptiles: In spite of remote common origin, each has large eyes, proximal nostrils, long rostrum, long neck, short back, large breastbone firmly attached to the pectoral girdle, many vertebrae articulating with the pelvic girdle, pneumatic bones, and other common features. Nevertheless, a different kind of airfoil is supported by a different digit, and the hind limbs (and almost certainly the internal organs) are dissimilar. A third example is found in a family of salamanders where several genera have lengthened the vertebral column by increasing the number of vertebrae, whereas another genus has instead accomplished the same result by increasing the length of the individual vertebrae.

Parallelism and convergence differ in the degree to which the development of corresponding structure in different lineages

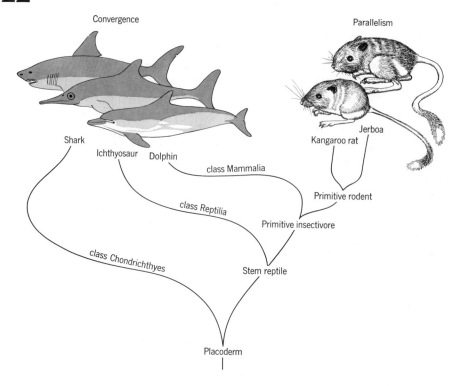

FIGURE 1-6
CONVERGENCE AND PARALLELISM both establish correspondence of structure in response to similar habits. They differ in the degree to which common
general plan extends to similarity of detailed structure.

has been facilitated by common body plan and by the degree to
which the common general plan extends to similarity of detailed structure. The concepts merge into one another. Each
process causes taxa to become more related by structure and
function than by phylogeny. Regardless of his emphasis (function or phylogeny) the morphologist will sometimes misin-
terpret his materials if he does not recognize and interpret parallelism and convergence.

Contributions from Embryology. Following the publication of
Darwin's *Origin of Species* in 1859, embryologists were quick to
endorse its new concept, and for as long as evolution was
taught as a theory to be proven, rather than a known process
inviting further interpretation, it was widely believed that embryology could provide sure evidence for the ancestral relationships of animal groups. It was thought that in the successive stages of its embryonic development, each kind of animal
repeats the stages of its evolutionary history. The former was
the newly discovered key to the latter. The author and foremost
champion of this doctrine was Ernst Haeckel (1834–1919), and

his three-word summary of the theory has been spoken by most biologists since that day: "Ontogeny Recapitulates Phylogeny." A list of the adherents of this attractive theory reads like a *Who's Who* of nineteenth century zoology: Charles Darwin, himself, Thomas Huxley, Gegenbaur, Lankester, Balfour, and Wiedersheim, to name some of the greatest.

We know now that Haeckel's interpretation of the relation between embryos and ancestors is only a partial truth. Further, although the concept of evolution now answers questions that would otherwise be riddles to the embryologist, embryology answers few questions for the evolutionist. The recapitulation theory has been corrected, expanded, and given a new name — biogenetic law — but has fallen into neglect.

Nevertheless, some important evolutionary problems *are* still best solved by embryology, and the unquestioned close relation between ontogeny and phylogeny remains one of the important concepts of biology. It behooves the vertebrate morphologist to know what that relation is. The classic little book by de Beer (see the References at the back of this book) is by no means the last word on the subject, but later contributions have added more technical terms than new ideas. De Beer has shown that ontogenies and phylogenies may have various relationships to one another. Several examples follow:

(*1*) Organs common to an ancestor and its descendant may come to develop earlier in the life history, or ontogeny, of the descendant. This relationship is called **acceleration.** Haeckel applied it to entire organisms, but it is preferable to apply it only to specific organs. Thus, the heart of most frogs develops after the external form of the tadpole is established, whereas the heart of a bird forms and begins to function outside of the still rudimentary body.

(*2*) Conversely, an organ which developed early in the embryo of an ancestor may develop later and later in the embryo of its descendant. In some instances development is not completed at all. Thus, some kinds of salamanders retain larval gills throughout adult life. This relationship is called **neoteny.** If reproductive structures mature quickly, becoming functional in the larva, and other structures do not mature, the related term **pedogenesis** is applied. The latter relationship is cited in theories on the origin of vertebrates and is mentioned in Chapter 2.

(*3*) Descendants commonly tend to be more complicated than their ancestors (though the reverse may also be true). For instance, land-dwelling vertebrates have limbs, lungs, and jaws, none of which was present in their remote ancestors. Consequently, embryos of descendants commonly add developmental steps not found in those of ancestors. This is **accumulation.**

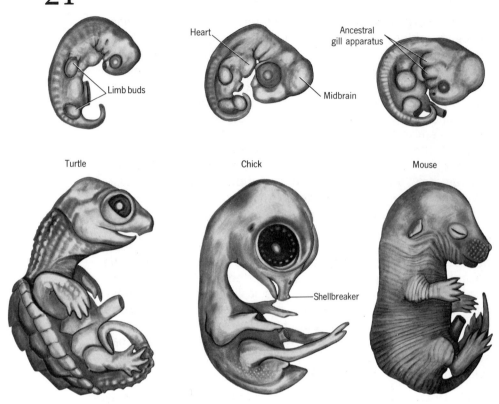

FIGURE 1-7
RELATIONS BETWEEN DEVELOPMENT AND EVOLUTION of the kinds numbered
3, 4, and 5 in the text are shown by embryos from three vertebrate classes.

(4)   Embryos of descendants are usually more like the embryos of their ancestors than the adult descendants are like the adult ancestors. It is as though the descendant embryos start developing according to an established plan but veer away from that plan as development proceeds. Thus, the gill pouches of a human embryo are similar to the gill pouches of larval fishes, but in humans the adult derivatives of those pouches are the thymus and parathyroid glands and parts of the ear, and these are certainly not similar to the gill clefts of adult fishes.

(5)   Some embryos need special structures to meet certain requirements of embryonic life, and, as evolution is resourceful, they come to have them even though the remote ancestors, not having the need, failed to develop the special structures. The shellbreaker of a chick is a sharp projection on the bill that permits the chick to break its shell at hatching. The remote ancestors had soft eggs that were laid in the water; a shellbreaker would not have been useful to them.

(6) The embryo of a descendant may start to form a certain organ early in development just as its ancestors had done, then, instead of completing the formation of that organ, may cause it to disappear again. The reptilian ancestors of birds had teeth, and we can be sure that their embryos had tooth buds. Duck embryos have tooth buds too, but no teeth develop from them.

Descendants do not inherit structures of a predetermined and fixed character in an established sequence. They inherit, instead, developmental mechanisms which they cannot refuse, but which they are free to modify by the slow process of natural selection. The human embryo constructs its parathyroid glands out of fish-like gill pouches instead of starting over with a direct method, because it inherits from fishes a developmental mechanism which includes the development of gill pouches having a high potential for forming glandular tissue. In life there is no genetic mechanism for starting over, only for making small deletions and for accumulating small alterations.

Surface-to-Volume Ratio. Many bodily functions depend on the relation between surface and volume: Oxygen is absorbed through a surface, but is used by the entire volume of the body; rough food is ground by the surfaces of teeth and absorbed through the surface of the gut, yet nourishes the entire body; the strength of muscles is approximately proportional to their cross-sectional areas, whereas their loads are proportional to the volumes of the structures moved.

Most adaptations required to sustain large body size derive from the fact that the surface-to-volume ratio of an object is size-dependent. It has been common for vertebrate giants, (examples are given in Chapter 19) to be 5–10 times as long as the ancestors of their respective phyletic lines. If such enlargement were not accompanied by alteration of form, their surface/volume ratios would be reduced 5–10-fold. Stated more generally, objects of similar form have surface areas that are proportional to the two-thirds power of their volumes. Clearly, as animals become larger, they fall behind in all requirements related to surface/volume ratios unless adaptations are made.

One way to adapt is to reduce the need for surface-related functions. The rate of gaseous exchange in the lungs, of absorption from the gut, and of excretion in the kidney all depend on the rate of metabolism. It is not surprising, therefore, that the basal metabolic rate of vertebrates is roughly proportional to the two-thirds power of body weight. The interpretation of this relationship is not simple; a direct or casual dependence of metabolic rate on body surface is not supported by the data.

Nevertheless, reduced demand for surface phenomena associated with metabolism adapt animals to gigantism.

Another way large animals adapt is by making structural modifications that increase surface areas so they can "keep up" with related volumes as overall size increases. Removal of wastes from the bloodstream occurs in the surface layer, or cortex, of the mammalian kidney. In order to provide enough cortex to remove enough wastes, the kidneys of large mammals are compound, each resembling a large cluster of small, smooth kidneys. Similarly, the grey matter of the mammalian forebrain is in the surface of that organ. In order to have enough grey matter, the large mammal has a much convoluted forebrain. Giant herbivores are never trim-bodied. Their bellies are large to accommodate the long intestines they need to absorb food. Further, the occlusal surfaces of the cheek teeth of large herbivores are disproportionately large and they have particularly intricate infolding of the enamel.

(Some adaptations of the skeleton for supporting great weight are given in Chapter 19.)

Some Trends in the Study and Teaching of Vertebrate Morphology

It is useful to contrast the former nature of a scientific discipline with its present nature. Useful, not because it shows those who are unsure which way to jump in order to be "modern," but because it enables participants and observers to assess its progress and vitality and to make short-range projections in regard to its materials, methods, and emphasis. Vertebrate morphology now differs from the antecedent comparative anatomy of several generations ago in various ways, though it is not implied here that the changes do or should characterize all work of all researchers and teachers.

Formerly, there was more speculation than there is today. Contrasting theories about, for example, the origin of paired appendages and of multicuspid teeth were elaborated and debated on meager evidence. Now, most morphologists prefer to work with tested, or at least testable principles. Many unsolved problems (e.g., the detailed homologies of the parts of various chondrocrania, and the origin of the vertebrate type of kidney) have largely been put aside because they are doubtfully answerable with present resources.

There is now less emphasis on phylogeny and more on function, less on general adaptations and more on specific adaptations, less on homology and more on analogy. Morphologists now stress differences instead of similarities. Descriptive studies remain of the utmost importance, yet seem less satisfactory if they do not support interpretation of form or function. Major taxa are studied less and minor taxa more. There is less emphasis on isolated structures and more on functional units. Examination of preserved materials is more often supplemented by study of living animals.

Morphology is now less narrow and more eclectic in regard to materials, methods, and the identification of problems. It merges into other disciplines. Variation is considered more important, so larger samples are used—when possible. Sampling techniques are of concern. Correlation and other quantitative methods are used for analysis. Superficial studies and generalizations from insufficient data are no longer acceptable.

Dissection, including micromanipulation, remains a very important, though too often neglected, technique. In recent years, however, morphologists have used much new instrumentation and many new techniques. These are examples: microscopy (light and electron), histochemistry, cinephotography (including high-speed recording, radiography, fluoroscopy), analytical film projection (variable direction and speed, remote control), artistic and graphic techniques, stereotaxic methods, telemetry, computer programming, analysis of television tape, electrophysiological techniques (including electromyography and sensors of nervous activity), and physical, electrical, and engineering data recording techniques (including osteometry, photoelastic and photostress analysis, use of force transducers and strain gauges, use of oscilloscopes and multigraph recorders).

Also important, particularly in teaching, are techniques for fixing, preserving, and embalming animals; preparing and demonstrating bony and cartilaginous skeletons by wet and dry methods; making dry bone-muscle demonstrations; injecting vessels, ducts, and cavities; dissecting and staining nervous tissue; air-drying hollow viscera; freeze-drying; embedding; molding; modeling; and casting. I have described and illustrated these techniques elsewhere (see References at the back of this book).

Anatomists formerly asked "What is it like?" Morphologists will long continue to ask the same question, but now also ask "How does it work?" The best research and teaching open new questions of these kinds as they answer old questions, and are ready with a satisfactory answer to the penetrating question, "So what?"

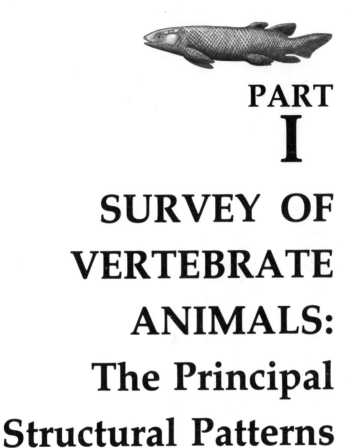

PART I

SURVEY OF VERTEBRATE ANIMALS: The Principal Structural Patterns

2

Nature, Origin, and Classification of Vertebrates

What is a vertebrate, and how did vertebrates originate? Some animal groups can be distinguished by one or two diagnostic characters: Any animal with feathers is a bird; any animal with mammary glands is a mammal. Similarly, any animal with a cranium is a vertebrate. However, since our objective is to understand and interpret the structure of the entire body, it is more helpful to use many characters in combination to describe vertebrates, selecting not only features that are unique to the group but also those that place vertebrates among related groups.

Remembering that VERTEBRATA is a subphylum of the phylum CHORDATA, let us start by describing vertebrates in

Relation of Vertebrates to Nonchordates

general terms that relate their phylum to others. Vertebrates are multicellular animals derived from embryos having three tissue (or germ) layers: ectoderm outside, mesoderm, and entoderm lining the gut tube. (The few embryonic terms used in this chapter are defined further on pp. 85–87.) The body has bilateral symmetry (right and left sides, anterior and posterior ends, dorsal and ventral surfaces). A body cavity, or coelom, is present and is lined by mesoderm. The gut is complete, which means that there are separate openings for mouth and anus. The anus is derived from an opening in the surface of the early embryo called the blastopore (or a point near the blastopore). There is an internal skeleton derived from mesoderm, and the mesoderm is formed at least in part from tissue derived from the embryonic gut.

These characters go far to describe vertebrates, yet they are shared by other chordates and by the phyla ECHINODERMATA (starfishes, sea urchins, sea cucumbers, etc.), HEMICHORDATA (worm-like, burrowing animals), and two lesser phyla (POGONOPHORA, CHAETOGNATHA). There are some departures: One group has no internal skeleton, another has no digestive tract, another lacks a coelom, and only the larvae of echinoderms have bilateral symmetry. In general, however, these features in combination set these animals apart from all others.

We conclude that chordates are more closely related to echinoderms, hemichordates, and members of similar phyla than to animals belonging to other groups. Because these animals do not form the mouth from the blastopore, they are collectively called DEUTEROSTOMIA (= second + mouth), a term sometimes called a superphylum but more often used without systematic rank.

Relation of Vertebrates to Other Chordates

There are three subphyla in the phylum Chordata: UROCHORDATA (called tunicates and sea squirts), CEPHALOCHORDATA (called amphioxus), and VERTEBRATA. These animals have various characteristics in common. Most distinctive is the **notochord** (= back + cord). This is a longitudinal rod of supportive tissue generally derived from the dorsal wall of the embryonic gut. Its turgid, vacuolated cells are unique. The cord is surrounded by sheaths of connective tissue. All chordates (as the name implies) have a notochord during early development. Cephalochordates and many vertebrates (but not urochordates) retain the notochord, or remnants thereof, as adults.

Hemichordates have a muscular proboscis in which there is a short organ considered by some to be a notochord or notochord homolog. For this reason hemichordates were formerly classified—and are still classified by some zoologists—as a fourth subphylum of Chordata. However, the structure is not supportive, has a cavity opening into the pharynx, and is questionably of entodermal origin. It is now usually called a stomochord instead of a notochord, and Hemichordata (also called Stomochordata) is usually considered a separate phylum.

A second chordate character is the **dorsal hollow nerve cord** derived from ectoderm by a folding process called neurulation. This feature is shared by hemichordates, but in them the central cavity, or neurocoel, is not continuous.

A third feature of importance is the presence, at least in early developmental stages and usually also in adults, of a **pharynx** (expanded anterior portion of the gut) which is perforated by numerous slits that permit water taken into the mouth to be passed out of the body. Among nonchordates, pharyngeal clefts are found only in hemichordates.

Another chordate character is a circulatory system having a **ventral heart** (or ventral pulsating vessel in cephalochordates) which drives blood up through the bars of the pharynx in vessels called aortic arches (these are secondarily modified in higher vertebrates) and thence caudad in a dorsal vessel. (Urochordates are exceptions in that the heart drives the blood in one direction for a time and then reverses to pump in the other direction.) The blood-vascular system of chordates is closed, i.e., blood remains in vessels and does not enter the tissue spaces.

Chordates tend to have their principal sense organs concentrated in a head, a condition spoken of as **cephalization.** This characteristic goes with the combination of bilateral symmetry and motility; urochordates, which are sessile, have no head. Two other chordate features which adult urochordates fail to exhibit are a **tail** extending posterior to the anus and **metamerism,** or segmentation of some features of the body.

The distinctive features of each subphylum should also be considered because differences as well as similarities are important to the establishment of relationships. UROCHORDATA inhabit coastal areas of all oceans. As already noted, only the free-swimming larva has a coelom, hollow nerve cord, and notochord. The notochord is prominent, but only in the tail ("urochordata" = tail + cord), where it is surrounded by muscle cells. Adults are sac-like or stalked, sessile, and often colonial. Colonies may be extensive, but individuals are small. The entire body is enveloped by a tough tunic. Between the tunic and the pharynx is a space called the atrium. Adult urochordates are filter-feeders. Water enters the pharynx by an incurrent siphon, seeps through complicated pores (one to thousands, depending on the species) in the pharynx to reach the atrium,

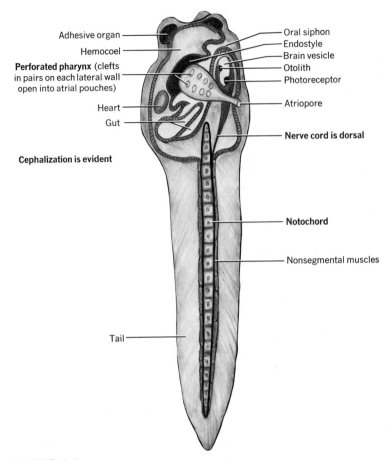

FIGURE 2-1
Stylized LARVAL UROCHORDATE drawn to illustrate internal structure. Characters of the phylum Chordata are shown by boldface labels;
characters of the subphylum are shown by standard labels.

and escapes by the excurrent siphon. Food particles are
trapped in sticky mucus which moves from the pharyngeal bars
to an endostyle and thence to the esophagus. The simple gut
loops within the body, and discharges into the atrium. Larvae
do not feed, and have a short gut. A heart pumps acellular
blood first in one direction and then in the other. Individuals
are hermaphroditic. Excretory cells may be grouped, but do not
form organs. Sex ducts are present. There is a nerve ganglion
and a nerve net. Urochordates are a large and diverse subphylum. Three classes (Ascidiacea, Thaliacea, Appendicularia)
are distinguished on the basis of number of pharyngeal slits,
colonial versus solitary habit, complete versus incomplete
metamorphosis, nature of tunic, etc.

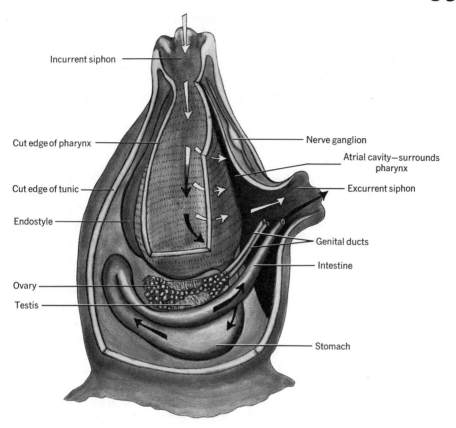

Incurrent siphon

Cut edge of pharynx

Cut edge of tunic

Endostyle

Ovary

Testis

Nerve ganglion

Atrial cavity—surrounds pharynx

Excurrent siphon

Genital ducts

Intestine

Stomach

FIGURE 2-2

Stylized ADULT UROCHORDATE with part of tunic and pharynx removed to show internal structure. Light arrows show path of respiratory current; dark arrows show path of food particles trapped on pharynx.

The subphylum CEPHALOCHORDATA has only two genera, yet is distributed over the world, especially in coastal areas having warm shallow water. The lance-shaped body measures 50–75 mm in length. There is a low continuous dorsal and tail fin and paired lateral body folds. The persistent notochord extends anterior to the nerve cord. Other skeletal elements support the fins, pharynx, and oral structures. There are light-sensitive cells on the floor of the nerve cord. Like urochordates, these animals are filter-feeders with a complicated pharynx surrounded by an atrial cavity, although details of their structure and development are unlike those of urochordates. Water enters the pharynx from the mouth and passes through about 150 slit-like oblique clefts to reach the atrium. The ciliated bars of the pharynx move a mucus strand into the straight digestive tract. Unlike a "true" liver, the hepatic diverticulum is hollow. The circulatory system lacks a heart but is otherwise closely similar to the vertebrate plan. Cephalochordates are unique and really quite peculiar in the asymmetry of various organs, which results from unusual developmental processes: The segmental muscles, pharyngeal slits, gonads, and spinal nerves alternate on the two sides of the body instead of lying in successive pairs. One other important feature departs from the vertebrate plan: The segmental excretory organs are derived from ectoderm (instead of mesoderm) and have clusters of flagellated cells such as are found in annelids.

What structures of VERTEBRATA distinguish these animals from all others? The nerve cord of larval urochordates enlarges anteriorly to form a vesicle, but only vertebrates have a true brain divided into several vesicles which serves to control and coordinate the nervous responses of the body. Further, only vertebrates have a skeletal structure, the cranium, which supports and protects the brain. Eyes, ears, and olfactory organs are present. Many other animals have light receptors and chemoreceptors, but the principal sense organs of vertebrates are not derived from others and rarely resemble them closely in structure. The concentration of brain, cranium, and sense organs in the head gives vertebrates a degree of cephalization that is approached only by some arthropods.

One would expect all vertebrates to have vertebrae. Most do, but some have only a few imperfectly formed vertebrae, and there is reason to doubt that the first vertebrates had them at all. The term Vertebrata became established before this was known. Only vertebrates have thyroid and pituitary glands. (Cephalochordates have a thyroid homolog without evidence of endocrine function; they have a structure thought by some to be a homolog of the pituitary, but this interpretation is questionable.) Most animals have accessory digestive glands and various of these are called "livers," but the solid liver of vertebrates is not homologous and only partly analogous to others. The related hepatic portal system and gall bladder are equally unique, as are the pancreas and spleen. A heart is present and chambered. The gonads are not segmental, and the kidneys are of mesodermal origin.

Vertebrates resemble the other chordate subphyla and hemichordates in having a perforated pharynx, but simple structure and predominately respiratory function of this organ are vertebrate characters. Complexity of the muscular system and attendant locomotor activities are also characteristic. Not all vertebrates have paired appendages, and some nonvertebrates have them, yet these are typical. Finally, although vertebrates include minnows, hummingbirds, and shrews, as a group they are relatively large animals.

This discussion started with a short definition of vertebrates, "any animal with a cranium is a vertebrate," and then ex-

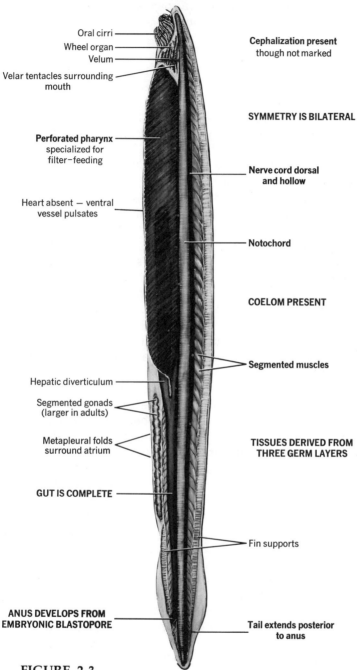

Oral cirri

Wheel organ

Velum

Velar tentacles surrounding
mouth

Cephalization present
though not marked

SYMMETRY IS BILATERAL

Perforated pharynx
specialized for
filter–feeding

**Nerve cord dorsal
and hollow**

Heart absent — ventral
vessel pulsates

Notochord

COELOM PRESENT

Segmented muscles

Hepatic diverticulum

Segmented gonads
(larger in adults)

Metapleural folds
surround atrium

**TISSUES DERIVED FROM
THREE GERM LAYERS**

GUT IS COMPLETE

Fin supports

**ANUS DEVELOPS FROM
EMBRYONIC BLASTOPORE**

**Tail extends posterior
to anus**

FIGURE 2-3

AMPHIOXUS, drawn from a cleared 37 mm specimen to
illustrate internal structure. Characters of the Deuterostomia are shown by boldface capital labels; characters of
the phylum Chordata are shown by boldface lowercase
labels; characters of the subphylum Cephalochordata are
shown by standard labels.

panded the concept for several pages. The student might find it instructive to phrase a definition in one paragraph.

Origin of Vertebrates

The preceding discussion of structure indicates that vertebrates are more closely related to cephalochordates than to urochordates, and that the phylum chordata is more closely related to echinoderms and, particularly, to hemichordates than to other phyla. Biochemical and immunological data also support these relationships.

These conclusions are accepted by most present-day zoologists. They are by no means self-evident, however, and in passing it should be said that many alternative theories have been proposed. As early as 1818, when the concept of evolution was but dimly perceived, the French naturalist St. Hilaire derived vertebrates from insects. Near the end of the nineteenth century the American biologist Patten, among others, proposed that vertebrates evolved from the king crab. Other arthropods were also suggested as ancestors. About the same time, many zoologists, including such prominent men as the Scottish embryologist Balfour (who proposed the phylum name "Chordata") and the German anatomist Gegenbaur, considered vertebrates to have evolved from one or another annelid worm. In order to get the nerve cord in the right place and the blood flow in the right direction, it was necessary to postulate that the annelid progenitor became inverted. Still other lineages were proposed, but none of these theories survived as knowledge of structure and development increased. Comparative embryology was particularly damaging to these notions and in 1894 led Garstang to the currently favored theory.

The current theory is no more than a theory, for the evidence is sufficient only to guide our speculations. Nevertheless, several tentative conclusions are worthy of notice.

No known echinoderm could be the chordate ancestor. Radial symmetry, water vascular system, nerve ring, modified coelom, and other features rule them out. It is the less specialized and bilaterally symmetrical larva which resembles the larvae of simple chordates. Further, no known hemichordate could be the chordate ancestor. Although hemichordates are probably closer to chordates than echinoderms, the structure of their circulatory system, proboscis, and collar could hardly be converted to the vertebrate body plan.

Similarly, adult urochordates are far too specialized to include the sought-for ancestor. One need only recall the loss of notochord, nerve cord, and coelom, and the presence of such nonchordate features as tunic, atrium, and siphons. Amphioxus is at first glance a possible ancestor, but what of its asym-

metry, atrium, unusual pattern of nerves, and annelid-like excretory organs?

One must conclude that vertebrates did not evolve from any
known animal, living or extinct. It follows that echinoderms,
hemichordates, and chordates must have diverged from a
common lineage not later than early Cambrian times some 600
million years ago. Since those remote days, each group has
gone its own way, retaining some of the common features over
the ages, but altering or deleting others and adding new structures according to their separate needs. Likewise, the subphyla
of chordates must have distinguished themselves soon thereafter. These tentative relationships are summarized by Figure 2-4.

There is an alternative theory in regard to cephalochordates.
Some investigators think that amphioxus is a degenerate and
specialized offshoot from a vertebrate ancestor. However, if
this were so, one would expect some trace of brain, cranium, or
eyes to be recapitulated in their embryonic development, and
none has been found.

Embryonic and larval structure must be stressed in any
discussion of vertebrate ancestry. Cleavage pattern, fate of
blastopore, and origin of mesoderm and coelom relate the
deuterostomes to one another. Hemichordates are linked to
echinoderms primarily by the similarity of their larvae, which
are called tornaria. Only larval urochordates resemble other
chordates, and cephalochordates are to be compared not with

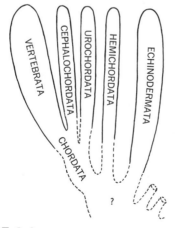

FIGURE 2-4
**TENTATIVE PHYLOGENY of the subphyla
of Chordata and of their closest nonchordate
relatives.**

adult vertebrates but with embryos and with ammocoetes, which is the larva of the eel-like lamprey.

It seems that each successive group evolved from the larva, not the adult, of the preceding group. Such an evolutionary relationship is called pedogenesis (see p. 23). Sexual development is accelerated and the development of other organ systems is arrested so the nonreproductive larva of the ancestor becomes the reproductive adult of the descendant.

The Construction and Interpretation of Classifications

When sorting many items it is desirable to establish categories of different rank arranged in a hierarchy. Thus, the postman can sort 6,250,000 addresses by arranging only 50 house numbers on each of 50 streets in each of 50 cities in each of 50 states. Similarly, the classifier of animals uses a number of categories arranged one within another. The general name he uses for a category is **taxon** (plural, taxa). The hierarchy of taxa assigned to every animal, in order from the most inclusive to the least inclusive, is **phylum, class, order, family, genus, species.** Either the "higher" (more inclusive) or "lower" (less inclusive) categories may be omitted from a particular classification if the scope of the study at hand does not require them. However, every animal is assignable to each of these categories. Every animal *has* a class, order, family, etc., whether it is stated or not. The classification of large assemblages of animals, such as the bony fishes, may require additional categories. These are established either by coining new words (e.g., cohort, branch, division, tribe), or by adding a prefix such as *sub-* (below or under), *infra-* (below or inferior), or *super-* (over) to the usual categories. In this book we will most often refer to the class and subclass. Several infraclasses and about two dozen orders will also be noted.

There remains the problem of deciding the degree of relationship required to associate animals at a particular level of the hierarchy. Do these new anatomical data justify the placing of these species in different genera? Does this new fossil belong in a previously named family, or should a new family be created? Only the species is an objective, or natural unit. It may be defined as a group of actually or potentially interbreeding natural populations that is reproductively isolated from other such groups. A species is not a specimen that is assigned to this rank on morphological criteria; it is a group of living animals, having morphological features in common, but assigned on the basis of the relationship of its breeding members. Although this definition emphasizes distinctions among contemporary animals, species (and also the other taxa) have a dimension in time. Over the ages species change to become new species. If the paleontologist had before him representative fossils from all the stages in such a transformation, he would need to designate an arbitrary boundary between the parent species and its descendant. Usually there are enough gaps in the fossil record so that he can conveniently let gaps separate the species he designates.

Definition of the other taxa is more arbitrary. Animals in nature are not arranged into categories which man, by his industry, tries to discover; they differ from one another in an almost infinite number of characteristics, both of kind and degree, which man, in his wisdom tries to understand. A genus is one or more lineages of related species. Similarly, each higher taxon is an assemblage of related lower taxa. Animals within one species, one genus, or one family tend to have a common adaptation such as climbing ability, fish eating, or resistance to cold; animals within a taxon above the family level tend to share basic structural patterns such as four legs, feathers, amnion, or gnawing teeth.

More and more effort is being made to use objective, numerical methods in the classification of animals. Nevertheless, the ordering of animals into categories has been based largely on the authority of specialists recognized as competent, and this seems likely to remain true, at least for the higher taxa.

Even competent authorities cannot be specialists on all kinds of vertebrates, usually cannot examine all relevant materials, and, like other people, sometimes make mistakes, so it is not surprising that even though we agree that evolution is to be our guide, still there are differences of opinion. Why, the student may ask, should I bother to learn any classification if the experts cannot agree among themselves? There are three good answers: First, the general pattern of vertebrate phylogeny is evident to all specialists; therefore, they are in agreement about the major features of their classifications. Second, any reasonably sound classification serves as the powerful tool we require to create order and express meaningful relationships among animals. Finally, since a classification is an abstraction that is designed to be useful, it is to be expected that classifications intended for different uses should be somewhat different.

The theoretical study of the principles, procedures, and rules of classification is called **taxonomy.** The application of names to the groups recognized is called **nomenclature.** There has been international acceptance of rules that regulate the kinds of names that may be assigned, the way new names must be announced, how priorities are established, and how mistakes can be corrected.

Since evolution has been a branching process, it is not possible to arrange animals in a linear sequence on a piece of paper and at the same time to accurately show all their phylogenetic relationships. Nevertheless, a classification that is based on evolution is consistent with our knowledge of phylogeny and tells us much about it. The points below apply to most vertebrate classifications, including the one described in the next section (which it may be well to consult as the points are considered).

(1) Animals grouped together in one of the higher, or more inclusive categories (e.g., a class or subclass) have relatively few characters in common; these are considered to delineate a structure pattern of broad evolutionary importance. The higher categories may be regarded as comprising the main channels in the stream of life. Conversely, animals grouped together in one of the lower, or less inclusive categories (e.g., genus or species) have relatively many characters in common.

(2) All the smaller categories within a common larger category share a common ancestry. The common ancestor may lie in one of these same smaller taxa or, alternatively, the common ancestor may lie in a taxon of equal or still smaller rank that is included in a different larger category. All the subclasses of Mammalia are derived from a certain smaller taxon (an order) in an entirely different class—the Reptilia.

(3) Any large taxon tends to be older in origin than the average age of the next smaller taxa that it includes. It follows that the characters shared in common by all members of a large taxon tend to be more primitive than the characters shared in common by the members of most of the included smaller taxa.

(4) Taxa that are considered to be relatively primitive are listed before others of equal rank that are regarded as being more advanced. Each animal group is usually placed adjacent to its closest relatives; proximity of listing is an indication of proximity of relationship. (Thus, amphibians are more closely related to bony fishes, which immediately precede them on the classification, than to either placoderms or cartilaginous fishes from which they are separated by other classes.) However, it was necessary to qualify the above statement because a written classification is of necessity linear. Thus, the classification indicates correctly that cartilaginous fishes evolved from placoderms, but is unable to show that at about the same time in earth history the bony fishes also evolved from the placoderms. Similarly, mammals did not evolve from birds as we might infer from the relative positions of these classes; mammals and birds each evolved from the reptiles.

Figure 2-5 shows the evolutionary relationships of the classes of vertebrates, their distribution in time, and the approximate relative abundance of the species in each. Figures

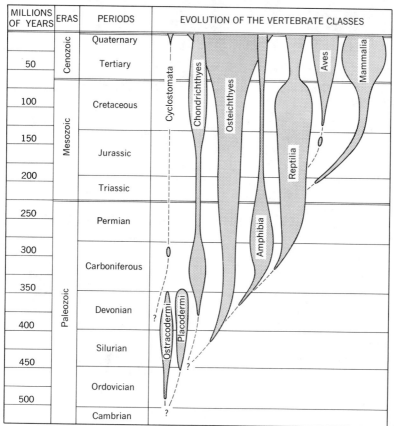

MILLIONS OF YEARS	ERAS	PERIODS	EVOLUTION OF THE VERTEBRATE CLASSES
50	Cenozoic	Quaternary	
		Tertiary	
100	Mesozoic	Cretaceous	
150		Jurassic	
200		Triassic	
250	Paleozoic	Permian	
300		Carboniferous	
350		Devonian	
400		Silurian	
450		Ordovician	
500		Cambrian	

FIGURE 2-5

OUTLINE OF VERTEBRATE EVOLUTION IN RELATION TO THE GEOLOGIC TIME SCALE.

2-6 and 2-7 summarize the evolutionary and systematic relationships of the principal taxa discussed in this book. Frequent reference to these figures may be desirable as subsequent chapters are studied.

A Classification of the Vertebrates

The classification given on p. 46 has been adapted to the needs of the remaining chapters. All classes and most subclasses are included. Categories below the subclass are included only if useful to this course of study. Technical names are listed first, followed by common names of familiar examples if such are available. Taxa preceded by an asterisk are extinct. It is suggested that the student memorize the classification as he studies Chapters 4–7.

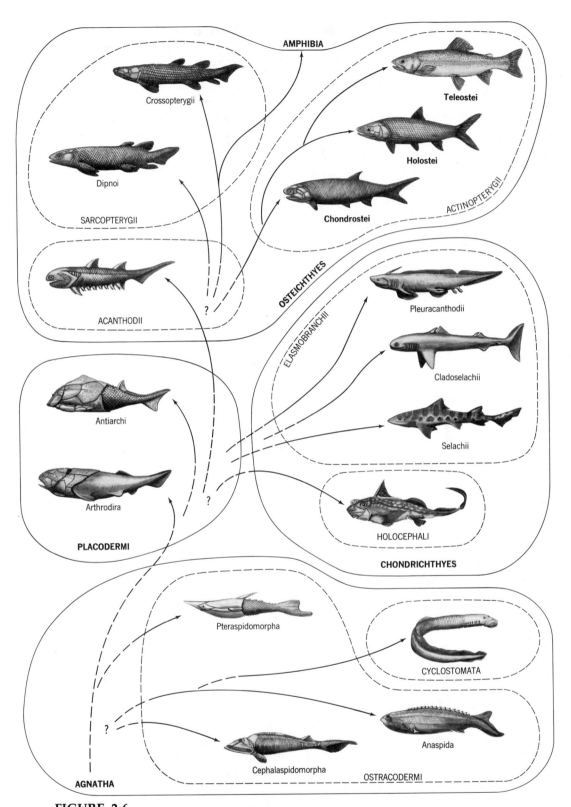

FIGURE 2-6

SYSTEMATICS AND PHYLOGENY OF FISHES. Classes are shown by boldface capital letters, subclasses by standard capitals, some infraclasses by boldface lowercase letters, and some orders by standard lowercase letters.

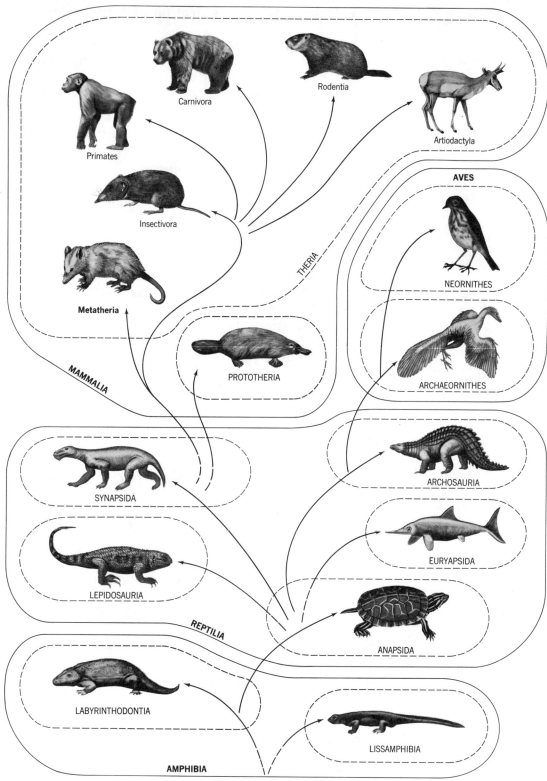

FIGURE 2-7
SYSTEMATICS AND PHYLOGENY OF TETRAPODS. Notations as for Figure 2-6.

Class **Agnatha:** jawless vertebrates
  *Subclass **Ostracodermi:** ostracoderms, including cephalaspids, anaspids, and pteraspids
  Subclass **Cyclostomata:** cyclostomes, including lampreys and hagfishes

*Class **Placodermi:** placoderms, including arthrodires, antiarchs, and shark-like forms

Class **Chondrichthyes:** cartilaginous fishes
  Subclass **Elasmobranchii:** elasmobranchs, including *pleuracanths, *cladodonts, and sharks and rays
  Subclass **Holocephali:** chimaeras

Class **Osteichthyes:** bony fishes
  *Subclass **Acanthodii:** acanthodians
  Subclass **Actinopterygii:** ray-finned fishes
    Infraclass **Chondrostei:** chondrosteans, including sturgeons and paddlefishes
    Infraclass **Holostei:** holosteans, including gars and the bowfin
    Infraclass **Teleostei:** teleosts
  Subclass **Sarcopterygii:** sarcopterygians or fleshy-finned fishes
    Infraclass **Dipnoi:** dipnoans or lungfishes
    Infraclass **Crossopterygii:** crossopterygians or lobe-finned fishes

Class **Amphibia:** amphibians
  *Subclass **Labyrinthodontia:** labyrinthodonts
  Subclass **Lissamphibia:** salamanders, frogs and toads, and caecilians

Class **Reptilia:** reptiles
  Subclass **Anapsida:** *stem reptiles and turtles
  Subclass **Lepidosauria:** lizards, snakes, amphisbaenians, and *Sphenodon*
  Subclass **Archosauria:** crocodilians, *dinosaurs, and *flying reptiles
  *Subclass **Euryapsida:** euryapsids, including plesiosaurs and ichthyosaurs
  *Subclass **Synapsida:** mammal-like reptiles

Class **Aves:** birds
  *Subclass **Archaeornithes:** ancestral birds
  Subclass **Neornithes:** familiar birds

Class **Mammalia:** mammals
  Subclass **Prototheria:** monotremes or egg-laying mammals
  Subclass **Theria:** mammals bearing "live" young
    Infraclass **Metatheria:** marsupials or pouched mammals
    Infraclass **Eutheria:** higher or "placental" mammals, including nearly all mammals not native to Australia

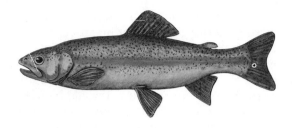

3

Fishes

In presenting the structure patterns that characterize vertebrate animals, it will be necessary to refer, again and again, to the major groups of vertebrates. The facing classification has merely presented group names; now it is desirable to relate those names to animals. A brief introduction to the more obvious characters serves this initial purpose. It is enough that the student visualize a representative of each group, recall several distinguishing group characteristics, and know the place of each group among the other vertebrates. If extinct and unusual animals seem to be stressed in this and the following chapter, it is not to give them special emphasis but because one needs little introduction to animals that are already familiar.

Introduction to the Introductions

Part II of the text is arranged by organ systems and Part III by adaptations; only in this part of the book are vertebrates arranged systematically. In separate chapters the student will learn something about (for instance) the mammary glands, pectoral girdle, urogenital system, and locomotor mechanisms of egg-laying mammals. But knowledge is useful in proportion to one's ability to apply it in different ways; therefore, as the end of the book is approached, it would be desirable for the student to see if he could bring together all he has learned about egg-laying mammals (or ray-finned fishes, frogs, or any other group).

Class Agnatha

All animals that have the general characters of the subphylum vertebrata (as reviewed in Chapter 2), but do *not* have jaws, belong to the class Agnatha. These animals are little known to laymen because all except the lampreys and hagfishes became extinct nearly 350 million years ago. Even lampreys and hagfishes are rarely seen by the public; they sometimes destroy fishes of commercial value, but otherwise are obscure.

The morphologist, however, should become acquainted with the animals in this class because of their important position in the phylogeny of vertebrates. They were the first vertebrates to evolve, and all other vertebrates trace their ancestry to the Agnatha. The most primitive body plan of a living vertebrate is that of ammocoetes, the larva of the lamprey.

The jawless vertebrates have advanced over their progenitors among the protochordates in the possession of several important characters: They have heads with a cranium, a brain, and paired organs of sight. Although incomplete by the standards of other vertebrates, vertebrae are at least represented by cartilaginous elements on the dorsal surface of the large persistent notochord. The internal skeletons of jawless vertebrates are mostly cartilaginous, but in the heavy scales and armor worn by the extinct Agnatha, that important material, bone, is present for the first time in earth history.

The Agnatha are also noteworthy for the lack of certain characters that became typical of vertebrates standing higher on the evolutionary ladder: They do not possess jaws, true teeth, girdles, or typical appendages. Pectoral spikes, folds, or lobes are often present; pelvic fins are never found. The gills are located in pouches.

Later chapters will refer to the jawless vertebrates in discussions of respiratory structures, tail shapes, the special senses, and the origins of scales, vertebrae, and fins.

Subclass Ostracodermi. The ostracoderms were small, fish-like animals that lived in the freshwater streams of several continents. They evolved nearly 500 million years ago and were most abundant at a time when other vertebrates were just making their appearance on earth. Any jawless vertebrate with scales or armor plates is an ostracoderm—the name means "shell skin."

Excellent fossils of some ostracoderms have been found; the patterns of nerves, blood vessels, and sensory structures of the head are known in considerable detail for certain genera. Nevertheless, the record of these long extinct creatures is fragmentary. Their classification and evolutionary relationships are debated by specialists. The subclass is commonly divided into three principal taxa which are here called orders: Cephalaspidomorpha, Anaspida, and Pteraspidomorpha. (Another group, the poorly known Coelolepida, is not described in this text.)

CEPHALASPIDOMORPHA are the only jawless vertebrates to have depressed heads and dorsal eyes—characters that indicate these queer creatures moved slowly along stream bottoms to nibble food found in mud or sand. Cephalaspids are also unique among the Agnatha for their lobe-like pectoral projections. Paired lateral and median dorsal sculptured areas on the head shield are probably sense organs. The mouth is small and ventral. The upper lobe of the tail is larger than the lower lobe.

ANASPIDA are the only jawless vertebrates having streamlined form and the only ostracoderms to have small, plate-like scales and no head armor ("anaspid" = without + shield). Their eyes are lateral. They may have been relatively active swimmers. They have pectoral spikes which probably relate to fin folds of unknown nature. Like cephalaspids they have many separate gill openings, but unlike that order, anaspids have lateral instead of ventral gill openings.

PTERASPIDOMORPHA appear in the fossil record 75 million years earlier than the other ostracoderms, yet they survived as long as any. Like cephalaspids, they have a heavy shield of armor covering the head and anterior part of the body. The rostrum usually projects over the mouth, and often there are bizarre spines on the shield or along the back ("pteraspid" = wing + shield). Other appendages are absent. These are the only ostracoderms to have the separate gill chambers covered by armor so there is a common exit from all the gill pouches. The head is not markedly flattened dorsoventrally as in cephalaspids, and the eyes are lateral—features by which one infers that pteraspids were slow swimmers but not habitual bottom feeders.

Subclass Cyclostomata. Lampreys (order Petromyzontiformes) and hagfishes (order Myxiniformes) are eel-like animals which comprise the subclass cyclostomata. They range in length from

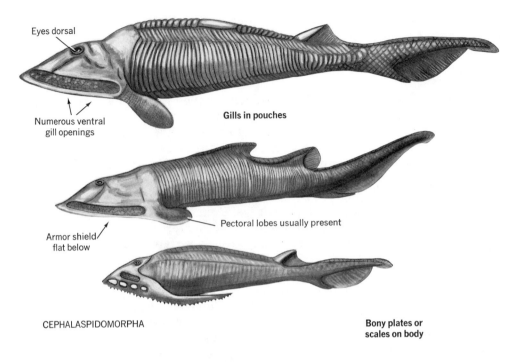

Eyes dorsal

Numerous ventral
gill openings

Gills in pouches

Pectoral lobes usually present

Armor shield
flat below

CEPHALASPIDOMORPHA

**Bony plates or
scales on body**

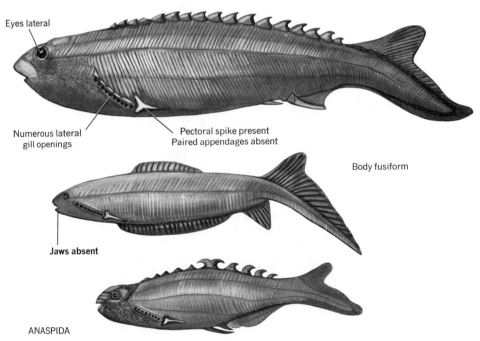

Eyes lateral

Numerous lateral
gill openings

Pectoral spike present
Paired appendages absent

Jaws absent

Body fusiform

ANASPIDA

FIGURE 3-1

Restorations of REPRESENTATIVE OSTRACODERMI illustrating some characteristics of the subclass (boldface labels) and of the orders Cephalaspidomorpha and
Anaspida (standard labels).

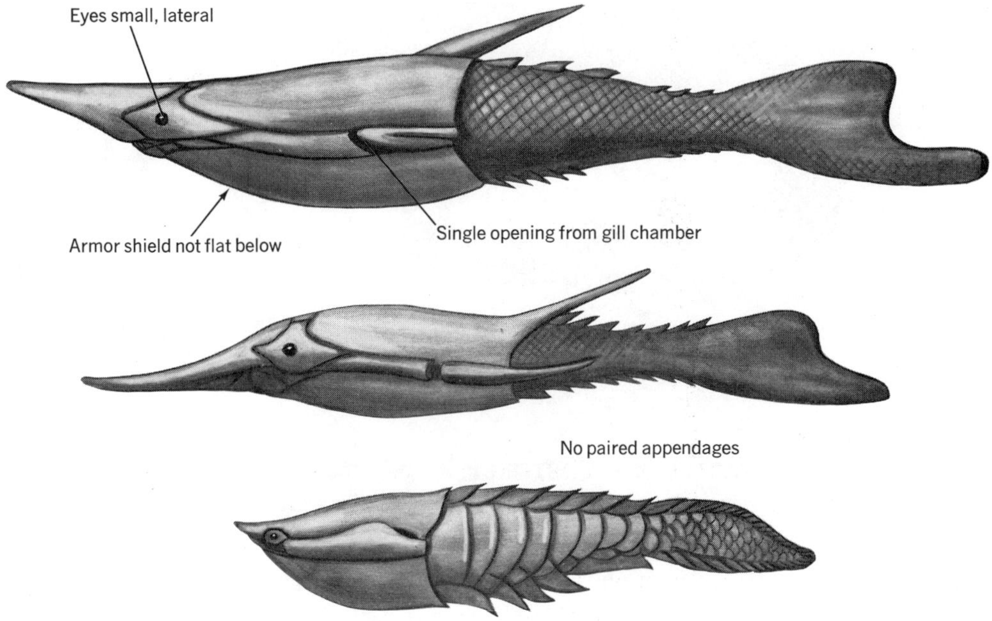

Eyes small, lateral

Armor shield not flat below

Single opening from gill chamber

No paired appendages

FIGURE 3-2

Restorations of REPRESENTATIVE OSTRACODERMI of the order Pteraspidomorpha.

10 to 90 cm, and they inhabit oceans and also freshwaters of several continents. The lamprey is commonly dissected in the comparative anatomy laboratory. We are fortunate to have this animal to show us something of the soft tissues of the class, even though we cannot tell how representative they may be. This book will refer to cyclostomes for the seemingly primitive nature of their axial musculature, digestive organs, circulatory system, release of sex cells, eye and nasal structure, and other parts.

The larva of the lamprey, ammocoetes, is even more primitive than the adult. This is particularly true in regard to mouth parts, pharynx, gonads, and some digestive organs. Ammocoetes becomes adult by metamorphosis.

Cyclostomes are semiparasitic upon bony fishes. Their mouths and "tongues" are adapted for holding onto their hosts and for rasping or cutting away their flesh ("cyclostome" = round + mouth). Sense organs, body form, and digestive structures are also modified in response to this way of life.

Cyclostomes are the only jawless vertebrates to have eel-like form, slippery naked skin, and a suctorial mouth. They are important because they retain such primitive characters as a persistent notochord and absence of complete vertebrae, paired appendages, and jaws. They also display specialized features (e.g., mouth parts) which the anatomist must avoid mistaking

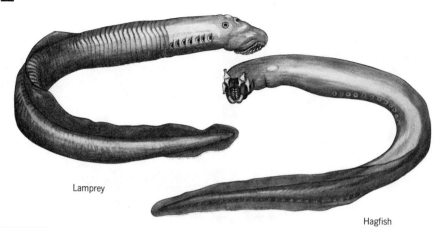

Lamprey

Hagfish

FIGURE 3-3

REPRESENTATIVE CYCLOSTOMATA showing attenuated form, rasping mouthparts, and absence of jaws, scales, and paired appendages.

for ancestral characters. The two orders differ markedly in the nature of their head skeletons, pouched gills, sense organs, and some other characters.

Most ostracoderms were heavily armored, and the others of which we have record had scales. Cyclostomes, on the other hand, have neither scales nor bone. For this reason, there is almost no fossil record of cyclostomes. However, one remarkable fossil impression shows that a lamprey-like cyclostome existed about 300 million years ago.

Jaw-Bearing Fishes

The most important evolutionary advance common to all the remaining fishes was the enlargement and adaptation of the first gill arch to function as jaws instead of gill supports. Chordates that lacked jaws could only filter microorganisms from water, grovel in mud, or rasp and cut soft flesh. The evolution of jaws permitted fishes and their descendants to utilize larger and harder food, and thus enabled them to become adapted to many new and diverse ways of living. This advance was of sufficient importance so that fishes and tetrapods are together called *gnathostomes* (= jaw + mouth) to set them apart from the agnatha (= without + jaw).

A second important advance common to all jaw-bearing fishes is the presence of paired appendages. True, some ostracoderms had pectoral folds (anaspids) or lobes (cephalaspids) which suggested incipient appendages, but it remained for jaw-bearing fishes to experiment with various types of paired appendages and perfect several models.

Most of the world's vertebrates—past and present—are jaw-bearing fishes. The diversity of their habits, form, and structure is enormous. They are important to the morphologist for the

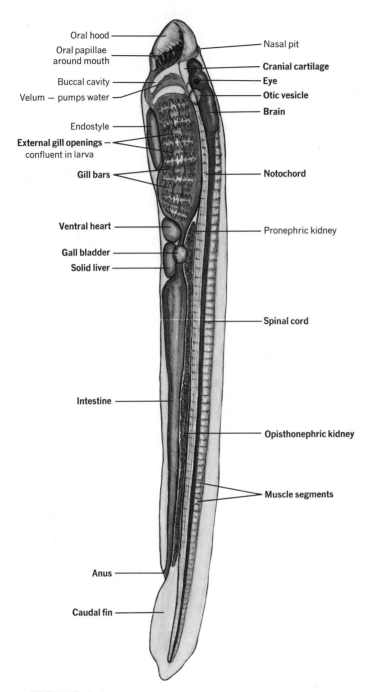

FIGURE 3-4

AMMOCOETES, larva of the lamprey, drawn from a cleared, 12 mm specimen to illustrate internal structure. Primitive and unspecialized vertebrate characters are shown by boldface labels; other characteristics are shown by standard labels.

advancements and specializations found in nearly every organ system. Three classes are recognized here.

Class Placodermi

Most placoderms have either bony scales or armor plates, particularly on the forward part of the body—the word "placoderm" means "plate skin." These fishes, like jawless vertebrates, have persistent notochords. The internal skeleton contains some bone. Placoderms swam first in the rivers and later also in the oceans of the world for 50 million years or more. They reached maximum abundance about the time amphibians evolved, and the last of their kind became extinct some 345 million years ago, before reptiles, birds, and mammals had made their appearance.

These remote creatures are not well-known to the general biologist, and few laymen have heard of them at all. There are good reasons, however, why the comparative anatomist should not entirely neglect the placoderms: Directly or indirectly, they probably include the ancestors of all of the more advanced classes. They were the first to evolve two of the most successful of all vertebrate structures—jaws and paired appendages. They pioneered gas bladders that were ultimately to become lungs. Features of their gill structure, scales, digestive organs, and sense organs were transitional between those of ostracoderms and higher fishes.

Placoderms are diverse in structure; the class may prove to be unnatural and polyphyletic. The evolutionary relationships within the class, and between this class and others, have by no means been worked out. The two best-known groups of placoderms, antiarchs and arthrodires, are commonly classified as orders. Each is world-wide in distribution and abundant in the fossil record. (Some less well-known groups having shark-like form and little or no armor are not included here, but may have been ancestral to cartilaginous fishes.)

ANTIARCHI are the most bizarre of placoderms because of their peculiar pectoral appendages ("antiarch" = opposite + arm). These long structures are hinged to the body and covered with bony plates. It is thought that they functioned as holdfasts against the currents of streams, or were used for creeping along stream bottoms. A bottom-living habit is also suggested by the dorsal position of eyes and olfactory organs and by the arched dorsal contour and flat ventral surface of the body. The small head and bulky thorax are encased in units of heavy armor which are hinged to each other. The jaws of antiarchs are small and weak. Overall body size is small. This group of fishes seems not to be ancestral to any other.

Typical ARTHRODIRA are the most spectacular of placoderms because most have large, predaceous jaws with serrated

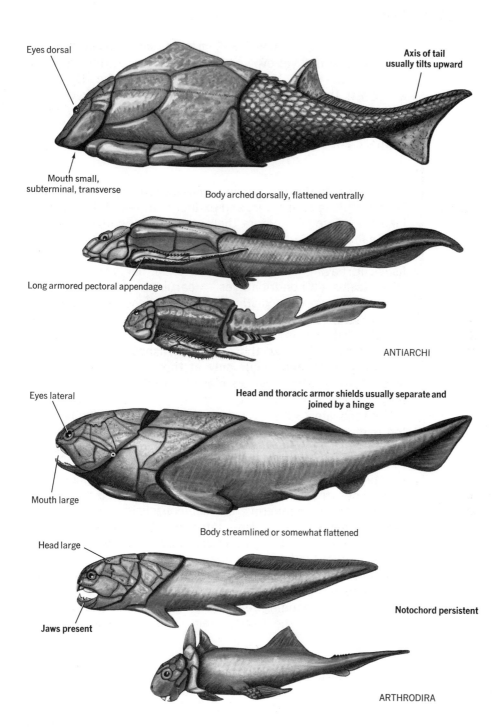

Eyes dorsal

Axis of tail usually tilts upward

Mouth small, subterminal, transverse

Body arched dorsally, flattened ventrally

Long armored pectoral appendage

ANTIARCHI

Eyes lateral

Head and thoracic armor shields usually separate and joined by a hinge

Mouth large

Body streamlined or somewhat flattened

Head large

Notochord persistent

Jaws present

ARTHRODIRA

FIGURE 3-5

Restorations of REPRESENTATIVE PLACODERMI illustrating some characteristics of the class (boldface labels) and of the two principal orders (standard labels).

margins, and some, being 6–9 m long, were the largest vertebrates that had yet evolved when they roamed the rivers and
oceans of the world. Most arthrodires have blunt heads and lateral eyes; they resemble antiarchs in having heavy cephalic
and thoracic shields that are hinged together ("arthrodire" =
joint + neck) and in having the gills usually hidden by the
cephalic armor.

Class Chondrichthyes This class includes the familiar sharks and rays, the less familiar but striking chimaeras, and extinct relatives of these fishes.
Like the arthrodires before them, the Chondrichthyes are pre-
dominantly marine and are of medium to large size. They differ
from their placoderm ancestors and also from most other
fishes in having no bone whatsoever, either internally or in
their scales ("Chondrichthyes" = cartilage + fishes). Cartilaginous fishes are also distinctive among fishes for their solid
braincase, fin structure, branching pattern of blood vessels associated with the gills, and small tooth-like scales (or none at
all). Their teeth, unlike those of other fishes, are anchored to
the integument and occur only at the margins of the jaws.
Most of these fishes have a series of external gill openings and
they lack a gas bladder.

Until the late 1930s it was thought that a cartilaginous skeleton is more primitive than a bony skeleton. Consequently, the
Chondrichthyes were accorded a key evolutionary position as
the ancestors of other fishes. As the fossil record grew, another
interpretation became more tenable: Both cartilaginous and
bony fishes (the next class to be described) evolved from (probably different) placoderms. Cartilaginous fishes appear in the
record some 30 million years later than bony fishes and have
not been ancestral to any other class. Evidence for this conclusion will unfold in later chapters.

Nevertheless, cartilaginous fishes do retain from their ancestors some features that are more primitive or unspecialized
than the corresponding features of bony fishes. Examples are
the structure of the heart, brain, and musculature of the
pharynx.

Subclass Elasmobranchii. Cartilaginous fishes are commonly
divided into two subclasses. Members of the first subclass
have slit-like external gill openings and hence are called
elasmobranchs (= plate + gills). Three orders of elasmobranchs—pleuracanths, cladodonts, and selachians—will be
recognized here; one or more additional orders are included in
some classifications.

PLEURACANTHODII are freshwater fishes, nearly a meter in

length, that became extinct about the time mammals evolved. A large spine borne on the back of the head is a handy recognition feature. (Few kinds of pleuracanths are known, however, and it is possible that fossils will be found that are assignable to this subclass, yet lack the spine.) The persistent notochord in this order and the next is mounted by somewhat calcified vertebral arches. Pleuracanths will be mentioned in later chapters for two other characteristic features that are unusual among fishes: Their paired fins have a jointed central axis, and the axis of the tail is straight all the way to the tip.

CLADOSELACHII were abundant in the Carboniferous period when the conquest of the land by reptiles was getting under way. These marine fishes look like large sharks except that the mouth is nearly terminal, the tail is nearly symmetrical externally, and the large pectoral fins have broad bases. This last feature is noteworthy because in the days when cartilaginous fishes were thought to be ancestral to bony fishes, the fin structure of cladodonts was believed to be the most primitive of any gnathostome.

The sharks and rays comprise the order SELACHII. All surviving cartilaginous fishes having a series of external gill slits and small abrasive scales are selachians. The first ancestral gill slit is reduced to a roundish opening called a **spiracle**. These are mostly marine fishes of medium to large size. The vertebrae have centra and contiguous arches around the spinal cord. Sharks and rays (commonly placed in separate suborders) differ markedly from one another in ways that relate to divergent habits. Many sharks are active predaceous fishes with fusiform bodies and large, strong tails. Their paired fins have restricted bases and their gill slits are lateral in position. Most rays, in contrast, spend much time resting on the bottom or swimming sluggishly along in search of shellfishes or other relatively inactive food. Their bodies are flattened dorsoventrally and the pectoral fins merge into the head and body. The mouth, being ventral, cannot take in water when the fish is at rest on sand or mud, so large spiracles perform this function instead.

Subclass Holocephali. Surviving members of this subclass are called chimaeras. Found only at sea, they rarely come to the attention of laymen and will be mentioned infrequently in the chapters that follow, yet they should not entirely escape notice because they include such peculiar fishes. These are the only cartilaginous fishes with a fleshy **operculum** covering the gills. There are few or no scales. The notochord is persistent, and there is no spiracle. Males have a unique club-shaped clasping

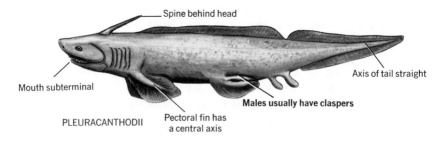

Spine behind head

Axis of tail straight

Mouth subterminal

Males usually have claspers

PLEURACANTHODII

Pectoral fin has
a central axis

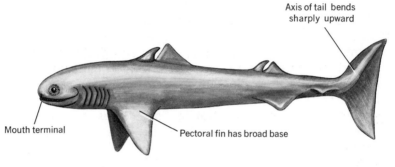

Axis of tail bends
sharply upward

Mouth terminal

Pectoral fin has broad base

CLADOSELACHII

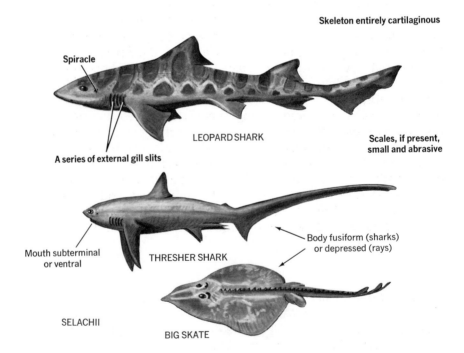

Skeleton entirely cartilaginous

Spiracle

LEOPARD SHARK

**Scales, if present,
small and abrasive**

A series of external gill slits

Body fusiform (sharks)
or depressed (rays)

Mouth subterminal
or ventral

THRESHER SHARK

SELACHII

BIG SKATE

FIGURE 3-6
**REPRESENTATIVE ELASMOBRANCHII illustrating some characteristics of the
subclass (boldface labels) and of the three orders (standard labels).**

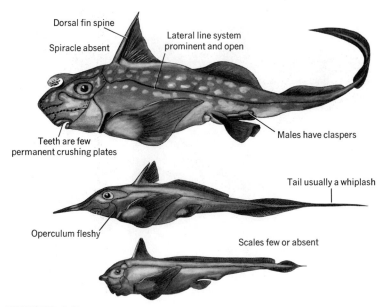

Dorsal fin spine

Spiracle absent

Lateral line system prominent and open

Teeth are few permanent crushing plates

Males have claspers

Tail usually a whiplash

Operculum fleshy

Scales few or absent

FIGURE 3-7
REPRESENTATIVE HOLOCEPHALI illustrating some characteristics of chimaeras.

organ on the top of the head. Large head and eyes, huge pectoral fins, whiplash tail, and striking colors give chimaeras a bizarre appearance. We will pay attention to the structure of their solid jaws, crushing teeth, and certain sense organs.

Bony fishes apparently evolved from placoderm ancestors at least 425 million years ago. For some 175 million years thereafter they were outnumbered, first by placoderms and then by cartilaginous fishes, but since the close of the Permian period, bony fishes have dominated the waters of the world. Since the end of the Mesozoic era, they have been the most abundant of all vertebrates. Their habits and structure are seemingly as diverse as common adaptation to an aquatic life will permit.

Most fishes of this class have bone in their skulls, vertebrae, girdles, fin supports, and scales. Some have secondarily substituted cartilage for much of the ancestral bone, yet even such fishes retain more bone in their internal skeletons than is present in other classes of fishes. Bony fishes are the only vertebrates having the gills of each side of the body in a common chamber covered by a movable, bony operculum. They have various sorts of fins, scales, and vertebrae, yet these structures nearly always differ from those of other classes. The pectoral girdle is joined to the skull by a chain of bones. A lung or gas bladder is usually present.

Class Osteichthyes

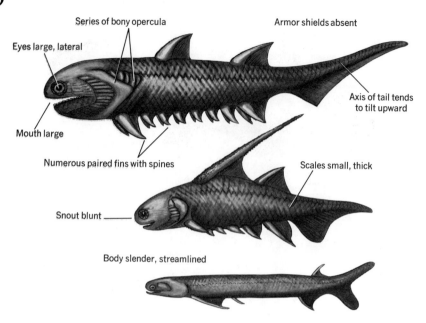

Series of bony opercula

Armor shields absent

Eyes large, lateral

Axis of tail tends to tilt upward

Mouth large

Numerous paired fins with spines

Scales small, thick

Snout blunt

Body slender, streamlined

FIGURE 3-8

Restorations of REPRESENTATIVE ACANTHODII illustrating some characteristics of the subclass.

Subclass Acanthodii. Of the three subclasses of bony fishes, the first to evolve, and the only one to become extinct is the acanthodians. They evolved a little before the known placoderms and outlived that class by about 80 million years. Their relationship to placoderms is unknown. They were apparently ancestral to the other bony fishes, but a clear story must await the discovery of further fossil evidence. Acanthodians have streamlined bodies, large lateral eyes, and wide mouths with numerous teeth. Their heads are bony and their small scales thick and hard, but they do not have the armor of most of their ostracoderm and placoderm contemporaries. We infer from these clues that these small, mostly freshwater fishes were active swimmers. The numerous fins of acanthodians are unique in that each has a thin membrane supported at its leading edge by a long stout spine. This feature gives the subclass its name; "acanthodian" means "spine form."

Subclass Actinopterygii. Most bony fishes belong to the subclass Actinopterygii, or ray-finned fishes. The membranes of the paired fins are supported by bony rays which radiate from the fin base. Consequently, the fins do not have fleshy stalks, as do the fins of fishes in the next subclass. The pattern of cranial bones and the nature of the venous system and reproduc-

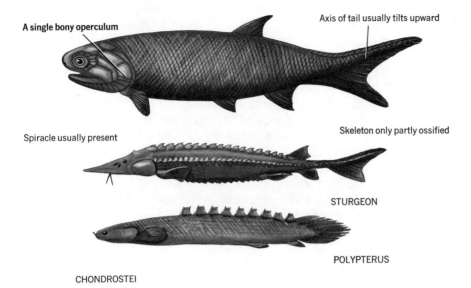

A single bony operculum

Axis of tail usually tilts upward

Spiracle usually present

Skeleton only partly ossified

STURGEON

POLYPTERUS

CHONDROSTEI

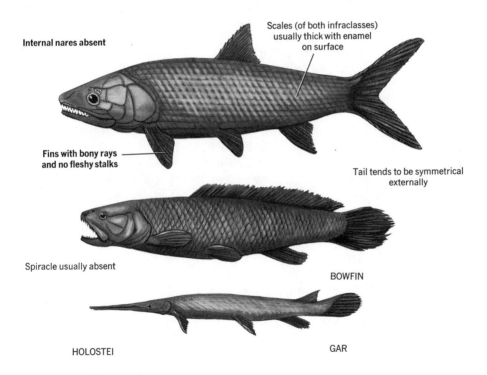

Internal nares absent

Scales (of both infraclasses)
usually thick with enamel
on surface

Fins with bony rays
and no fleshy stalks

Tail tends to be symmetrical
externally

Spiracle usually absent

BOWFIN

HOLOSTEI

GAR

FIGURE 3-9

REPRESENTATIVE CHONDROSTEI AND HOLOSTEI illustrating some characteristics of the subclass Actinopterygii (boldface labels) and of these two infraclasses (standard labels). Larger drawings are of extinct fishes, others are of surviving fishes.

tive ducts are also distinctive and clearly show that these fishes do not include the ancestors of land vertebrates.

Ray-finned fishes are classified into three infraclasses termed CHONDROSTEI, HOLOSTEI, and TELEOSTEI. These are regarded as forming an evolutionary sequence in the order named. They differ from one another in regard to degree of ossification of the skeleton; presence or absence, mobility, or pairing of certain bones of the skull and pectoral girdle; shape of tail fin; presence or absence of spiracle; and certain features of the digestive system. Chondrosteans (which may not be an entirely natural group) thrived in the Triassic period. They are now represented by the sturgeons and paddlefish, which are, in some respects, not representative of their infraclass, and by two related genera native to Africa which are also atypical (and hence variously classified) but of special interest because of the primitive nature of their scales and gas bladder. Of these latter two fishes, *Polypterus* is the better-known. Holosteans were most abundant in the Jurassic and Cretaceous periods. The gars and bowfin are among the few survivors. Aquarium, sport, and food fishes belong to the large infraclass Teleostei. About 30,000 species and subspecies are arranged in 400–600 families and are grouped variously in 6 (conservative) to 30 or 40 (usual) or more orders. Teleosts are recognized as being either primitive or advanced on the basis of mouth structure, position of pelvic fins, number and stiffness of dorsal fin rays, and other characters.

Subclass Sarcopterygii. This subclass was formerly termed Choanichthyes (= nostril + fishes), but the term was not apt and has been replaced by Sarcopterygii (= flesh + fin) because the fins of these fishes have fleshy stalks. Sarcopterygeans thrived in the Devonian period, 200 million years before teleosts evolved, and became scarce after the Triassic period. Several somewhat degenerate representatives of the subclass survive today.

DIPNOI, or lungfish, comprise the first of the two infraclasses. These are mostly freshwater fishes of moderate size and either "normal" or elongate form. They have atypical **internal nares** (nostrils opening into the mouth), functional lungs, and advanced circulatory systems. For these reasons they were once regarded as being ancestral to tetrapods. This view is no longer tenable because the other infraclass also possesses (or doubtless formerly possessed) these characters and is less specialized in regard to other features.

The paired fins of typical (extinct) dipnoans have a fleshy stalk and a jointed skeletal axis resembling that of pleuracanths. The few teeth are peculiar fan-shaped plates adapted for crushing. Braincase and vertebral column are poorly os-

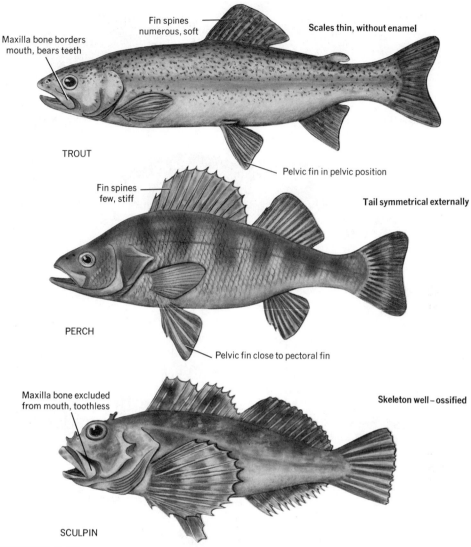

Fin spines
numerous, soft

Scales thin, without enamel

Maxilla bone borders
mouth, bears teeth

TROUT

Pelvic fin in pelvic position

Fin spines
few, stiff

Tail symmetrical externally

PERCH

Pelvic fin close to pectoral fin

Maxilla bone excluded
from mouth, toothless

Skeleton well – ossified

SCULPIN

FIGURE 3-10

REPRESENTATIVE TELEOSTEI illustrating some characteristics of the infraclass (boldface labels) and differences between more primitive (above) and more advanced species (center and below).

sified. The scales of surviving dipnoans are simplified in structure and sometimes degenerate.

The respiratory and circulatory systems of lungfishes are at the same time advanced for fishes and primitive for tetrapods. Unfortunately, modern lungfishes are rarely dissected in the classroom because of lack of availability and the aberrant nature of much of their anatomy.

The other infraclass, CROSSOPTERYGII, is of particular importance. Many ancient crossopterygians had internal nares,

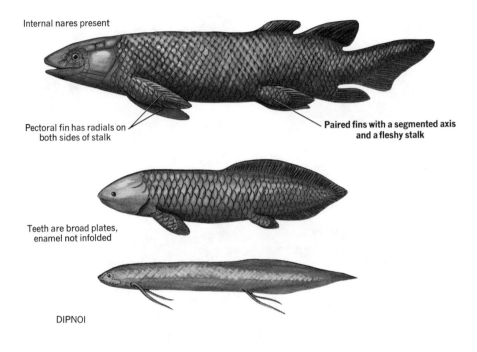

Internal nares present

Pectoral fin has radials on
both sides of stalk

Paired fins with a segmented axis
and a fleshy stalk

Teeth are broad plates,
enamel not infolded

DIPNOI

A functional lung is
usually present

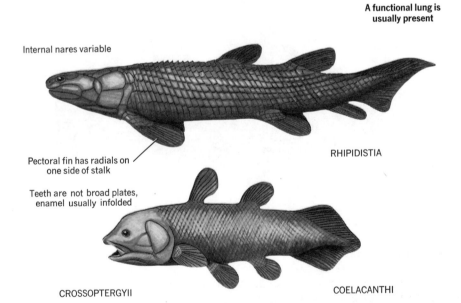

Internal nares variable

Pectoral fin has radials on
one side of stalk

Teeth are not broad plates,
enamel usually infolded

RHIPIDISTIA

CROSSOPTERGYII

COELACANTHI

FIGURE 3-11

REPRESENTATIVE SARCOPTERYGII illustrating some characteristics of the
subclass (boldface labels) and of the two infraclasses (standard labels). The most
slender fish is the African lungfish; the other examples shown here are extinct.

and it is probable that they also resembled dipnoans in having functional lungs and advanced circulatory systems. They differed from dipnoans, however, and resembled early amphibians in the pattern of their cranial bones and in having numerous conical teeth with complicated internal structure. Their fin structure is also regarded as being closer to that of the tetrapod limb than is that of dipnoans. Crossopterygians fall into two groups. Only one of these, the Rhipidistia, is ancestral to amphibians. The other group, Coelacanthi, lacks internal nares and has tail fin and scales that are more modified from the primitive condition.

Coelacanths were long thought to be extinct. Then, in 1938, and again in 1952, somewhat decomposed remains of a large coelacanth were recovered off the coast of South Africa. It was a significant and exciting discovery. Numerous good specimens of this large fish, *Latimeria,* have since been obtained in deep water near Madagascar. Some of its features (scales, teeth, gas bladder) do not seem to retain their ancestral characters, whereas others (cranial mechanics, circulatory system) are apparently primitive and very instructive.

4

Tetrapods

Tetrapods are simply vertebrates having four legs (or at least four legs in their ancestry). It is instructive, however, to avoid this tautology by defining tetrapods instead as vertebrates that dwell on land (or that had land-dwelling ancestors), and then to consider the changes that enable descendants of fishes to live terrestrial lives.

Out of water, the body usually no longer benefits from being streamlined. A neck becomes advantageous because the head now can turn to facilitate feeding and vision without affecting the mechanics of locomotion. Median fins are no longer useful, and paired fins are converted to limbs. Deprived of the buoyancy of water, the body must be supported by the limbs and this necessitates having appendages that are strong, girdles that are more firmly related to the axial skeleton, and a vertebral spine that can better resist bending.

Lungs and a pulmonary circulation, pioneered by air-breathing fishes, are usually retained to replace gills, which would be damaged by exposure to dry air. Gill covers can therefore be dispensed with. The superficial layer of the skin becomes cornified to resist abrasion and drying. The eye, ear, and nose also must be modified to function in air instead of water. Oral glands are needed to moisten food that is now dry. Eggs and delicate larvae formerly were supported by water and could pass the waste products of their metabolism directly into the environment. Before becoming able to reproduce completely away from either water or moist environments, tetrapods had to accomplish the seeming miracle of evolving eggs with shells and fetal membranes to protect their embryos from desiccation and mechanical harm and to receive metabolic wastes. Some other structural changes were also necessary to make possible the shift to terrestrial life, and physiological and behavioral changes were also needed. It is not surprising, therefore, that the first tetrapods, the amphibians, did not fully accomplish the change.

Class Amphibia

Amphibians evolved from crossopterygian ancestors (of the group called Rhipidistia) some 50 million years after bony fishes evolved. The class reached its greatest expansion after another 100 million years, in the Upper Carboniferous period, but continued to abound until the end of the Triassic period. Relatively few kinds of amphibians have survived to the present, yet they are distributed in tropical and temperate areas of all the world, and their habits and habitats are diverse.

The skin of modern amphibians is unable to withstand long exposure to dry air, and since fetal membranes are lacking, eggs must be layed either in water or in damp places on land. In this sense the class is not completely terrestrial ("amphibian" = both kinds + life), though many species do not utilize open water, and several live in remarkably arid places.

Adult amphibians have large mouths and a fleshy tongue which is attached near the front of the lower jaw. One of the bones that supported the jaws of the piscine ancestor is converted to an ear ossicle, which usually contributes to hearing in air. Lungs are usually present (they are secondarily lost by one large group of salamanders), and some respiration occurs also through the skin and lining of the mouth and throat. Eyelids and glands to moisten the eyes have evolved.

Relationships within the class are still being worked out, but three subclasses are commonly recognized. One of these (Lepospondyli) comprises slender aquatic forms, some of which retain characters of skull, gills, and girdles that resemble those of their fish ancestors. They are all extinct and will not be discussed further in this book. The other two subclasses are of concern here.

Subclass Labyrinthodontia. Labyrinthodonts, extinct for 175 million years, comprise most of the amphibians that have ever lived, including the ancestors of reptiles. Different authorities recognize two or three orders. Some labyrinthodonts were entirely aquatic whereas others appear to have been true land animals with strong limbs, robust bodies, and probably dry skins. Several kinds were as large as alligators, and many had rather large, flat heads. The complicated structure of their teeth (from which the name of the subclass is derived), and the varied structure taken by their vertebrae are among the subjects to which we shall return in subsequent chapters.

Subclass Lissamphibia. All surviving amphibians are lissamphibia. Most are less than 30 cm in length. Their moist skin has abundant mucous glands and only rarely supports scales ("liss" = smooth). The outer cornified layer of the skin is shed

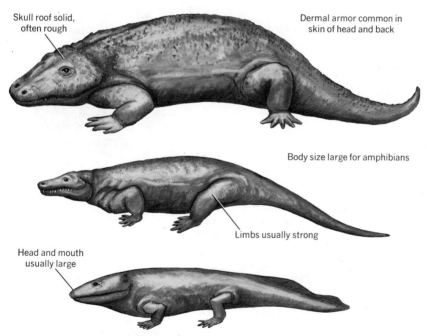

Skull roof solid, often rough

Dermal armor common in skin of head and back

Body size large for amphibians

Limbs usually strong

Head and mouth usually large

FIGURE 4-1

Restorations of REPRESENTATIVE LABYRINTHODONTIA illustrating some characteristics of the subclass.

periodically. Parts of the skeleton, particularly of the feet, are commonly cartilaginous, and several ancestral bones have been lost from the braincase. There are only four toes in the hand. Teeth are never complicated as in labyrinthodonts and are absent from some groups.

There are three orders of lissamphibia. ANURA (= without + tail) includes frogs and toads. Salamanders belong to the URODELA (= having a tail). Finally, the APODA, commonly called caecilians, are, as the name tells us, legless. They are obscure animals which will merit mention in Part II of this book because of their primitive excretory organs and their scales (unusual in surviving amphibians), and in Part III because of their burrowing habits.

Class Reptilia   Reptiles evolved from labyrinthodont amphibians some 50 million years after amphibians evolved. From the Permian through the Cretaceous periods, they were the most abundant of vertebrates. This was the first class of tetrapods to have all the structures noted at the beginning of this chapter as requisite to fully terrestrial life, including fetal membranes and an in-

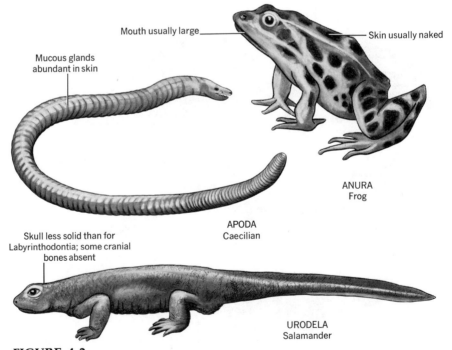

Mouth usually large

Skin usually naked

Mucous glands abundant in skin

ANURA
Frog

APODA
Caecilian

Skull less solid than for Labyrinthodontia; some cranial bones absent

URODELA
Salamander

FIGURE 4-2
REPRESENTATIVES OF THE THREE ORDERS OF LISSAMPHIBIA showing some characteristics of the subclass.

tegument that is resistant to drying. During the "age of reptiles," the different genera ranged from small to gigantic, from herbivorous to carnivorous, and from sluggish to swift. Several groups independently reverted to aquatic habitats, becoming highly skillful swimmers, and one group even invaded the air. No class of vertebrates had theretofore been so diverse in habits, and only the mammals have matched them since. Today, reptiles remain an important part of the faunas in tropical and temperate regions but are less numerous than bony fishes, birds, or mammals.

Reptiles are covered with horny scales. Excepting such specialized forms as snakes, most of them have claws, ribs that are used in drawing air into the lungs (amphibians instead use the mouth and throat as a force pump), and a vertebral column that is more differentiated into regions and more firmly attached to the pelvic girdle than in their amphibian ancestors. However, none of these characteristics is unique to the class. There are features of the heart and related blood vessels that *are* unique to reptiles, yet most other single structures that are typical of the class are not sufficiently distinctive to separate it from other vertebrates.

Why, one may ask, does this class have so few distinctive characters? First, it includes all the animals that made the first true invasion of land and radiated out into its varied habitats. Hence, it is a large and diverse class. Such a group can have less in common than one, such as birds, whose members have a common adaptation. Second, reptiles were ancestral to members of two other classes, birds and mammals. They have correspondingly many relatives from which to be distinguished. For these very reasons, Reptilia is a key class in vertebrate evolution and one to which the morphologist must often refer.

The class is variously grouped in 17 to about 23 orders commonly arranged in 5 or 6 subclasses. Only 4 orders survive to the present.

Subclass Anapsida. The term "anapsida" (not to be confused with "anaspida") refers to the absence of openings in the bones that roof over the temporal region of the skull, a feature which distinguishes this subclass from the others. The group is divided into two principal orders.

COTYLOSAURIA are small, lizard-like creatures that were abundant in the Permian period but have long been extinct. We will rarely refer to them in later chapters, yet they are important in that they include the most primitive of reptiles and because most, and possibly all, other subclasses of reptiles evolved from them. Accordingly they are called **stem reptiles.**

The CHELONIA comprises the turtles and tortoises, which the layman easily recognizes by their broad, armored bodies. A

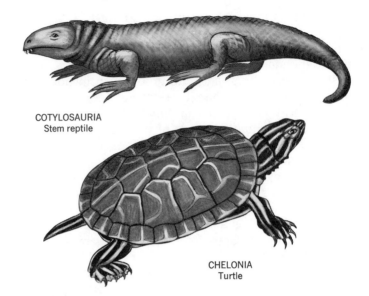

COTYLOSAURIA
Stem reptile

CHELONIA
Turtle

FIGURE 4-3
REPRESENTATIVES OF TWO OF THE ORDERS OF ANAPSIDA. The principal characteristic of the subclass is the absence of openings in the temporal area of the skull.

turtle is often dissected in the laboratory as a representative of its class because of its availability and suitable size. Students should realize that the turtle's "shell," ribs, spine, toothless mouth, and pectoral girdle are highly specialized, and therefore not typical of the class. The skull, limbs, and (so far as we can judge) soft parts remain primitive.

Subclass Lepidosauria. This was one of the first subclasses to evolve after the anapsids, but it did not reach maximum abundance until after the passing of the dinosaurs. The taxon is defined by cranial characters. There are two openings in the roofing bones of the temporal area (unless secondarily modified), openings among the bones of the palate, and usually teeth on the roof of the mouth as well as at the margins. Three or more orders are recognized, including one (Eosuchia) which is extinct and not useful to this course of study.

The only survivor of the order RHYNCOCEPHALIA is the tuatara, genus *Sphenodon,* which looks like a large robust lizard. It has probably survived longer (135 million years) than any other single genus of tetrapods. It is of particular interest for the primitive nature of its skeletal and circulatory systems. The one remaining species is confined to islands off the coast of New Zealand. It is rigidly protected and never available for class dissection.

The remaining surviving lepidosaurs are placed in the order

RHYNCOCEPHALIA
Tuatara

Lizard

SQUAMATA

Snake

FIGURE 4-4
REPRESENTATIVES OF TWO OF THE SURVIVING ORDERS OF LEPIDO-SAURIA. The principal characteristics of the subclass are cranial features noted in Chapter 7.

SQUAMATA. Most of the 3000 species of lizards have legs and a tail. Amphisbaenians have small forelimbs only or (usually) no limbs at all. They live underground. Snakes probably evolved from burrowing, lizard-like ancestors. They have lost almost all traces of limbs and generally have modified the skull to allow them to swallow prey of diameter equal to or greater than that of their bodies.

Tetrapods

Subclass Archosauria. This largest and most spectacular of reptilian subclasses is also characterized by two temporal openings and usually by an open bony palate. Often there are openings in the skull in front of the orbit and in the side of the lower jaw. All teeth are marginal.

Five or more orders are recognized. The only survivors of the subclass, crocodiles and their relatives, are in the order CROCODILIA. Two orders comprise the beasts popularly known as dinosaurs: Some SAURISCHIA were carnivorous; most ORNITHISCHIA were herbivorous. These animals, which tended to gigantic size and sometimes to bipedalism, give the subclass its name: "archosauria" = ruler + lizard. Flying reptiles are in the order PTEROSAURIA (= wing + lizard). This leaves only a short-lived order (Thecodontia) which we need mention only as being ancestral to other archosaurs and to birds.

Subclass Euryapsida. The three or four orders grouped in this subclass all have a single temporal opening and only marginal teeth. The subclass is diverse, however, and may well prove to be unnatural. We will note only two of the orders, which included some highly specialized marine animals mentioned in Chapter 23. Aquatic members of the order SAUROPTERYGIA are called plesiosaurs. They had broad bulky bodies, tapering tails without lobes, paddle-like limbs, blunt heads, and necks that were sometimes very long. ICHTHYOSAURIA, by contrast, had dolphin-like body contours, fish-like tails, large eyes, and a large rostrum.

Subclass Synapsida. Members of this extinct subclass were usually terrestrial carnivores of moderate size. There is one temporal opening. Unlike archosaurs, they never tended to bipedalism. Passing over the more primitive of the two orders (Pely-

FIGURE 4-5
The alligator is a REPRESENTATIVE OF THE CROCODILIA, the only surviving order of the subclass Archosauria.

Roof of skull has two vacuities

SAURISCHIA
Carnosauria

Hindquarters and tail
usually relatively large

SAURISCHIA
Sauropida

Bones common
within the skin

ORNITHISCHIA
Stegosauria

Body size moderate
to gigantic

ORNITHISCHIA
Ceratopsia

FIGURE 4-6

Restorations of REPRESENTATIVE ARCHOSAURIA of the orders Saurischia
and Ornithischia and of the suborders and infraorders indicated. Some characteristics of the terrestrial members of the subclass are shown.

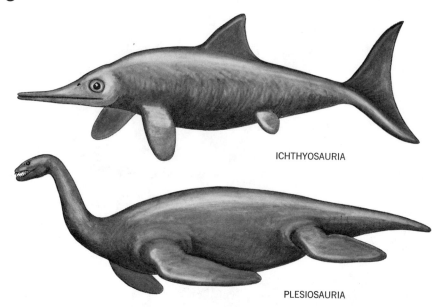

ICHTHYOSAURIA

PLESIOSAURIA

FIGURE 4-7
Restorations of REPRESENTATIVE EURYAPSIDA of two dissimilar aquatic groups. The principal characteristics of the subclass are cranial features noted in Chapter 7.

cosauria) brings us to the THERAPSIDA, which are important as the ancestors of mammals. Commonly called **mammal-like reptiles,** they tended to have robust legs placed relatively close to the center line of the body and rooted teeth which were specialized, according to position in the mouth, for biting, tearing, or chewing. The architecture of the deep skull, palate, ear, and jaw came to resemble the corresponding features of mammals.

Class Aves   No other locomotor adaptation requires so much structural specialization as that of flight, and all birds fly or are descendants of flyers. In striking contrast to reptiles, birds are, therefore, the most homogeneous and distinctive of all tetrapod classes. However, for all their unique characteristics among the living fauna, birds are not very different from the particular reptiles from which they evolved. Those small Mesozoic archosaurs tended to be bipedal and therefore to have robust hind limbs with elongate feet. Like birds, they had long necks, and their pelvic bones and skulls approached avian structure. We may safely infer that their urogenital systems, fetal membranes, and sense organs were virtually like those of birds.

Feathers are of particular importance, for birds are the most expert of flyers and the only vertebrates ever to achieve the highly successful combination of flight with bipedalism. Such

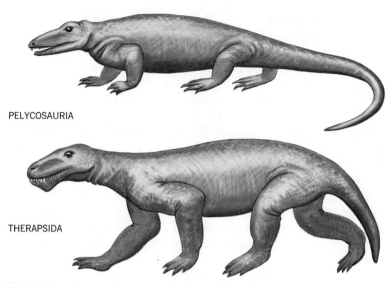

PELYCOSAURIA

THERAPSIDA

FIGURE 4-8
Restorations of REPRESENTATIVE SYNAPSIDA of the two orders.
Principal characteristics of the subclass are slender form, strong limbs,
and features of the cranium noted in Chapter 7.

flight is mechanically dependent on feathers. Further, sustained flight probably requires the high metabolic rate made
possible by an elevated body temperature. Contrary to former
belief, many reptiles do have considerable control of body temperature, but lack of feathers or other insulation puts them at a
disadvantage in some respects. We must not be overly sure
that the flying reptiles could not stay aloft for long periods, and
there are indications that some had hair-like insulation, but it is
unlikely that they could match the birds that replaced them
during the Cretaceous period.

Subclass Archaeornithes. The small, light bones of birds are
not easily fossilized; the ancestry of no other class of comparable size is so poorly documented. However, fossils of Jurassic
birds are known from Germany, and they are "missing links" of
utmost value. The Archaeornithes (= ancient + birds) were fully
feathered, but unlike present-day birds had the tail feathers arranged in a row along each side of a long lizard-like tail. These
arboreal climbers could already fly, but various volant adaptations of the wing skeleton, spine, and breastbone were absent
or incomplete by Cenozoic standards. The skull had the large
orbits and beak-like rostrum of a bird but was reptilian in other
respects, including having teeth in its jaws.

Subclass Neornithes. All remaining birds are placed in the subclass Neornithes (= recent + birds). In contrast to the preceding subclass, this one is characterized by tail feathers arranged like a fan at the end of a tail having a short bony axis, and by fusions of bones in the spine, braincase, lower leg, and "hand." A system of air sacs is usually present, and air spaces are found within most of the bones. Unless the power of flight has been secondarily lost, as in the ostrich, the breastbone has a large keel from which the flight muscles take their origins.

Although the subclass is relatively homogeneous, it is also large, and it has been convenient to establish 32 orders—as many as for the more diverse mammals and half again as many as for all reptiles, living and extinct. It follows that the characteristics that distinguish the orders and three superorders are relatively minor.

The superorder ODONTOGNATHAE has long been extinct. These toothed birds were specialized for swimming. The superorder PALEOGNATHAE includes the ostrich, emu, cassowary, and their relatives. Features of the pelvis and palate unite the group. However, all but one member (the tinamou) have secondarily lost the power of flight, and many of the characters they have in common (including large size, strong legs, reduced pectoral girdle) are apparently due to convergent evolution. Nearly all surviving birds belong to the remaining superorder, NEOGNATHAE.

Class Mammalia

Mammals, like birds, are familiar and distinctive. Children learn that only mammals have hair and mammary glands which give the class its name. Succeeding chapters will present unique features of the cranium, jaw, teeth, ear, pectoral girdle, pelvis, muscles, brain, and other structures. Mammals are also numerous. Some 3000 genera are known (2000 are extinct).

Unlike the transition from reptiles to birds, which is documented by only several fossils, the transition from reptiles to mammals is well-recorded—so well, in fact, that the conventional boundary based on the structure of the jaw, ear, and cranium is somewhat arbitrary.

Some Mesozoic Mammals. Mammals evolved late in the Triassic period several million years before the origin of birds, yet for 120 million years thereafter—for two-thirds of their entire existence—they remained few in number, small in size, and subordinate to the great reptiles with which they were contemporaneous. The fossil record of these inconspicuous creatures is scanty and consists mostly of teeth and jaw fragments. About five orders are recognized which are variously classified as to subclass and infraclass by different paleontologists. Brief refer-

Teeth present

Three free, clawed
digits in wing

ARCHAEORNITHES

Skeletal axis
of tail long

Teeth
usually absent

Digits of wing
less free, rarely
clawed

Skeletal axis
of tail short

NEORNITHES
Neognathae

NEORNITHES
Paleognathae

FIGURE 4-9

**REPRESENTATIVE AVES showing some characteristics of the two sub-
classes and examples of the two surviving superorders of Neornithes.**

ence to these obscure mammals will be made only in relation to
the evolution of teeth.

Subclass Prototheria.   The most primitive of surviving mammals
are the Prototheria (= first + beasts), which are not represented as fossils before the Pleistocene epoch. The single
order, MONOTREMATA, includes a few animals of diverse hab-

FIGURE 4-10

REPRESENTATIVES OF THE SUBCLASS PROTOTHERIA AND THE IN-FRACLASS METATHERIA. These taxa comprise, respectively, the orders Monotremata and Marsupialia.

its: the aquatic platypus and the insect-eating echidnas. These odd creatures are rare both in zoos and in their native Australia and New Guinea. If known only by the pectoral girdle, they surely would be classified as reptiles. Further, they are oviparous (egg-laying), which is equally unique among mammals. The young are nourished by milk, however, and the presence of hair and a single bone in each half of the lower jaw qualify monotremes as mammals.

Subclass Theria. All familiar mammals belong to the subclass Theria. They are viviparous (give birth to live young) which sets them apart from Prototheria. Two infraclasses survive to the present. METATHERIA, includes the single and long-enduring order MARSUPIALIA. Opossums, bandicoots, pha-

INSECTIVORA
Shrew

CHIROPTERA
Bat

PRIMATES
Chimpanzee

EDENTATA
Armadillo

RODENTIA
Marmot

CETACEA
Dolphin

CARNIVORA
Bear

PERISSODACTYLA
Rhinoceros

ARTIODACTYLA
Pronghorn

FIGURE 4-11
REPRESENTATIVES OF NINE ORDERS OF THE INFRACLASS EUTHERIA.

langers, wombats, and kangaroos are marsupials. They give birth to tiny embryonic young which are nourished in the pouch ("marsupium" = pouch) of the mother until they are able to walk about.

The other surviving infraclass, EUTHERIA (= true + beasts), comprises the animals commonly known as "placental mammals." The term is misleading: A placenta is an organ that accomplishes physiological exchange between mother and fetus. Some reptiles and even several fishes and amphibians have placentas, and so do all marsupials. The placenta of marsupials is always vascularized on the fetal side by a membrane called the yolk sac; that of eutherian mammals is usually vascularized by the allantoic membrane. However, some marsupials have both allantoic and yolk sac circulations, and some eutherian mammals have no allantoic circulation. Numerous features of the skeletal, reproductive, and nervous systems provide more technical but more exact ways to distinguish the infraclass Eutheria.

There are 16 surviving and 10 extinct orders of eutherian mammals. Diagnosis of ordinal characters must be left for textbooks of mammalogy. However, in Part II, and particularly in Part III, of this book, reference will be made to representatives of all surviving and several extinct orders. It will be useful, therefore, to identify those that are most common.

INSECTIVORES (order Insectivora) includes shrews, moles, and hedgehogs. All are small animals with numerous sharp teeth. This is the oldest and most primitive order in the infraclass. BATS (Chiroptera) are the only mammals capable of sustained flight as distinguished from gliding. PRIMATES (technical name the same as the common name) include lemurs, monkeys, apes, and man. As the name indicates, the EDENTATES (Edentata) have simple teeth or none at all. They are the anteaters, sloths, and armadillos.

The largest order of all is the RODENTS (Rodentia). Squirrels, beavers, rats, mice, porcupines, and a host of other small mammals with gnawing incisors belong to this group. WHALES and dolphins of the order Cetacea are the most specialized of mammals. Bears, dogs, weasels, raccoons, civets, cats, hyenas, seals, and most other mammals with large canine teeth and a taste for flesh are CARNIVORES (Carnivora). ELEPHANTS (Proboscidea) were once more numerous, both in numbers and kinds. Horses, tapirs, and rhinoceroses comprise the order Perissodactyla. The term means odd-toed and distinguishes them from the Artiodactyla (= even-toed) such as pigs, camels, deer, antelopes, and cattle. The PERISSODACTYLS and ARTIODACTYLS include most (not all) hoofed mammals and are collectively known as UNGULATES ("unguis" = hoof), a useful term of no systematic rank.

Other orders accommodate the rabbits, sea cows, hyraxes, pangolins, and aardvark.

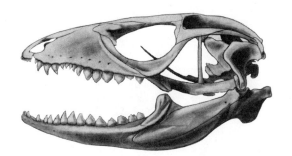

PART
II
THE PHYLOGENY
OF STRUCTURE:
Evolution in
Relation to Time
and Major Taxa

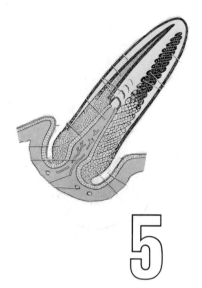

5

Early Development and Integument

The embryonic development of the various organ systems will be reviewed, albeit rather briefly, in the separate chapters of this part of the book. However, before describing the development of the first organ system, the skin and its derivatives, it is desirable to review the early development of the entire embryo which establishes the general plan and symmetry of the body and the tissues from which all organ systems are formed.

Recall that the chordate **zygote,** or fertilized egg, is transformed by a process called **cleavage** into a multicellular embryo called a **blastula.** The structure of the blastula is related to the amount of yolk present (Figure 5-1). In amphioxus, which has eggs with virtually no yolk, it is a small, hollow ball of cells of

Development of the Early Embryo

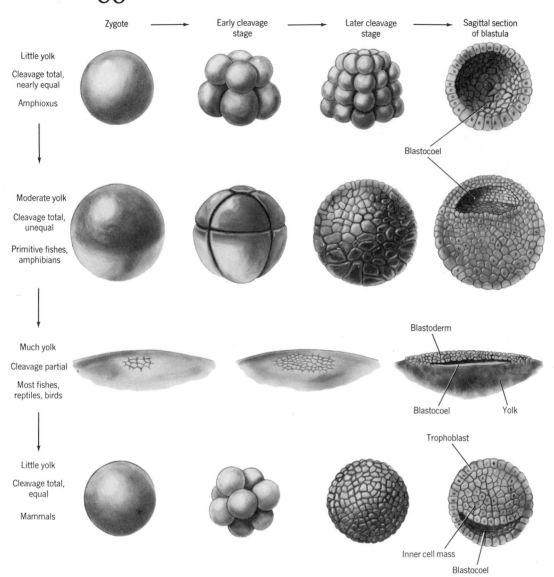

FIGURE 5-1

PRINCIPAL TYPES OF CHORDATE CLEAVAGE. Membranes, shells, and polar bodies omitted. Relative sizes only approximate.

nearly equal size. In amphibians and several fishes, which have moderate amounts of yolk, it is a hollow ball of cells of unequal size. In most fishes, reptiles, and birds, the blastula is a disk of cells which rests on the relatively enormous uncleaved yolk. In mammals, which have secondarily lost the ancestral yolk, the blastula is again a small mass of cells. Regardless of size and shape, the blastula consists of a single tissue layer made up of several hundred cells which have a polarity that relates to the axes of the future body.

The blastula is converted to an embryo called a **gastrula** by various processes, collectively called **gastrulation,** which include the folding of tissues and the migration of certain cells relative to others (Figure 5-2). The gastrula has at first two, and then three, tissue or **germ layers.** The outer layer, called **ectoderm** will form the outer part of the skin, all of the nervous system, and parts of the eye and ear. The inner layer, called **entoderm,** will form the linings of the respiratory tract and of the digestive tract and associated organs. The middle layer, or **mesoderm,** is the last to develop. It will form the skeletal, muscular, circulatory, and urogenital systems and the inner part of the skin. The figure shows that the cavity of the gastrula, which is different from the cavity of the blastula, becomes the cavity of the gut, and communicates with the outside of the embryo by an opening, the **blastopore,** which in chordates and other deuterostomes either becomes the anus or comes to lie near the anus.

The gastrula is converted to a **neurula** by processes which are together called **neurulation** (Figure 5-3). The embryo lengthens. The mesoderm in the middorsal line is now called **chordamesoderm** because it is destined to become the rod-like notochord. Chordamesoderm has the important function of influencing or inducing the overlying ectoderm to roll up into two

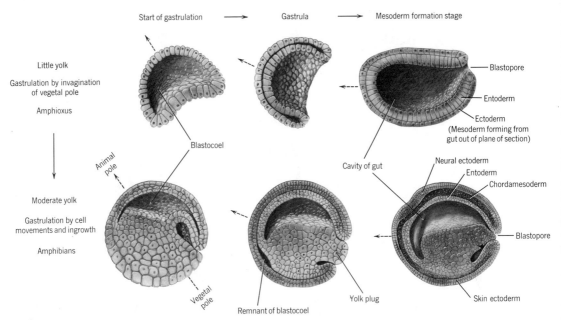

FIGURE 5-2
SOME TYPES OF CHORDATE GASTRULATION that are uncomplicated by a large yolk mass. Sagittal sections.

longitudinal neural folds. These subsequently fuse along their crests, thus establishing the **neural tube,** or primordium of the central nervous system. The tube sinks somewhat and is closed over by surface ectoderm.

The remaining mesoderm (after the formation of the notochord) divides into a dorsolateral portion called the **epimere,** which flanks the neural tube and notochord; a small lateral portion called the **mesomere;** and a ventrolateral **hypomere.**

The epimere is segmented into blocks of tissue termed **somites** which are paired on the right and left sides of the body. Each somite, in turn, differentiates into an internal **sclerotome,** which will surround the notochord; a middle **myotome;** and a lateral **dermatome,** as shown in Figure 5-4. The sclerotome

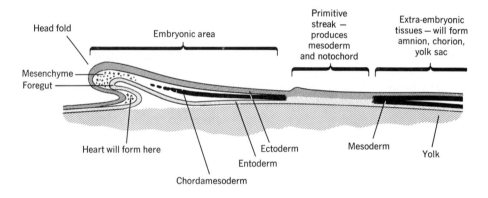

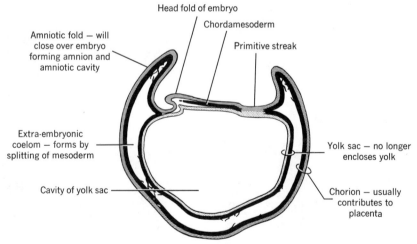

FIGURE 5-3

DIAGRAMS OF SAGITTAL SECTIONS OF EMBRYOS AND ASSOCIATED MEMBRANES OF AMNIOTES AS MESODERM IS ESTABLISHED. Above, reptile or bird with much yolk; below, mammal with no yolk. (The drawings are representative; there is variation within the classes.)

forms vertebrae, the myotome becomes muscles, and the dermatome contributes to the skin. The dermatome, myotome, and, to a lesser extent, hypomere all have derivatives at some distance from the primordia just described. Their cells become

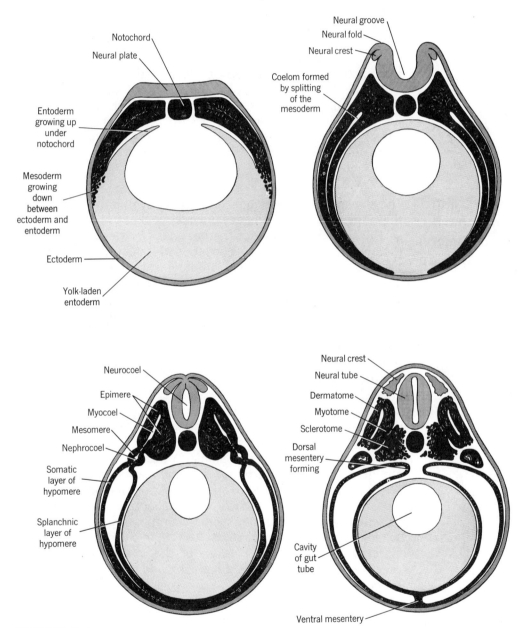

FIGURE 5-4

DIAGRAMS OF NEURULATION AND EARLY DIFFERENTIATION OF THE MESODERM as seen in cross-sections of an embryo having a moderate amount of yolk.

loosely associated into a kind of mesoderm called **mesenchyme** and migrate away to their destinations.

The mesomere is segmented anteriorly but not posteriorly. It forms urogenital structures. The sheet-like hypomere is not segmented. It is split by the developing coelom into a lateral **somatic layer** which will line the body wall, and a medial **splanchnic layer** which forms the heart and contributes to the digestive and respiratory systems.

As the neural tube is formed, ectodermal **neural crest** cells are pinched off between the neural tube and the surface ectoderm (see again Figure 5-4). Most of these inconspicuous cells migrate from their place of origin as **ectomesenchyme** (mesenchyme derived from ectoderm) and form varied and important derivatives including parts of the head skeleton, some ganglia of the nervous system, and pigment cells and other parts of the skin.

Nature, Development, and General Structure of Skin

The skin and its appendages perform many functions: Together they may (according to need) provide mechanical protection to soft tissues below, restrict the passage of water in and out of the body thus preventing desiccation and disturbance of water balance, help regulate heat transfer with the environment, guard against the entry of injurious organisms and materials, provide adaptive coloration, and facilitate locomotion in various ways. The skin may also function in respiration, secretion, excretion, sense reception, and the storage of fat and glycogen.

Considering these many functions, it is not surprising that the various vertebrates have diverse kinds of integuments: Contrast those of the fish, frog, lizard, armadillo, and rhinoceros! The integuments of the different vertebrates have in common only some general features and the source of their diversity, which is contact with the environment. Because this is so, the integument tells the morphologist something of the habits and environments of animals, and enables the systematist to identify most vertebrates. Some mammals can be identified by single hairs, and ornithologists commonly preserve only the skins of the birds they collect. Because the integument is variable and responsive to the environment, it has so far told the morphologist less than most other organ systems about phylogeny. Study of the integument is now a very active field for research, however, and new insights can be expected.

The skin of all vertebrates has two principal layers, a superficial **epidermis** and a deeper **dermis.** The epidermis is derived from the ectoderm on the surface of the embryo. The dermis is derived from the dermatome supplemented by contributions from the lateral and ventral somatic mesoderm. Cells from these sources migrate as mesenchyme to distribute themselves evenly under the ectoderm. Some neural crest cells also invade the developing dermis.

The epidermis is stratified into two or more layers. The deepest layer rests on the dermis and consists of closely packed, discrete cells. It is called the **stratum germinativum** because its daughter cells push outward and mature to become the more superficial cells of the skin. The layer or layers of the epidermis that are superficial to the stratum germinativum are exceedingly varied according to taxon. Most are (or will become, or have been) secretory in nature and, as classified by W. B. Quay, (see Maderson, 1972), fall into two general categories, mucous cells and proteinaceous cells, The former produce various types of mucus, some kinds of poisonous secretions, and, in some fishes, **photophores,** (light-producing cells). The proteinaceous line of epidermal cells may produce slime, poisons, substances eliciting alarm reactions, enamel, and possibly some photophores. The principal product of this cell line, however, is the horny material **keratin,** which is the main constituent of feathers, hair, claws, reptilian scales, and also the dead outermost layer, or **stratum corneum** of the dry skin of tetrapods ("keratos" and "cornu" both = horn). Two molecular types of keratin, designated α- and β-keratin, are recognized. The α-keratin tends to be more flexible, though the functional difference between the two kinds is not well-known.

The dermis is usually thicker than the epidermis. It has relatively fewer kinds of cells and is characterized by a meshwork of fibers. Most abundant are **collagenous fibers** which are tough, straight, and made up of fibrils. The fibrils are oriented in a three-dimensional network. They are linked to a ground substance which resembles plasma protein and contains mucopolysaccharides. **Elastic fibers,** which are fewer in number, are wavy, nonfibrillar, and branched.

The dermis commonly has an outer, vascular, **stratum spongiosum** and a deeper, thicker, **stratum compactum.** These merge with one another and also bridge across to secure the skin to the connective tissue covering the muscles of the body wall. Smooth muscle fibers may be present in the dermis. Fat is commonly deposited between the skin and the body. The glands of the skin are derived from the epidermis but usually penetrate into the dermis which then adds supportive tissue to them.

Pigment cells are called **chromatophores.** They are derived from neural crest cells and occur, according to taxon, in any level of the skin, but tend to concentrate near the epidermal–dermal boundary. Chromatophores of the epidermis are particularly characteristic of **homeotherms** (animals maintaining constant temperature) and are of one type called **melanophores.** They have numerous migratory organelles, termed melanosomes, which contain the pigment melanin. Melanin is black, brown, or red. Color imparted by these cells may be constant or may be responsible for **morphological color change,** which is seasonal, age-related, or otherwise relatively slow change.

Chromatophores of the dermis are found almost exclusively in **poikilotherms** (animals having variable body temperature.) They may maintain constant color, cause morphological color change, or cause **physiological color change,** which is relatively rapid, as when a fish or lizard adapts its color to that of an altered substrate. There are three types of dermal chromatophores: Melanophores are similar to those of the epidermis. **Iridophores** have organelles called reflecting platelets which are oriented in stacks and contain crystalline deposits (chiefly of guanine) which scatter or reflect light. These cells are commonly large. **Xanthophores,** which are yellow, and **erythrophores,** which are red, have their pigments (pteridines and carotenoids) in organelles called pterinosomes.

These various kinds of chromatophores can be structurally and physiologically interrelated in the achievement of certain color effects. Their complex control, which is currently under intensive study, may include influence by hormones of the pituitary, thyroid, gonads, and adrenals, and in some poikilotherms by the nervous system as well. Color is "used" by the various vertebrates for concealment, for making themselves conspicuous (e.g., as warning, social releaser, or sexual attractant), for control of heat absorption and conservation, for protection of the nervous system or gonads from light, and for control of the synthesis of vitamin D.

Approaches to Study of the Integument

Some years ago it was considered convenient to describe separately skin derivatives thought to be developed entirely from the epidermis (e.g., horny scales, feathers, hair) and those thought to be developed from the dermis (e.g., fish scales). It is true that some derivatives are formed by one or the other layer, and it may still be convenient to sort them on that basis, but it is now recognized (1) that many derivatives (e.g., teeth, fish scales) have structural contributions from each layer, and (2) that neither layer ever forms any derivative without the presence and influence of the other. In the absence of underlying dermis, the embryonic epidermis degenerates, and experiments show that the kind of epidermal derivative formed (whether scale, feather, hair, or other) is controlled by the nature of the associated dermis. The influence of the epidermis on the dermis seems to be less universal, yet epidermis apparently triggers the dermal contributions to teeth and fish scales (of which more below). These interactions make the study of skin derivatives more interesting but complicate their classification.

A step toward order is a distinction stressed by Maderson between the fundamental patterning of the integument and appendages of the integument. The ancient vertebrate integument was probably soft and smooth like that of cyclostomes, holocephalians, eels, and modern amphibians. Some descendants developed hard inflexible integuments like the armor shields of ostracoderms and placoderms. Many other vertebrates evolved integuments which are more or less hard, yet flexible because they have serial, patterned folds which can move over one another. Such folds are called scales. (Unfortunately, the word "scales" is also used to refer to the horny or bony parts of such folds.) Soft skins, armor shields, and scales are all fundamental patterns of the generalized integument.

In addition, the integument usually has complex, localized, and specialized derivatives such as multicellular glands, dermal denticles, feathers, and hair. These are appendages of the integument. The scales and nonglandular integumentary appendages of fishes are largely of dermal origin, whereas those of tetrapods are largely of epidermal origin.

Integument of Fishes: Emphasis on Dermal Derivatives

Soft Structures. The soft part of the integument of OSTRACODERMS and PLACODERMS is, of course, unknown. The epidermis of CYCLOSTOMES is thin and has several kinds of glands, all of which are unicellular. Most numerous are **mucous glands** of two types. **Club glands** produce slime of fibrous protein. **Granular glands** probably discharge at the surface of the body, but their function is not yet known. Keratin is absent. A thin noncellular **cuticle** covers the epidermis. The dermis, which may be thinner than the epidermis, consists largely of a fibrous layer which contains collagenous fibers but no elastic fibers. There is no trace of scales.

The skin of jawed FISHES is usually thin and glandular. It fits tightly over the body. With rare exceptions, keratin is entirely absent. The replacement of worn epidermis is constant. Mucous glands of one or another type are nearly always abundant. They are usually unicellular but may also be multicellular. The slimy mucus they secrete cleans the body and produces a cuticle which prevents the entry of foreign material, assists in osmoregulation, and reduces resistance as the fish swims. Granular and club glands are also common. Some fishes have **poison glands** associated with fin spines; others have multicellular light organs which may even be provided with tiny lenses and reflectors. The dermis, though still thin, is divided into a stratum compactum and a stratum spongiosum (except when it covers the fins, where it is reduced to a basal membrane).

Development and Structure of Hard Tissues. The most complex derivatives of the integument of fishes are hard scales and denticles of various kinds. Before studying their nature and phylogeny it is desirable to know about the tissues of which they are constructed.

In historical perspective, the description and classification of the scales and hard integumentary appendages of fishes have been complicated by various factors: (*1*) Fish scales—particularly those of fossil fishes—include many kinds of hard tissues. (*2*) The various hard tissues grade into one another and combine in many ways. (*3*) Certain virtually identical tissues seem to develop from different germ layers. This throws doubt on their homology and has led to repeated use of such vague designations as "enamel-like dentine." (*4*) Some of the terms applied to scales describe gross shape (e.g., cycloid, rhomboid), others identify one of the tissues present (ganoid, cosmoid), and still others indicate a particular combination of tissues (palaeoniscoid, lepidosteoid). This has resulted in a confusing lack of parallel terms.

These difficulties are being overcome, and the phylogeny of scales promises to emerge as one of the more complete stories of vertebrate morphology. Much remains to be done, however, and the study of hard tissues is an active field of research; biochemical analysis, physiological studies, and electron microscopy are revealing details in the mechanisms of deposition and resorption of hard tissues.

Significance is attached by many morphologists to the embryonic origin and interactions of the precursors of hard tissues. Some neural crest cells join mesenchyme from the dermatome and contribute to the dermis of the skin and the tissues of the gums. Specifically, they form papillae which in-

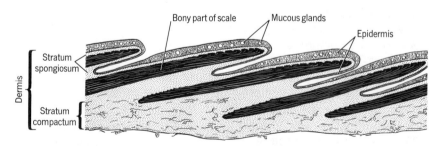

FIGURE 5-5
SECTION OF THE SKIN OF A TELEOST.

duce (cause to be formed) enamel organs from the overlying ectoderm. These organs, in turn, induce the papillae to form dentine if this substance is to develop at all. Bone may also be induced in this way. Finally, dentine induces the enamel organ to produce enamel (these tissues are defined below). If dentine is not deposited, then the enamel organ (or its equivalent – the term is not apt in this instance) may form horny scales or any of the other derivatives of the ectoderm described earlier in this chapter. There is, therefore, a basic similarity in the mechanism of development of teeth and all skin derivatives regardless of ultimate hardness and principal germ layer of origin. Proponents of these views believe that continuous armor could be formed under the influence of an extensive interaction between neural crest derivatives in the dermis and ectoderm which has the potential to form various hard tissues.

Meanwhile, the Scandinavian paleontologists Ørvig and Stensiö have used new techniques to reveal minute structural detail of scales and fragments of fossil armor. They postulate that the ancestral vertebrate, and also the young of ostracoderms and placoderms, had tiny scales, each called a **lepidomorium** (= scale + division). Larger scales and armor were formed, they believe, by the aggregation of lepidomoria edge to edge (producing cyclomorial scales) or in onion-like layers (producing synchromorial scales). Many paleontologists believe that the sequence of fossils does not support this theory, but the two general hypotheses need not be entirely mutually exclusive.

It is generally agreed that there are three principal kinds of hard tissue: enamel, dentine, and bone. **Enamel** is the hardest tissue of the body. It is shiny, translucent in thin sections, and composed of elongate crystals of hydroxyapatite $[3(Ca_3PO_4)_2 \cdot Ca(OH)_2]$. In therian mammals it is prismatic. Internal cells and tubules are absent. Only about 3% of the tissue is organic. Enamel occurs only on teeth and superficial denticles, scales, or armor plates, and is usually external to any other hard tissues present. It is produced only by ectoderm – even at the back of the mouth where the ectoderm has migrated into position. Growth is by accretion on the inner surface of an enamel organ. Hence, enamel cannot be altered or replaced once it has been deposited. **Ganoine** is probably an enamel, though the germ layer of its origin is questioned. It is characterized by thick deposition in successive waves of growth which create a laminar structure.

Dentine is harder than bone and usually softer than enamel. The chemical composition of its inorganic salts is the same as that of enamel, but the content of organic fibers is typically about 30%. The generative cells usually, but not invariably, remain external to the hard tissue; their processes then penetrate the dentine via dentinal tubules. Dentine occurs only in teeth, denticles, scales, and external armor. It is present unless secondarily lost, and lies internal to enamel and usually external to bone where those tissues are also present. It is produced only by the outer surface of a mesodermal papilla which, in turn, occurs adjacent to the boundary of mesoderm with ectoderm. Dentine can be altered little, and only at its generative surface. Various types of dentine are recognized. **Osteodentine** is organized in **osteons** (columns having cylinder-within-cylinder construction) which are usually interspersed in a matrix of bone. (Osteons that occur in dentine are also termed denteons.) **Orthodentine** has no true osteons or bony matrix but instead is either laid down in a superficial compact layer (pallial dentine) or in layers that are concentric around a central pulp cavity (circumpulpar dentine). **Cosmine** (not a parallel term) is dentine having characteristic tufts of tiny canals that radiate upward and outward from a succession of small vascular centers distributed more or less in a plane that is parallel to the surface of a scale. Very hard dentine is termed **enamel-like** or **enameloid,** and may be difficult to distinguish from enamel. Other terms, many of them confusing, have been applied to the types of dentine.

Bone has about the same organic content as dentine, usually occurs internal to dentine if each is present, and develops in a deeper and less restricted part of the dermis. It usually has internal bone cells (osteocytes) located in small vacuities (lacunae) which intercommunicate by small canals (canaliculae). Acellular bone is common, however, in pteraspids and teleosts. The ancestral relationship between cellular and acellular bone is still in dispute. Bone (like osteodentine) is characterized by osteons (here also called Haversian systems), but when it is adjacent to internal and external surfaces it is usually deposited in laminar sheets (like orthodentine). Bone may be compact or vascular and spongy. It may have few intrinsic collagenous fibers or many which are more or less layered and make the bone soft and flexible. Other collagenous fibers (Sharpey's fibers) may penetrate bone or dentine from adjoining connective tissue to bind scales together and anchor them to the substrate. The lamellae and osteons of bone can be destroyed and replaced at any internal or external surface. Growth includes reorganization as well as accretion.

Hard tissues not found in the integument are described in Chapter 7.

Phylogeny of Bony Scales and Their Derivatives. Early in the twentieth century it was assumed that the most primitive ver-

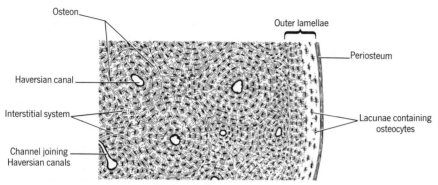

FIGURE 5-6
CROSS-SECTION OF COMPACT BONE.

tebrates had cartilaginous skeletons and either lacked hard tissues entirely or had only small scales. Body armor was considered to be a secondary development, and Chondrichthyes were accepted as the most primitive of living fishes. Then, as the fossil record became more complete, it became apparent that the situation was quite the reverse. The earliest known fishes and jawless vertebrates are nearly always *more* heavily armored than the descendants in their respective lineages. Enamel, dentine, and bone were all present in the Ordovician fragments of armor which are the earliest known vertebrate fossils. Primitive armor may have served as an ion reservoir, for protection, for osmotic control, or perhaps to make the body heavy.

The theory that hard tissues are primitive for vertebrates fits the record but leaves one puzzled as to the origin of the heavy and complicated armor of ostracoderms from the (apparently) naked integument of protovertebrates. Surely there must have been an intermediate step. The lepidomorial theory, and the theory that extensive dermal papillae underlying extensive enamel organs could have formed large armor plates, are attempts to reconstruct that step.

Armor shields of ostracoderms and placoderms (cephalaspids, pteraspids, arthrodires, antiarchs) differ only in size from the coarse scales found elsewhere on their bodies. In section, armor shows three principal layers. The surface is composed of dentine which is often capped with enamel (or, according to some authorities, enamel-like dentine) completing surface projections called **denticles.** A middle layer is composed of bone which is riddled with anastomosing channels for small blood vessels and sensory pits. The basal layer is lamellar bone with fewer vascular channels.

Anaspids have no armor shields, and their scales have regressed in that only the basal layer is retained. It is generally assumed that the naked skin of cyclostomes resulted from fur-

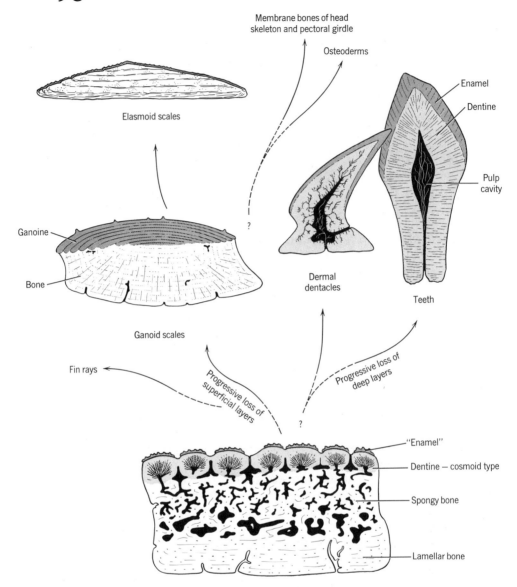

Membrane bones of head
skeleton and pectoral girdle

Osteoderms

Enamel

Dentine

Pulp
cavity

Elasmoid scales

Ganoine

Bone

Dermal
dentacles

Teeth

Ganoid scales

Fin rays

Progressive loss of
superficial layers

Progressive loss of
deep layers

"Enamel"

Dentine — cosmoid type

Spongy bone

Lamellar bone

Armor and heavy ancestral scales

FIGURE 5-7
STRUCTURE AND RELATIONSHIPS OF DERMAL SCALES AND DERIVATIVES.

ther degeneration and that their ancestors had armor plates.
However, the record does not exclude the possibility that their
ancestors had small scales — or none.

Cosmoid scales differ in no fundamental respect from the ancient armor just described — the same basic layers are present.
The term has gained wide usage, however, and is helpful for
describing somewhat more advanced scales that are typically

smaller, thinner, and characterized by having dentine of the cosmoid type. The surface of the scale is usually sculptured by the enamel of the denticles ("cosmoid" = ornamented). The scales may be **cycloid** (roundish in outline) or **rhomboid** (a parallelogram in outline). Cycloid scales are usually imbricated (overlapping). Rhomboid scales often overlap at their inner margins, but are fitted edge to edge at their outer surfaces. Cosmoid scales are found on the posterior parts of the bodies of some placoderms and on crossopterygians and early dipnoans. (The modern crossopterygian, *Latimeria,* has large cycloid scales with reduced surface layers; modern dipnoans have thin scales without cosmine.)

Ganoid scales are thick rhomboid structures which evolved from cosmoid scales. There are two principal types. The more primitive is called a **palaeoniscoid scale.** The surface is thickened during successive periods of growth by laminations of the "enamel" called ganoine. Cosmoid dentine is retained under the ganoine. The base of the scale is lamellar bone pierced by vascular canals. This type of scale is found on primitive extinct actinipterygians and on the surviving chondrostean, *Polypterus.* The second type of ganoid scale was derived from the first and is called a **lepidosteoid scale.** The ganoine is the same. The cosmine is deleted. The bony base is acellular, and the canals, though present, are no longer vascular. This type of scale is found on more recent chondrostean and holostean ray-finned fishes (but not on modern chondrosteans which have degenerate scales) and also, surprisingly (since this is a rather advanced scale), on fishes of the extinct subclass Acanthodii.

Elasmoid scales were derived from ganoid scales of the lepidosteoid type. They are restricted to teleosts. The basal layer, which now forms the bulk of the scale, remains acellular but is laced by collagenous fibers coursing in various directions. The resulting bone (called isopedine) is somewhat flexible and soft. The ancestral ganoine is absent, and in its place is a thin surface glaze derived from the enamel organ. Elasmoid scales are thin, imbricated, and cycloid or **ctenoid** (having comb-like projections on the exposed margins).

Isolated **dermal denticles** or placoid "scales" are confined to elasmobranchs and some aberrant placoderms. They evolved from cosmoid scales or possibly from the integument of a placoderm stock that sidestepped armor plates and heavy scales entirely. They lack the bony basal layer of scales and are always small and usually isolated. A central pulp cavity is surrounded by dentine, and this is capped by a tissue of debatable nature. Some researchers consider it to be enamel-like dentine, but Moss is sure that it is true ectodermal enamel.

Early Development and Integument

FIGURE 5-8
OSTEODERM FROM
THE NECK SKIN OF A
LARGE CROCODILE.
Actual size 7½ × 10½ cm.

Various hard structures of tetrapods are derived from the bony scales of fishes. Caecilians have bony ossicles buried in the skin. **Osteoderms** are plates of bone which are derived from scales and located under the horny scutes of crocodilians, some lizards, and some extinct labyrinthodonts and reptiles. Some bones in the shells of turtles probably had the same origin. (Other bones in the shells of turtles are flattened ribs.) **Gastralia** are slender, splint-like bones, also seemingly derived from dermal scales, which lie in the muscles of the ventral abdominal wall of various reptiles. (The bones in the shells of armadillos are of similar nature, but they evolved secondarily, long after their ancestors had lost all ossifications in the skin.)

Other hard structures have also evolved from scales. Teeth certainly evolved about as dermal denticles evolved. Various bones of the roof of the skull and pectoral girdle represent armor plates that lost their enamel and dentine and sank below the skin to join bones originating in the internal skeleton. The fin rays of bony fishes are also regarded as derivatives of scales. These structures will be described in later chapters.

Integument of Tetrapods: Emphasis on Epidermal Derivatives

Skin of Amphibians, Living and Extinct. The epidermis of living amphibians is thin (typically five to eight cell layers), but in response to contact with the air it has a particular muccopolysaccharide which apparently helps control desiccation, and a stratum corneum with α-keratin. Only the outermost cell layer is dead, however, and this is shed every few days, sometimes in large patches. The molting is under hormonal control.

Amphibians have two kinds of multicellular, alveolar (flask-shaped) glands which originate from the epidermis and grow down into the dermis. Their products reach the surface by ducts. Abundant mucous glands secrete continuously and spontaneously to clean and lubricate the skin and to keep it moist so that cutaneous respiration will be possible. Granular glands are under nervous or hormonal control. They secrete an acrid milky fluid which is distasteful, and in some instances very toxic to predators. Granular glands are grouped together in the "warts" of toads. The amphibian dermis is two-layered and may be provided with lymph spaces and muscle fibers.

Terrestrial amniotes are able to withstand abrasion and desiccation largely because of keratinized derivatives of the epidermis. Having only a thin layer of dead keratinized cells, modern amphibians must instead seek moist habitats or use behavioral adaptations to avoid drying out. It is probable that the skin of terrestrial labyrinthodonts was thicker, drier, and more like that of some modern reptiles. Many had bony ossicles in the skin, and these are usually associated with a heavily keratinized epidermis.

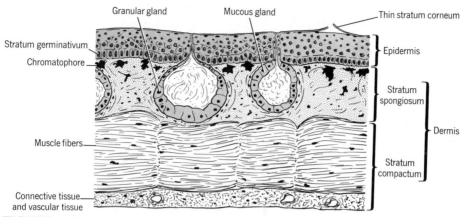

FIGURE 5-9

SECTION OF THE SKIN OF AN AMPHIBIAN.

Skin of Reptiles: Horny Scales. The first of several ways that tetrapods utilize keratin to "air-proof" the integument is seen in reptiles. Here the adaptation involves the fundamental patterning of the skin in that the distinctive epidermis forms a complete body covering of horny scales. Joints between scales are merely regions where the horny material is thin and folded (Figure 5-10). The epidermis of lepidosaurs, as described chiefly by Maderson, is of particular complexity and interest. In these animals an entire "generation" of the epidermis is molted as a single unit. This occurs at least several times a year. Let us enter the cycle just after such a molt in what is termed the resting stage. The epidermis now consists of the stratum germinativum and an **outer epidermal generation** which characteristically has five layers. From the outside inwards there is first a thick, dead, acellular layer heavily keratinized by β-keratin. The surface of this layer is called the *oberhautchen* and has microscopic spicules. Under this β layer is a thin mesos layer of unknown significance and then a moderately thick layer of loose, dead, anucleate material having α-keratin. Below this are two layers of living cells: an outer layer which will later be taken into the α layer, and an inner layer which will later become clear and create the separation leading to molting.

At the end of the resting stage, the germinal epithelium rapidly proliferates the various layers of an **inner epidermal generation.** As these mature, they separate from the innermost layer of the outer epidermal generation, and molting follows.

The keratinous plate on the outer surface of a large flat scale is called a **scute.** The scutes of crocodilians and chelonians are not shed. Growth adds keratinized material over the entire inner surface of a scute, thus compensating for wear. Each

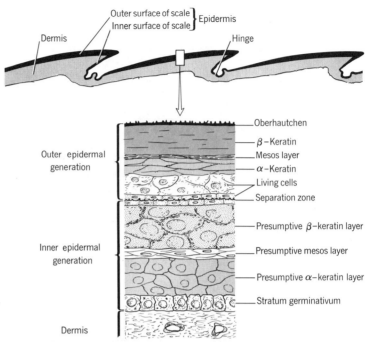

FIGURE 5-10
SECTION OF THE SKIN AND EPIDERMIS OF A SQUAMATE REPTILE shortly before a molt.

FIGURE 5-11
CARAPACE OR SHELL OF A DESERT TORTOISE, *Gopherus*, **SHOWING SCUTES WITH GROWTH LINES.**
Dorsal view; anterior to left.

wave of growth extends beyond the previous margin of a scute to form the familiar concentric rings of the turtle shell.

The reptilian dermis is thin. Mucous glands are absent, as they are also from the skins of the other truly terrestrial tetrapods. Scent glands of various types (generation glands, preanal glands, femoral pores, etc.) occur variously on the tail (some lizards), cloacal area (most squamates), thighs (lizards), and under the jaws (crocodilians). Their secretions influence social behavior. Some bones found in the skin of reptiles were mentioned above on p. 100.

Integument of Birds: Thin Skin and Feathers. Following a different strategy, birds have over most of the body a thin, weakly keratinized skin which is loosely joined to the underlying tissues. It is appendages of the skin—the feathers—which are heavily keratinized. The lower leg and toes, however, are covered by horny scales similar to those of archosaurs. These are not shed. The **beak** is also heavily keratinized. The **shellbreaker** of birds and some reptiles is an elevation on the beak or rostrum which helps the hatchling to break out of its shell (Figure 1-7). (The egg tooth of snakes and lizards serves the same purpose, but is a real tooth.) The **spurs** of gamecocks are horny spines covering bony cores.

With rare exceptions, glandular derivatives of the avian skin are restricted to a large, branched, alveolar **uropygial gland** above the tail ("uropygium" = tail + rump) which secretes an oil used by the bird to preen its feathers (Figure 9-9). It is most developed in water birds.

Although intermediate stages are still largely speculative, there is biophysical, developmental, and anatomical evidence that feathers evolved from the epidermal scales of reptilian ancestors. They contain β-keratin, which is the same type of keratin that occurs in the outer surface of the scales of archosaurs. The keratin of avian skin, and also of the thinner, undersurfaces of archosaurian scales, is α-keratin. It is probable that feathers evolved to provide insulation and only later became adapted as airfoils.

There are several principal kinds of feathers and various intergrades. Most familiar and complex are the **contour feathers** which give the bird its external form and provide airfoils for flight. Few animal structures are so exquisite in design. The axis has a hollow, proximal (toward the body) **quill** and a solid, distal (away from the body) **shaft** or rachis. The vane is made up of **barbs,** which branch from opposite sides of the shaft, and smaller **barbules,** which branch from the barbs. The barbules on the distal side of each barb have on their edges **hooklets** which engage the proximal barbules of the adjacent barb (Figure 5-12). The resulting web is strong, light, and flexible. If disturbed, the elements of the vane may separate, but they do not break. The integrity of the feather is restored by preening with the modified edges of the bill, which reengages the hooklets.

Wing feathers (remiges) and tail feathers (rectrices) are enlarged and stiffened contour feathers. (Primary remiges are attached to the manus; secondary remiges are attached to the arm.) A variation of the contour feather lacks hooklets and therefore has no firm vane but is instead fluffy. Many birds, including the more primitive orders, have double contour feathers—a principal feather joined at its base by a shorter, softer feather called the **aftershaft.**

Contour feathers are evenly distributed over the bodies of several kinds of birds (probably the primitive condition), but usually are restricted to feather tracts called **pterylae.** The feathers spread from the pterylae to cover the intervening areas. The conformation of pterylae is of use to systematists.

Down feathers have little or no shaft. Long barbs branch from the base of the feather and there are no hooklets. The resulting feather is small and soft. Down feathers, hidden by the contour feathers, are widely distributed and not restricted to pterylae. Their function is insulation: No material is known that is superior to the down of the eider duck as filling for lightweight sleeping bags.

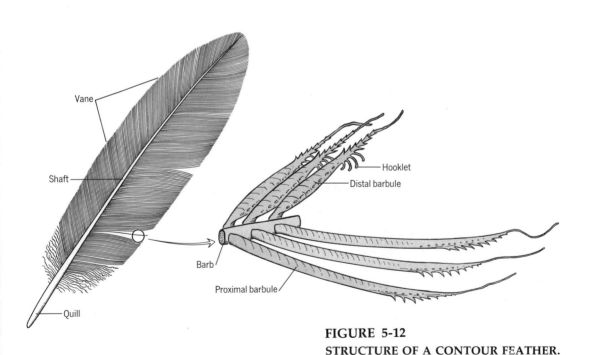

FIGURE 5-12
STRUCTURE OF A CONTOUR FEATHER.

Bristles are derived from contour feathers by the partial or complete deletion of the vane. They are short stiff feathers which may screen foreign objects from the nostrils (hawks, blackbirds), increase the effective gape of the mouth (flycatchers), or form eyelashes (ostrich).

The colors of feathers come from two sources. Yellow, orange, red, brown, and black are the result of specific pigments introduced into the feather during its development. White results entirely from the microstructure of the feather. Blue, green, and iridescent hues result from a combination of black, yellow, or other pigment with microstructure that reflects only part of the light.

Feathers are molted and replaced once or (less commonly) twice a year. Most species shed the feathers one at a time so that function is not impaired, but ducks and some other birds lose most of the flight feathers at one time.

The development of a feather starts with a hummock of mesoderm, the **dermal papilla,** which is covered by ectoderm. This structure sinks into the skin, thus forming a narrow depression, or **feather follicle,** all around its base (Figure 5-13). The feather is formed only by ectoderm, but the ectoderm not only must be nourished by the vascular mesoderm, but also activated by it. Experiments show that in the absence of mesoderm, no feather can form, and that in the presence of a papilla, ectoderm that does not normally form a feather may do so.

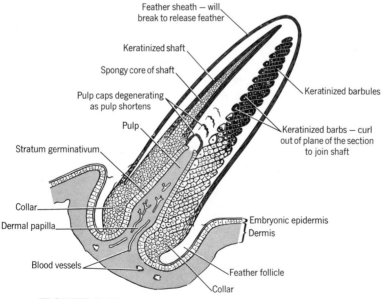

FIGURE 5-13
SECTION OF A DEVELOPING CONTOUR FEATHER.

A superficial keratinized **feather sheath** surrounds erupting feathers and subsequently sloughs away. At the base of the follicle the germinative layer forms a **collar.** The barbs of a down feather grow straight upward from the collar within the sheath. An early step in the formation of a contour feather is the development of a shaft as an outgrowth from one point on the collar. Barbs form first as branches of the shaft, and then as outgrowths of the collar itself which migrate onto the base of the shaft as it lengthens. When the sheath breaks away from the maturing feather the barbs unfold to the right and left of the shaft to change the cylindrical, embryonic feather into the flat, mature feather.

Skin, Scales, and Integumentary Glands of Mammals.   The skin of mammals is relatively thick — particularly the dermis, from which leather is made. However, the thickness varies greatly according to species and location on the body (and sometimes also according to season). The epidermis thickens where the hair is sparse and also in areas subject to pressure and abrasion, such as footpads, the kneepads of camels and warthogs, and pads on prehensile tails. Between the stratum germinativum and stratum corneum there may be one or more transitional layers, the most common of which is a thin **stratum granulosum.** Bundles of smooth muscle in the dermis are related to hair follicles.

The stratum corneum may form horny scales, as on the tails of opossums and beavers. **Claws** are strong keratinized structures which wrap around the tapering terminal bones of the digits. The tip and upper and lateral parts comprise the **unguis** and are harder than the underside (subunguis). Hooves are derived from claws. The unguis of the horse's hoof is built up of compacted horny tubules. The entire hoof spreads somewhat under the impact of a footfall. The shell of the armadillo has a heavily keratinized epidermis as well as bony dermal ossicles.

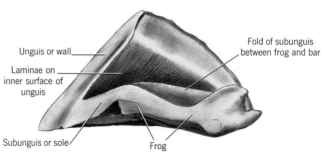

Unguis or wall

Laminae on inner surface of unguis

Subunguis or sole

Fold of subunguis between frog and bar

Frog

FIGURE 5-14
HALF OF THE HOOF OF A HORSE seen from the medial (cut) surface.

The unique pangolin (mammalian order Pholidota) has markedly overlapping scales on its dorsal surface (Figure 21-2). These scales, which may be more than an inch long, are shed one at a time and replaced in larger sizes as the animal grows. The **baleen plates** of whales are lath-like outgrowths of the buccal epithelium which serve as strainers during feeding (as explained further in Chapter 25).

Sweat glands (also called sudoriferous glands) are unique to mammals. Many species have a million or more of these small glands distributed over the entire body. Others have fewer and restrict them to the muzzle or soles of the feet. Still others, including whales and sea cows, which have no use for them, have none. Sweat glands are tubular, simple (not lobulated), and coiled at their inner ends. There are two kinds which differ somewhat in structure and nature of secretion. They develop in the embryo from cords of ectoderm which sink into the dermis. The evaporation of sweat from the surface of the skin helps to prevent overheating of the body and opposes slipping of foot pads over the substrate. Salt, urea, and some other wastes are excreted in the sweat. Glands in the eyelids (Moll's glands) which open near the eyelashes, and the wax glands of the external ear are considered to be enlarged and modified sweat glands.

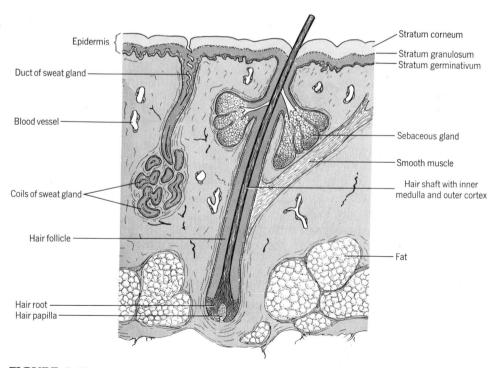

FIGURE 5-15
SECTION OF THE SKIN OF A MAMMAL.

Sebaceous glands are also limited to mammals. One or more of these branched alveolar glands drains into each hair follicle. They also occur without relation to hair on the nipples, lips, and genitalia. Their oily secretion dresses the hair and prevents excessive drying of thin skin. Lanolin, used as a base for cosmetics, is refined from the sebaceous secretion of sheep. Modified sebaceous glands (Meibomian glands) occur in the eyelids, where their secretion films over the eyeball and normally prevents overflow of the tears.

Many mammals have **scent glands.** There is wide variation in their nature and distribution. They may serve for defense, recognition, or sexual attraction. The glands may be located in the anal region (weasel family), on the face (bats, antelopes), on the back (kangaroo rat), on the feet (some artiodactyls), or indeed, on any other part of the body. Some scent glands are said to be derived from sebaceous glands and others from sweat glands.

Only mammals have **mammary glands** which secrete the milk to suckle the young. The first indication of the development of the glands is the appearance in the embryo of a pair of epidermal ridges, the **milk lines,** which extend lengthwise from the chest to the inguinal region. At intervals along the lines where the adult mammae will ultimately form, ectoderm sinks into the dermis and branches into solid cords. In females these cords enlarge at maturity, pushing under the skin and becoming compound (lobulated) and alveolar. Much of the human breast is fat. The gland becomes active at parturition under the influence of ovarian and pituitary hormones.

The number of mammae correlates with the number of young per litter and varies from one pair to about a dozen pairs. The mammae may be on the chest (primates, elephants, bats, sea cows), in the inguinal area (ungulates), or at intervals in between (rodents, carnivores). Mammary glands appear to be phylogenetic derivatives of primitive sweat glands.

Each milk gland sends numerous ducts to the surface. In monotremes these emerge in depressions whence the young lap up the milk. This probably approaches the primitive condition. Usually the site of emergence of the ducts is elevated into a **nipple** which the young can hold in the mouth and suck. In ungulates it is the skin circling the point of emergence that is elevated forming a hollow **teat.**

Hair.   The phylogenetic origin of hair is obscure. There is no indication that hair evolved from reptilian scales. Where hair and scales occur together (as on tails of rats, shells of armadillos, and backs of pangolins), the hair grows between the scales, the pattern of the scales imposing a pattern on the distribution of the hair. A similar pattern of hairs often occurs where scales

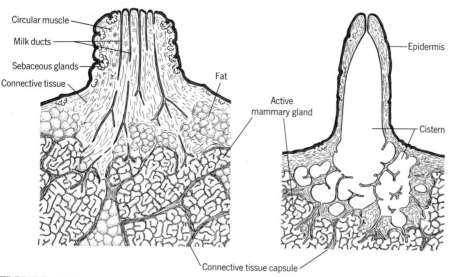

FIGURE 5-16
SECTION OF THE NIPPLE AND ASSOCIATED TISSUES OF A PRIMATE (left)
AND THE TEAT OF AN ARTIODACTYL (right). Not drawn to quite the same
scale.

are lacking. Evidence is presented by Maderson to support the
hypothesis that hairs arose from reptilian sensory appendages
of the mechanoreceptor type that were located between scales
and contributed to thermoregulatory behavior. It is postulated
that such structures multiplied sufficiently to become useful as
an insulatory body covering. Some mammal-like reptiles may
well have reached such a stage of evolution.

A typical hair has an expanded root and a shaft which, below
the skin, is hidden in an epidermal sheath or **hair follicle** (see
Figure 5-15). One or more sebaceous glands usually drain into
the cleft between the hair shaft and adjacent tissues. The
follicle slants at an angle to the skin surface, and a tiny smooth
muscle runs downward from the outer part of the dermis to insert on it at such an angle that contraction elevates the follicle
and its hair. This merely gives a human "goose flesh," but for
most mammals it deepens the fur coat, thus increasing the effectiveness of the fur in displays and as an insulator.

In section, hairs are seen to be made up of two or three
layers. Of greatest structural importance is the **cortex,** which is
relatively dense and contains the pigment of the hair. Around
the outside are microscopic scales which form the **cuticle.** The
size, shape, and overlapping pattern of these scales vary from
species to species. Mammalogists sometimes take advantage
of this individuality when they wish to identify hairs from owl
pellets or droppings of carnivores. Coarse hairs also have a cen-

tral pith or **medulla** consisting of shrunken dead cells and air spaces.

Guard hairs are the relatively long straight hairs that give a pelt its apparent color and texture. These hairs are often grouped by twos and threes. Parting the contour hairs of most mammals (particularly of "fur-bearers") reveals shorter hairs which are very fine and numerous. These are wool hairs or **underfur.** They are usually somewhat flattened in cross-section, and this makes them wavy. They often occur in groups of a dozen or more about the base of each guard hair. Underfur traps innumerable air pockets which provide insulation and may prevent water from penetrating to the skin.

Extra long and coarse hairs comprise eyelashes and manes and are found on the tails of ungulates. Whiskers, or **vibrissae,** are even coarser hairs that are specialized as tactile organs. The stiff shaft of a whisker serves as a lever which pivots at the surface of the skin to translate the slightest movement to the root. The bulb at the lower end of the root is surrounded by erectile tissue rich in nerve endings. The heaviest "hairs" of all are **quills.** Quills are hollow, but can be very stiff. The cuticle at the tip of a porcupine quill is modified to produce tiny, effective barbs.

Most mammals molt once or twice a year, the winter and summer coats often being different in density, quality, and color. The new pellage usually comes in first at one or several locations and spreads over the body in a pattern characteristic of the species.

Mesoderm plays a less prominent role in the formation of a hair than it does in the formation of a scale or feather. A solid cord of ectoderm sinks into the dermis. The walls of the cord become the double-layered **root sheath.** A small dermal papilla forms at the enlarged base of the cord. The ectodermal cells over this papilla proliferate to form the hair itself, which pushes outward through the sheath cells to emerge from the skin.

Horns and Antlers.   The horns and antlers of tetrapods are of various types. **Rhinoceros horn** is composed of keratinized fibers about $1/2$ mm in diameter which are compacted into a solid structure tough enough (as some animal collectors have learned) to punch holes in army trucks. Growth is from the epidermis, where it covers many small dermal papillae at the base of the horn. The horn is evergrowing and is not shed.

Giraffe "horns" are merely knob-like projections from bones of the skull. They are permanently covered by skin.

The **antlers** of the deer family are also bony outgrowths of the skull, but they are shed and replaced each year. The hard, compact bone is covered by skin ("velvet") only during growth. When full size is attained, the circulation to the velvet is cut

True horn
Sectioned to show core

Pronghorn

Antler
After shedding   In velvet

FIGURE 5-17
SOME HORNS AND ANTLERS.

off, thus causing its death and ultimate sloughing. At the end of the breeding season, the bone at the base of an antler, just below a rough expansion called the burr, is weakened, and the antler is shed. Antlers are of varied shapes, often large, and usually branched in mature animals.

Pronghorns are limited to the American artiodactyls of that name. Again, there are bony projections from the skull which are covered by skin. However, instead of producing hair, the skin forms horn. The bony core is permanent and the horny cap is shed and regrown each year.

The **true horns** of cattle and antelopes (and of some dinosaurs and chameleons, among reptiles) have bony cores which are vascular and may contain extensions of the frontal sinuses. Over these cores are permanent, horny sheaths of epidermal origin. The horny substance is not filamentous like that of the rhinoceros horn and horse's hoof. The growth is by internal deposit so that the core keeps slipping outward, and growth rings may be seen around the base.

Rhinoceros horn and the sheaths of true horns are alike in that each grows only by the accumulation at the basal end of hard material that cannot subsequently change its shape. In this they resemble claws, tusks, and mollusk shells. When such structures grow at equal rates on all sides, they grow straight. More often the rate of growth is unequal around the base. If the point of minimum growth lies opposite the point of maximum growth, then the structure always forms a logarithmic (equiangular) spiral. The rhinoceros horn is an example. If the point of minimum growth does not lie opposite the point of maximum growth, then a helical (corkscrew) spiral in space is superimposed on the flat logarithmic spiral. The ram's horn is an example.

Phylogeny? As indicated earlier in this chapter, a reasonably satisfactory phylogeny can now be prepared for the various kinds of bony scales and related structures. Attempts have also been made to construct a phylogeny of other integumentary structures, but the task is made difficult by multiple origins, evolutionary plasticity, parallelism, and convergence. Considerations of paleontology, development, innervation, and function have been of little help. Quay believes that the various mucous glands of the aquatic vertebrates probably originated at various times, evolved along separate lines, and are not homologs of any glands of reptiles, birds, and mammals. Similarly, it is not known to what extent the granular glands of cyclostomes, fishes, and amphibians are related. The general correspondence of the keratinized skin derivatives of terrestrial vertebrates seems more evident, but truly homologous integumenary structures may not exist above the class level in modern vertebrates.

6

Teeth

Teeth have an importance for vertebrate morphology that is out of proportion to their contribution to the bulk of the body. This is true for several reasons: First, their durability has made them a significant part of the fossil record. Second, they are so adaptive that the diet of most animals can be approximated from their teeth. Third, great variation of structural detail among kinds of vertebrates combined with relative stability of structure within kinds makes teeth invaluable in systematics; experts can identify most species of mammals by a single cheek tooth. Finally, in spite of their adaptation to diet, teeth can often be used to trace the general course of evolution within, and among, genera, families, and orders. For these various reasons the teeth have been the subject of much study.

113

Teeth

The origin of ancient integumentary armor was discussed briefly in Chapter 5. Such armor had surface denticles of enamel (or enameloid, according to some authorities) and dentine which merged into bone below. Teeth appear to have evolved from denticles released from armor near the margins of the mouth as ossification in the integument was gradually reduced.

The part of a mature tooth that is above the root and ultimately subject to wear is the **crown**. The **root** is hidden below the gum and is usually anchored to a jawbone. The **pulp cavity** within contains blood vessels and nerves. The bulk of a tooth usually consists of orthodentine. The bases of the teeth of some fishes contain a vascular osteodentine which may merge into the jawbone. Unworn crowns are covered by enamel which rarely exceeds 2 mm in thickness even in large animals. The nature of these hard tissues was described in Chapter 5. Roots

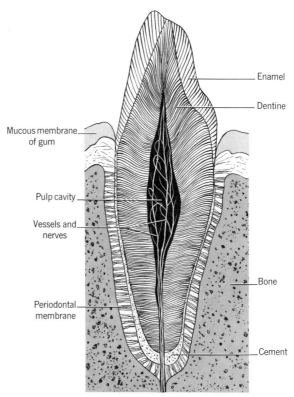

FIGURE 6-1
LONGITUDINAL SECTION OF MAMMALIAN INCISOR TOOTH.

of teeth that are held in sockets are covered by a thin layer of **cement.** (Some teeth that are specialized for grinding also have cement on the crowns.) Cement is a nonvascular bone that has no osteons and is usually acellular. It is rich in collagenous fibers and is softer than dentine.

Development

Knowledge of the process of tooth development facilitates an understanding of the mechanism of replacement and of the origin of complicated teeth from simple ones. The first step in the formation of teeth occurs in the embryo when a fold of ectoderm forms along the margins of the mouth and penetrates into the underlying mesoderm of the gums as a double-layered wall called the **dental lamina** (Figure 6-2). At intervals along the dental lamina, hummocks of tissue push into its inner edge causing it (as seen in section across the jaw) to look like an inverted goblet. This tissue is regarded as mesoderm, but is derived from neural crest cells which have migrated from their site of origin (see p. 90). The double-walled bowl of the goblet now constitutes the **enamel organ.** The cells of its inner layer (ameloblasts) will form enamel. Each mesodermal hummock, or **dermal papilla,** develops on its surface the cells (odontoblasts) that will form dentine. The entire unit is called a **tooth bud.**

There is a reciprocal induction between the enamel organ and its papilla: each is necessary for the proper function of the other. The shape of the crown of the future tooth is determined by the conformation of the interface between enamel organ and papilla at the time the hard tissues are deposited. Until that time, the interface is gradually sculptured by differential

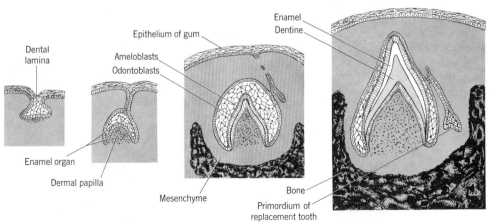

FIGURE 6-2

Diagrammatic sections showing DEVELOPMENT OF A THECODONT TOOTH.

pressures and growth rates of the various parts of the tooth bud. If there is to be more than one cusp, deposition of enamel and dentine starts at what will be the tip of the principal cusp. As maturation proceeds, the developing tooth slowly climbs up toward the surface of the gum. The enamel organ regresses ahead of the emerging crown, and no further enamel can then be deposited. The formation of dentine continues after the tooth becomes functional. Cement forms only in the presence of dentine. The papilla becomes the pulp.

Attachment and Replacement

The teeth of cartilaginous fishes are anchored to the skin by collagenous fibers (Sharpey's fibers) which run into the dentine from the dermis. The teeth of most vertebrates, however, are more or less fixed to the bones of the jaws. Often the outer margin of each jawbone forms a thin wall having on its inner (lingual) side a series of hollows to accommodate the teeth. Each tooth touches the bone only with the outer (buccal) surface of its root. It may be joined to the jaw by collagenous fibers or by cement. This mode of attachment, which may be the primitive one, is called **pleurodont** (= side + tooth) (Figure 6-3).

Some other teeth scarcely have any roots and abut against the rim of the jawbones to which they are joined by a continuum of hard tissue. This form of attachment, which has evolved independently several times, is called **acrodont** (= summit + tooth). Still other teeth have their roots held in sockets (alveoli) in the jawbones. This is the **thecodont** (= sheath + tooth) condition. There are intergrades among these kinds of attachment.

Replacement of teeth is necessary to provide for growth and to compensate for wear and accidental loss. Even before a first tooth is fully functional, a new tooth bud forms to initiate the development of its replacement. As the second tooth matures, the root of the first is resorbed, thus causing it to loosen and fall away. Replacements for pleurodont teeth form either just lingual or just anterior to the roots of the old teeth and move into position after the old teeth are shed. Replacements for thecodont teeth form directly under the roots of the old teeth.

Most vertebrates replace their teeth continuously, generation following generation, for as long as they live. Such animals are said to be **polyphyodont** (= many + to grow + tooth). Most mammals have only two generations of teeth and are said to be **diphyodont**. Some acrodont teeth, and some that fuse to form large tooth plates, are not replaced. However, many of the animals that seem to have only one generation of teeth do have one or more other generations that are shed before birth or resorbed in the embryo.

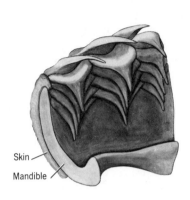

Skin

Mandible

Teeth anchored to skin

Pleurodont

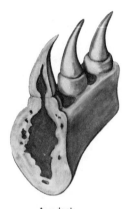

Acrodont

Thecodont

FIGURE 6-3
Sections of jaws showing **TYPES OF TOOTH ATTACHMENT.**

Polyphyodont teeth are not replaced at random, seemingly because that could result in temporary impairment of function. Some 375 million years ago natural selection evolved a mechanism for replacing teeth which minimizes loss of function by maintaining two conditions. First, the teeth are in two sets, the even-numbered teeth constituting one set and the odd-numbered teeth another, and replacement in one set is out of phase with replacement in the other. Adjacent teeth are at different stages of the growth cycle, so vacant positions are flanked by functional teeth. Second, neighboring teeth in a set (alternate teeth in the mouth) are at slightly different developmental stages. Every tooth is a little more mature than the next anterior tooth of its set. It follows that waves of growth can be traced along each set of the tooth row (upper drawings of Figure 6-4), and when the sets are combined (lower drawing), few tooth positions are vacant—and these are not adjacent. The length of one wave of growth varies from species to species; commonly five to nine teeth are involved.

This model was developed largely by Edmund, using paleontological and developmental studies in combination. It seems true and useful, yet nature is usually less precise than the model, and the mechanism may be difficult to observe in a given jaw. Analysis is difficult if there are two or more parallel rows of teeth in each jaw (as in some fishes), and there may be other complications.

Comparative Anatomy of Teeth

From Denticulate Armor to Heterodonty.   AGNATHOUS VERTEBRATES do not have true teeth. Cyclostomes have epidermal tooth analogs on the oral funnel and on the "tongue" (Figure 3-3). Anaspids had on the "tongue" (not a tongue homolog) a bony plate which was roughened by surface denticles and must have been an effective rasping organ. Some cephalaspids apparently had a weaker but similar mechanism on the tongue and at the margins of the mouth. Pteraspids had a series of small plates at the ventral border of the mouth which were probably somewhat movable and may have been partial analogs of both teeth and jaws.

Teeth probably evolved among the PLACODERMS, but the

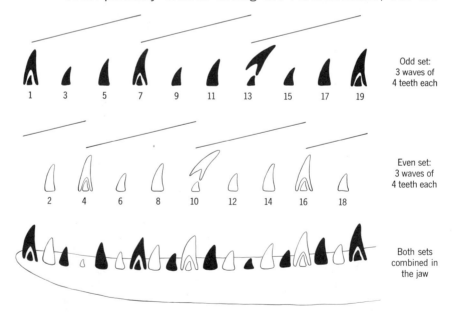

Odd set:
3 waves of
4 teeth each

1 3 5 7 9 11 13 15 17 19

Even set:
3 waves of
4 teeth each

2 4 6 8 10 12 14 16 18

Both sets
combined in
the jaw

FIGURE 6-4
Stylized diagram of the METHOD OF TOOTH REPLACEMENT IN AMPHIBIANS AND REPTILES.

first teeth have not been identified. The predaceous arthrodires had on the margins of their jaws large bony plates with jagged edges which sometimes contained dentine (Figure 3-5). These made a formidable shearing mechanism. The weak jaws of antiarchs were supported by small bony plates devoid of dentine and enamel.

CARTILAGINOUS FISHES usually have numerous teeth carried only at the margins of the mouth. Their structure resembles that of dermal denticles. Typically, they have several large, sharp cusps (cladodonts, pleuracanths, some selachians), but they may also be blade-like with serrate margins (Figure 6-5). Less commonly, they are blunt or flat. Chimaeras have only several crushing plates, each formed by the aggregation of numerous ancestral teeth. The teeth of a cartilaginous fish are usually all of about the same size and shape and are said to be **homodont** (= same + tooth). They are anchored to the skin of the gums and are not fixed to the skeleton of the jaws. Replacement teeth form in continuous series on the lingual side of the tooth row whence they migrate up onto the margins of the mouth (Figure 6-3). Several teeth of one series may be functional at a time. Old teeth are ultimately pushed to the outside of the gums, where they are resorbed and torn off.

As one would expect, the teeth of bony fishes are exceedingly diverse. Nevertheless, the major taxa have characteristic dentitions. ACANTHODIANS had numerous, sharp, conical teeth distributed along the margin of the lower jaw or arranged in whorls. RAY-FINNED FISHES usually have numerous teeth (the number ranges from zero to more than 10,000!) carried by the margins and roof of the mouth, the fifth gill arch, and the tongue. They are usually simple and conical, but may also be serrate, blunt, or flat. These fishes usually have homodont teeth. It is thought that primitively they were probably pleurodont and polyphyodont. Most, however, are now somewhat or entirely acrodont, and such teeth often are not shed.

Adult DIPNOANS have several, large, fan-shaped tooth plates, each of which is derived by the fusion of many separate teeth. These distinctive plates are adapted for crushing and are not replaced.

The teeth of CROSSOPTERYGIANS resemble those of ray-finned fishes in most respects. However, they have one distinctive and important characteristic: the enamel and dentine are folded so as to form complicated patterns as seen in cross-section (Figure 6-5). This structure, called **labyrinthodont,** strengthens the tooth and makes it resistant to wear. Labyrinthodont teeth survived for 100 million years among terrestrial descendants of crossopterygians, but were not retained by *Latimeria,* the lone survivor of the infraclass.

Labyrinthodonts (the subclass of amphibians) had conical teeth on the margins and roof of the mouth. These teeth were

CLADOSELACHII

PLEURACANTHODII

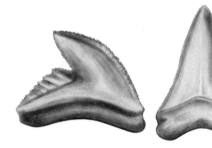
SELACHII

DIPNOI

Salmonid fish

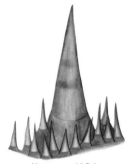
Hexagrammid fish

ACTINOPTERYGII

Stylized labyrinthodont tooth characteristic of
CROSSOPTERYGII
LABYRINTHODONTIA   COTYLOSAURIA

Boid snake

Varanid lizard

Iguanid lizard

LEPIDOSAURIA

FIGURE 6-5
CHARACTERISTIC TEETH OF VARIOUS TAXA.

numerous, but they tended, like those of the reptiles to follow, to be fewer than the teeth of fishes. They were attached to shallow pits or grooves in the jaws. They were homodont and — as the name of the group tells us — labyrinthodont. MODERN AMPHIBIANS tend to have fewer teeth (none in toads) which are small, simple, pleurodont, and no longer labyrinthodont. They are supported by pedicels of dental origin to which they are attached by zones of soft tissue (Figure 6-6). The phylogentic and functional interpretation of this unique pattern remains to be clarified.

Many REPTILES are homodont, though there may be some variation in the size of teeth in different parts of the mouth. Various reptiles, in several orders but particularly among mammal-like reptiles, have teeth of different form and function in different parts of the tooth row (see Figure 7-12). Such dentition is **heterodont** (= different + tooth). Only stem reptiles (extinct anapsids) retained the labyrinthodont condition. All forms of tooth attachment are observed in the class: Some snakes and many archosaurs are (and were) thecodont, various reptiles (including the primitive lepidosaur, *Sphenodon*) are acrodont, and many lizards are at least partly pleurodont. Most reptiles are polyphyodont, but acrodont teeth often are not replaced. One major group — the turtles — is **edentate** (toothless).

FIGURE 6-6 DIVISION OF TEETH OF LISSAMPHIBIA INTO CROWNS AND PEDICELS shown by the caecilian, *Gymnopis.*

ARCHAEORNITHES had thoroughly reptilian teeth, a condition retained by birds at least until the Cretaceous period. Modern birds are edentate, but embryos of ducks have dental laminae in their jaws.

Some Consequences of Chewing. Although exceptions will be noted in Chapter 25, most nonmammals use the teeth only for securing and holding food which is then quickly swallowed. MAMMALS also secure and hold food, but in addition they usually shear, crush, or grind their food. These new functions are of the utmost importance because they speed digestion and greatly increase the diversity of foods — particularly plant foods — that can be eaten. In order to be sheared, crushed, or ground, it is desirable for food to be retained in the mouth and chewed. This necessitates (1) modification of the ancestral jaw articulation and palate (discussed in other chapters), (2) cheeks to hold food in the mouth, (3) a tongue capable of positioning the food, and (4) several kinds of teeth. Mammals are heterodont unless they have secondarily reverted to homodonty in response to a diet of fish or insects (e.g., toothed whales, armadillos). Other distinctive features also relate to heterondonty: Mammalian teeth are thecodont because such teeth can best withstand shearing forces without being loosened. They are carried only at the margins of the jaws, possibly because elsewhere they would occlude with less force and precision. Specialization of the teeth leads to increased size, complexity, strength, and wearing ability. These, in turn, correlate with reduction in number of teeth functional at one time and reduction in number of sets of replacement teeth.

Most mammals are diphyodont, but this statement needs qualification. Typical mammals have a first set of teeth which consists of the temporary milk teeth plus the permanent molars. The fact that all these teeth belong to one generation is obscured by the circumstances that they erupt in sequence and may or may not be deciduous. The second generation of teeth includes all permanent teeth except the molars. Some marsupials replace only one tooth in each jaw, and moles replace none.

Numbers and Kinds of Teeth.   Typical mammalian dentitions have three or four kinds of teeth. **Incisors** are adapted for securing food and sometimes for grooming. They may be conical spikes for holding insects or flesh, or simple blades for cutting plant stems. They are relatively small teeth and are single-rooted. Their function requires that they be forward in the mouth; the upper incisors are rooted in the more anterior of the two bones of the jaw (the premaxilla). Numbers of teeth are expressed in terms of one side of the mouth only and are commonly written as a fraction, the count of upper teeth forming the numerator and the count of lower teeth forming the denominator. Incisors of eutherian mammals are always 3/3 unless secondarily reduced. This count is exceeded by some marsupials.

The **canines** are next posterior in the mouth. They are simple spike-like teeth with single roots. If not secondarily modified, they are long and strong and serve for holding and piercing in relation to both feeding and fighting. Canines always number 1/1 if not reduced. The alveolus for the upper tooth lies in, or just posterior to, the suture between the two bones of the upper jaw.

All replacement teeth behind the canines are called **premolars,** and all nondeciduous teeth of the first generation are called **molars.** Molars tend to be larger than premolars and to have more cusps and roots. The numbers of premolars and molars among ancestral therian mammals were probably 4/4 and 7–8/7–8, respectively. The primitive counts for eutherian mammals are considered to be 4/4 and 3/3.

The distinction between premolars and molars is sometimes unsatisfactory. As a practical matter it may not be possible to tell from an adult skull which teeth have been replaced. The entire tooth row forms a series such that teeth tend to resemble their immediate neighbors. Specialization may make the kinds of teeth indistinguishable: Artiodactyls have an in-

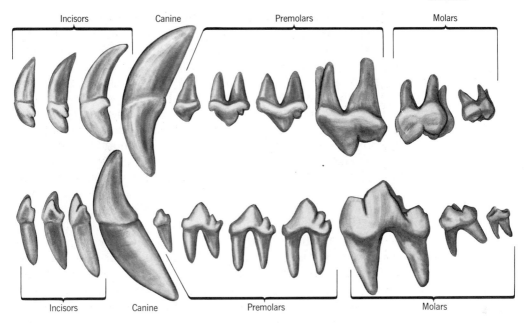

FIGURE 6-7
HETERODONT MAMMALIAN DENTITION illustrated by the teeth of the dog.

cisiform lower canine and horses have molariform premolars. Further, the distinction between premolars and molars breaks down if the replacement generation of teeth is incomplete or absent. For these reasons it is often convenient to speak of these teeth collectively as **cheek teeth.**

The numbers and kinds of teeth are expressed by a dental formula which consists of the numerical fractions already noted written in sequence starting with the incisors. Thus, the formula for the wolf is 3/3, 1/1, 4/4, 2/3; for the deer is 0/3, 0/1, 3/3, 3/3; and for man is 2/2, 1/1, 2/2, 3/3.

More About Cheek Teeth. The cheek teeth of primitive mammals (Jurassic mammals, marsupials, insectivores) and of some more advanced orders, as well, have several cusps. The upper molars usually have three principal cusps arranged in a triangle; the lower molars commonly have five principal cusps. Accessory cusps may also be present. The cusps are intricately arranged to interlock when the mouth is closed. Such teeth are admirably adapted for diets of flesh or invertebrates.

Speculation about the origin of multicuspid teeth from unicuspid teeth has been the subject of much debate. The theory of Bolk is now of only historical interest. His **concentration theory,** presented in a series of papers in the first decade of this century, postulated a fusion of teeth derived from two different replacement generations. The **concrescence theory** of Kükenthal and Röse was formulated in the 1890s. It holds that multi-cuspid teeth originated by the fusion of adjacent unicuspid teeth. This does actually happen in some fishes, but does not appear to have been a usual event. The favored theory, called the **differentiation theory,** was presented in the 1880s by Cope and Osborn and was revised and extended some years later by Gregory. This theory, for which there is abundant paleonto-logical and embryological evidence, states that the single cusp (called the protocone) of the ancestral tooth gradually became supplemented by two secondary cusps (paracone and meta-cone) which formed from the side walls of the tooth crown. Additional smaller cusps may form from a ridge (cingulum) that encircles the base of the crown. The patterns of cusps are various and intricate. Further, cusps may become joined in various ways by ridges (lophs). All cusps and lophs are named, and the terms distinguish between corresponding cusps of upper and lower teeth. Unfortunately, some of the terms as-signed by Cope and Osborn and now in wide use are not apt because they do not express true homologies between upper and lower teeth. Since it is impossible for the nonspecialist to recognize the specific cusps and lophs of the various animals, the terminology will not be presented here.

The description and functional analysis of the principal kinds of cheek teeth is a particularly interesting subject. It will be postponed until Chapter 25 because it relates to adaptation to diet rather than to the broad sweep of evolutionary change among major taxa, which is our concern in this part of the book.

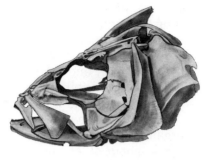

7

Head Skeleton

The internal, jointed skeletal system of vertebrates is unique in the animal kingdom and is the most important of all organ systems in the study of vertebrate morphology. It is conservative enough in general pattern to show the broad outlines of vertebrate phylogeny: Homologous bones and evolutionary trends are readily demonstrated in skeletons characterizing the successive major taxa. Further, the skeleton plays a central functional role. Wide variations have always been superimposed on the gradually evolving general pattern, and the skeleton has been sufficiently plastic to respond to the particular habits of the various animals. Therefore, it also provides reliable information about the specific adaptations of vertebrates:

Importance of the Skeleton to Morphology

125

Postures and locomotor adaptations are accurately revealed, and other adaptations are sometimes indicated.

Because of its hardness and durability, the skeleton (including the teeth) becomes fossilized relatively often, and this contributes virtually (though not quite) all of our knowledge of past vertebrate life. Paleontology can be regarded as the comparative anatomy and morphology of extinct animals. No other science has contributed so much to our knowledge of vertebrate evolution, and it is, of course, limited almost entirely to the study of hard parts. Fortunately for the paleontologist, of all organ systems the skeletal system tells the trained observer the most about *other* organ systems: Most muscles take their origins and insertions on bones, and often leave tuberoscities or scars to show the positions and extent of these contacts; the important cranial nerves reveal their sizes and courses by the foramina they traverse in the skull; the relative development of the different parts of the brain may be revealed by the braincase; nasal chambers, orbits, and otic cavities give some information about the sense organs they house; the nature and distribution of sensory canals on the head may be shown in detail; even some blood vessels leave traces on the skeleton.

Further, of all organ systems this one is the easiest to preserve, store, and demonstrate. It is, therefore, a good one to teach and learn about. Students would have to spend hours merely finding the circulatory system of one animal; they can devote the same time to studying the prepared skeletons of a variety of animals.

More About Hard Tissues

Hard tissues have been described as we have come to them: horn, enamel, dentine, and bone in Chapter 5 and cement in Chapter 6. As we turn to the internal skeleton, another hard tissue, cartilage, must be added and a little more should be said about the development of bone.

Cartilage is a tissue that combines hardness with elasticity. It is present in several invertebrates but is particularly characteristic of vertebrates. Its cells (chondrocytes) are dispersed in spaces (lacunae) in a matrix of a glycoprotein (chondromucin) and fibers. Chondrocytes form at the surface membrane, or perichondrium, and gradually round up, grow, and become wider spaced as new matrix pushes them farther apart. **Hyaline cartilage** has relatively few contained fibers. It is translucent in thin sections, and its surfaces are exceedingly smooth. **Elastic cartilage** is rich in elastic fibers and consequently has flexibility and resilience. **Fibrous cartilage** contains a heavy meshwork of collagenous fibers which make it very tough. **Calcified cartilage** contains deposits of calcium salts (the same that occur in bone) which make it relatively hard and firm. Unlike bone, it is not remodeled once it has formed. There are intergrades among these types of cartilage.

Cartilage is usually derived from mesoderm, but, curiously, in part of the head and in virtually all of the gill region it may also be derived from neural crests, which are initially ectodermal.

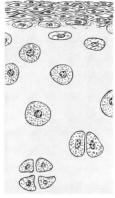

In terms of ontogeny, there are two kinds of bone. As explained in Chapter 5, bone that is associated with the integument ossifies directly from the dermis. Bones of the ancestral integument that come to lie below the skin also ossify directly from somewhat deeper mesoderm. This kind of bone is called **dermal bone** or **membrane bone.**

The other kind of bone does not ossify directly from membranous tissue but instead gradually replaces cartilage which has formed earlier. Such bone is called **replacement bone** or **endochondral bone.** (Its development is described further on p. 178.) Replacement bone does not occur in classes below the bony fishes. Once the two kinds of bone have formed, they are histologically identical. Some bones form by a combination of the two methods or even alternatively by either method, and it is known that replacement bones do rarely evolve into membrane bones. The distinction is, in general, useful, however, for establishing the origins and homologies of various elements of the skeleton.

FIGURE 7-1
HYALINE CARTILAGE.

Segmental Nature of the Head

The head skeleton is the oldest part of the skeleton. Except for the brain it is perhaps the most instructive part of the entire body. Certainly morphologists have given it the most attention. Zoology students learn in introductory courses that internal segmentation is a characteristic of the phylum Chordata. Segmentation is rarely as conspicuous in this phylum as it is for annelids or arthropods, yet it is evident enough for the vertebrae, the serial spinal nerves, and the body musculature of fishes. It is not so evident that the vertebrate head was initially also a segmental structure, at least in part. Evidence of its fundamentally serial nature is found (albeit with difficulty) in the embryology of the cranial skeleton and muscles of certain animals—particularly of sharks.

It is not an easy matter to decide how many body segments are represented in the head and which structures, if any, are derived from each. Since the series of segments is never complete, and since the number of segments represented has bearing on the interpretation of various head structures, it is desirable to learn how many somites (embryonic segments of the mesoderm) have been lost. These problems constituted a principal field of research among morphologists from the turn of the eighteenth century through the first decades of the nineteenth century. The German poet–naturalist, Goethe, a pioneer in comparative anatomy who introduced the word "morphology," supported a philosophical concept called "unity of type." He believed the vertebra to be a fundamental "type," or structure pattern of the body, and saw the skull as a fusion of six enlarged and modified vertebrae. Later, the Swiss-American naturalist, Louis Agassiz; the English evolutionist, Thomas Huxley; the Scottish anatomist, Balfour; and the German anatomists, Wiederscheim, Gegenbaur, Fürbringer, and others; showed this concept to be false and developed instead variations on the idea that although only the posterior part of the skull can be homologized with vertebrae, most or all of the head is fundamentally segmental. Certain assumptions of these investigators had to be corrected, chiefly by Goodrich, to bring the idea of the segmentation of the head to its present status. The concept will contribute to what follows in this chapter and will be developed further in the chapters on muscles and the nervous system.

Components of the Head Skeleton

Since the head skeleton is complicated, both as to origin and structure, it is helpful to find parts that can be described one at a time. The logical components to select are the chondrocranium, visceral skeleton, and dermal elements, because the structures that these contribute to the complete skull are at first rather distinct, both in their phylogeny and their ontogeny.

The **chondrocranium** supports the brain and organs of special sense. The **visceral skeleton** supports the gill arches and their derivatives. The **dermal elements** complete the relatively superficial framework of the skull. The chondrocranium and visceral skeleton may remain cartilaginous or become bony, but they are always present. The dermal skeleton is always bony and is usually present, but has been secondarily lost by several major vertebrate groups. Authorities are divided on the relative antiquity of these components. Regardless of which may have evolved first, the earliest known jawless vertebrates already had all three of them. After learning about these three sources of the many elements of the head skeleton, we will survey the general changes that have characterized the evolution of the skull among the classes.

The term "skull" is sometimes used for all of the skeleton of the head. However, it is also used—and will be used here—for the single unit that forms the braincase and upper jaw and houses the nose and ear. The word is useful, but inexact.

Chondrocranium and Derivatives. Most organs of the body can continue to function when subjected to moderate pressures by contacts with the environment or by locomotor or digestive activities. This is not true, however, of the central nervous system or organs of special sense. The spinal cord of primitive vertebrates derived some protection from the stiff notochord. The larger brain was shielded above by bony scales or plates that formed from the dermis of the skin and was supported below and on the sides by a trough of hyaline cartilage. Capsules and rods of cartilage that protected the sense organs and stiffened the rostrum merged with this brain trough to form the chondrocranium (= cartilage + cranium). This structure has persisted in all vertebrates ever since its origin some 550 million years ago. Animals that do not have bony heads (cyclostomes and cartilaginous fishes) have relatively complete and heavy chondrocrania. Most vertebrates that do have bony heads retain a completely cartilaginous chondrocranium only in larval or fetal life. In adult life these animals replace at least part of the more delicate cartilage by bones that ossify within the cartilage. These bones are not spoken of as a chondrocranium but can be thought of as a derivative or modification of the chondrocranium rather than as a substitute for it.

The chondrocranium has evolved steadily through the ages. It is not to be regarded as a degenerate structure merely because it is absent from adult tetrapods. Homologies of its major features have been established by noting their relations to such conservative landmarks as the notochord, hypophysis, and cranial nerves and blood vessels. Homologies of its numerous and complicated little rods and vacuities are more difficult to establish; seemingly, some will never be deciphered. Embryonic recapitulation has not been very helpful. The homologies of the bones that ossify in the chondrocranium are more satisfactory.

The chondrocranium is a single unit. It is composed, however, of numerous elements which are distinct in the embryo. We will arrange these in five groups that are relatively large and constant.

The **notochord** lies within, just above, or just below the base of the developing chondrocranium. It may remain free or be obliterated, but usually its sheaths become cartilaginous and join the chondrocranium. The contribution it makes is small but important because its presence in the head may predate that of all other skeletal structures there, and because the constant position of its anterior end, just posterior to the hypophysis, serves as a reference point.

Anterior to the notochord is a pair of bars called the **trabeculae** (= small beams). The hypophysis is located between their posterior ends. Sometimes, in lower vertebrates having broad heads, the two bars remain free and widely separated (the platytrabic condition); sometimes they are joined by a plate of cartilage; and usually they fuse anteriorly to form a narrower Y-shaped structure having its leg pointing forward (the tropitrabic condition). The trabecular part of the chondrocranium is related to the forebrain, orbits, and rostrum. In older embryos it may be complicated by curved outgrowths which form an elaborate framework. This portion of the chondrocranium is probably older in origin than any other part except the notochord. It is flanked by mesodermal structures that are initially segmented, but is apparently not itself of segmental origin.

Behind the trabeculae, and flanking the notochord, another pair of cartilages form which are called (because of their position) **parachordal cartilages.** These soon merge above, below, or around the notochord to form the **basal plate.** The anterior part of the plate commonly encloses a large vacuity, and its lateral margins are pierced by foramina for the exit of cranial nerves. The side walls of the chondrocranium are usually incomplete. They consist of varied and sometimes complicated, pillars and rods that fuse to the basal plate at its lateral margins. All but

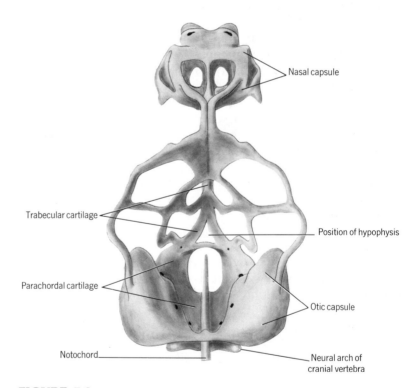

FIGURE 7-2

COMPONENTS OF THE VERTEBRATE CHONDROCRANIUM seen in dorsal view. Slightly simplified from the 25 mm stage of development of the lizard *Lacerta.*

the most anterior part of the basal plate is of segmental origin from the sclerotomes of embryonic head somites. This part of the chondrocranium is considered to be derived from, or at least to be in series with, the bases of vertebrae. The number of segments in the head varies but is often in the range of three to five. At the back of the basal plate are one or two projections called **occipital condyles** which articulate with the first free vertebra of the spine. Their interesting origin from vertebral elements will be explained in Chapter 8.

A fourth contribution to the chondrocranium consists of one or more pairs of arches that rise up from the posterior angles of the basal plate to flank or encircle the spinal cord just where it enters the skull. These are clearly the **neural arches of cranial vertebrae.** In embryos of some fishes they sometimes even bear short ribs. The separate arches merge in late embryos and are then known collectively as the **occipital arch** (occiput = back part of the skull).

A final contribution to the chondrocranium is the cartilaginous **sense capsules** that house the nasal chambers and inner ear. The nasal capsules join the anterior ends of the trabeculae. The otic, or auditory, capsules join the margins of the basal plate just in front of the occipital arch. The eyes are also enclosed in capsules of cartilage. These remain free, however, because if they joined the chondrocranium, the eyes could not be moved independently of the head.

Bones typically develop in the chondrocrania of five of the eight classes of vertebrates (all except Agnatha, Placodermi, and Chondrichthyes). The names and positions of these bones are best learned from specimens and from Figure 7-3. No one list of such bones serves for all vertebrates having ossified skulls; most of the bones are rather constant, but some (including the mesethmoid and orbitosphenoids) are of variable occurrence. In the higher vertebrates, the otic bones and certain of the basal bones tend to fuse together. The sclerotic bones that form in the optic capsules of some animals will be discussed along with the eyes in another chapter.

Visceral Skeleton and Derivatives. The **pharynx** is a somewhat expanded part of the digestive tube lying between the oral cavity and the esophagus or stomach. In the earliest stages of chordate evolution the pharynx became perforated laterally by paired gill slits which served in feeding and respiration. A part of the pharynx still contributes to the feeding mechanism of all the vertebrates that have jaws, and the respiratory function of the pharynx is retained by all jawless vertebrates, fishes, and larval amphibians. When tetrapods ceased to use the pharynx for respiration, much of its musculature and skeleton became available for other uses and were adapted for the control and

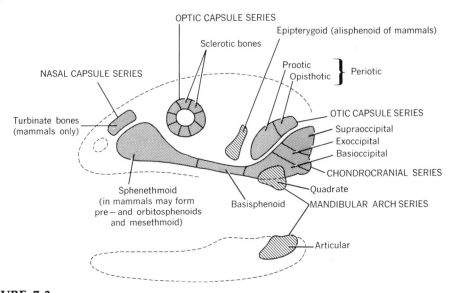

FIGURE 7-3
REPLACEMENT BONES OF THE SKULL AND MANDIBLE AT THE EARLY TET-RAPOD STAGE OF EVOLUTION (except as noted). Ossifications of the chondro-cranium and sense capsules are shaded; those of the mandibular arch are hatched.

support of the tongue, vocal apparatus, and related structures. Thus, the pharynx has had a long and varied history.

The first protochordates probably had at least a dozen pairs of gills. The number was increased to as many as 100 by some protochordates (amphioxus), so that food particles could be entrapped on sticky gill bars as water was strained through the pharynx. The earliest vertebrates may have also included some filter-feeders, but the most ancient well-known vertebrates subsisted on larger food and did not require such a specialized pharynx. They reduced the count to 5–15 pairs of gill slits. By the time jaws evolved, the number had been further reduced and stabilized at about 6 (more in some sharks and fewer in some tetrapods).

Between each typical gill pouch, and also anterior to the first and posterior to the last, are bars of tissue that consist of skeletal elements (largely derived from neural crests), muscles, and respiratory filaments together with the nerves and vessels serving these structures. Each such bar is called a **visceral arch** —"visceral" because it forms largely from a specialized part of the gut tube. The basic number of visceral arches for jaw-bearing vertebrates is one more than the primitive number of pouches, or 7.

The relationship between the segmentation of the visceral arches and the segmentation of the head (discussed above) is of interest. The mesodermal part of the digestive tube forms in the embryo from the hypomere. Unlike the somites above it, the hypomere is not segmented. It follows that the muscles and skeleton of the visceral arches do not exhibit the primary segmentation of the body. The presence of serial gills, however, imposes on the derivatives of the pharynx a secondary segmentation which comes to be functionally integrated with the primary segmentation of the cranial nerves.

It is convenient to number the arches and slits in sequence from anterior to posterior. Details of the composition and distribution of the cranial nerves of certain fishes led nineteenth century anatomists to surmise that at least one anterior arch and two pouches had been lost very early in vertebrate evolution. The subsequent discovery of cephalaspids and the description of the internal structure of their heads in astonishing detail by the Scandinavian paleontologist Stensiö has verified the prediction. However, since most vertebrates lack evidences of such structures, premandibular arches and pouches are ignored when numbering the series. The first constant arch is numbered 1, and each slit, or pouch, is given the same number as the arch that lies just anterior to it.

No other part of the vertebrate skeleton, except possibly the notochord itself, is as ancient in origin as the visceral skeleton. Each gill bar of jawless vertebrates is supported by a single cartilaginous rod. It may angle and bend, and usually joins its fellows above and below the gill slits, but it is not jointed. The skeleton of each visceral arch of gnathostomes is jointed. The basic number of paired segments is seemingly four per arch. Of these, the middle two, called the **epibranchial** (above) and the **ceratobranchial** (below), are relatively important for our story. There are also unpaired, midventral segments between the lower ends of the gill bars, which are considered, on somewhat slim evidence, to have been derived by remote ancestors from the paired gill bars.

The first visceral arch enlarges to become the jaws, if jaws are present, and is then called the **mandibular arch.** Its epibranchial forms the upper jaw and takes the name **palatoquadrate.** The ceratobranchial presumably forms the lower jaw and is called the **mandibular cartilage.** The second visceral arch, as we shall see, usually has a special function and, therefore, also has a special name—the **hyoid arch.** The epibranchial is its key element and is called the **hyomandibula.** Succeeding arches that serve primarily in respiration are called **branchial arches** (branch = gill). Thus, the third visceral arch may also be the first branchial arch. This terminology is clarified by Figure 7-4, which may be compared to Figure 7-8.

Ossifications that form in the visceral cartilages are, like those of the chondrocranium and nearly all of the postcranial

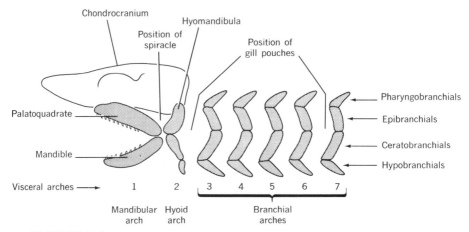

FIGURE 7-4

PRIMITIVE VISCERAL SKELETON represented by a stylized elasmobranch.

skeleton, replacement bones. As noted in the following section, most of the bones that become functionally related to the jaws are not replacement bones and, therefore, are not derived from the visceral skeleton. The true first arch bones are usually small, yet no other bones of the body have a more engrossing and unexpected history. More of this later in the chapter.

Contributions from the Integument. The rigid scales or heavy armor of most ostracoderms, placoderms, dipnoans, and crossopterygians was the most complete and solid over the head where it supported the teeth and shielded the roof of the brain, the delicate gills, and other soft tissues. At first these integumentary bones were variable as to size and pattern: Anaspids had small scales; cephalaspids had huge solid shields; arthrodires had large plates composed of individual elements joined by sutures. By the time the bony fish stage of evolution was reached, the elements tended to stabilize as medium-sized pieces. The crossopterygians developed a general pattern of bones that was to persist throughout tetrapod history. The various classes have modified the basic pattern, however, by fusions, deletions (particularly where the gills were lost, at the back of the head, and between head and pectoral girdle), and sinking the bones below the skin and even underneath certain muscles.

In order to establish the homologies of these numerous bones it has been necessary for morphologists to use as many clues as possible. The paleontological sequence of various changes has indicated trends. Attention has been given to the relation of these bones to ossifications in the chondrocranium and to nerves, vessels, sense organs, and the opening often present on the top of the head that accommodates the pineal organ. The sensory canals of aquatic forms have been helpful because they follow a definite pattern and usually leave grooves, pits, or tunnels in the underlying bones. The number of ossification centers and the sequence of their appearance in the young animal have also been useful because a particular bone usually ossifies from a given number of centers which appear in constant sequence in relation to the centers of other bones. The characteristics of some centers are known to vary somewhat, however, so this clue must be interpreted with caution. In these various ways it has been possible to establish with confidence the homologies of nearly all the bones derived from the integument. The names and positions of the more constant bones are shown in Figure 7-5.

The recognition of these components of the head skeleton is more than a mere convenience. As we have seen, they may retain a degree of structural independence in the adult. This is particularly true of anamniotes. The skeleton of the gills must be discrete as long as the gills are functional, and the membrane bones often remain superficial where they meet edge to

Relations of the Cranial Components

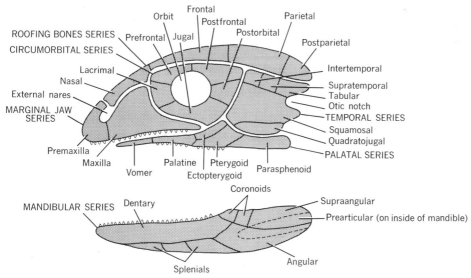

FIGURE 7-5

PRINCIPAL DERMAL BONES OF THE SKULL AND MANDIBLE AT THE EARLY TETRAPOD STAGE OF EVOLUTION.

edge to form a continuous covering. Further, in the evolution of the various groups of animals, the bones of the skull have sometimes all become more and more strongly ossified and, in contrast, have sometimes all regressed. Other times, however, derivatives of the several components have evolved independently. Thus, it appears that the chondrocranium and visceral skeleton were emphasized by cartilaginous fishes and cyclostomes at the same time that the dermal skeleton was regressing. Conversely, the replacement bones of the skull, and these only, tended to regress in late cephalaspids, lung fishes, some primitive ray-finned fishes, and late labyrinthodonts. Evidently, replacement and membrane bones may respond differently to evolutionary forces even though they are visually identical.

In spite of original and potential independence, the replacement and membrane bones of higher vertebrates become intimately associated. Enlargement of the brain forces the chondrocranial ossifications outward, whereas enlargement of jaw musculature forces many membrane bones inward. These components meet and jointly form a single, firm unit.

When the first visceral arch was converted to jaws it became mechanically necessary to brace the arch more firmly than had been the case while it functioned only for respiration and filter-feeding. This was first accomplished by attaching the palatoquadrate cartilage to the chondrocranium. Later, the hyomandibula of the second arch became a strut to further brace the jaw. Subsequently, the hyomandibula served alone, or the membrane bones covering the palatoquadrate joined other membrane bones to provide a firm anchor. The number, position, and firmness of the various points of attachment of the jaws to the remainder of the skull have been various. In fact, there are more than a dozen named patterns of jaw suspension. To further complicate the story, similar types of suspension have evolved independently in several instances. Properly interpreted, the relatively simple terminology proposed long ago by Thomas Huxley is still adequate. The terms, together with the animal groups to which they apply and the relationships among them, are shown in Figure 7-6. Some aspects of the story are amplified in the following pages.

Comparative Anatomy of the Head Skeleton

Jawless Vertebrates: Innovations and Variety. The chondrocranium of ostracoderms was a solid structure that usually remained cartilaginous throughout life and is, therefore, not well-known. In cephalaspids it was sometimes partly ossified and often was covered on all surfaces by a thin veneer of bone.

The visceral skeleton is a single continuous unit of cartilage.

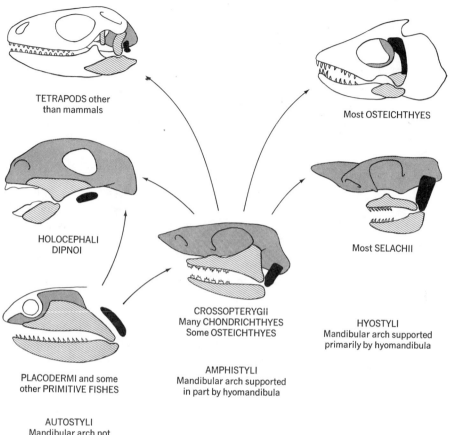

TETRAPODS other
than mammals

Most OSTEICHTHYES

HOLOCEPHALI
DIPNOI

Most SELACHII

CROSSOPTERYGII
Many CHONDRICHTHYES
Some OSTEICHTHYES

HYOSTYLI
Mandibular arch supported
primarily by hyomandibula

AMPHISTYLI
Mandibular arch supported
in part by hyomandibula

PLACODERMI and some
other PRIMITIVE FISHES

AUTOSTYLI
Mandibular arch not
supported by hyomandibula

FIGURE 7-6

**PRINCIPAL TYPES OF JAW SUSPENSION. Mandibular arch and derivatives
are hatched, hyomandibula and derivatives are black, chondrocranium is shaded,
teeth and extent of membrane bones are indicated in outline.**

All the arches are branchial in function. Their number is variable but numerous relative to other vertebrates. A premandibular arch is present in some ostracoderms. The individual arches tend to join each other above and below the gill slits and also to join the chondrocranium. The specialized mouth parts of cyclostomes are supported by cartilages which are not derived from the visceral skeleton and are not represented in other vertebrates.

The dermal skeleton of ostracoderms usually consisted of heavy, superficial scales. Cephalaspids and pteraspids had an armor shield covering the anterior part of the body. The bones comprising these shields cannot be homologized with the cranial bones of other vertebrates. Anaspids had small scales covering the head. Cyclostomes have no dermal skeleton.

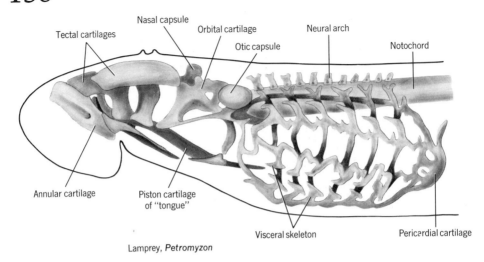

Tectal cartilages

Nasal capsule

Orbital cartilage

Otic capsule

Neural arch

Notochord

Annular cartilage

Piston cartilage of "tongue"

Visceral skeleton

Pericardial cartilage

Lamprey, *Petromyzon*

FIGURE 7-7
HEAD SKELETON OF A CYCLOSTOME seen in left lateral view.

Placoderms: Enter Jaws. The chondrocranium of placoderms was similar to that of agnathous vertebrates in being a solid structure, which was entirely cartilaginous or partly ossified.

Important advances had been made in the visceral skeleton. Most significant, the first arch had made the transition from gill support to jaws. Some of the predaceous arthrodires had formidable jaws indeed. Replacement bones often formed in the large palatoquadrate. This structure was **autostylic** (= self-supported), as it was attached to the chondrocranium only by ligaments. The second arch remained a typical branchial arch. As in higher vertebrates, the branchial arches of at least some placoderms were each supported by several skeletal elements.

The dermal skeleton usually consisted of heavy head and thoracic shields that were joined by hinges (arthrodires and antiarchs). Individual dermal bones of placoderms cannot be homologized with those in the armor or skulls of other vertebrates.

Cartilaginous Fishes: Specialization and Regression. Cartilaginous fishes have lacked bones throughout their long history. To provide the protection that dermal bones gave to the brains of remote placoderm ancestors, the chondrocranium became unusually solid, with complete side walls and a roof. Although the chondrocranium never ossifies, it is sometimes so hardened with granules of calcium salts that it cannot be cut with a knife.

Jaw suspension is usually amphistylic (pleuracanths, cladodonts, some sharks). It may also be hyostylic, however (other sharks), and in Holocephali, as that term implies, jaw suspension is strengthened for a shellfish diet by fusing firmly with the braincase—a form of autostyly. There are usually six, but as many as eight, postmandibular arches. Typical branchial arches have four paired skeletal elements, and there are also unpaired median ventral cartilages.

The segmentation and configuration of the visceral skeleton of cartilaginous fishes are considered to be generally primitive for all fishes. Loss of the dermal skeleton and specialization of the chondrocranium indicate, however, that the cartilaginous fishes do not include the ancestors of higher vertebrates. Biology classes often study the chondrocranium of the shark, but it is not typical of that structure for vertebrates in general.

Bony Fishes: Diversity and Complexity.   In no other class is the skull so varied and complex as in the bony fishes. This is not unexpected, since the class is so large, yet other parts of the body do not exhibit as much diversity. Much of the variety

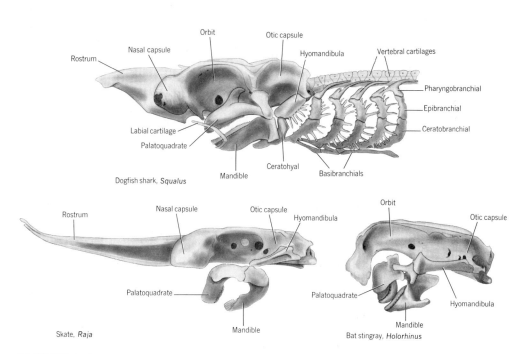

FIGURE 7-8

HEAD SKELETONS OF SELECTED CARTILAGINOUS FISHES seen in left lateral view. Above, chondrocranium, visceral skeleton, and anterior vertebrae; below, chondrocranium and jaws with their support.

results from adaptation to many diets, body shapes, and habits. These differences will, for the most part, not be discussed here. Some of the variety shows evolutionary trends, and these we will note. It is important that the three subclasses have somewhat different skulls, for only one of them can include the ancestors of tetrapods, and the skull helps us to recognize which subclass this is.

The chondrocrania of the earliest known bony fishes were relatively well-ossified, often, seemingly, in one unit without sutures. From this start the dipnoans had already regressed by Mesozoic times so that only the exoccipital bones ossify in an otherwise cartilaginous braincase. Crossopterygians also came to have a rather cartilaginous chondrocranium that always has a striking and unique feature: It is in two pieces. An anterior (or ethmosphenoid) unit is derived from the trabeculae and supports the orbits and rostrum. Joined to this by a movable hinge is a posterior (oticoccipital) unit which is derived from the parachordal cartilages and supports the brain. This crossopterygian trademark relates to the function of their jaws (see p. 629). Typical ray-finned fishes have braincases which are more or less completely ossified with separate bones.

The visceral skeleton is more nearly the same among the three subclasses. Dermal bones support the teeth, hence the mandibular arch usually regresses. Only in dipnoans does the palatoquadrate fuse to the braincase to become autostylic. In this they resemble the holocephalians, but the condition was acquired independently because of an analogous diet of shellfish. Several replacement bones may form in the palatoquadrate of bony fishes. Of these, the quadrate has the most evolutionary significance. It forms the upper part of the hinge of the jaw of bony fishes, amphibians, reptiles, and birds. The only replacement bone in the lower jaw is the small articular which forms the lower part of the hinge.

The second arch is not branchial in function. Its large hyomandibula usually supports the jaw (except in dipnoans) which, therefore, is hyostylic (ray-fins) or amphistylic (acanthodians and crossopterygians). We noted that some sharks are also hyostylic, but these two fish lineages probably evolved the condition independently. There are usually five branchial arches. Sometimes the more posterior of these are quite small. The gill arches may be bony (most ray-fins) or cartilaginous (dipnoans) and do not differ importantly from those of placoderms and cartilaginous fishes.

With few exceptions, the bony fishes have complete dermal skeletons of small to medium-sized bones. A series of bones joins the pectoral girdle to the skull. A movable bony operculum covering the gills is distinctive. Each of the major taxa of bony fishes has its own pattern of dermal bones. With the ex-

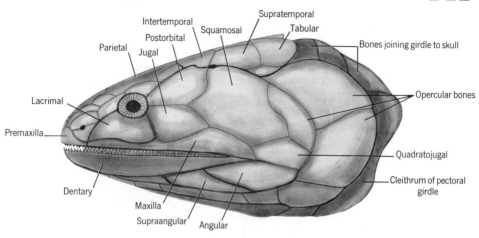

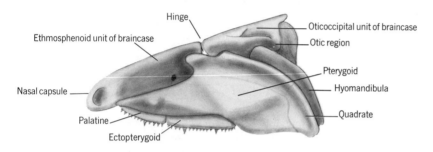

FIGURE 7-9

HEAD SKELETON OF *Eusthenopteron*, **A CROSSOPTERYGIAN OF THE RHIPIDIS-TIAN GROUP. Superficial dermal bones are shown in the upper drawing; the hinged chondrocranium and some other deep parts of the skeleton are shown in the lower drawing.**

ception of some small bones on the forward part of the snout, the pattern of the crossopterygians, and only this pattern, can be homologized with the pattern of amphibians. This is of primary importance and is one of several lines of evidence supporting crossopterygian ancestry for all tetrapods. The dermal bones of the other major groups of fishes are assigned names that are also applied to the bones of tetrapods, but for them the relationships are, in fact, quite uncertain.

Amphibians: Conservativism or Regression.   When the progenitors of amphibians crawled out of the water they "quickly" (over a period of several million years?) lost from their skulls such piscine characters as gill bars and operculum. Evolution of the tetrapod skull was then by no means finished—important changes were still to come to the palate, skull roof, and jaw mechanism—yet a trend toward stability of form had been initiated by crossopterygians that was to continue through all

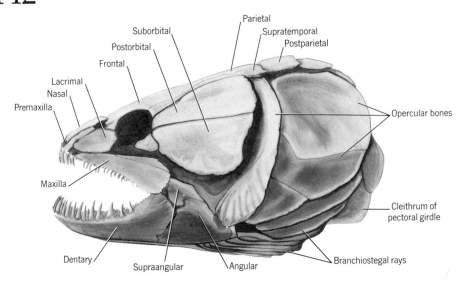

Bowfin, *Amia*

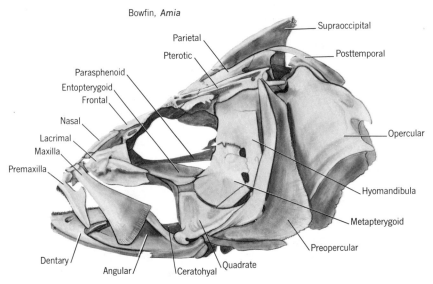

Bass, *Roccus*

FIGURE 7-10
HEAD SKELETONS OF RAY-FINNED FISHES of the infraclass Holostei (above) and Teleostei (below).

their descendants. There is less variation of skull structure in all the tetrapods together than in the bony fishes alone. Labyrinthodonts bequeathed the crossopterygian skull to the first reptiles without making striking changes. Surviving amphibians, on the other hand, have somewhat specialized skeletons.

The skulls of extinct labyrinthodonts are more within the main stream of evolution than are those of most modern amphibians.

Ancestral amphibians had a nearly full complement of replacement bones derived from the chondrocranium; only the supraoccipital was missing. Some of these bones were lost by late labyrinthodonts. Lissamphibia lack even such usually prominent bones as the basioccipital and basisphenoid. The exoccipitals are retained by these animals largely to provide the paired occiptal condyles. The more primitive labyrinthodonts had a single occipital condyle formed by the basioccipital.

The most significant changes that occurred in the skull during the transition from fish to amphibian concerned the visceral skeleton. The quadrate of the upper jaw now articulated with the squamosal without an "assist" from the hyomandibula. Thus, jaw suspension had become autostylic (again), as it was to remain throughout subsequent vertebrate evolution. Most extinct amphibians probably rested their large heads on the ground and detected vibrations in the substrate by bone conduction through the jaw and other structures to the inner ear. This mechanism was improved by incorporating a reduced, stubby hyomandibula into the line of sound transmission. The bone was then an ear ossicle and is assigned a new name, **stapes,** or columella. (The stapes of Lissamphibia also incorporates another element.)

Ventral elements of the hyoid arch support the tongue as they do also in fishes, but the tongue of amphibians is larger and more muscular. The more posterior arches function as branchial arches only in larvae and in those few species that retain gills as adults. Otherwise, they are reduced in number to three and converted to help support the tongue and newly evolved larynx. The tracheal rings may have the same origin. The complex of bones that moves the tongue and suspends the larynx from the base of the skull is called the **hyoid apparatus.**

Of the membrane bones there is less to say. The operculum is lost, and the pectoral girdle is no longer joined to the skull. There has been some tendency to simplify the pattern of bones, but in general, amphibians have not introduced innovations.

Reptiles and Birds: Variations on the Basic Plan. Reptiles are important to our story for modifications introduced by one or more of their lineages in relation to the structure of ear ossicles, palate, and jaw mechanism. The skulls of birds are sufficiently similar to those of their reptilian ancestors, so that we can here consider birds to be merely specialized reptiles.

In these classes, the replacement bones that are derived from the chondrocranium vary widely in configuration, but a full complement is typical. In birds, all of them fuse with one

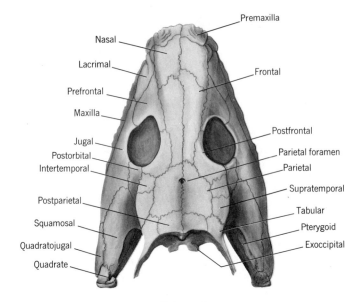

Palaeogyrinus

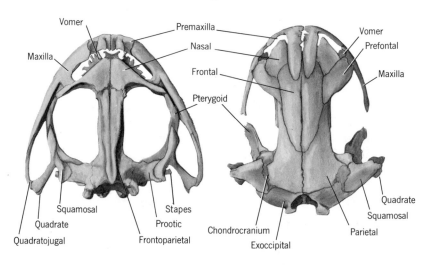

Bullfrog, *Rana*   Tiger salamander, *Ambystoma*

FIGURE 7-11

AMPHIBIAN SKULLS OF THE SUBCLASS LABYRINTHODONTIA (above), AND ORDERS ANURA (below, left), AND URODELA (below, right).

another and with other bones of the braincase before, or soon after, hatching, leaving no trace of sutures in the adult. Reptiles and birds have one occipital condyle.

The visceral skeleton of birds and most reptiles remains essentially the same as for amphibians: Quadrate, articular, and

Comparative Anatomy of the Head Skeleton

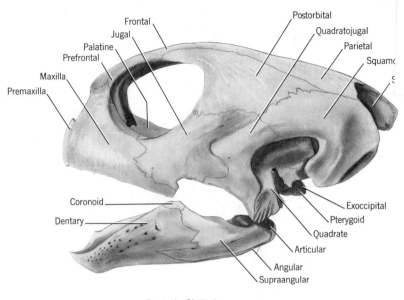

Sea turtle, *Chelonia*

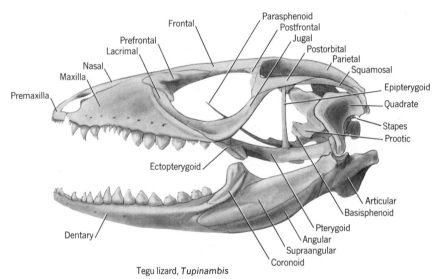

Tegu lizard, *Tupinambis*

FIGURE 7-12

REPTILE SKULLS AND MANDIBLES OF THE SUBCLASSES ANAPSIDA (above) AND LEPIDOSAURIA (below).

some cartilage form from the first arch; stapes from the second arch; hyoid apparatus chiefly from the second arch, but also from the third and sometimes fourth arches; larynx and tracheal rings probably from the sixth and seventh arches.

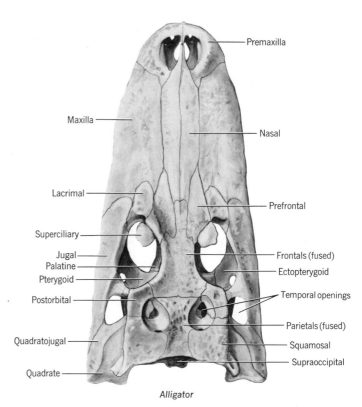

FIGURE 7-13
REPTILE SKULL OF THE SUBCLASS ARCHOSAURIA.

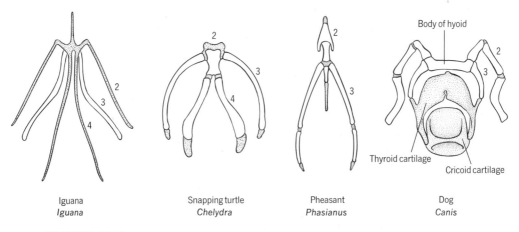

FIGURE 7-14
VISCERAL SKELETONS OF SELECTED TETRAPODS shown in ventral view. Numbers indicate visceral arch of origin. Stippled elements are cartilaginous.

   Mammal-like reptiles modified the first arch derivatives in a
way that proved to be significant. As the dentary bone of the
lower jaw became ever larger, the articular became smaller and
was displaced until it finally lost its position as the lower element in the hinge of the jaw. Likewise, the quadrate became
smaller and at last lost its position as the upper element of the
hinge. A forward shift of the hinge helped to release these
bones and create a new jaw articulation between the dentary
and squamosal. The articular and quadrate now became ear ossicles as the hyomandibula had done 100 million years before.
The articular is given the new name **malleus** and the quadrate
the new name **incus.** The stapes now articulates laterally with
the incus and medially with the inner ear. This classic evolutionary sequence was first pieced together from elusive clues

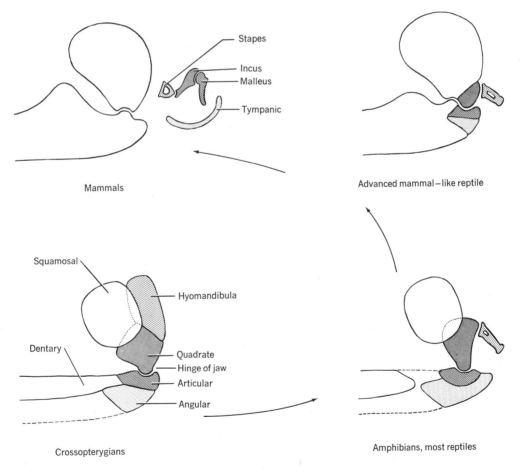

FIGURE 7-15
PHYLOGENY OF THE ARTICULATION OF THE JAW AND ASSOCIATED
STRUCTURES.

derived largely from embryology and comparative anatomy. More recently, abundant paleontological evidence has confirmed even the details of the story. These transitions are of particular interest because they relate to the recognition of the first mammals. Conveniently, but arbitrarily, mammals were long distinguished from their immediate reptilian ancestors on the basis of having an ossicle called the malleus instead of a jawbone called the articular. This distinction is no longer satisfactory for some fossils that are on the borderline.

The membrane bones of reptiles vary somewhat in number and may be heavy or delicate. Variations in configuration and relationship are more striking, however, and differences in the relation of bones on the roof and side walls of the skull to the muscles of the jaws are the basis for the naming of the subclasses of reptiles. Amphibians and fishes accommodate their jaw muscles beside the braincase and under the superficial roof bones of the skull which may be continuous or notched. This arrangement persisted in the most ancient group of reptiles (cotylosaurs, or stem reptiles — long since extinct). Subsequent reptilian lineages developed more powerful jaws. Presumably in response to the resultant increased and altered stresses, the cranial vault strengthened in some places and weakened in others. Ultimately, one or two pairs of openings formed in the temporal region of the skull, and jaw muscles then moved out through them and took origin in part from their margins. It is evident that this process occurred independently several times, because the fenestration (fenestra = window) evolved at different places on the skull in different groups.

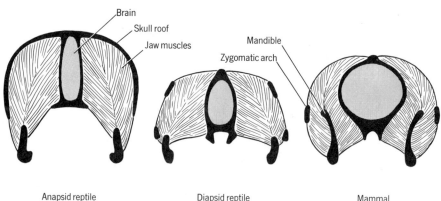

FIGURE 7-16
SOME RELATIONSHIPS BETWEEN JAW MUSCLES AND THE HEAD SKELETON shown by cross-sections of the head at the epipterygoid–alisphenoid level of the skull. Diagrammatic, but based on the sea turtle, *Chelonia*, the tuatara, *Sphenodon*, and the dog, *Canis*.

Names for the various skull types consist of the combining form "apsid," meaning arch, plus a prefix meaning "not," "two," "wide," etc., as appropriate.

Turtles are related to the ancestral, **anapsid** stem reptiles, and like them have no true temporal openings. Sea turtles demonstrate this condition well, but other turtles have formed an emargination into the skull roof bones from behind. Synapsida and Euryapsida each have one pair of temporal openings, but in each it is bordered by different bones. The mammalian condition was derived from the **synapsid** arrangement of mammal-like reptiles merely by enlarging the opening. **Diapsid** skulls have the same opening as synapsid skulls and also another that is dorsal to it. This pattern is found in extinct archosaurs, surviving archosaurs (crocodilians, which have small openings, but must usually be used to illustrate the condition to students), and also in the primitive lepidosaur, *Sphenodon.* Other surviving lepidosaurs and birds have modified the primitive diapsid pattern by the deletion of one or more of the arches of bone from below or between the original openings, as shown by Figure 7-17.

Another change in cranial architecture that was introduced by certain reptiles and later became universal among mammals is likewise related to the feeding mechanism. Amphibians and most reptiles bolt their food. It is unimportant to them, therefore, that respired air comes through the nares into the forward part of the mouth. Other reptiles, and particularly mammal-like reptiles, tore, crushed, or even chewed their food before swallowing. To avoid interrupting their breathing, it became necessary to deliver inspired air into the pharynx behind the chewing mechanism. This was doubly important for those that were becoming endotherms and were therefore increasing their respiratory rates. The paired vomers, which formerly had bordered the internal nares behind, gradually merged, migrated posteriorly, and moved dorsally above the new, more posterior internal nares. The parasphenoid was lost, thus "getting out of the way" of the relocated nares. The pterygoids shortened to the rear. Meanwhile, shelf-like processes of the maxillas, and later also of the palatines, grew to the midline in front of the shifting nares to form a new, or **secondary palate.** Actually, secondary palates evolved independently several times. Of animals other than mammal-like reptiles, some turtles have an incomplete secondary palate and crocodilians have a complete one.

The palate of crossopterygians and early labyrinthodonts was complete and solid. The palates of later labyrinthodonts, surviving amphibians, most reptiles, and birds have lateral vacuities of larger or smaller proportions which lighten the skull without much loss of strength and, for some of these animals,

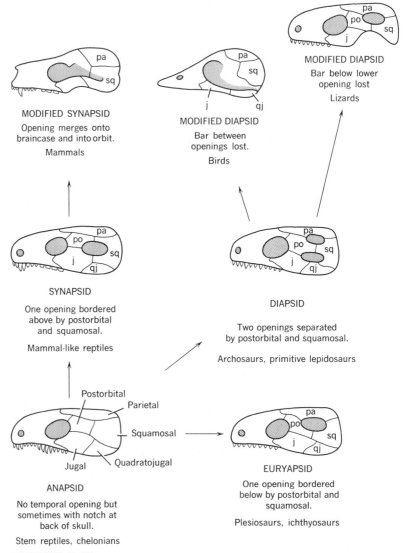

MODIFIED SYNAPSID

Opening merges onto braincase and into orbit.

Mammals

MODIFIED DIAPSID

Bar between openings lost.

Birds

MODIFIED DIAPSID

Bar below lower opening lost

Lizards

SYNAPSID

One opening bordered above by postorbital and squamosal.

Mammal-like reptiles

DIAPSID

Two openings separated by postorbital and squamosal.

Archosaurs, primitive lepidosaurs

Postorbital

Parietal

Squamosal

Jugal   Quadratojugal

ANAPSID

No temporal opening but sometimes with notch at back of skull.

Stem reptiles, chelonians

EURYAPSID

One opening bordered below by postorbital and squamosal.

Plesiosaurs, ichthyosaurs

FIGURE 7-17

PHYLOGENY OF TEMPORAL OPENINGS AMONG REPTILES AND THEIR DESCENDANTS.

enable protruding eyes to be withdrawn into the head on occasion to escape harm or assist in swallowing.

Birds and some reptiles are able to move the palate forward and backward on the braincase by pivoting the quadrates to and fro and by providing a hinge in the roof of the skull in front of the orbits. This action elevates the upper jaw, which increases the gape of the mouth. Skulls that have this mechanism are said to be kinetic and are described further in Chapter

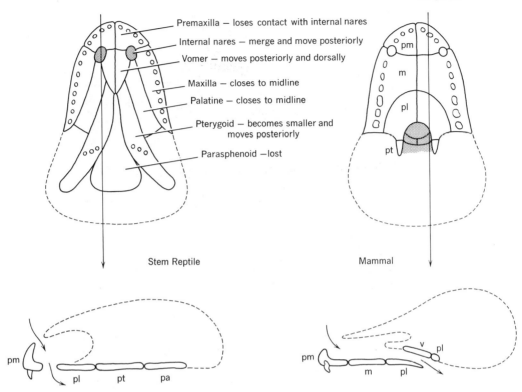

Premaxilla — loses contact with internal nares
Internal nares — merge and move posteriorly
Vomer — moves posteriorly and dorsally
Maxilla — closes to midline
Palatine — closes to midline
Pterygoid — becomes smaller and moves posteriorly
Parasphenoid — lost

Stem Reptile

Mammal

FIGURE 7-18
EVOLUTION OF THE SECONDARY PALATE. Palatal views above; parasagittal sections below. Dashed lines indicate parts of skull not involved in this evolutionary sequence. Lower arrows show path of inspired air.

25. Ornithologists divide the birds into four or more groups on the basis of the relationships of the palatal bones involved in this mechanism.

Mammals: Some Further Modifications. The mammalian skull is at the same time exceedingly variable in regard to adaptive features such as strength and proportions, and conservative in regard to basic plan. Temporal architecture, jaw suspension, ear ossicles, and secondary palate remain about as inherited from the most advanced synapsid reptiles. The brain is, of course, larger than for other vertebrates, a factor that contributes to complete functional integration of the replacement and membrane bones of the braincase. There are again two occipital condyles as for most amphibians. Cranial sutures are nearly always more in evidence than for birds and usually less in evidence than for reptiles. Certain combinations of bones

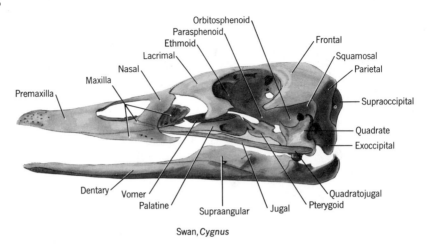

Orbitosphenoid
Parasphenoid
Ethmoid
Frontal
Lacrimal
Squamosal
Nasal
Parietal
Maxilla
Premaxilla
Supraoccipital
Quadrate
Exoccipital
Dentary   Vomer
Quadratojugal
Palatine
Supraangular   Jugal   Pterygoid

Swan, *Cygnus*

FIGURE 7-19

SKULL AND MANDIBLE OF A BIRD. Sutures of the braincase are obliterated in the adult.

are particularly prone to fuse early in life. Common fusions include the postparietal with the supraoccipital, basioccipital with the exoccipitals to form a single occipital, the otic bones (prootics and opisthotics) with the squamosals to form temporals, and the four sphenoid bones (basi-, pre-, ali-, orbito-) with one another to form a single sphenoid.

Mammalian nasal structure is distinctive. The anterior bony nares have merged to form a common opening. A relatively large nasal chamber is in part filled by delicate scrolls of bone, the conchae, or turbinates, which are outgrowths from the inside walls of the maxillas, nasals, and ethmoids (Figure 10-4). These are covered with epithelium and serve to warm and clean inspired air before it reaches the lungs. Their presence is believed to be associated with endothermy. It is, therefore, of great interest that some mammal-like reptiles possessed incipient turbinates *before* the articular became a malleus!

A trend toward simplification of the skull by the loss of bones runs all through tetrapod evolution. Bones that are characteristic of amphibians and reptiles and are not represented in mammals include the prefrontals, postfrontals, postorbitals, ectopterygoids (probably retained by monotremes), quadratojugals, parasphenoid, and all membrane bones of the lower jaw except the angular, prearticular, and dentary (which is retained as the single jawbone.) The prearticular contributes a process to the malleus; the angular moves over from the angle of the jaw to contribute to a uniquely mammalian structure, the tympanic bulla, which helps to enclose the middle ear. Details of the construction of the bulla are used in the classification of carnivores.

Reptilian bones that occur in mammals under new names may be summarized as follows:

Reptilian bone	Mammalian homolog
articular + prearticular	malleus
quadrate	incus
angular	tympanic
epipterygoid	alisphenoid
sphenethmoid	presphenoid + orbitosphenoid
vomer (paired)	vomer (unpaired)

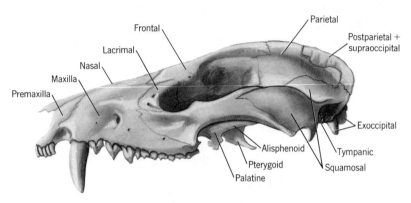

Opossum, *Didelphis*

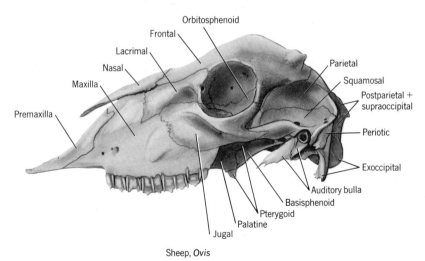

Sheep, *Ovis*

FIGURE 7-20

MAMMALIAN SKULLS OF THE PRIMITIVE ORDER MARSUPIALIA (above) AND THE ADVANCED ORDER ARTIODACTYLA (below).

It is doubtful that mammals have evolved any really new bones, but the entotympanic which contributes to the tympanic bulla seems to be a candidate.

Features of the Skull

This chapter has emphasized evolutionary trends of the isolated skull. But the skull is not isolated. To be fully understood it must also be related to other organ systems. The vertebrate skull is so complex and the range of its variation is so great that its named features are legion. Student and professor alike must limit himself to the major parts of skulls representing the major taxa, leaving the details to specialists. I have selected for illustration (Figure 7-21) some features of the mammalian skull and mandible that are relatively prominent and constant and that correlate with structures to be presented elsewhere in this book. Figures 7-22 and 7-23 show processes and crests in relation to muscles, and foramina in relation to cranial nerves.

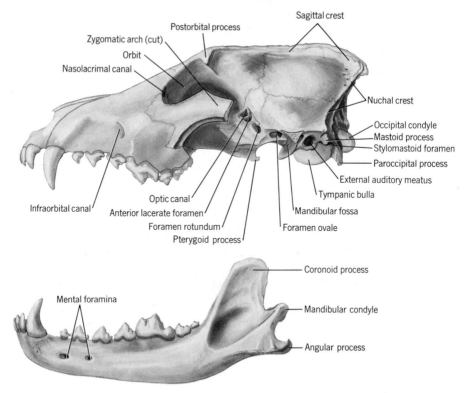

FIGURE 7-21
SOME FEATURES OF THE SKULL AND MANDIBLE OF A WOLF.

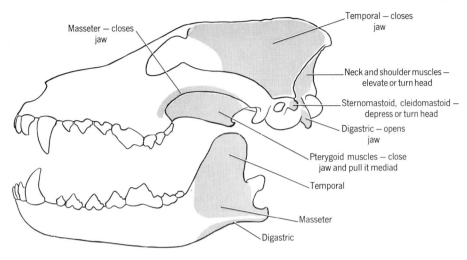

FIGURE 7-22
RELATIONS OF THE SKULL AND MANDIBLE OF A WOLF TO VARIOUS MUSCLES.

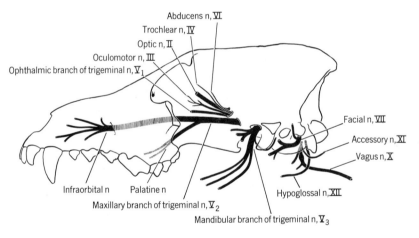

FIGURE 7-23
RELATIONS OF THE SKULL OF A WOLF TO VARIOUS NERVES.

Finally, although the elements present in a skull and the basic relationships of these parts are determined by a long evolutionary history, the relative sizes and configurations of these parts are related to specific adaptations of (chiefly) the feeding mechanism, brain, and sense organs. The importance of these factors to the architecture of the skull will be noted in other chapters.

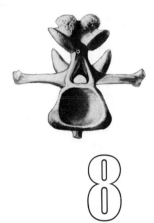

8

Body Skeleton

The vertebral column is older than any other part of the postcranial skeleton except the notochord. Nevertheless, it is not so ancient as the major features of the soft organ systems, and is virtually absent in the oldest known vertebrates. Its structure and functions diversified slowly. We shall start with an account of the features of a typical vertebra to gain the terminology necessary to discuss the evolution of vertebrae.

Structure and Development of Vertebrae

General Structure. The principal part of a vertebra is the spool-like body or **centrum** which lies just below the spinal cord, where it surrounds, restricts, or replaces the notochord. The centrum may consist of one or two elements (rarely more—in

157

fishes). If a tetrapod centrum has two elements (many extinct amphibians, few amniotes), then the more anterior is called an **intercentrum** and the more posterior is called a **pleurocentrum.** If a tetrapod has only one central element, it may be the intercentrum (some extinct amphibians) or the pleurocentrum (amniotes). As explained below, it is not yet clear if these terms are appropriate for fishes.

Extending dorsally from the centrum are pillars which flank the spinal cord, one on each side, and join above the cord to complete the **neural arch.** A **neural spine** may extend from the summit of the neural arch. Of frequent though less universal occurrence is a similar **hemal arch** which extends ventrally from the centrum to surround blood vessels. It may be continued by a **hemal spine.** Neural and hemal spines are related to muscles that move the axial skeleton. Of like function are a variety of processes that project, according to species, from walls of the vertebra. Centra may accommodate the **capitulum,** or head of a rib, either with a process (parapophysis) or with a concavity. The part of a rib called the **tuberculum** (Figure 8-1) is supported by a **diapophysis.** Any lateral process is a **transverse process.** The most prominent of these is usually in the position of the diapophysis but on vertebrae lacking ribs. Adjacent vertebrae always articulate by their centra (if the centra are complete), and in tetrapods they also articulate by processes carried by the neural arches and called **zygapophyses** (= joining + processes). The **prezygapophyses** on the anterior aspect of one vertebra articulate with the **postzygapophyses** on the posterior aspect of its neighbor. Prezygapophyses face upward or inward, whereas postzygapophyses face downward or outward. This is useful to remember for determining which end of a single vertebra is which. (A ventrolateral process may be called a basapophysis, and a median ventral process a hypapophysis. Some mammals have a metapophysis on the neural arch close to the prezygapophysis, or an anapophysis near the postzygapophysis.)

The shapes of the articulatory surfaces at the ends of the centra are of evolutionary, functional, and systematic importance. If each surface is concave, the centrum is said to be **amphicoelous** (= on both sides + hollow). Such centra actually touch only at the periphery of the intervertebral joint. Limited motion is permitted in any direction. Within the joint is a space filled by connective tissue, cartilage, or remnants of the notochord. Adjacent spaces may be joined by a perforation through the interior of the centrum. Other kinds of centra are concave anteriorly and convex posteriorly, the bulge of one vertebra fitting into the hollow of the next. Such a centrum is **procoelous** (= in front + hollow). Conversely, an **opisthocoelous** (= behind + hollow) centrum is concave posteriorly and convex

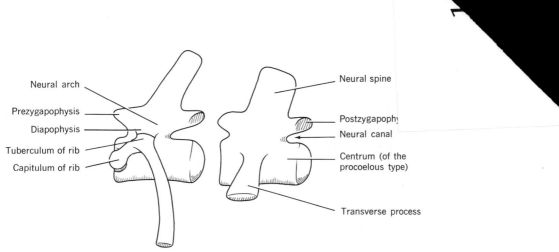

FIGURE 8-1
SOME FEATURES OF VERTEBRAE.

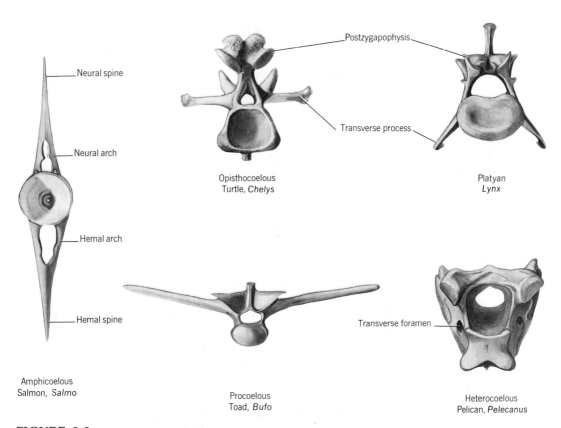

FIGURE 8-2
VERTEBRAE SHOWING VARIOUS SHAPES OF CENTRUM and other features as seen in posterior view.

anteriorly. Joints between procoelous or opisthocoelous vertebrae permit motion in any direction (except as modified by the zygapophyses) and they resist dislocation. Some centra have flat ends and are said to be **platyan** (or acoelous). These centra withstand compression and limit motion (unless the intervertebral joints are provided with thick fibrous disks). Still other centra have saddle-shaped ends. They are called **heterocoelous** (= different + hollow). They allow vertical and lateral flection, but prevent rotation around the axis of the spine. There are intergrades among these types, combinations, and occasionally doubling of the concave and convex surfaces. The functional aspects of vertebral structure are interpreted further in Part III.

Evolution of Vertebrae: A New Look at Old Theories. Vertebrae are complex structures, and the variety of their construction is seemingly endless. Long ago it became evident that a unifying theory was needed that could explain their diversity in terms of development and evolution. Such a theory was presented in some detail in 1896 by the anatomist Hans Gadow. He postulated that the primitive vertebra was double: In adults as well as embryos it had cranial and caudal halves in each segment. Further, each half had dorsal and ventral parts. Thus, each vertebra had four pairs of elements. Collectively, they were called **arcualia** and each was named. Nearly all known vertebrae—fossil and modern—were considered to be derivable from this primitive pattern by deletions and combinations. Partial exceptions were modern amphibians: Anura and Urodela were described as atypical, unlike one another, and very different from Labyrinthodontia. Evidence for the theory came initially from the embryology and comparative anatomy of (chiefly) elasmobranchs and primitive bony fishes and from the paleontology of amphibians. Subsequently, other workers contributed additional embryological evidence.

Gradually Gadow's theory of the arcualia became more and more strained to accommodate new data. Following careful studies by Mookerjee in the 1930s, and a critical review by Williams in 1959, the theory was found to be unacceptable. Elasmobranchs are no longer considered ancestral fishes, as they were in Gadow's day. Tetrapod embryos never have arcualia. The ontogeny of fish vertebrae and the relevance of paleontological material has been reevaluated. The phylogeny of tetrapod vertebrae is now one of the best-documented stories of vertebrate evolution. Modern amphibians remain atypical yet are not considered as divergent as was once thought. The evolution of fish vertebrae is still somewhat in doubt. The long-sought-for unifying theory is taking shape but remains incomplete.

Development of Vertebrae. The differentiation of the embryonic somite into dermatome, myotome, and sclerotome was described in Chapter 5 (see Figure 5-4). Vertebrae develop from the sclerotomes (= hard + sections). Sclerotomes are segmentally arranged by pairs all along the axis of the body. A small and transitory cavity (the sclerocoel) is usually present inside each sclerotome and divides it into cranial and caudal halves. The halves are distinct from one another (even in the absence of a sclerocoel) because the cells of the caudal half are more densely packed. (The significance of this difference is not clear because the adult tissues derived from the two halves are histologically indistinguishable.) Intersegmental blood vessels develop between sclerotomes, and spinal nerves form at the level of the cranial half sclerotomes.

The next event (except in elasmobranchs) is the condensation of mesenchyme adjacent to the notochord to form a continuous cylinder (perichordal tube) enclosing the chord. This tube has segmental thickenings at the center of each sclerotome and is flanked intersegmentally by less dense tissue (central zones).

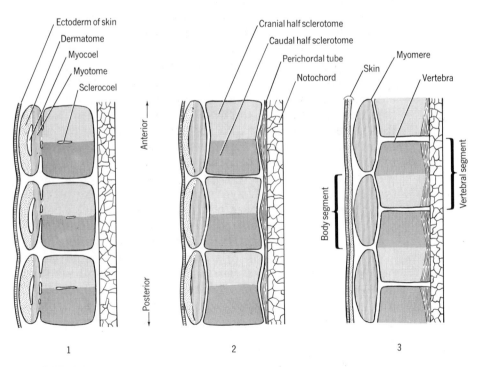

FIGURE 8-3

ONTOGENY OF TETRAPOD VERTEBRAE. Frontal sections of three body segments on one side of the body at the level of notochord and developing centra.

Body Skeleton

Next (with emphasis on tetrapods), the neural arches form chiefly (but not exclusively) from the caudal half sclerotomes, and the hemal arches form from mesenchyme derived from each half sclerotome. Only at this stage does chondrification occur. The cartilage is continuous for a time but then breaks are established which separate individual vertebrae. It is important to note that the breaks come *within* (not between) the original sclerotomes. It follows that vertebrae are intersegmental, the anterior part of each being derived from the posterior part of a sclerotome. Since the musculature of primitive vertebrates remained segmental, this developmental mechanism provided that muscles would hold adjacent vertebrae together and move one on another.

Finally, the cartilage of the vertebrae is replaced by bone. (Some membrane bone may be added to the replacement bone of the neural arches.)

Comparative Anatomy of the Spine

Beginnings: Notochord with Supplemental Cartilages. The notochord is persistent in adults of jawless vertebrates, placoderms, pleuracanths, cladodonts, chimaeras, acanthodians, dipnoans, crossopterygians, and most members of the lower orders of ray-finned fishes. In contrast to the embryonic notochord (more often seen in the laboratory), the adult chord is large and resilient (3 × 3.8 cm in diameter in *Latimeria*). It has an outer elastic sheath which merges with a tough, inner, fibrous sheath. The stiffness of the notochord enables muscles of the body wall to flex the body (as required for the vertebrate manner of swimming) rather than shorten it. It has been shown that the notochord of amphioxus is itself a contractile organ. Its contraction increases its stiffness, particularly during fast swimming. It is probable that the notochords of ancestral vertebrates had the same capacity.

All these vertebrates (except hagfishes) have neural arches (see Figure 7-7); many have hemal arches, at least in the tail; and some have arch bases or other elements which flank, but do not restrict, the notochord. Limited ossification of these elements is present in some arthrodires and bony fishes.

This kind of spine is primitive for vertebrates in general. (However, some of the fishes mentioned probably have secondarily reduced the complexity of their vertebrae.) It is possible that the various cartilages evolved where tendons inserted on the notochordal sheaths and then enlarged to extend the areas of insertion.

Advanced Fishes: The Spine Takes Over. Selachians among cartilaginous fishes, teleosts among ray-finned fishes, and also

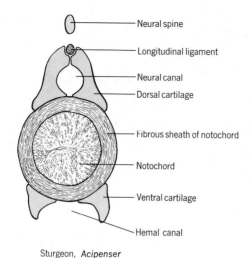

Neural spine

Longitudinal ligament

Neural canal

Dorsal cartilage

Fibrous sheath of notochord

Notochord

Ventral cartilage

Hemal canal

Sturgeon, *Acipenser*

FIGURE 8-4
CROSS-SECTION OF THE NOTOCHORD AND SPINAL CARTILAGES OF A CHONDROSTEAN FISH.

some other bony fishes evolved vertebrae with firm centra that articulate with one another. The deeply amphicoelous centra restrict or interrupt the notochord, thus destroying its integrity and function. Such a spine is stronger than the notochord and provides greater anchorage for muscles.

Variation of structure among the fishes is considerable. SE-LACHIANS are unique in having not only the centra but also the arches in continuous contact. Such an arrangement would virtually immobilize a bony spine, but the cartilaginous nature of the selachian spine provides the slight flexibility that is needed. Each vertebral centrum may have several separate pieces (the arcualia of Gadow), and there are two principal dorsal arches, a neural plate and an intercalary plate, or inter-neural. The centrum chondrifies in the notochordal sheaths (a **chordal centrum**). Calcification of the centra is common, but instead of being continuous, the hard tissue is arranged in concentric cylinders around the central axis and in pillars in the positions of the primitive arch bases. Details of the pattern are various and have systematic value.

The vertebrae of BONY FISHES usually have one central element, a neural arch with spine, and, in the tail, a hemal arch with spine (Figure 8-2). The centrum usually ossifies directly from mesenchyme surrounding the notochord (a **perichordal centrum**), but may also form from cells that invade the chordal sheaths or from cartilaginous precursors.

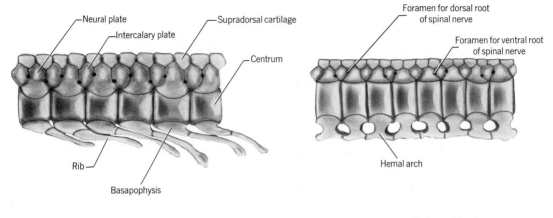

Trunk region

Caudal region (diplospondylous)

Leopard shark, *Triakis*

FIGURE 8-5

SECTIONS FROM THE TRUNK (left) AND CAUDAL REGIONS (right) OF THE SPINE OF AN ELASMOBRANCH in left lateral view. As the nerve foramina show, the caudal section is diplospondylous.

Angel shark
Squatina

Leopard shark
Triakis

FIGURE 8-6

CROSS-SECTIONS OF THE CENTRA OF TWO ELASMOBRANCHS.

There is little regional differentiation in the vertebral column of either cartilaginous or bony fishes. The first vertebra may be slightly modified for articulation with the skull. Behind the coelomic cavity a caudal region is distinguished by the absence of rib facets and the presence of enlarged hemal arches with spines. Some bony fishes (including the bowfin commonly seen in the laboratory) and many cartilaginous fishes have a curious doubling of vertebrae in the caudal region, giving two (sometimes even more) complete vertebrae per muscle segment and per pair of spinal nerves (Figure 8-5). The origin and significance of this condition, called **diplospondyly,** have been much discussed but remain obscure. Perhaps it increases flexibility. Rarely the centrum is single and the neural arches are doubled. Arches may be fused to the centra or free.

Amphibians: Varied Solutions to New Problems. Few organs were so affected by the change from aquatic to terrestrial life as was the spine. Formerly it resisted only the stresses imposed by strong axial muscles; now the axial musculature gradually became reduced but there were new stresses, imposed by gravity, which acted largely in a different plane. Formerly the paired appendages had not been related to the spine; now the slowly strengthening limbs transmitted their support to the axis of the body. Formerly it was sufficient for the spine to be nearly uniformly flexible throughout its length; now it had to resist bending in some places and provide new mobility elsewhere. The requirements were for vertebrae with firm centra, intervertebral joints that could facilitate or restrict motion according to location, processes that could increase the leverages of muscles, and a more intimate relation with the girdles. All this took many millions of years to accomplish. Indeed, it remained for the amniotes to complete the changes. Nevertheless, the credit for most of the transformations, and nearly all of the evolutionary experimentation that tested the innovations, must be given to the extinct amphibians. Various taxa of labyrinthodonts are named and largely defined by the nature of their vertebrae.

The story of the evolution of amniote vertebrae starts with the particular crossopterygians (Rhipidistia) that were ancestral to tetrapods. Their vertebrae were nearly identical with those of the earliest amphibians (Ichthyostegalia). The notochord was persistent. There was a neural arch, a large intercentrum encircling the notochord, and small paired pleurocentra. The spine was still fish-like in function, but it contained the three ossifications that were to feature in all subsequent vertebrate evolution (Figure 8-7).

Two principal lineages diverged from this common origin. One led first to a group of labyrinthodonts (Rhachitomi; = vertebra + cut) that replaced the notochord by a large intercentrum and a pair of small, unfused pleurocentra, and thence to a group (Stereospondyli; = solid + vertebra) that retained only the intercentrum. This lineage had no descendants. The other lineage either increased the size of the pleurocentrum to equal status with the intercentrum (the condition in the Embo-

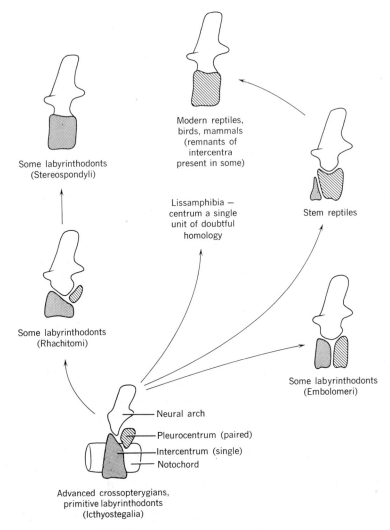

Some labyrinthodonts
(Stereospondyli)

Modern reptiles,
birds, mammals
(remnants of
intercentra
present in some)

Lissamphibia —
centrum a single
unit of doubtful
homology

Stem reptiles

Some labyrinthodonts
(Rhachitomi)

Neural arch

Pleurocentrum (paired)

Intercentrum (single)

Notochord

Some labyrinthodonts
(Embolomeri)

Advanced crossopterygians,
primitive labyrinthodonts
(Icthyostegalia)

FIGURE 8-7
**PHYLOGENY OF TETRAPOD VERTEBRAE as seen in
left lateral view.**

lomeri; = wedge + parts), or enlarged the pleurocentrum to the
near or complete exclusion of the intercentrum (amniotes).

The vertebrae of modern amphibians remain an enigma. They
have the same intersegmental position and mesenchymal primordia as vertebrae of other tetrapods. However, evidence for
segmentation and then resegmentation of the sclerotome is
not good, and the pattern of development gives no useful information about the homologies of vertebral elements. The centra
of frogs, salamanders, and caecilians (which always consist of a
single unit) are probably broadly homologous to each other and
to those of other tetrapods.

Amphibians, like other tetrapods, have zygapophyses to strengthen the spine and control its flexibility. Centra are amphicoelous, procoelous, or opisthocoelous. Regional differentiation of the spine is minimal for tetrapods but exceeds that of fishes. The first (and sometimes the second) vertebra is modified to give increased mobility to the head. A **cervical vertebra** is somewhat set off by the reduction or absence of ribs. **Trunk vertebrae** bear ribs (Anura excepted). A single **sacral vertebra** is enlarged to articulate (via its fused rib) with the pelvic girdle. Typical tail, or **caudal vertebrae** lack zygapophyses, and have hemal arches. Anura, however, have no free caudal vertebrae but have instead a rod-like **urostyle** which is probably derived by the fusion of several caudal vertebrae.

Amniotes: Strength and Specialization. The centrum of amniotes is a pleurocentrum. Stem reptiles inherited vertebrae with large pleurocentra and small intercentra. Subsequent amniotes retain only one functional intercentrum (in the first vertebra of the neck). Some reptiles and several mammals have vestigial intercentra elsewhere along the spine. Amniotes have strong zygapophyses (rarely they are doubled for added strength) except in the caudal region. The first two cervical vertebrae are specialized to support the skull (of which more under the next subheading).

Because most REPTILES have more distinct necks than amphibians, they have more distinct cervical regions; and because their limbs are stronger, they have two (sometimes more) sacral vertebrae instead of one. Centra are usually procoelous but also take other shapes. Hemal arches are retained only in the caudal region where they are separate bones shaped like Y's. The ancestral intercentra with which they once articulated having disappeared (or merged with them?), they articulate with the spine at the intervertebral joints. Such arches are called **chevron bones** because of their shape.

Because BIRDS have a locomotor specialization in common—flight—they have more specialized and more uniform spines than other tetrapods. The spine of Archaeornithes is transitional between those of their reptilian ancestor and Neornithes. No other class has so many cervical vertebrae—commonly, 15–20 (10 in Archaeornithes)—and these vertebrae are distinctive for being heterocoelous. Fused to the 2 sacral vertebrae of the reptilian ancestor are 10–20 trunk and caudal vertebrae (5–6 in Archaeornithes) and the pelvises to form a solid unit called a **synsacrum.** The remaining trunk vertebrae number only 4–6, and there may even be fusions among some of these. The result is a short, rigid back. The free caudal vertebrae of Neornithes have been reduced to (usually) 6 or 7; Archaeornithes had a long bony axis to the tail. At the end of

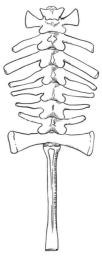

Toad, *Bufo*

FIGURE 8-8
VERTEBRAL COLUMN OF AN ANURAN showing in dorsal view the single cervical vertebra, short trunk, absence of free ribs, single sacral vertebra, and fusion of caudals to form a urostyle.

the tail of modern birds is a unique blade of bone called the **pygostyle** which supports the tail feathers. It represents the fusion of 4–7 vertebrae. Even more embryonic somites form and disappear again in recapitulation of the long ancestral tail. Some of these features are seen in Figure 24-15.

MAMMALS are unique in forming bony, plate-like caps, or **epiphyses,** at the ends of their centra posterior to the first intervertebral joint. These usually fuse to the centra when the animal is mature. They are derived from the segmental thickenings of the mesenchyme (perichordal tube) that surrounds the embryonic notochord. Their function relates to the growth process. With rare exceptions (some edentates and sirenians), mammals have 7 cervical vertebrae whether the neck is short (dolphin) or long (giraffe). There are about 20 trunk vertebrae. Unlike the condition in other classes, these are sharply divided, by presence or absence of ribs, into anterior **thoracic vertebrae** and posterior **lumbar vertebrae,** (see Figure 22-9). To assign isolated vertebrae to region, look for rib facets. Lumbar vertebrae tend to have larger centra, shorter and stouter neural spines, and longer transverse processes than thoracic vertebrae. (We shall see on pp. 556 and 557 that from a functional viewpoint there are other ways to divide the trunk vertebrae.) Mammals have 3 or more sacral vertebrae fused to form a **sacrum.** Chevron bones are usually restricted to the base of the tail.

Cranio–Vertebral Joint. The occipital process of the skull of fishes usually looks about like the end of a typical centrum. The head seldom has more mobility on the spine than one vertebra has on another. The cranio–vertebral joint is like an intervertebral joint.

Amphibians are the first vertebrates to have a neck, however short, and linked with this advance is a more flexible cranio–vertebral joint. The second vertebra is unchanged or only somewhat enlarged.

Reptiles and birds have a ball-and-socket joint between the small first vertebra, now called the **atlas,** and the single occipital condyle. The second vertebra, called the **axis,** is enlarged, and the joint between the atlas and axis is specialized. (The remaining cervical vertebrae of various reptiles—particularly turtles—are also quite complex.)

The atlas of mammals has large zygapophyses, wing-like transverse processes, scarcely any centrum, and no neural spine. Its articulation with the two condyles of the skull permits hinge-like up and down motion. The axis has a blade-like neural spine and a large centrum. The joint between atlas and axis permits side-to-side motion and rotation around the axis of the spine.

In adults of some extinct amphibians, in reptiles, and in the embryos of many tetrapods, small structures with no apparent function are found between the skull and the neural arch of the atlas. These form the **proatlas.**

The story of how these changes came about is of particular interest. It has been pieced together from many embryological, paleontological, and comparative anatomical studies. In Chapter 7 it was noted that several ancestral vertebrae were incorporated into the vertebrate chondrocranium. It follows that the cranio–vertebral joint falls within the primordial spine, not beyond its anterior limit. At some time (or times?) in tetrapod evolution one vertebra at the cranio—vertebral boundary became reduced in size and partly fused to adjacent units of the series. This was the proatlas. Its neural arch remains free in some labyrinthodonts and some reptiles. Otherwise, the parts of the proatlas may fuse to other vertebrae or to the skull, though there is no evidence of such fusion in modern amphibians.

The atlas of amniotes always retains its "own" neural arch and intercentrum. The axis not only retains its own neural arch and pleurocentrum, but also incorporates into its odontoid process the pleurocentrum of the atlas and even that of the proatlas if that element has not joined the skull (mammals, some reptiles). The intercentrum of the axis is lost (mammals) or also incorporated into the composite centrum of the axis.

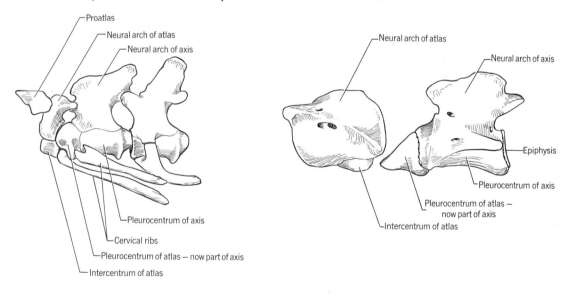

Crocodile, *Crocodylus*   Deer fawn, *Odocoileus*

FIGURE 8-9

ORIGIN OF THE ELEMENTS OF THE ANTERIOR CERVICAL VERTEBRAE of a reptile and a mammal shown in left lateral view.

Ribs Ribs are intersegmental splints of cartilage or replacement bone which articulate with the vertebrae. Originally they extended the entire length of the spine to increase the direct contact of axial muscles with the skeleton. As they lengthened they came to protect underlying viscera while retaining requisite flexibility and minimum weight. Only in amniotes are they specialized to contribute to the breathing mechanism.

There appear to be two sets of ribs among the different vertebrates. Each set evolved in the septa of connective tissue that occur between successive muscle segments of the primitive body wall. The set called **dorsal ribs** evolved where these septa intersect with the partition between dorsal (epaxial) and ventral (hypaxial) muscle masses. **Ventral ribs** form between the hypaxial muscles and the lining of the coelom. These occur only in the trunk region.

The first ribs known are already complete and independent structures; paleontology gives no clue of their origin. It was formerly thought that ribs originated as outgrowths from the sclerotomes. The sclerotomes probably do contribute to the heads of ribs, but it is now thought, on the basis of their ontogeny, that they are primarily new structures not derived from the spine. The oldest known tetrapod ribs had two heads, and this condition is retained by most ribs that are strong. Weaker ribs (often more posterior on the trunk) may have a single head as the result of either fusion or loss. The more ventral head, named the **capitulum,** articulates with the intercentrum, if that piece is present, and otherwise with the pleurocentrum near the intervertebral joint. The dorsal head, or **tuberculum**, articulates with the diapophysis of the neural arch (Figure 8-1).

Jawless vertebrates and placoderms have no ribs. The functions of ribs were performed in part by the armor of some of them, and none has centra with which ribs could articulate. With few exceptions (e.g., Holocephalia) other fishes have ribs. There is disagreement as to whether the short cartilaginous ribs of pleuracanths and selachians are of dorsal or ventral origin (Figure 8-5). The long bony ribs that enclose the body cavity of ray-finned fishes are ventral ribs. Various primitive but unrelated fishes (e.g., salmon, *Polypterus)* have both ventral and smaller dorsal ribs. The ribs of tetrapods are in the position of ventral ribs; however, consensus now is that they are the ancestral dorsal set.

The short cervical ribs of tetrapods may articulate firmly with the vertebrae (labyrinthodonts, many reptiles), fuse with the vertebrae (birds, mammals), or be lost (turtles). Free caudal ribs are retained by most labyrinthodonts and some reptiles but otherwise are either lost or indistinguishably fused to the transverse processes. The same is true of the lumbar ribs of mammals. Some Anura have free ribs, but many retain only

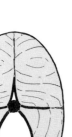

FIGURE 8-10
CROSS-SECTION OF A BONY FISH SHOWING POSITION OF DORSAL AND VENTRAL RIBS in relation to muscle masses and body cavity.

fused sacral ribs. Urodela have short ribs with unique articulations. The ribs of amniotes are in two pieces — a principal ossified segment and a shorter **sternal rib** which ossifies in birds but is otherwise usually cartilaginous. The more anterior thoracic ribs of most amniotes articulate with the sternum (breastbone) via their sternal segments (Figure 22-9). Ribs of birds and of some reptiles have **uncinate processes** to provide anchorage for shoulder muscles (Figure 24-15).

Sternum

The **sternum** is a midventral skeletal element that usually articulates with the more anterior thoracic ribs. Its functions are to strengthen the body wall, help protect the thoracic viscera, accommodate muscles of the pectoral limb, and, in some amniotes, aid in ventilating the lungs. Most of these functions are not relevant to fishes, and only tetrapods (and not all tetrapods) have a sternum. There is wide structural variation among species and relatively great individual variation among animals of a kind. The sternum is rarely present in fossils of amphibians and reptiles because it was cartilaginous. It contributes virtually nothing to our knowledge of vertebrate evolution.

The sternum forms from either paired or midventral primordia (or both) which are now generally regarded (following considerable debate) as new structures not derived from the pectoral girdle or ribs. (The rib ends of mammalian embryos inhibit ossification of the sternum.)

It is probable that the most primitive tetrapods had no sternum. It is a simple cartilaginous plate in most living Urodela and is missing in others. In Anura it ranges from poorly developed to well-ossified. The sternum is cartilaginous, and is often large in lizards and crocodilians. It is absent in snakes and turtles. Modern birds have a huge ossified sternum with a prominent keel to give origin to flight muscles (see again Figure 24-15). Mammals are unique in having the sternum divided into a linear series of about half a dozen bony segments.

Median Fins

The **dorsal fin**(s), located along the middorsal line; the **anal fin**(s), between anus and tail; and the **caudal fin** comprise the median fins. They occur in nearly all jawless vertebrates and fishes. (Larval amphibians, adults of many Urodela, and some amniotes that are highly specialized for aquatic life have secondarily evolved analogous structures. Some of these will be mentioned in Chapter 23 but are not described here.) The origin of median fins is considered to be related to the origin of paired fins and will be discussed in the next section.

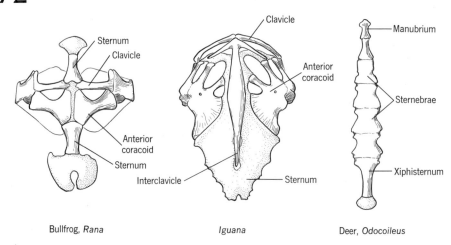

Clavicle

Sternum

Clavicle

Anterior
coracoid

Manubrium

Sternebrae

Anterior
coracoid

Sternum

Interclavicle

Sternum

Xiphisternum

Bullfrog, *Rana*

Iguana

Deer, *Odocoileus*

FIGURE 8-11
VARIED STERNUMS AND RELATED BONES OF THE PECTORAL GIRDLE
seen in ventral view.

Dorsal and Anal Fins. Dorsal and anal fins function to prevent the body from yawing (turning around the vertical axis) and rolling (turning around the longitudinal axis).

The primitive condition probably was for each fin to be supported within the contour of the body by a series of rod-like **radials,** or **pterygiophores** (= fin + bearer of), arranged one per body segment. Commonly, the number is reduced (or increased) and segmental arrangement is lost. Each pterygiophore is usually divided into two or more pieces. The proximal piece is often conspicuously larger than those more distal and is then called a **basal.** Pterygiophores sometimes articulate with neural and hemal spines, but they are not derived from vertebrae.

The exposed membrane of the fin may originally have been supported only by dermal scales in the covering skin—those on the leading edge being larger than the rest. The fins of cephalaspids, placoderms, and acanthodians apparently were of this nature. The fins of more advanced fishes are supported internally by a series of slender **fin rays.** Fin rays of cartilaginous fishes are slender, unsegmented, and horny and are called **ceratotrichs** (= horn + hair). Those of bony fishes are slightly broader, segmented, and bony and are called **lepidotrichs** (= scale + hair). Lepidotrichs are considered to be derived from dermal scales that have sunk below the surface. Higher teleosts have in the dorsal fin only six or fewer lepidotrichs which have enlarged and become rigid. The leading edge of one or more median fins of many fishes (pteraspids, acanthodians, most cartilaginous fishes) is stiffened by a stout spine which serves as a cutwater and sometimes also (or instead) for de-

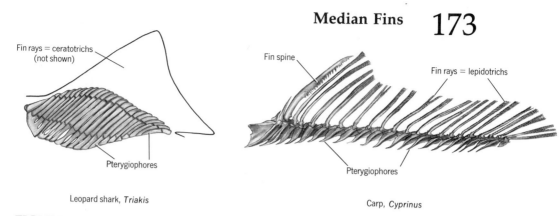

Fin rays = ceratotrichs
(not shown)

Pterygiophores

Leopard shark, *Triakis*

Fin spine

Fin rays = lepidotrichs

Pterygiophores

Carp, *Cyprinus*

FIGURE 8-12

DORSAL FIN SKELETONS OF A CARTILAGINOUS FISH (left) AND A BONY FISH (right).

fense or display. This spine may be an enlarged lepidotrich, but in cartilaginous fishes it is derived from one or more dermal denticles. Dorsal and/or anal fins may be long and continuous (some cyclostomes, pleuracanths, some teleosts), single (most ray-finned fishes), double (many selachians and sarcopterygians), multiple (*Polypterus*), or absent.

Caudal Fin. If the spine is straight to the tip of the tail, then dorsal and ventral lobes of the tail are about equal and the fin is said to be **diphycercal** (= double + tail). If the spine tilts upward and enters the dorsal lobe, then the dorsal lobe is longer than the ventral lobe, most of the fin membrane is ventral to the axis of the tail, and the tail is termed **heterocercal.** If the spine enters a larger ventral lobe, the tail is **hypocercal.** If all the fin membrane is posterior to the spine, then dorsal and ventral lobes are about equal and the tail is said to be **homocercal.** Several intergrades and modifications are also recognized.

The ancestral tail may have been diphycercal, but the most primitive tails of record are heterocercal (cephalaspids, placoderms, most Chondrichthyes, the more primitive Osteichthyes of each subclass). Few vertebrates have hypocercal tails (anaspids). From the heterocercal tail there evolved the (secondarily?) diphycercal tail (cyclostomes, pleuracanths, later sarcopterygians, *Polypterus*) and homocercal tail (nearly all teleosts). These tail shapes can be identified in the illustrations for Chapter 3.

The caudal fin of ray-finned fishes, unlike the other median fins, is supported within its fleshy base by several modified neural arches and spines called **epiurals** (= upon + tail), and more numerous modified hemal arches and spines called

hypurals (= below + tail). These and related structures are further defined by Figure 8-13. (Some bony fishes have small radials in the caudal fin, but these have been lost by nearly all teleosts.) The membrane of the fin is stiffened by fin rays corresponding in structure to those of the dorsal and anal fins of the same fish. Lepidotrichs of the caudal fin are usually branched.

Paired Fins and Their Girdles

The Origin of Appendages. Speculation over the origin and evolution of appendages dates back nearly to the publication of the *Origin of Species*. One fanciful theory, later disproved by evidence from both embryology and paleontology, derived paired appendages from gill arches. In the early 1880s several anatomists (Balfour, Mivart, Thatcher) postulated that the primitive vertebrate had continuous paired lateral fins from the gills to the anus, and that from there a single continuous fin ran in the midline around the tail and forward along the back to the head. It was considered that the fins of familiar fishes represent such portions of the ancestral continuous fins as remained after intervening deletions were established. The long lateral body folds of amphioxus, the broadly based paired fins

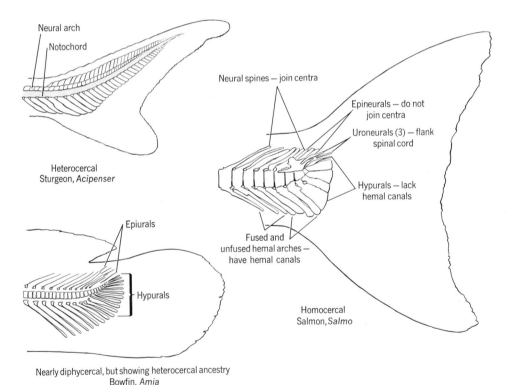

Neural arch

Notochord

Heterocercal
Sturgeon, *Acipenser*

Neural spines — join centra

Epineurals — do not
join centra

Uroneurals (3) — flank
spinal cord

Hypurals — lack
hemal canals

Epiurals

Hypurals

Fused and
unfused hemal arches —
have hemal canals

Homocercal
Salmon, *Salmo*

Nearly diphycercal, but showing heterocercal ancestry
Bowfin, *Amia*

FIGURE 8-13
SHAPE AND STRUCTURE OF THE TAILS OF SOME BONY FISHES.

of cladodonts, and the embryology of musculature in the fins of sharks were cited in evidence. Some of this support must now be abandoned: The lateral folds of amphioxus are not considered homologous with true fins; cladodonts are no longer regarded as close to the ancestral vertebrate stock. However, some anaspids may have had fin folds. Modern knowledge of fossil fishes, and recent experiments on the induction of accessory limbs in amphibian larvae, support the conclusion that appendages do always tend to form where the continuous fin folds were postulated to be. In truth, we still do not know just how appendages evolved. We hope for the discovery of fossils that will provide answers, but the complete story may be forever lost in antiquity.

Comparative Anatomy of Paired Fins. The function of paired fins is usually to prevent the body from pitching (turning around the transverse axis) and rolling, and (particularly in higher fishes) to brake forward motion. The function and position of fins is discussed further in Chapter 23.

Among AGNATHOUS VERTEBRATES, cyclostomes have no trace of paired appendages, and the same was true of pteraspids (Figures 3-2 and 3-3). Anaspids had lateral folds or pectoral spikes of questionable relationship with fins. The spikes were of dermal origin, superficial, and not motile. Cephalaspids had pectoral lobes behind the lateral wings of the cephalic armor. They were muscular, but seem to have had no internal skeleton.

PLACODERMS and ACANTHODIANS experimented with newly acquired paired appendages (Figures 3-5 and 3-8). Arthrodires had stiff fins with fin rays and advanced musculature. Antiarchs had large pectoral appendages which joined the body by a unique ball-and-socket joint. A complex dermal skeleton was always present, and a cartilaginous internal skeleton is sometimes recorded. Acanthodians had numerous pairs of fins along the sides of the body. Each had a stout fin spine and weak fin rays.

To this point in the evolution of fins, the dermal skeleton was emphasized over the internal skeleton. Subsequent evolution reversed this relationship. It is postulated that the ancestral internal skeleton consisted of a series of parallel radials. These were probably segmented, with relatively heavy basals.

Paired fins of CHONDRICHTHYES (except pleuracanths) have cartilaginous radials which are parallel or radiating, and fewer basals derived by fusions among the proximal ancestral radials. (If there are three of these, they are termed pro-, meso-, and metapterygia.) Details of the pattern vary widely. The distal part of the fin is supported by ceratotrichs.

The ACTINOPTERYGII are named according to fin structure;

actinopterygium = ray + fin. There is a proximal row of bony radials and a distal series of lepidotrichs which, like those of the caudal fin, are branched and segmented.

SARCOPTERYGII are also named according to fin structure; **sarcopterygium** (= flesh + fin) refers to the fleshy nature of the bases of these fins. The ancestral fin skeleton is modified — probably by reorienting the row of basals — to establish an axis. There are two principal kinds of sarcopterygium. In dipnoans and pleuracanths (the latter are not classified as Sarcopterygia but have similar fins), the radials are biserial; there is a series of radials on each side of a median axis. This fin was once thought to be the most primitive of all and accordingly was named an **archipterygium.** The term remains but is no longer apt. In Crossopterygii the radials are uniserial; there is a series on one side of the axis. This fin, called a **crossopterygium** (= fringe + fin), was unquestionably ancestral to the tetrapod limb.

Comparative Anatomy of the Girdles of Fishes. Placoderms had small pelvic fins at best, and a pelvic girdle has not been described. The pelvic fins of other fishes are weakly supported by a single skeletal element on each side of the body. They are bony except in cartilaginous fishes and dipnoans. The two

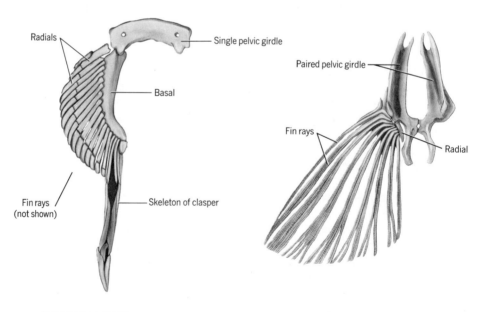

♂ Leopard shark, *Triakis*

Carp, *Cyprinus*

FIGURE 8-14

PELVIC GIRDLE AND LEFT FIN SKELETON OF A CARTILAGINOUS FISH (left) AND A BONY FISH (right) seen in dorsal view.

pieces are usually separate but may overlap or articulate with one another, and in cartilaginous fishes are joined across the midventral line by a bridge of cartilage (Figure 8-14).

The pectoral girdle is older, larger, and more complicated than the pelvic girdle. It includes one or more elements of cartilage or replacement bone and several membrane bones derived from ancestral scales or armor plates.

PLACODERMS illustrate the initial stages in the evolution of the pectoral girdle. A cartilaginous fin base was related to overlying plates of the dermal skeleton. The large pectoral girdle of CARTILAGINOUS FISHES is distinctive in two respects: Dermal elements are absent (presumably because of regression), and right and left halves of the girdle are fused in the midline. The result is a U-shaped girdle of one piece called the **scapulocoracoid.**

In BONY FISHES the replacement element, also termed scapulocoracoid, may ossify in one or several units. The membrane bones are identified in Figure 8-15. They join the girdle to the skull. This anchors the girdle in a manner which is not available to cartilaginous fishes (because they lack the requisite bones),

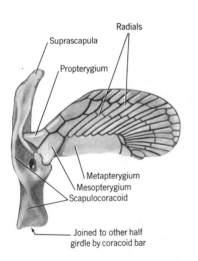

Radials
Suprascapula
Propterygium
Metapterygium
Mesopterygium
Scapulocoracoid
Joined to other half
girdle by coracoid bar

Leopard shark, *Triakis*

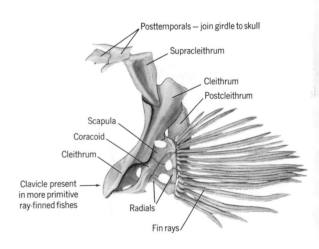

Posttemporals — join girdle to skull
Supracleithrum
Cleithrum
Postcleithrum
Scapula
Coracoid
Cleithrum
Clavicle present
in more primitive
ray-finned fishes
Radials
Fin rays

Rockfish, *Sebastodes*

FIGURE 8-15
**LEFT PECTORAL GIRDLE AND FIN SKELETON OF AN ELASMOBRANCH (left)
AND A TELEOST (right).**

and hence few bony fishes follow cartilaginous fishes in having the halves of the girdle joined in the midventral line. The **cleithrum** is the basic dermal element. The **clavicle** is lost in higher teleosts. The bones between the cleithrum and skull vary in number.

Limbs and Girdles of Tetrapods

Development and Growth of Long Bones. The larger bones of the appendages are called **long bones.** A typical long bone has a cylindrical shaft called the **diaphysis** (= through + growth) which contains the marrow cavity, and at each end an enlargement called the **epiphysis** which articulates with adjacent bones. Epiphyses may be cartilaginous or bony. If bony, they are spongy within. Bony epiphyses usually fuse with their respective diaphyses at maturity.

The first embryonic primordium of a long bone is a condensation of mesenchyme. This then forms a one-piece, cartilaginous model of the future bone. Ossification of the diaphysis begins as a thin veneer of intramembranous bone deposited around the center of the model by the limiting membrane, or **periosteum.** Spicules of bone penetrate the cartilage and replace the matrix, which is gradually destroyed. Soon the incipient marrow

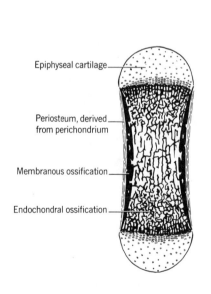
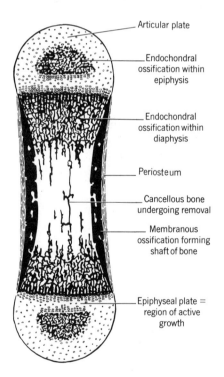

Epiphyseal cartilage

Periosteum, derived from perichondrium

Membranous ossification

Endochondral ossification

Articular plate

Endochondral ossification within epiphysis

Endochondral ossification within diaphysis

Periosteum

Cancellous bone undergoing removal

Membranous ossification forming shaft of bone

Epiphyseal plate = region of active growth

FIGURE 8-16

HISTOGENESIS OF A MAMMALIAN LONG BONE at two developmental stages as seen in longitudinal section.

cavity is surrounded by solid lamellae of bone arranged like the layers of orthodentine. The smallest mammals may retain this structure, but usually longitudinal canals are eroded within the developing bone tissue and the cavities reossify with the cylinder-within-cylinder construction of osteons. Only at the surface of the bone is lamellar structure retained into full maturity. Epiphyses, by contrast, ossify (if at all) from one or more internal centers which expand outward to replace the cartilage. Their contained spicules are largely lamellar.

Bones grow by a complicated and wonderful process which is best known for reptiles and mammals. The young bone does not merely expand like a grain of rice swelling in water. The diaphysis increases in length only at the cartilaginous plates that separate it from its epiphyses. If (as in mammals) these plates are obliterated at maturity, increase in length stops. Growth in the transverse diameter of the shaft is accomplished by the deposition of bone at its outer surface and erosion of its inner surface to enlarge the marrow cavity. Some tendonous insertions must migrate to maintain their proportionate positions as a bone grows. The periosteum stretches like a sheet of rubber as the bone enlarges within; it must slip over the surface of the bone since the latter grows without stretching. Some further points are shown in Figure 8-17.

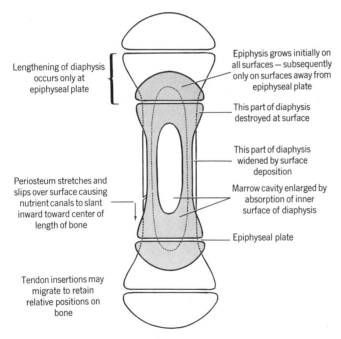

FIGURE 8-17
MORPHOGENESIS OF A MAMMALIAN LONG BONE between two developmental stages as seen in longitudinal section.

Origin and Evolution of Limbs. The transformation from fin to limb is another subject that has been long debated. More than a dozen leading anatomists have discussed the evidence, but final conclusions cannot yet be drawn. An early theory suggests that fishes evolved limbs to crawl between pools of water in times of drought. Another author believes that certain fishes adapted their fins for burrowing in mud to escape drought, and that such fins were only secondarily "discovered" to be useful also as crude limbs. Another disagrees, saying the first tetrapods lived in humid tropical areas where water was abundant. Still another view is that certain fishes first learned to "walk" on the bottom with strong stalked fins—the flexible fins of modern dipnoans are used in this way—and occasionally crawled into shallow water and onto damp shores to escape aquatic enemies or find terrestrial food.

There is no question that limbs evolved from a crossopterygium, and there is general agreement about the homologies between the proximal segments of the fin axis and the proximal limb bones. Opinion differs, however, on which digit represents the continuation of the fin axis (the fourth or fifth digit is favored), the number of segments of the ancestral axis that have been retained (probably four or five), which fin radials became what foot bones, and (for the pectoral appendage) which way the fin twisted to assume the tetrapod stance. The transition shown in Figure 8-18 is an educated guess.

The paired fins of fishes always have about the same functions and vary in basic structure according to class or subclass. The pectoral fins are the stronger and more firmly related to the axial skeleton. The limbs of tetrapods, by contrast, have a variety of functions but retain a unity of basic structure. The pelvic limbs are the stronger and more firmly related to the axial skeleton. The basic skeletal structure of limbs is indicated in Figure 8-18. (The terminology shown is used worldwide by comparative anatomists, but other terms are used by human anatomists and mammalogists to identify the particular variations of the foot bones of their materials. Some of their terms are noted below.) The bones of the wrist comprise the **carpus;** those of the ankle comprise the **tarsus.** Carpal and tarsal bones are known collectively as **podials.** Of the podials, the centralia are least constant. The forefoot is called the **manus** and the hind foot the **pes.** Metacarpal and metatarsal bones are known collectively as **metapodials.** The skeletal patterns of the various tetrapod feet are derived from the primitive pattern by deletions and fusions which can usually be verified by embryonic development. The derivations of some patterns representative of major taxa are shown in Figures 8-19 and 8-20. Some other modifications are analyzed in Part III.

AMPHIBIANS (other than the legless caecilians) usually have short limbs splayed to the sides of the body. The trunk is

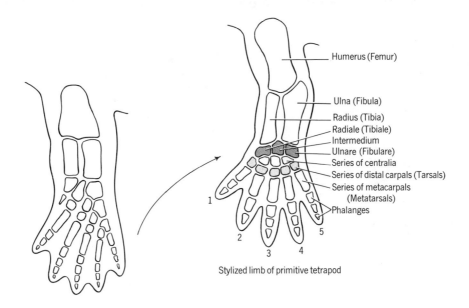

Humerus (Femur)

Ulna (Fibula)
Radius (Tibia)
Radiale (Tibiale)
Intermedium
Ulnare (Fibulare)
Series of centralia
Series of distal carpals (Tarsals)
Series of metacarpals
   (Metatarsals)
Phalanges

1

2

3   4

5

Stylized limb of primitive tetrapod

Transitional stage — hypothetical

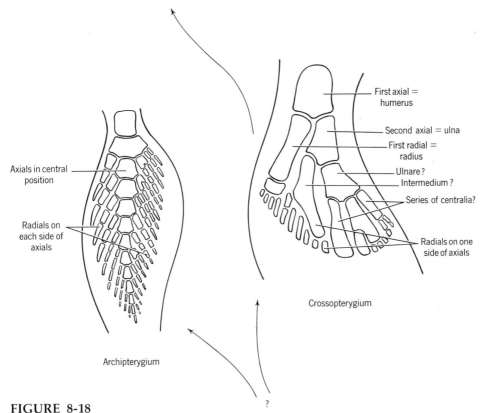

Axials in central
position

Radials on
each side of
axials

Archipterygium

First axial =
humerus

Second axial = ulna
First radial =
radius

Ulnare?
Intermedium?

Series of centralia?

Radials on one
side of axials

Crossopterygium

?

FIGURE 8-18

PHYLOGENY OF THE PECTORAL APPENDAGE OF SARCOPTERYGII AND TET-RAPODS. Left appendage in dorsal view. Corresponding terminology for the hind leg of a primitive tetrapod shown in parentheses.

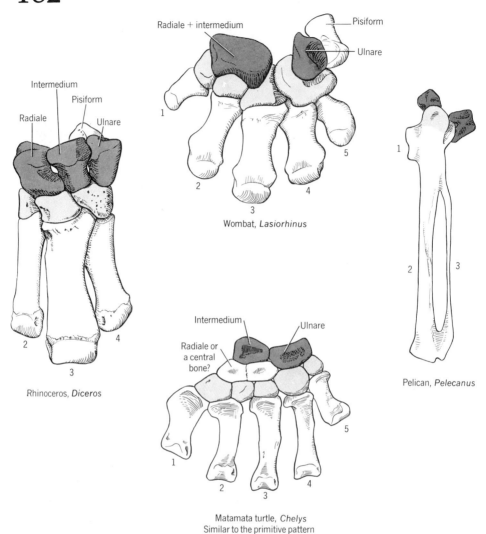

Radiale + intermedium

Pisiform

Ulnare

Wombat, *Lasiorhinus*

Intermedium

Pisiform

Radiale

Ulnare

Rhinoceros, *Diceros*

Pelican, *Pelecanus*

Intermedium

Ulnare

Radiale or
a central
bone?

Matamata turtle, *Chelys*
Similar to the primitive pattern

FIGURE 8-19

LEFT CARPUS AND METACARPUS in dorsal view. Distal carpals shown by light shading, proximal carpals by dark shading, metacarpals, centralia, and pisiform by no shading. There is doubt regarding the homologies for birds.

usually lifted from the ground when the animal walks, but only with difficulty. In urodeles undulations of the spine may be used to twist the girdles, thus helping to advance the limbs. Epiphyses are of hyaline cartilage and fit like corks into the ends of the bony shafts (most amphibians) or are calcified and fit over the ends of the shafts like match heads (Anura). The podials of modern amphibians are often cartilaginous. The principal joint of the foot is between the podials and met-

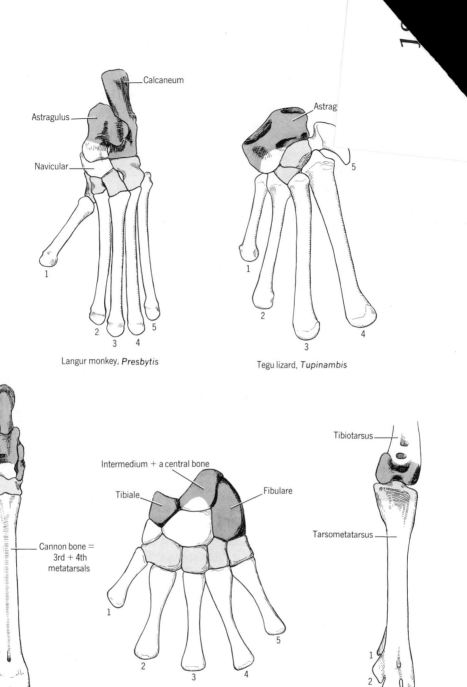

Langur monkey, *Presbytis*

Tegu lizard, *Tupinambis*

Artiodactyl, *Pudu*

Salamander, *Salamandra*
Similar to the primitive pattern

Pelican, *Pelecanus*

FIGURE 8-20
LEFT TARSUS AND METATARSUS in dorsal view. Distal tarsals shown by light shading, proximal tarsals by dark shading, metatarsals and centralia by no shading.

Body Skeleton

FIGURE 8-21
FORELEG SKELETON
OF A MATAMATA
TURTLE, *Chelys.* **Dorsal**
view of left leg.

apodials. There are only four digits on the hand and four or five on the foot. One to three phalanges are present in each toe. The marrow cavities of the long bones of amphibians and higher vertebrates produce blood cells—a function not performed by the skeleton of fishes.

Many REPTILES still have the limbs positioned far to the sides of the body, but some dinosaurs and mammal-like reptiles placed their feet quite as far under the body as mammals do. Disregarding legless reptiles, the limbs are usually stronger than for amphibians, and the hind limbs are often disproportionately large. Epiphyses are usually cartilaginous but may ossify in lizards. An "extra" bone, the **pisiform,** is added to the outside of the carpus, and the tibiale is no longer a free bone in the tarsus. The joint of the foot is between podials. The **phalangeal formula** (number of phalanges per digit starting with digit 1) is 2·3·4·5·3 for the manus and 2·3·4·5·4 for the pes, if segments and digits have not been lost through specialization or degeneration.

The limb structure of BIRDS is uniform and specialized, as shown in the figures. Epiphyses are cartilaginous in immatures and virtually absent in adults. The principal digit of the wing is customarily identified as number 2, although there is some embryological evidence that it is number 3, and the question is not yet resolved. The phalangeal formula of the foot is 2·3·4·5·0.

MAMMALS have bony epiphyses on each end of the long bones, on the distal ends of the metapodials, and on the proximal ends of all but the terminal phalanges. In the carpus the radiale is commonly called the **scaphoid,** the intermedium the **lunar,** and the ulnare the **cuneiform.** It is hardly worthwhile for the nonspecialist to learn the mammalian terms for the remaining bones because deletions and fusions among them are the rule. The pisiform is retained. In the tarsus the fibulare is called the **calcaneum.** This is a large bone which extends behind the ankle joint (as it did also in mammal-like reptiles) to form a heel which is important as a lever arm for muscles that straighten the foot on the shank. The tibiale joins the intermedium, and the resultant large bone is called the **astragulus.** This bone lies partly over the calcaneum, and the ankle joint is between the astragulus and tibia. The ancestral articulation between fibula and calcaneum is reduced or lost. The prominent centrale, which articulates with the astragulus, is called the **navicular.** Fusions among the tarsals are common. The basic phalangeal formula is 2·3·3·3·3. Features of the appendicular skeleton are identified in Figures 8-23 and 8-24, and are related to articulations and muscles. The **olecranon process** of the ulna is particularly characteristic of mammals.

FIGURE 8-22
HIND LEG SKELETON
OF A MATAMATA
TURTLE, *Chelys.* **Dorsal**
view of left leg.

Comparative Anatomy of the Girdles of Tetrapods.   The **pelvic girdle** of tetrapods is much enlarged over that of fishes. Each half of the girdle is a single cartilaginous unit in the embryo, but three bones are constant in the adult. These are a dorsal **ilium** which articulates with one or more sacral vertebrae, an anterior **pubis,** and a posterior **ischium.** The bones of one side

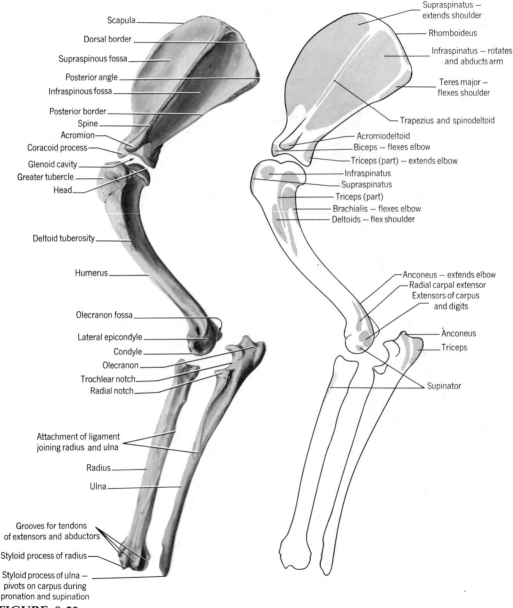

Scapula

Dorsal border

Supraspinous fossa

Posterior angle

Infraspinous fossa

Posterior border

Spine

Acromion

Coracoid process

Glenoid cavity

Greater tubercle

Head

Deltoid tuberosity

Humerus

Olecranon fossa

Lateral epicondyle

Condyle

Olecranon

Trochlear notch

Radial notch

Attachment of ligament
joining radius and ulna

Radius

Ulna

Grooves for tendons
of extensors and abductors

Styloid process of radius

Styloid process of ulna —
pivots on carpus during
pronation and supination

Supraspinatus —
extends shoulder

Rhomboideus

Infraspinatus – rotates
and abducts arm

Teres major –
flexes shoulder

Trapezius and spinodeltoid

Acromiodeltoid

Biceps – flexes elbow

Triceps (part) – extends elbow

Infraspinatus

Supraspinatus

Triceps (part)

Brachialis – flexes elbow

Deltoids – flex shoulder

Anconeus – extends elbow

Radial carpal extensor

Extensors of carpus
and digits

Anconeus

Triceps

Supinator

FIGURE 8-23

FEATURES AND FUNCTIONS OF THE FORELEG SKELETON OF THE DOG. Lateral view of left leg. Not all muscle attachments are shown.

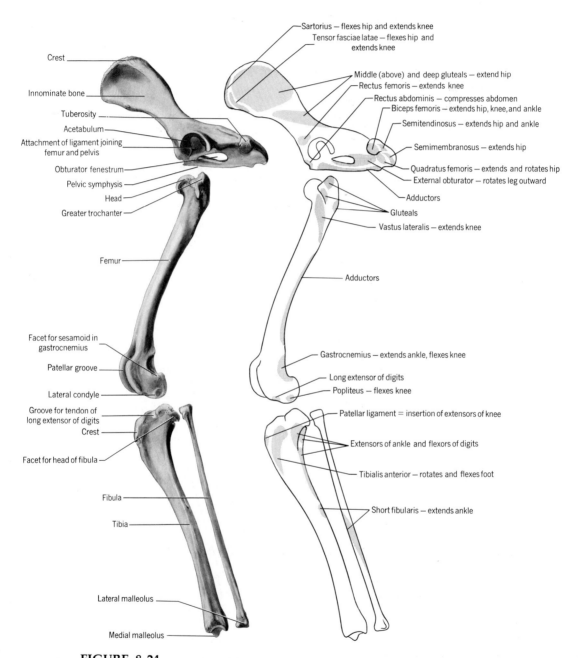

FIGURE 8-24
FEATURES AND FUNCTIONS OF THE HIND LEG SKELETON OF THE DOG. Lateral view of left leg. Not all muscle attachments are shown.

commonly fuse in the adult to form the **innominate bone.** One or both of the ventral bones of the two sides usually articulate or fuse across the midventral line; the contact is called the **pelvic symphysis.**

Primitive AMPHIBIANS had a solid girdle shaped like a triangle with the ilium forming the apex. The pubis can be distinguished from the ischium by having a foramen (obturator foramen) which accommodates a nerve. The atypical girdle of frogs has a long, anteriorly inclined ilium. The pubis of modern amphibians is cartilaginous.

The girdle of REPTILES takes various shapes but is like that of labyrinthodonts in basic plan. The contact with the spine is firmer. A large pubo-ischiadic fenestrum is usually present between the two ventral bones.

The pelvic girdle of BIRDS is distinctive. It is large and firmly attached to the synsacrum. The long ilium extends both anterior and posterior to the socket for the femur, or **acetabulum.** The pubis is turned backward below the ischium. There is no symphysis.

MAMMALS have a long and expanded ilium which extends only forward from the acetabulum. The large obturator fenestrum represents both the obturator foramen and the pubo-ischiadic fenestrum of the ancestor. A symphysis is nearly always present. Monotremes and marsupials have **epipubic bones** of doubtful origin which articulate with the pubic bones and extend forward in the ventral body wall.

The **pectoral girdle** of PRIMITIVE AMPHIBIANS differed from that of fishes in that the replacement bones were larger and the membrane bones reduced. All bones dorsal to the cleithrum were lost (except in one transitional group of primitive labyrinthodonts); thus the contact with the skull was broken and the head was freed to turn on the evolving neck. Some dipnoans and crossopterygians had a small **interclavicle** which united the two half girdles in the midventral line. This bone enlarged in labyrinthodonts, probably to compensate for loss of anchorage of the girdle to the head. There were two replacement bones: a dorsal **scapula** and a ventral bone which for the moment I shall call the coracoid. Among MODERN AMPHIBIANS, Urodela have no membrane bones at all in this girdle, and Anura have no interclavicle and usually lack the cleithrum.

STEM REPTILES, SYNAPSIDS, and MONOTREMES are alike in having a full complement of bones. Interclavicle and clavicle are present, and the cleithrum is present (for the last time) in the more primitive reptiles. The scapula is large and there are two coracoids. It is agreed that the single coracoid of amphibi-

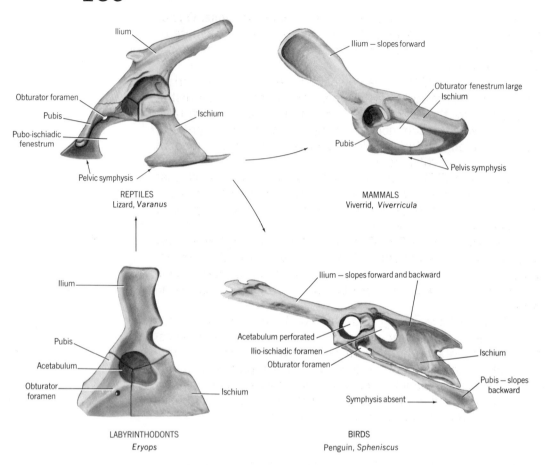

FIGURE 8-25

PHYLOGENY OF TETRAPOD PELVIC GIRDLE. Evolutionary stages identified in capital letters; examples in lowercase letters. Left lateral views.

ans did not split; instead, a new bone was added behind the original coracoid. We refer, therefore, to an **anterior coracoid** and a **posterior coracoid.** (Unfortunately these bones have also been called, respectively, procoracoid and coracoid, procoracoid and postcoracoid, and coracoid and postcoracoid.) OTHER REPTILES, including turtles, lepidosaurs, and archosaurs, usually lose the posterior coracoid and at least some of the membrane bones.

BIRDS have a blade-like scapula which is oriented parallel to the spine. The anterior coracoid is large and articulates firmly with the sternum. The posterior coracoid has been lost. The two clavicles fuse ventrally to form the wishbone or **furcula.** The interclavicle may be incorporated into the furcula of some birds, but probably is often absent.

When only one coracoid is present (labyrinthodonts, turtles, crocodilians, lizards, birds), it is usual to refer to it as *the*

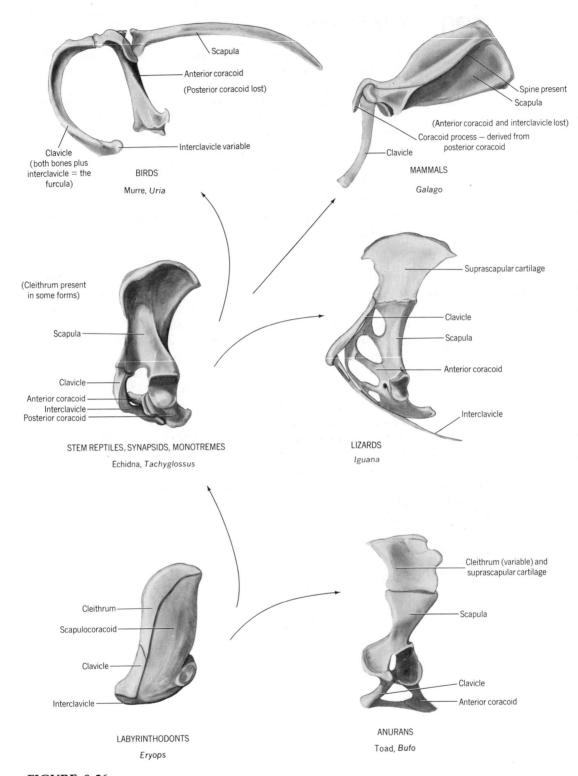

FIGURE 8-26
PHYLOGENY OF TETRAPOD PECTORAL GIRDLE. Evolutionary stages identified in capital letters; examples in lowercase letters. Left lateral views.

coracoid without qualification. We have seen, however, that with respect to tetrapod phylogeny it is the anterior coracoid.

The clavicle is the only membrane bone retained by MAMMALS, and even this bone may be missing. It is the *anterior* coracoid that is completely lost this time. The posterior coracoid ossifies independently in the fetus and then fuses to the scapula to form the **coracoid process** of that bone. The scapula is unique in having a **spine.** The spine represents the anterior border of the ancestral bone, so in fact it is the anterior border of the mammalian scapula that is new. The ventral end of the spine is continued as the **acromion process** to articulate with the clavicle.

Some Miscellaneous Bones

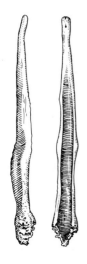

Nodules of bone tend to form where tendons play over joints. These are called **sesamoid bones** ("sesamoid" = having the shape of a seed). The largest sesamoid is the **patella,** or kneecap. The central tendons of certain kinds of muscles tend to ossify—particularly in birds (where they may be encountered in a drumstick). A bone called the **baculum** (or os penis) is present in the penis of carnivores, bats, insectivores, rodents, colugos and some primates. Its size and shape vary widely among species and it is therefore useful in systematics. A corresponding but much smaller bone may be present in the female clitoris. Additional small bones are found here and there among tetrapods: in the eyelids of crocodilians, in the crest of a bird, in the snout of pigs, at the base of the external ear of some rodents, at the base of the aortic arch of some artiodactyls, and so on. These are of more functional than evolutionary significance.

FIGURE 8-27
BACULUM of the maned wolf, *Chrysocyon,* in left lateral and ventral views.

Muscles and Electric Organs

The muscular system (when studied in conjunction with the skeleton) is of primary importance in the analysis of locomotor mechanisms. It is of secondary importance for establishing phylogenetic relationships. Homologies are nearly always evident among muscles of related orders and lower taxa, and can usually be established among related classes, but there are exceptions, and the muscular system is not often useful for verifying phylogenies worked out on the basis of other organ systems.

**Function and
Structure of
Muscles**

General Function and Gross Structure.   Nearly all functions of the body are in part muscular. Without muscles, vertebrates could not move, their tissues soon would be starved or poisoned, and the products of their glands could not be distributed. Humans could not read, speak, or write to communicate their thoughts.

Muscles accomplish all this by doing only one thing — by creating tension along the axis of their fibers, which tends to shorten their substance. Active muscles usually do shorten, thus moving a bone or constricting a space. They may also prevent motion by opposing gravity or the pull of other muscles. With little or no motion, they may function to cause a part to become more rigid. Further, they may offer controlled resistance to extrinsic forces that tend to stretch them out.

The muscular system also has secondary functions: It contributes importantly to the maintenance of the body temperature of homeotherms and, because of its bulk, distributes the weight of the body, influences the contours of the body, and offers protection to some of the viscera.

There is much variation in the shapes of muscles. Since each muscle fiber contracts only along its length, it is usually most effective for the fibers of one muscle to lie approximately parallel, thus establishing a longitudinal axis for the muscle as a whole. A long cylinder might be taken, therefore, as a fundamental shape. Several muscles of the throat are usually of this nature. If one end of a muscle is tapered to insert on the skeleton, the shape of the muscle becomes that of a teardrop. Several muscles of the hip, thigh, and upper arm are commonly of this sort. Most muscles of the limbs are tapered at each end and hence are spindle-shaped. One end may be divided into two or more parts (or "heads") and the fleshy part of the muscle is seldom symmetrical because it must fit against other muscles. Some muscles are spread out as sheets (most abdominal muscles), and others are flat at one end and gathered at the other, thus becoming fan-shaped (various shoulder muscles). Muscles that surround orifices have curved fibers and are washer-shaped; muscles that surround spaces (stomach, uterus) have fibers oriented in various directions and are more or less hollow spheres.

Some spindle-shaped muscles (particularly of birds) do not have their fibers oriented in the long axis of the muscle but instead have fibers that slope inward to insert on a central tendon. In longitudinal section such muscles may look like feathers and hence are called **pinnate**.

Where muscles attach directly to the skeleton, the connective tissue that surrounds and pervades them is continuous with that which surrounds the bones. Where muscles do not impinge directly on the skeleton, they are usually joined to bones by **tendons,** which are tough cords of closely packed, parallel, collagenous fibers. (**Ligaments** join bone to bone. Their collagenous fibers are somewhat less regular and they include elastic fibers.) Some muscles (e.g., several abdominal muscles of mammals) do not attach to bones but instead distribute their forces over broad areas by means of strong flat sheets of connective tissue. These sheets are called **aponeuroses.** The loose connective tissue that binds muscle to muscle, and skin to muscle, is called **fascia.**

When a muscle contracts, it pulls equally on each end. Commonly, one attachment is relatively free to move and is then called the **insertion.** The relatively fixed attachment is the **origin.** These terms must be used with caution, however; tables of muscle origins and insertions can be misleading. Which end of a muscle is the more movable depends on posture, the activity of other muscles, and contacts with the environment. When a man lifts an object from a table, the proximal end of his biceps is its origin; when he chins himself on a bar, the distal end is the origin. During the propulsive phase of a limb's motion, its muscles move the body on the limb, not the limb on the body; contrary to usual terminology, it is then the distal ends of the muscles that are relatively fixed. Often neither end of a muscle is fixed, and sometimes neither moves.

It is necessary to have a vocabulary for describing the actions of muscles. **Flexors** are commonly defined as muscles that reduce the angle between adjacent bones; **extensors** increase the angle. These definitions usually serve but can be misleading. Considerations of position, phylogenetic origin, and innervation (as explained below) make it clear that flexors of a hind limb swing joints to the rear, whereas extensors swing them forward. At the hip, such muscles may increase or decrease the angle between adjacent bones, depending on posture. Also, since the elbow and knee bend in opposite directions, the situation is reversed for the forearm, flection bends the elbow forward, not to the rear. Further, some muscles of the back that straighten (hence extend) the spine from a hunched position also, on further contraction, arch (hence flex) the spine. If confusion threatens, it is well to avoid one-word designations of muscle function. **Adductors** move parts inward toward the sagittal plane of the body or axis of a limb; **abductors** move parts outward away from the body or axis. Opening and closing the fingers and clapping the hands use these sets of muscles alternately. **Levators** raise and **depressors** lower such parts as the jaw or shoulders. Paired fins may alternatively be said to be depressed or adducted. **Protractors** push a part (such as the tongue or an entire limb) away from its base and **retractors** draw it back. **Sphincters** constrict openings (mouth, duct orifices) and **constrictors** compress spaces (pharynx, abdomen); they are opposed by **dilators. Rotators** turn parts about their axes (spine, limbs). The rotators that turn the soles of the hands or feet upward are **supinators**; those that turn them downward are **pronators.**

For every action there is an opposing or restoring action. Opposing muscles are called **antagonists.** Rarely does one muscle contract alone. Muscles that supplement one another are called **synergists.**

Muscles are given names that describe their actions (levator maxillae, flexor digitorum, adductor mandibulae), shapes (biceps, rhomboideus, trapezius), positions (temporalis, pectoralis, gluteus), or attachments, the origin being named before the insertion (geniohyoid, sternomastoid, cleidobrachialis). Most muscles were first named for the human; the terms are not always apt when applied to homologs in other animals. Nevertheless, it is easier to remember the terms when their derivations are understood.

Histology and Contractility.   Three types of muscle are distinguished by differences in histology and physiology. **Smooth muscle** is found in the skin, blood vessels, urogenital system, respiratory channels, and alimentary tract and the ducts of its derivatives. Smooth muscle tissue is described briefly on p. 225. **Cardiac muscle** is restricted to the heart and will be described with that organ. The remaining muscles of the body are **skeletal** (or **striated**) **muscle.**

Each skeletal muscle is surrounded by a tough envelope of connective tissue, the **epimysium** (= upon + muscle), which is continuous with tendons and with such fascia as may be present. The epimysium also merges with septa, collectively called the **perimysium,** which penetrate the muscle and divide it into bundles of fibers. Such bundles may be distinct, making the muscle stringy, as for the rhomboideus of many mammals, or may be virtually absent. The perimysium, in turn, is continuous with a net of connective tissue, the **endomysium,** which surrounds the sarcolemma or limiting membrane of the individual muscle fibers. This continuous system of connective tissue gives muscles their shape and strength, and transmits their forces to origin and insertion.

If a fragment of muscle is macerated and pulled apart with fine needles, the hair-like **muscle fibers** can be seen. Although scarcely visible to the naked eye, they may be as long as the whole muscle. They have multiple, peripheral nuclei and are cross-banded (hence the term "striated" muscle). The muscle fibers are penetrated by transverse tubules which open to the outside and by tiny anastomosing canals called the sarcoplasmic reticulum. These function in both transport and transmission.

When muscle fibers are stained and viewed with a light microscope it is seen that each is made up of dozens of finer strands, called **myofibrils,** each of which is about 0.001 mm (1μ) in diameter. When enlarged some 400,000 diameters by the electron microscope it is seen that each myofibril is made of many filaments of two kinds. **Thick filaments** are about 0.01 μ in diameter and consist of a protein called myosin. **Thin filaments** have half the diameter of the other filaments and consist of another protein called actin. The striations of the muscle result from a regular pattern of interlocking of these filaments. In the presence of the energy source adenosine triphosphate, the two kinds of filaments slip through one another, like the bristles of two hairbrushes that are pushed together, to cause contraction. The same reaction has been identified where contraction occurs in the absence of muscles (Protozoa, sperm tails) and has been produced experimentally in the absence of life.

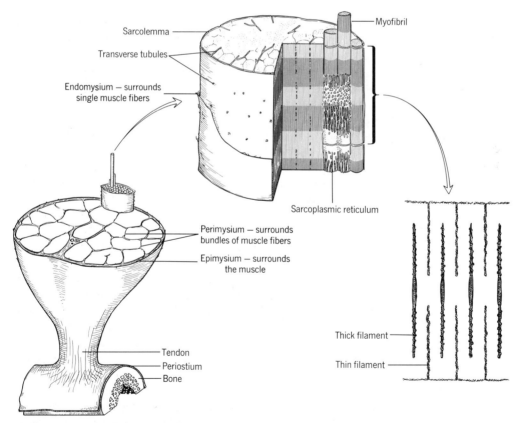

FIGURE 9-1
STRUCTURE OF SKELETAL MUSCLE. Nuclei and mitochondria omitted.

**Major Categories
of Muscles**

It is both convenient and instructive to arrange the many muscles into groups for study. Several methods of grouping suggest themselves, and the student can use each to his profit. First, all muscles of one region of the body can be studied together. Thus, the muscles of the spine, forelimb, head and neck, etc., can be studied in turn. What muscles would be seen in a cross-section of the thigh? What muscles would be cut in passing from the breast to the lung? What muscles have origin or insertion on the scapula? This approach is efficient at the operation table and dissection table, particularly with large animals.

A second method groups together muscles of like function. What muscles extend(protract) the forelimb? Or turn the head in a specified way? Or maintain standing posture? One discovers that a single muscle (e.g., pectoralis) can have several actions, and that some actions (e.g., swinging the thigh to the rear) can be accomplished alternatively by two or more muscles (though not with identical efficiency). This approach is practical for the functional morphologist, behaviorist, and physical therapist.

A third method is of significance for comparative studies. The various muscles can be arranged in major categories on the basis of embryonic origin. Such categories have somewhat independent phylogenies, relate to the positional and functional groups noted above, and can be distinguished by nervous innervation. Let us identify them by tracing their origins and innervations.

The mesoderm of the early embryo is differentiated into a dorsolateral, segmented epimere, a small mesomere, and a ventrolateral, unsegmented hypomere (Figure 5-4). As has been explained in previous chapters, the epimere further differentiates into dermatome, myotome, and sclerotome. The sclerotome forms no muscles. The dermatome forms much of the skin, including any intrinsic smooth muscles which may be present there. The myotome and the hypomere are the sources of virtually all other muscles of the body.

As development proceeds, the myotomes behind the head and pharynx form much of the musculature of the body wall, or **axial muscles.** In most fishes, the axial muscles of each side of the body are clearly separated by a membranous partition, the **lateral septum,** into dorsal **epaxial muscles** and ventral **hypaxial muscles.** On the trunk, but not on the tail, it appears (on the basis of experimental studies done chiefly on the chick) that in higher vertebrates the hypomere also contributes importantly to the formation of hypaxial muscles. Epaxial muscles dorsoflex the spine and are innervated by dorsal rami of spinal nerves. Hypaxial muscles ventroflex the spine and support the body wall. They are innervated by ventral rami of spinal nerves.

When epaxial and hypaxial muscles of one side of the body contract together, the spine is flexed to that side.

The epaxial parts of three pairs of head myotomes form the **extrinsic eye muscles.** These myotomes are identified with the premandibular, mandibular, and hyoidean segments of the ancestral head. The extrinsic eye muscles are innervated by the third, fourth, and sixth cranial nerves. The intrinsic muscles of the eye will be discussed in Chapter 16. The remaining head myotomes lie behind the ear and are transitory or absent in embryos of surviving vertebrates above the cyclostomes.

Below the pharynx, from the pectoral girdle to the jaw, are muscles which were derived phylogenetically by the forward migration of hypaxial muscles originally located on the trunk. Because of their position they are called **hypobranchial muscles.** Their innervation by the twelfth cranial nerve and by ventral rami of cervical nerves indicates their posterior origin.

There is evidence from the embryology of sharks that the muscles of the fins, or **appendicular muscles,** form as extensions from the hypaxial muscles of the body wall. This may be the primitive conditon. In tetrapods, however, the appendicular muscles form in place from mesenchyme derived, at least in

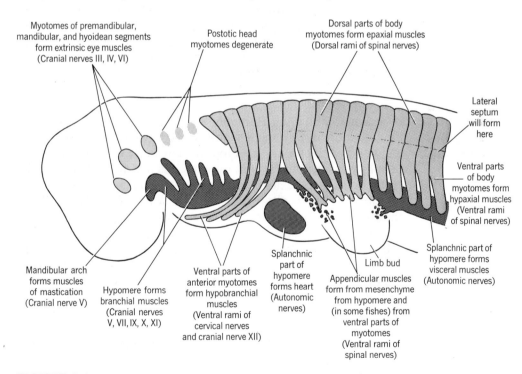

FIGURE 9-2

STYLIZED EMBRYO SHOWING THE DIFFERENTIATION OF MUSCLE GROUPS FROM THE MYOTOMES AND HYPOMERE as identified in Figure 5-4. Ultimate innervation of the muscle groups is given in parentheses.

part, from the hypomere. Innervation is by ventral rami of spinal nerves.

The hypomere has important muscular derivatives other than the hypaxial and appendicular muscles already noted. At the level of the trunk it splits to enclose the coelomic cavity. The outer (somatic) layer forms no further muscle. The inner (splanchnic) layer forms the **heart** and the **muscles of the viscera.** At the level of the pharynx, there is no persistent coelom and the hypomere does not split into layers. It is instead interrupted by the pharyngeal slits and becomes associated with the visceral skeleton. The **branchial muscles,** which develop from this portion of the hypomere, are striated and voluntary, but are, nevertheless, in series developmentally with the smooth muscle of the gut. Branchial muscles are innervated by cranial nerves, the fifth nerve serves the mandibular arch, the seventh nerve the hyoidean arch, and the ninth nerve the first branchial arch. The remaining arches are served principally by the tenth nerve, but the eleventh nerve and ventral rami of cervical nerves may also innervate branchial muscles.

In summary, the major categories of muscles are the axial musculature (which has epaxial and hypaxial divisions), extrinsic muscles of the eye, hypobranchial muscles (which are derived from hypaxial muscles), appendicular musculature (which has dorsal and ventral divisions), muscles of the gut, and branchial muscles (which are serially related to the visceral skeleton).

Comparative Anatomy of Muscles

Bases for Establishing Homologies. In order to trace the evolution of individual muscles one must have criteria for recognizing homologous muscles in different taxa. Within orders it is usually possible, and within families it is nearly always possible, to recognize equivalent muscles on the basis of position and relationships with other muscles and with the skeleton. Thus, the supraspinatus muscle of mammals always occupies the supraspinous fossa of the scapula and inserts on the greater tubercle of the humerus. In some instances, however, the criterion of positional relationships fails: Reptiles do not have a supraspinous fossa; the attachments of a muscle may change with evolution to such an extent that the action of the muscle is materially altered; adjacent muscles of similar action may fuse; a muscle may disappear; and ancestral muscles (unlike bones of the skeleton) tend to split in the course of evolution to become several muscles.

Paleontological evidence of homology is sometimes provided by a series of fossil bones having muscle scars that evince the migration, fusion, or loss of a particular muscle. The progressive change of certain muscles in the feet of extinct horses has been learned in this way. However, the paleontologist must be careful to avoid making unjustified assumptions and is usually dependent on the muscles of surviving animals to guide his analyses.

It was shown above that embryology is useful for establishing major categories of muscles. Since the pattern into which the initial muscle masses split tends to be less specialized than that of the adult, embryology is also useful for homologizing specific muscles. This approach has been applied to birds and some other vertebrates and deserves further study.

The criterion of muscle homology that has received the most attention is that of nerve supply. In the last decade of the nineteenth century, the German anatomist Fürbringer postulated an invariable relationship between peripheral nerves and the muscles they innervated: Homologies could be established by nerve–muscle dissections, and apparent exceptions were interpreted to mean simply that previous assumptions were in error. This theory was immediately attacked, and the developmental mechanism which Fürbringer had postulated as the basis for the nerve–muscle relationship was shown to be unfounded. Various authors have now studied nerve–muscle relationships in detail (notable among them have been Howell, Romer, and Haines), and it is agreed that nerve supply remains a useful criterion of homology but that numerous instances are known of muscles that have evolved nerve relationships differing from those of their evolutionary precursors.

The comparative myologist is well-advised to use as many criteria of homology as are available to him.

Muscles of Primary Swimmers.   The muscular system of CYCLOSTOMES (particularly of lampreys) is more simple and more primitive than that of other vertebrates. A lateral septum is lacking, so the prominent axial musculature is not divided into epaxial and hypaxial divisions. The segmentation of the body is clearly evident: Each myotome contributes one muscle segment, or **myomere.** An axial skeleton being virtually absent, the short fibers of the myomeres insert on partitions of connective tissue, the **myosepta,** which lie between successive myomeres. Myomeres and septa are thrown into gentle folds which are scarcely more complicated than those of amphioxus. The ventral portions of those myomeres lying close behind the pharynx turn somewhat forward, foreshadowing the hypobranchial musculature.

Appendicular muscles are, of couse, absent, and, since jaws are lacking and the visceral skeleton is constructed in one unit, related branchial muscles are not prominent. There is an elaborate musculature associated with the specialized mouth and tongue. These are dissimilar in lampreys and hagfishes and cannot be homologized with muscles of higher vertebrates.

Extrinsic eye muscles develop from head myotomes as in other vertebrates. However, their number and position are such as to make it difficult to homologize them with the analogous muscles in other classes.

The musculature of JAWED FISHES is more advanced, but remains less complex than that of tetrapods. Axial muscles are divided into epaxial and hypaxial portions by a lateral septum. Dorsal ribs, if present, lie in this septum. The myomeres, although straight in the embryo, become more angled than in cyclostomes. In some fishes, strands of connective tissue extend from myoseptum to myoseptum, thus forming tendons which may be stiffened by bones. Strong longitudinal tendons are found in the tails of certain fishes.

Strap-like hypobranchial muscles extend from the pectoral girdle to the visceral arches. One of them, the **coracomandibularis,** assists in opening the jaw. The hypobranchial muscles have become distinct from the hypaxial muscles from which they evolved, but they retain the longitudinal orientation imposed by their forward migration.

The girdles of fishes lie firmly anchored within the axial musculature. Appendicular muscles have evolved with the fins

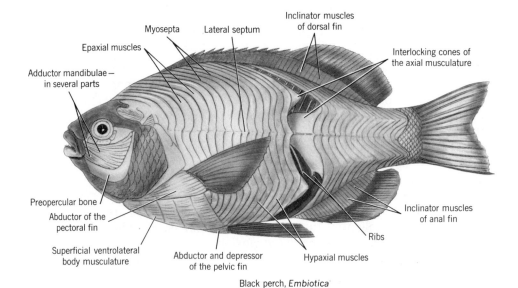

Black perch, *Embiotica*

FIGURE 9-3

MUSCULATURE OF A TELEOST with two myomeres removed to show the shape of the myosepta.

and are divided into a dorsal mass of extensors (or abductors, or levators—all these terms are used) which move the fins upward or forward, and a ventral mass of flexors (adductors, depressors) which move them downward or backward. These muscles, which may be somewhat subdivided, have origins on the girdles and adjacent fascia and insert on the radials of the fins.

The pharyngeal morphology of some sharks suggests that ancestral fishes, having homogeneous visceral arches, had simple and serial branchial muscles. Superficially, sheet-like **dorsal** and **ventral constrictors** were apparently nearly continuous over the gill area. These compressed the pharynx. A series of **levators** ran from fascia above the pharynx to the several arches and served to lift the gill bars. (Hypobranchial muscles performed the complementary role of depressing the pharyngeal skeleton.) Deeper series of **adductors** closed the internal angles of each ⪵-shaped visceral arch. The adductor closing the middle angle between the epibranchial and ceratobranchial elements probably was the most important and the ventral adductor the least important. The regularity of this ancestral pattern is much altered among the various surviving fishes according to type of jaw suspension, feeding mechanism, and presence or absence of spiracle and operculum. Muscles of the first two arches are the most specialized. The levator is retained to lift the palatoquadrate cartilage (except in some autostylic fishes). The middle adductor of the mandibular arch is much enlarged to become the **adductor mandibulae,** which closes the jaws. The ventral constrictors of the mandibular and hyoid arches form the sheet-like **intermandibularis** muscle, which lies between the mandibles and raises the floor of the mouth. A muscle of the second arch, the **depressor mandibulae,** opens the jaw, and other second arch muscles move the operculum in bony fishes. Muscles of the branchial arches are relatively unspecialized in Chondrichthyes. The levators, however, tend to mass over the more posterior gills as the **cucullaris,** or **trapezius** muscle (Figure 11-4). Branchial muscles of the gills of Osteichthyes are usually reduced to remnants of the ventral constrictors.

Fishes have six extrinsic eye muscles. Four **rectus muscles** (anterior, posterior, superior, inferior) have their origins close together deep in the posterior part of the orbit. These rotate the eye around the longitudinal and vertical axes of the head. The superior and inferior **oblique muscles** have their origins deep in the forward part of the orbit. They rotate the eye around its optical axis (the transverse axis of the head). Four of the muscles (anterior, superior, inferior recti, and inferior oblique) are derived from the premandibular myotome and are innervated by the third cranial nerve. The superior oblique is

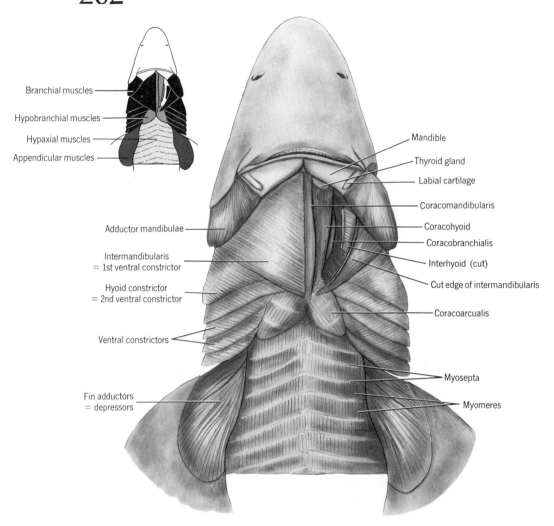

Branchial muscles
Hypobranchial muscles
Hypaxial muscles
Appendicular muscles

Mandible
Thyroid gland
Labial cartilage
Coracomandibularis
Adductor mandibulae
Coracohyoid
Coracobranchialis
Intermandibularis
= 1st ventral constrictor
Interhyoid (cut)
Cut edge of intermandibularis
Hyoid constrictor
= 2nd ventral constrictor
Coracoarcualis
Ventral constrictors
Myosepta
Fin adductors
= depressors
Myomeres

FIGURE 9-4

**ANTERIOR VENTRAL MUSCULATURE OF AN ELASMOBRANCH shown by the
shark,** *Squalus.*

derived from the mandibular myotome and is innervated by the
fourth cranial nerve. The posterior rectus is derived from both
mandibular and hyoidean myotomes and is innervated by the
sixth cranial nerve.

The shark is particularly favorable for the study of the
muscles of fishes.

Axial and Hypobranchial Muscles of Tetrapods. Several general trends are evident in the evolution of the axial musculature
of tetrapods. In fishes, these muscles, being the propulsive
muscles, are the most massive of the body. As limbs take over
the propulsive role, their muscles enlarge and the axial muscu-

lature diminishes. The axial skeleton of tetrapods, in contrast, becomes firmer in order to play a new supportive role, and, concomitant with this trend, the remaining axial musculature becomes more intimately related to the skeleton and adds to its functions dorsoflection and ventroflection of the spine, which are rarely marked in fishes. Myosepta regress and disappear, and many muscles develop long fibers which span two to many vertebrae. Further, certain muscles tend to form sheet-like layers, and others become associated with the pectoral girdle.

The axial musculature of AMPHIBIANS is transitional between that of fishes and reptiles. Epaxial muscles, now collectively called the **dorsalis trunci,** are conservative; myosepta are still present and are nearly vertical instead of angled, as in fishes. Hypaxial muscles are equally primitive on the tail; on the trunk they are more advanced and are commonly classified in three groups. The **subvertebral group,** located below the transverse processes of the vertebrae, is small in mass and flexes the spine. The **rectus abdominis** muscle (or group) runs lengthwise along the ventral body wall between the two girdles. (It extends anterior to the pectoral girdle as the rectus cervicis in all living groups. This is probably a secondary condition resulting from reduction of that girdle.) The rectus abdominis ventroflexes the body and supports the viscera. The **lateral group** is located on the flanks between the other groups and is the most differentiated part of the hypaxial musculature. It breaks into (usually) three sheet-like layers, each having its fibers oriented in a different direction. Together they support and compress the body wall. The layers, in order from outside in, are the **external oblique** muscle, the **internal oblique,** and the **transversus.**

The pelvic girdle of amphibians is stronger than that of fishes, but since it gains an articulation with the spine it does not require muscular support. The pectoral girdle, in contrast, no longer articulates with the head and does not establish an articulation with the spine. Several muscles, probably derived from part of the external oblique, evolve to hold this girdle to the trunk.

The hypobranchial muscles of amphibians are, in general, like those of cartilaginous fishes. The muscles of the fleshy tongue are, however, added derivatives which are of the utmost importance to the biology of the class.

Amphibian muscles are seen to best advantage in Urodela, such as *Ambystoma, Taricha,* or *Notophthalmus.* They are not seen to advantage either in Anura or the commonly studied *Necturus,* where they are specialized.

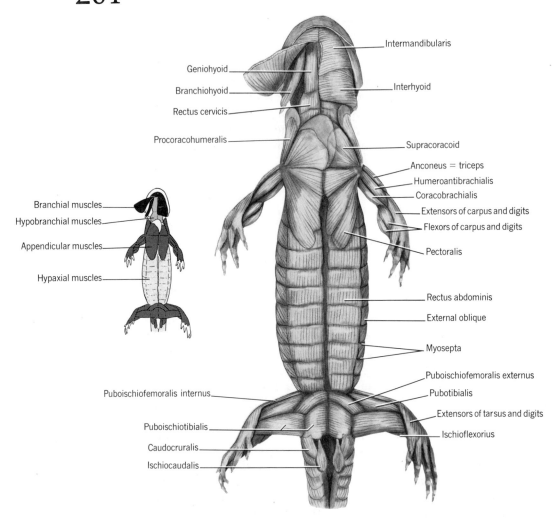

Intermandibularis
Geniohyoid
Branchiohyoid
Rectus cervicis
Procoracohumeralis
Interhyoid
Supracoracoid
Anconeus = triceps
Humeroantibrachialis
Coracobrachialis
Extensors of carpus and digits
Flexors of carpus and digits
Pectoralis
Branchial muscles
Hypobranchial muscles
Appendicular muscles
Hypaxial muscles
Rectus abdominis
External oblique
Myosepta
Puboischiofemoralis externus
Pubotibialis
Extensors of tarsus and digits
Ischioflexorius
Puboischiofemoralis internus
Puboischiotibialis
Caudocruralis
Ischiocaudalis

FIGURE 9-5
VENTRAL MUSCULATURE OF A URODELE shown by the tiger salamander, *Ambystoma.*

The axial musculature of REPTILES and of MAMMALS are alike in basic design. Epaxial muscles of the tail, though without myosepta and much reduced in bulk, are about as in amphibians; those of the cervical region are also relatively unspecialized but do tend to form layers behind the head. Epaxial muscles of the trunk become exceedingly complex and varied in detail. Three groups are distinguished: a large **longissimus dorsi** group which has long bundles and fills all but the deepest part of the space between the neural spines and transverse processes; an **iliocostalis** group which is more lateral and has attachments to the ilium and ribs; and a smaller **transversospinalis** group which lies under the longissimus dorsi and has

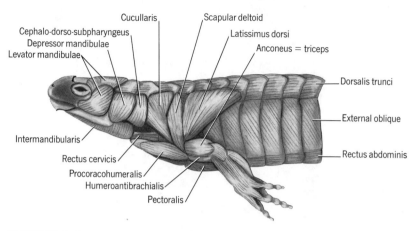

FIGURE 9-6

ANTERIOR LATERAL MUSCULATURE OF A URODELE shown by the tiger salamander, *Ambystoma*.

a complex of short bundles which lace among the centra, transverse processes, and bases of the neural spines of adjacent or nearby vertebrae.

The hypaxial musculature of the cervical and caudal regions is similar to that of amphibians, though often with longer tendons. On the trunk, the subvertebral group of hypaxial muscles is much reduced, except in the lumbar area, and the rectus abdominis is excluded from the thoracic region. Behind the thorax the lateral group remains essentially as for amphibians. More anteriorly, however, the ribs, now enlarged, penetrate and alter the lateral group of muscles. The transversus is excluded from the thorax and the external and internal obliques become, respectively, the external and internal **intercostal muscles** which contribute to the new function of ventilation of the lungs. In various reptiles—particularly turtles—ventilation is assisted by other muscles that are unique.

Muscles of the shoulder derived from the lateral group of hypaxial muscles are more prominent in amniotes than in amphibians. Principal among them are the **serratus**, which runs from the ribs to the scapula to suspend the thorax, sling fashion, from the girdle; the **levator scapulae**, which draws the shoulder forward; and (in mammals and some reptiles) the **rhomboideus**, which elevates the scapula and binds it to the thorax.

The muscular **diaphragm**, found only in mammals, is apparently also of hypaxial origin. Its nerve, the phrenic nerve, branches from ventral rami of cervical nerves. The embryonic diaphragm originates quite anterior to the adult position.

Hypobranchial muscles remain much as in amphibians, but are somewhat altered by the regression of the gills and compli-

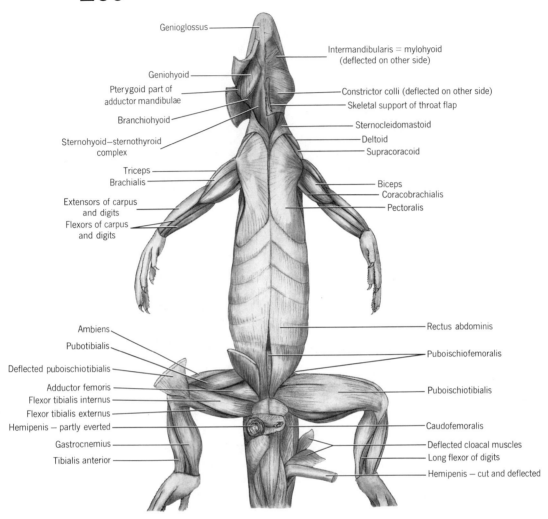

Genioglossus

Geniohyoid

Pterygoid part of
adductor mandibulae

Branchiohyoid

Sternohyoid—sternothyroid
complex

Triceps
Brachialis

Extensors of carpus
and digits
Flexors of carpus
and digits

Intermandibularis = mylohyoid
(deflected on other side)

Constrictor colli (deflected on other side)
Skeletal support of throat flap

Sternocleidomastoid
Deltoid
Supracoracoid

Biceps
Coracobrachialis
Pectoralis

Ambiens
Pubotibialis
Deflected puboischiotibialis
Adductor femoris
Flexor tibialis internus
Flexor tibialis externus
Hemipenis — partly everted
Gastrocnemius
Tibialis anterior

Rectus abdominis

Puboischiofemoralis

Puboischiotibialis

Caudofemoralis
Deflected cloacal muscles
Long flexor of digits
Hemipenis — cut and deflected

FIGURE 9-7
VENTRAL MUSCULATURE OF A SQUAMATE REPTILE shown by the iguanid, *Iguana*.

cated by the enlargement of the larynx.

Large lizards (but not turtles) are favorable for the demonstration of the muscles of reptiles.

The trunk of BIRDS is short and relatively rigid as an adaptation to flight. The tail is short and the tongue is usually bony. Consequently, the axial and hypobranchial musculature is much reduced.

Muscles of the Pectoral Limb. Muscles of the pectoral limb of tetrapods have three general sources. First, one or several muscles are contributed by the branchial musculature. These are innervated by cranial or cervical nerves and are identified under the next subheading. Second, as noted under the previous subheading, several muscles are contributed by the axial musculature. These are innervated by ventral rami of spinal nerves that do not join the network of nerves at the base of the limb called the **brachial plexus.** Finally, most appendicular muscles of tetrapods are derived directly from appendicular muscles of fishes. These are also innervated by ventral rami of spinal nerves, but these nerves each join the plexus before entering the appendage.

When the appendicular nerves of fishes emerge from their plexuses, they tend to be arranged in a more dorsal group which runs to the dorsal mass of fin elevators and a ventral group which runs to the ventral mass of fin depressors. The appendicular muscles of adult tetrapods are numerous and complex, yet in the embryo they differentiate from dorsal and ventral masses in recapitulation of the ancestral piscine condition, and in the adult (with some exceptions in the hind limb) the many individual muscles can still be identified as derivatives of the dorsal or ventral mass by their relationship to nerves emerging from the dorsal or ventral part of the respective plexus. In some instances, however, the ancestral functions of extension (for dorsal mass derivatives) and flection (for ventral mass derivatives) become reversed.

Homologies of the muscles of the pectoral limb are well-established. The larger and more constant of the **dorsal group muscles** are as follows: (*1*) The **latissimus dorsi,** present in all

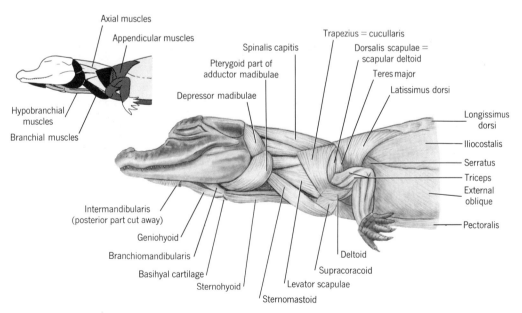

FIGURE 9-8
ANTERIOR LATERAL MUSCULATURE OF A REPTILE shown by the crocodilian, *Caiman.*

tetrapods and large in mammals, extends from the humerus to the body wall behind the girdle. It flexes the humerus. (2) A muscle or muscle complex that lies between the girdle and body wall of amphibians, reptiles, and birds becomes the strong **subscapularis** of mammals. This muscle rotates, adducts, or flexes the humerus according to position at the start of the action. The **teres major** muscle, which runs from the posterior angle of the scapula to the humerus and swings the limb to the rear, is believed on embryological evidence to be derived from the subscapularis by splitting. (It is also stated to be derived from the latissimus dorsi.) (3) The **dorsalis scapulae** of amphibians is an arm abductor. It persists in higher classes and usually forms two or more muscles called **deltoids**. (4) In all tetrapods the **triceps** is the powerful extensor of the lower arm.

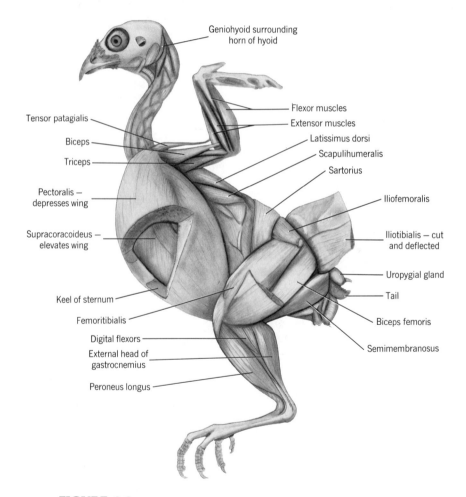

FIGURE 9-9
LATERAL MUSCULATURE OF A BIRD shown by the Japanese quail, *Coturnix.*

It extends from the shoulder to the ulna where, in mammals, it inserts on the olecranon process. (5) The **extensors of the manus and digits** are numerous and will not be named individually. They take origin largely from the lateral epicondyle of the humerus. Their actions are to elevate the wrist and straighten the digits.

The principal **ventral group muscles** of the pectoral limb are the following: (1) The **pectoralis,** present in all tetrapods and sometimes somewhat divided, inserts on the humerus and fans out on the chest. This powerful adductor is the largest flight muscle of flying vertebrates. (2) The **supracoracoideus** muscle of the lower classes is importantly altered in birds and mammals. In birds it shifts to the sternum, under the pectoralis, and inserts over a bony pulley onto the upper surface of the head of the humerus, thus serving to elevate the wing. In mammals the insertion on the humerus is retained, but as the coracoid bone regresses, the origin of the muscle shifts to the scapula. The embryonic muscle grows out on each side of the spine of the scapula to become the **supraspinatus,** which extends and abducts the limb, and the **infraspinatus,** which rotates and sometimes flexes the arm according to position when the action starts. (3) The **coracobrachialis,** large in amphibians, becoming smaller in mammals, is an arm flexor or adductor. (4) The flexors of the lower arm are the **brachialis** and (in amniotes only) the **biceps.** (5) The numerous **flexors of the manus and digits** have origin largely on the medial epicondyle of the humerus. In general, they depress the wrist and curl the digits, though in primates, at least, they may also stablize the hand in subtle ways.

Muscles of the Pelvic Limb.   With minor exceptions, the muscles of the pelvic limb of tetrapods are all derived from appendicular muscles of piscine ancestors. Dorsal and ventral group muscles are again recognized, and each group is commonly classified into prehip and posthip portions. In some tetrapods each of the four resulting units has its own nerve, but often the nerves combine or invade each other's "jurisdictions." Homologies between muscles of reptiles (particularly Lepidosauria) and those of mammals are relatively satisfactory; those for Lissamphibia and birds are more provisional than the commonly adopted terminologies indicate.

Dorsal group muscles include the following: (1) In reptiles a strong group of muscles runs from pubis and ischium to the femur. Because of changes in posture, only small derivatives remain on mammals (including the psoas and iliacus). (2) The **iliofemoralis** of reptiles probably becomes the important **gluteal muscles** of mammals. If so, the action changes from abduction to flection of the thigh. (3) The extensors of the shank run from

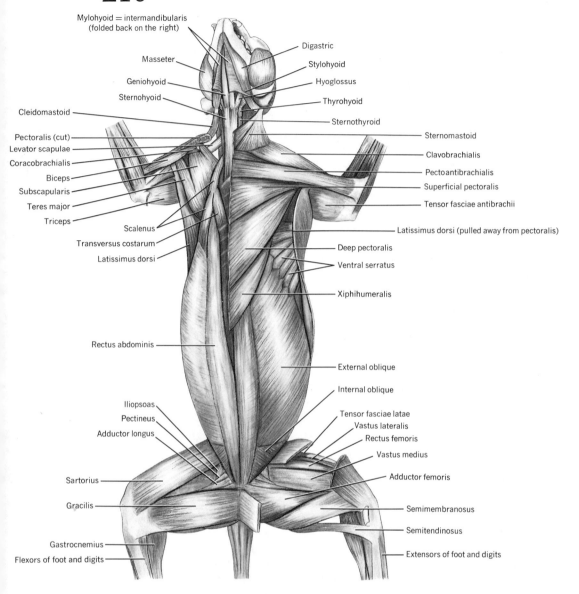

Mylohyoid = intermandibularis (folded back on the right)
Masseter
Geniohyoid
Sternohyoid
Cleidomastoid
Pectoralis (cut)
Levator scapulae
Coracobrachialis
Biceps
Subscapularis
Teres major
Triceps
Scalenus
Transversus costarum
Latissimus dorsi
Rectus abdominis
Iliopsoas
Pectineus
Adductor longus
Sartorius
Gracilis
Gastrocnemius
Flexors of foot and digits

Digastric
Stylohyoid
Hyoglossus
Thyrohyoid
Sternothyroid
Sternomastoid
Clavobrachialis
Pectoantibrachialis
Superficial pectoralis
Tensor fasciae antibrachii
Latissimus dorsi (pulled away from pectoralis)
Deep pectoralis
Ventral serratus
Xiphihumeralis
External oblique
Internal oblique
Tensor fasciae latae
Vastus lateralis
Rectus femoris
Vastus medius
Adductor femoris
Semimembranosus
Semitendinosus
Extensors of foot and digits

FIGURE 9-10
VENTRAL MUSCULATURE OF A MAMMAL, as seen in the cat. Sternomastoid, pectoralis complex, and tensor fasciae antibrachii removed on the right. See also Figures 7-22, 8-23, and 8-24.

the girdle and femur to insert by a tendon on the crest of the tibia, with the patella, where present, guiding the tendon over the knee joint. The individual muscles can be homologized among the amniotes; collectively they are called the **quadratus femoris**. (*4*) The **extensors of the pes and digits** lie on the lateral and anterior surface of the shank.

Ventral group muscles include the following: (*1*) A group of muscles that extends from the pubis and ischium to the femur includes, in mammals, the **obturator muscles** which lie immediately inside and outside of the pelvic fenestrum of that name. They are adductors and rotators. (2) A large group of muscles, variously subdivided, runs between the ventral part of the girdle and the femur and is called the **femoral adductors.** (3) Powerful flexors of the thigh and shank extend from the ischium to the femur or tibia. They include the **gracilis, semimembranosus, semitendinosus,** and **biceps femoris** of mammals and their homologs (not clear in each instance) in the other classes (4) The **caudofemoralis** is an important flexor of the thigh in reptiles, but with reduction of the tail, whence this muscle has its origin, only several small derivatives remain in mammals. (5) **Flexors of the pes and digits** lie on the medial and posterior surface of the shank. Most of them send long tendons around the ankle joint to insert under the foot, but in mammals the strong **gastrocnemius** inserts on the newly evolved heel bone (calcaneum).

Branchial Muscles of Tetrapods. Branchial muscles of the first (mandibular) arch are essentially as for fishes. The ancestral adductor (or levator) mandibulae pushes to the surface of the skull (as described on p. 148) and becomes complexly divided. Thus, in mammals, it splits into one or more **pterygoid muscles** which run from the inside of the mandible to the pterygoid region of the skull, a **temporalis** which runs from the coronoid process of the jaw to the temporal area, and a complex **masseter** which runs from the outside of the mandible to the zygomatic arch and adjacent parts of the skull. The variations of these muscles are closely related to feeding habits. The intermandibularis, called **mylohoid** in mammals, is retained from the ventral constrictor series of fishes. The levator of the first arch is retained in birds and some reptiles having kinetic skulls (movable upper jaws), but is otherwise lost.

The principal muscle of the second arch of all tetrapods except mammals is the **depressor mandibulae** which supplements or replaces hypobranchial muscles as the opener of the jaw. In mammals this muscle is lost and the mouth is opened by a new muscle, the **digastric,** which is derived from the ventral constrictors of both the first and second arches. Accordingly, it is innervated by both the fifth and seventh cranial nerves. Another second arch muscle of interest is the **stapedial muscle.** This tiny muscle controls the motion of the stapes — a second arch bone.

The trapezius (or cucullaris) of fishes is a muscle of moderate size derived from the more posterior gill levators. This muscle persists in tetrapods but enlarges and splits to become several muscles which assume the new functions of moving the scapula on the thorax and turning the head. In mammals these

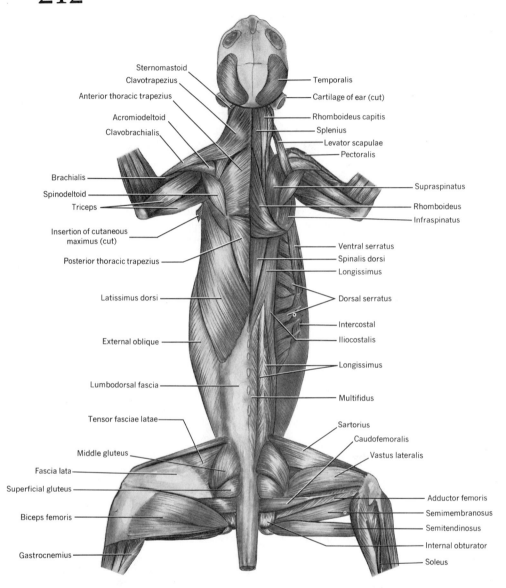

FIGURE 9-11

DORSAL MUSCULATURE OF A MAMMAL as seen in the cat. Trapezius muscles, clavobrachialis, latissimus dorsi, lumbodorsal fascia, tensor fasciae latae, and biceps femoris removed on the right. See also Figures 8-23 and 8-24.

muscles include at least two muscles that retain the name **trapezius,** and also the **sternomastoid** and **cleidomastoid.** (In man, at least, the branchial origin of these muscles is no longer evident in ontogeny.) The other tetrapod muscles of branchial origin are small. They are the muscles of the larynx and constrictors of the throat.

Extrinsic Muscles of Skin and Eye in Tetrapods. Muscles that run from underlying tissues to insert on and move the skin are not found in fishes or amphibians and are rare in reptiles. Snakes are a notable exception; their locomotor apparatus may include separate muscles to move each ventral scute. Birds have a muscle to tense the skin on the leading edge of the wing. Extrinsic skin muscles are most characteristic of mammals. A derivative of the second arch constrictor of reptiles is the **constrictor colli,** a thin superficial muscle over the ventral and lateral parts of the neck. In mammals this muscle becomes a complex of facial muscles collectively known as the **platysma.** Facial muscles reach their highest development in man. The ancestral innervation, by the seventh cranial nerve, is retained.

A second muscle of the skin, the **cutaneous maximus,** is derived from the latissimus dorsi and pectoralis. Although vestigial in man, this is often an extensive muscle over the trunk where it may serve to curl the body (spiny anteater) or become subdivided for flicking insects from the skin (horse).

The six extrinsic eye muscles of fishes are retained in tetrapods with remarkably little variation. However, the eyeball usually can no longer be rotated around its optical axis and one or more additional muscles evolve by splitting from one of the preexisting muscles. From the posterior rectus develops a **retractor bulbi** of from one to four parts which pulls the eyeball deeper into its socket. This action is protective and may also aid in swallowing. The muscle is marked in amphibians and some reptiles, but is lacking in many mammals. From the superior rectus there develops a muscle to lift the upper lid, and other muscles form to move the nictitating membrane.

Electric Organs

Electric organs are found in some 500 species of fishes belonging in seven families of Chondrichthyes and Osteichthyes. It is possible that richly innervated areas on the head shields of cephalaspids were also electric organs, though this interpretation is rejected by many paleontologists. The organs may be on the tail (electric skate, some teleosts), on the fins (electric ray), under the skin (a teleost), behind the eye (star-gazer), or over much of the trunk (electric eel). They are usually derived from muscle cells (hence their inclusion in this chapter), but origin from glandular and nervous tissue is not ruled out in every instance. Diversity of occurrence, location, structure, and also physiology, indicate that electric organs are ancient specializations which evolved independently several times and have undergone convergence.

Many fishes are only weakly electric. The electric ray, however, can develop 50 A (the ampere is a measure of the amount of current delivered) and the electric eel can produce more than 500 V (the volt is a measure of the driving force of the current). Shocks of 2000 W have been recorded (the watt, a measure of power, is the product of current flow times force).

Communication, orientation, and the detection of prey are the most common functions of electric organs—particularly for fishes living in murky water. The organs of some species serve also for offense or defense; even large fishes can be electrocuted by the more powerful discharges. Electric fishes emit constant discharges and are highly sensitive to the disturbances that objects produce in the electric fields near their bodies. It has been shown that one fish can detect a potential gradient of only 0.03 μv/cm. The sense organs that monitor the electric field are derived from the lateral line system and are located at the bases of deep pits in the skin (see p. 397).

The functional unit of an electric organ is the **electroplax,** a large, coin-shaped, multinucleated cell. Usually one flat surface is minutely folded; mitochondria concentrate under this membrane. The other flat surface is richly innervated. Hundreds or thousands of electroplaxs are stacked to form a column, and many columns are commonly present in one organ. In the resting state, an electric potential develops between the inside (negative) and outside of each electroplax. When the organ is fired by its nerve, the potentials are momentarily reversed (at least in some species) so the current exceeds

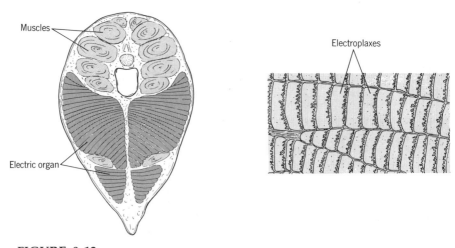

FIGURE 9-12
ELECTRIC ORGAN OF THE ELECTRIC EEL, *Electrophorus.* **Left, cross-section of the body; right, columns of electroplaxes.**

the resting potentials. The organ may either be "wired" largely in series (+ pole of one cell to − pole of adjacent cell, which gives maximum voltage as is desirable for freshwater species) or largely in parallel (+ pole to + pole, which gives maximum amperage).

Electric organs continue to be much studied, not only for their inherent interest, but also as a means of investigating general electric phenomena of nerve and muscle.

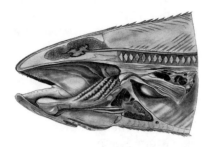

Digestive System
and Coelom

The digestive system has been less significant for comparative morphology than have most other organ systems. Its structure is at times useful in systematics, and it reveals general dietary habits, but some of the organs are too constant in structure to offer a phylogenetic story, whereas others vary without evident correlation with phylogeny or function. However, correlations of its fine structure with function and systematics deserve further attention.

The coelom has been of even less importance to the vertebrate morphologist because its structure is too simple and constant to contribute importantly to functional or evolutionary analysis.

217

Mesenteries correlate with systematics at the class level and also with marked postural differences (e.g., of dog, man, sloth). The detailed configurations of mesenteries are often too complicated to decipher without embryological analysis.

Function and General Structure

Coelomic cavities are the spaces that surround the heart, lungs, digestive system, and certain urogenital organs. The coelom functions to allow these structures to move freely and to change their relative sizes and positions. It also transmits eggs from ovaries to oviducts. (In Agnatha it carries metabolic wastes from blood stream to excretory tubules.) The coelom, unlike cavities of the nervous and respiratory systems, occurs within tissues of mesodermal origin. Its lining, which covers the body wall and envelopes the viscera, is a **serous membrane** composed of flat cells that secrete a fluid to lubricate the organs so they will slip easily past one another.

Mesenteries extend across the coelom from body wall to viscera. They are sheets of serous membrane strengthened by thin layers or bands of collaginous and elastic fibers. Mesenteries that join one viscus to another are called omenta or "ligaments" (the latter is an unfortunate term, since these are not true ligaments). Mesenteries support the internal organs, without restricting function, and transmit nerves and vessels. In mammals they are commonly sites of fat storage.

The digestive system functions to (1) receive ingested food, (2) store it temporarily, (3) reduce it physically, (4) further reduce it chemically, (5) absorb the products of digestion, and (6) hold temporarily and then eliminate undigested wastes.

The general structure of the gut relates to contours of the body (particularly in fishes) and to the general nature of the functions noted. Food is taken into the mouth with the aid of teeth and jaws, tongue (some agnaths and tetrapods), forefeet (some mammals), lips (some mammals), beak (birds), or pharyngeal suction (many fishes, amphibians, and turtles). The nature of these mechanisms relates closely to feeding habits in ways described in Chapter 25.

Most vertebrates are intermittent feeders; when food is available it must be taken into the body faster than it can be digested. If the food is bulky, or if feeding and drinking are infrequent and rapid, quantities of food and water must be stored temporarily. The principal storage organ is the stomach. Some birds have a storage sac off the esophagus. Many rodents, and some other mammals, have internal or external cheek pouches (opening, respectively, inside or outside the lips). If digestion is slow (as for diets of coarse vegetation), much digesting food must be retained at one time and the storage capacity of the entire tract is greatly increased.

Physical reduction of food—especially of roughage—is necessary to release nutrients from undigestable components and to increase the surface contact of food with digestive juices. Physical reduction is accomplished by the (*1*) chewing, rasping, or grinding of oral teeth, pharyngeal teeth (some fishes), or stomach (gizzard of many birds); (*2*) moistening, softening, and dissolving of food by fluids of the mouth, stomach, and intestine; (*3*) churning and mixing by peristalsis (anterior to posterior waves of contraction), reverse peristalsis, and segmentation (dividing motions) of the stomach and small intestine; and (*4*) emulsification of fats by secretions of the liver.

Chemical reduction of food is accomplished principally in the stomach and small intestine by enzymes produced in those organs or in the pancreas. Since the chemical nature of foodstuffs eaten by the various animals is similar, it is not surprising that the enzymes provided and the glands secreting them are also similar in the different vertebrates. Animals that employ bacterial fermentation as an aid to digestion (ungulates, some marsupials) must provide long storage in the stomach or caecum.

Absorption of the end products of digestion requires great surface contact between the digested food and the intestinal epithelium. This is accomplished by (*1*) a long intestine, (*2*) folds or pleats in the lining of the gut, (*3*) microscopic villi on the lining of the tract, and (*4*) smaller microvilli (revealed by the electron microscope) in portions of the tract.

Development of the Gut, Coelom, and Mesenteries

The developmental process of gastrulation provides the early embryo with an inner germ layer, the entoderm, which is nearly spherical if there is little yolk and sheet-like if there is much yolk (Figures 5-2 and 5-3). In either event, as the embryo lengthens, the entoderm is drawn out into a tube by processes shown in Figure 10-1. This tube becomes the lining of the gut. At each end it breaks through to the ectoderm, thus establishing oral and anal openings. The midgut of embryos provided with a large yolk mass (fishes, reptiles, birds) is continuous with a fetal membrane, the **yolk sac,** which envelopes and gradually absorbs the yolk.

Initially the gut tube is straight, or nearly so, but soon it folds and coils and establishes outgrowths, or diverticula, which become the lining and secretory cells of associated organs. The complicated pouches that form in the oral cavity and pharynx will be described in other chapters. Close behind the pharynx of tetrapods a ventral diverticulum foreshadows the respiratory system. Several diverticula develop posterior to the expanding stomach. These become the liver, gall bladder, pancreas, and their ducts. Near the posterior end of the gut of amniotes, a

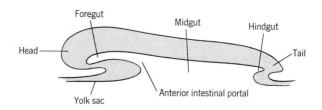

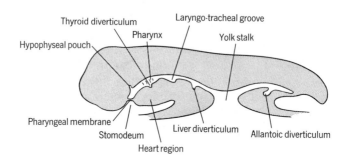

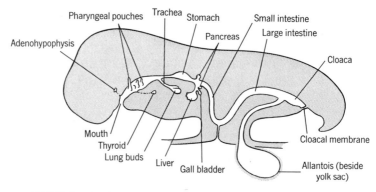

FIGURE 10-1

DEVELOPMENT OF THE GUT TUBE AND ITS DERI-
VATIVES IN AN AMNIOTE. Other organs and coelom
are not represented. The first stage shown here follows
the last stage shown in Figure 5-3. A later stage is shown
in Figure 10-13.

ventral diverticulum grows rapidly to become the fetal membrane called the **allantois.**

The muscular and connective tissue associated with the gut
and related organs (and therefore most of their bulk) is of
mesodermal origin. The differentiation of this mesoderm is
related to the ontogeny of the coelom.

In amphioxus, echinoderms, and other invertebrate deuterostomes, the coelom forms from a series of pouches that pinch

off from the dorsolateral wall of the initial gut tube. This process, called **enterocoely** (= gut + hollow), links echinoderms to chordates and may have been the ancestral method of coelom formation in vertebrates. All surviving vertebrates form coelom by the cavitation or splitting of initially solid mesoderm (Figure 5-4). This process is called **schizocoely** (= split + hollow).

Early embryos of vertebrates may have small and transitory coelomic cavities in the sclerotomes (these are called **sclerocoels**), myotomes **(myocoels)**, and mesomeres **(nephrocoels)**. Some nephrocoels have adult derivatives (described in Chapter 13); the other spaces have no known functional significance. The coelom of the hypomere is the **splanchnocoel,** but since it is large and persistent and gives rise to all the coelomic cavities of the adult, it is usually called simply the coelom.

This coelom splits the hypomere into an inner **splanchnic layer** and an outer **somatic layer.** As the coelom expands, right and left splanchnic layers move toward one another. They either come together in the midsagittal plane of the body or encounter the entoderm of the gut tube and its diverticula. Where they come together dorsal to the gut, they form the **dorsal mesentery;** ventral to the gut they form the **ventral mesentery,** and between the gut and its derivatives they form omenta. Parts of the ventral mesentery quickly degenerate, thus causing right and left coelomic cavities (from right and left hypomeres) to become confluent.

The parts of the splanchnic layer of the hypomere that encounter the gut tube and its derivatives form the smooth mus-

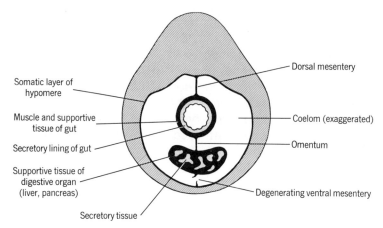

FIGURE 10-2

DERIVATIVES OF THE HYPOMERE IN RELATION TO THE GUT AND COELOM. This developmental stage follows the last shown in Figure 5-4.

culature, connective tissue, and serous membranes of the various organs. The somatic layer of the hypomere remains in contact with the body wall. It forms the serous membrane that lines the outer part of the coelom and, as noted in Chapter 9, may form mesenchyme which contributes to hypaxial and appendicular musculature.

Structure and Comparative Anatomy of the Digestive System

Mouth and Oral Cavity. The complicated structure of mouth, oral cavity, and pharynx cannot conveniently be discussed under the heading of any one organ system. The teeth have been described (Chapter 6) as has the evolution of the secondary palate in relation to chewing and breathing (Chapter 7). Respiratory and glandular derivatives of the pharynx will be presented in Chapters 11 and 17, and various specializations of the digestive tract will be noted in Chapter 25.

The ANCESTRAL VERTEBRATE was likely a filter-feeder. Like amphioxus and the larva of the lamprey, it probably had a small mouth, virtually no oral cavity, and a large pharynx specialized to remove microscopic food particles from water by trapping them in mucus covering numerous gill bars (Figures 2-3 and 3-4). The mucus was probably moved to the intestine by cilia.

The AGNATHA have given up filter-feeding, but because they have no jaws or true teeth they ingest small or soft food and have correspondingly small oral cavities. Mouth parts may be adapted for nibbling (most ostracoderms) or modified for clinging to a host and abrading its flesh (cyclostomes). Anaspids, some cephalaspids, and cyclostomes have a rasping organ on the floor of the oral cavity which is called a tongue but is not homologous with the tongue of higher vertebrates.

The oral cavity of antiarchs resembled that of ostracoderms. The oral cavity of other PLACODERMS was much enlarged in relation to the development of jaws.

The mouths and oral cavities of CARTILAGINOUS and BONY FISHES are extremely varied. Although fleshy lips are absent, the mouth parts may be highly protrusible or otherwise specialized. The oral cavity and pharynx are usually distensible. Gill bars are often provided with food strainers, grinding mills, or teeth. The basal elements of the visceral skeleton support a firm tongue which is little movable, yet may be provided with teeth and is the partial homolog of the tetrapod tongue. Since the food of fishes is always wet, no further lubrication need be provided and oral glands are restricted to scattered mucous cells.

TETRAPODS have oral cavities of moderate or large size, depending on feeding habits. A tongue is present and is supported by derivatives of the second and third (sometimes also

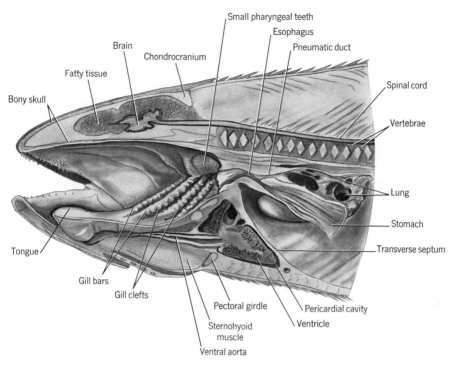

FIGURE 10-3
**SAGITTAL SECTION OF THE HEAD AND ANTERIOR BODY OF A HOLOS-
TEAN FISH, the bowfin,** *Amia*.

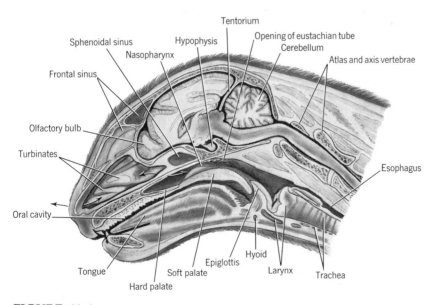

FIGURE 10-4
**SECTION OF THE HEAD AND NECK OF THE CAT cut just beside the
midsagittal plane.**

fourth) visceral arches. The tongue is usually fleshy and highly movable, but is relatively firm and fixed in many birds and some reptiles. It may function in securing food, manipulating food during chewing, swallowing, and sound control. Mammals have lips which are moved by derivatives of the platysma muscle. Lips may assist in taking food into the mouth.

Tetrapods have multicellular oral glands with ducts. They are named according to position as labial, lingual, palatine, nasal, maxillary, mandibular, pharyngeal, etc. However, the number, distribution, and detailed structure of these glands is diverse, and correspondence of name or position does not necessarily indicate homology. All tetrapods that are not secondarily aquatic require the secretions of oral glands to lubricate dry food. For some tetrapods the oral secretions seem to have no other function, but they facilitate tasting in frogs, and in some mammals (and to a lesser degree in certain other forms) a starch-digesting enzyme is present. Further, various animals have evolved special functions for oral glands. Certain glands of some snakes, lizards, and shrews become poison glands, blood-feeding bats secrete an anticoagulant, and the nasal glands of some marine birds and reptiles migrate to the orbits where they function in salt excretion.

Esophagus and Stomach. The fine structure of the alimentary canal is basically similar throughout its length. A general account will therefore precede brief remarks on the histology of the esophagus and stomach.

The gut is constructed in layers. The innermost principal layer is the **mucosa.** It consists of a surface **epithelium,** a deeper

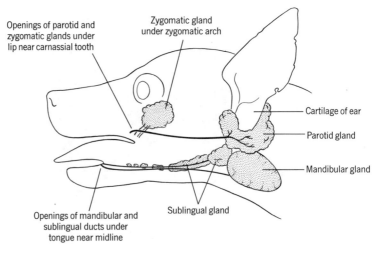

Openings of parotid and zygomatic glands under lip near carnassial tooth

Zygomatic gland under zygomatic arch

Cartilage of ear

Parotid gland

Mandibular gland

Sublingual gland

Openings of mandibular and sublingual ducts under tongue near midline

FIGURE 10-5
ORAL, OR SALIVARY, GLANDS OF THE DOG.

lamina propria, and a **muscularis mucosae.** Cells of the epithelium may be squamous, but usually are columnar with basal nuclei. Interspersed among the less specialized cells are mucus-secreting goblet cells and (according to region of the tract) unicellular or multicellular glands which secrete digestive juices. The epithelium has many folds when the tract is empty. These flatten out as the gut is distended. Microscopic projections called villi may be present on the folds, and numerous microvilli sometimes further increase the surface of the cells. The lamina propria is a network of loose tissue that underlies the epithelium and fills the cores of the villi. Next is the muscularis mucosae (not always present), a thin layer of smooth muscle which controls motions of the lining of the gut that are independent of the gut tube as a whole.

The second principal layer is the **submucosa,** which is a conspicuous stratum of loose connective tissue containing nerves, capillaries, lymphatic ducts and nodules, and ganglia of the parasympathetic nervous system. The larger crypts and glands of the epithelium may subtend into the submucosa.

Outside the submucosa is the third prominent layer, the **muscularis externa.** It consists of **smooth muscle,** which differs markedly from striated muscle. Its cells (or fibers) are spindle-shaped. Myofibrils are present but lack striations and are difficult to demonstrate. Each cell has a central oval nucleus.

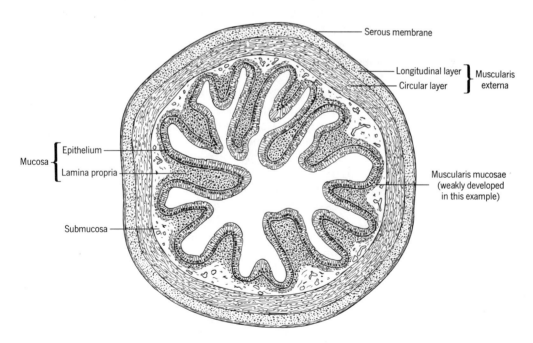

FIGURE 10-6
GENERAL STRUCTURE OF THE GUT TUBE as seen in cross-section.

Smooth muscle cells may be scattered individually in the dermis, but elsewhere are arranged in layers of cells of like orientation. In the gut, an inner portion of the muscularis externa is called the circular layer because its fibers are arranged in a tight spiral around the gut tube. Their contraction lengthens and constricts the gut. An outer portion is called the longitudinal layer. Its fibers are nearly longitudinal and serve to shorten the tract. The coordinated action of the two layers of the muscularis accomplish peristalsis and segmentation. Smooth muscle contracts and relaxes relatively slowly. It is involuntary and subject to control by the autonomic nervous system.

The pharynx, rectum, and part of the esophagus are bound directly to adjacent structures; other parts of the tract lie in the coelom and are enveloped in serous membranes.

The histology of the esophagus is distinctive in two ways. The much folded and highly distensible epithelium consists of stratified squamous cells which are cornified in animals that swallow coarse food. The muscularis externa of the anterior part of the esophagus (particularly in mammals) may have striated (hence voluntary) instead of smooth muscle fibers.

The lining of the stomach is divided into several regions which are distinctive in fine structure and function. They may be sharply demarcated or may merge. The most anterior, called the **esophageal region,** is of variable occurrence. Its fine structure is identical to that of the esophagus, but origin from the esophagus is doubtful. Like the esophagus, its only secretion is mucus. A **cardiac region** is present in mammals only. It likewise secretes only mucus, but the cells of its epithelium are columnar. The digestive region of the stomach is the **fundus.** Its lining is thickened by a dense layer of straight tubular gastric glands oriented perpendicular to the surface. Their open mouths form microscopic pits. The columnar cells of the glands are of several kinds. They secrete an enzyme which initiates protein digestion; hydrochloric acid, which provides the acidity necessary for the action of this enzyme; and sometimes also a fat-splitting enzyme. The necks of the glands secrete mucus. The stomachs of mammals also secrete rennin, which coagulates milk. The most posterior region of the stomach is the **pyloric region.** Its coiled tubular glands secrete mucus.

The muscularis externa of the stomach is relatively thick. Circular and longitudinal fibers are supplemented by oblique fibers. At the anterior end of the stomach, circular fibers form the **cardiac sphincter;** at the posterior end, the **pyloric sphincter** controls passage of chyme (digesting food) into the intestine.

Neither esophagus nor stomach can be identified in filter-feeders. They are also absent in CYCLOSTOMES. It is probable that OSTRACODERMS had little or no development of these organs.

The esophagus of FISHES is usually short and sometimes merges into the stomach. Commonly it has distinctive pleats or papillae directed toward the stomach. It is ciliated in most selachians. The stomach of fishes is usually either straight or bent into a J or U shape. It is particularly large in selachians. Esophagus and stomach have secondarily become vestigial or absent in holocophalians, dipnoans, and several kinds of bony fishes.

The esophagus of AMPHIBIANS is short, ciliated, and well-supplied with mucous glands. The stomach is straight or nearly so and relatively simple.

REPTILES tend to have a longer esophagus because of the increased length of the neck and more developed lungs. The esophagus tends to be ciliated if soft food is eaten, but is cornified in some turtles. The stomach remains simple and straight or gently curved in most reptiles, but is rounded and very muscular in crocodilians. It may have been highly modified in some extinct herbivorous archosaurs.

The esophagus of BIRDS is long. Its lining is usually cornified. At least some members of many families of birds have a permanent dilation of the lower part of the esophagus to serve as a storage organ called a **crop.** The dilation is usually ventral to the esophagus and sharply set off, but it may be dorsal instead and less distinct. The stomach of birds is in two parts. The anterior part, derived from the fundus but called the **proventriculus,** is very glandular and produces digestive enzymes. The posterior part corresponds to the pyloric region and is called the **ventriculus,** or **gizzard.** It may be exceedingly muscular for grinding coarse food (sometimes with the aid of pebbles eaten by the bird). Proventriculus and ventriculus are least distinct in carnivorous birds and most sharply demarcated in granivorous species. The ventriculus is usually a firm, somewhat rounded or lenticular chamber having adjacent openings into proventriculus and intestine. Rarely the ventriculus is vestigial.

The esophagus of MAMMALS is long, devoid of cilia, and cornified in such roughage eaters as artiodactyls, perissodactyls, and rodents. The stomach may be simple and sac-like (man, many rodents, some insectivores, carnivores) or complexly compartmentalized (artiodactyls, most whales, sloth, many marsupials, sea cows). An esophageal region is often present and is sometimes extensive. The entire stomach of monotremes is cornified, and most of the stomach of ruminants (cud

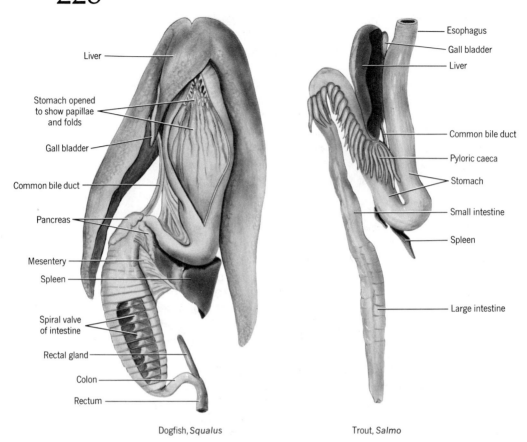

Liver

Stomach opened
to show papillae
and folds

Gall bladder

Common bile duct

Pancreas

Mesentery

Spleen

Spiral valve
of intestine

Rectal gland

Colon

Rectum

Esophagus

Gall bladder

Liver

Common bile duct

Pyloric caeca

Stomach

Small intestine

Spleen

Large intestine

Dogfish, *Squalus*   Trout, *Salmo*

FIGURE 10-7
**DIGESTIVE TRACTS OF AN ELASMOBRANCH (left) AND A TELEOST (right) that
feed, respectively, largely on flesh and insects. Ventral views.**

chewers) is of the esophageal type. A cardiac region, found
only in mammals, is characteristic but not universal. The thickened and glandular fundus is usually conspicuous and often
set off as a somewhat separate pouch or compartment. The
pyloric region is usually well-differentiated, but is less distinctive than in birds.

Intestine and Caeca.   Digestion is completed and foodstuffs absorbed in the anterior part of the intestine. Here the columnar
epithelium forms folds and microscopic villi which, in turn,
have microvilli on their exposed surfaces. Both unicellular and
tubular multicellular glands are abundant in the intestinal
lining. Their secretions contain a variety of enzymes, but seemingly more in higher vertebrates than in fishes. Hormones (secretin, pancreozymin, cholecystokinin, enterogastrone) may
also be secreted which influence the activities of stomach,

liver, and pancreas. Secretions of the liver and pancreas empty into a portion of the gut close behind the stomach which is called the **duodenum.** The combined intestinal juice is alkaline.

Water is absorbed and feces formed in the shorter posterior part of the intestine. Villi and microvilli are usually absent from this region. Mucous cells are abundant, and lymph nodules are often present in the submucosa. The part of the intestine that passes out of the coelom and through the pelvic girdle is called the **rectum.** The posteriormost end of the gut of many vertebrates is a **cloaca,** or common chamber for wastes of the digestive and urinary systems and for products of the gonads.

Anterior and posterior regions of the intestine are relatively distinct in tetrapods where they are called, respectively, the **small** and **large intestines** because of the larger diameter of the latter. Further, tetrapods have one or two pouch-like diverticula at the juncture between small and large intestines. These are

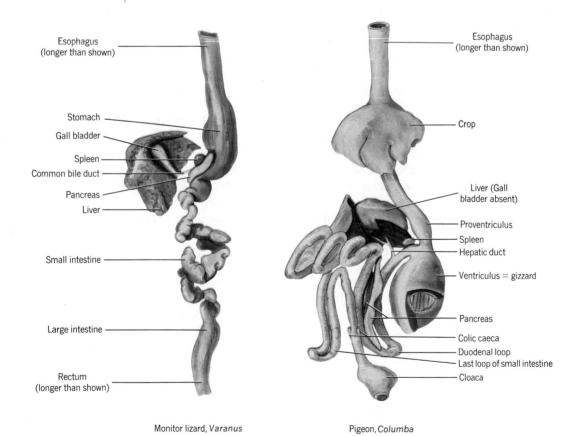

Monitor lizard, *Varanus*   Pigeon, *Columba*

FIGURE 10-8
DIGESTIVE TRACTS OF A REPTILE (left) and a GRANIVOROUS BIRD (right). Ventral views.

called **colic caeca.** Histologically the caeca are distinctive but similar to the large intestine. Primitively their function may have been merely to increase the surface of the gut; now they function variously for storage, fermentation, or vitamin concentration.

The length, pattern of looping, specializations, and fine structure of the intestine are exceedingly varied. The intestine is relatively long in very large animals (because of surface–volume relationships as explained in Chapter 1). In general, carnivorous animals have short intestines, long villi, and small caeca, whereas herbivores have long intestines, shorter villi, and large caeca. Exceptions are many, however, particularly among fishes and birds. Much remains to be learned about the functional morphology of the digestive system.

The intestine of CYLOSTOMES runs straight from pharynx to cloaca. Regional differentiation is slight, only a short rectum being set off from a nearly uniform intestine. These features, seen also in amphioxus, are doubtless primitive for vertebrates. In lampreys, a single marked fold of the intestinal mucosa runs lengthwise of the gut in a gentle spiral. This is the analog, if not also the homolog, of the spiral valve described below.

The nature of the digestive tract of SELACHIANS is relatively constant. The gut, now too long to run straight through the coelom, is N-shaped. One angle lies in the long stomach and one in the intestine. The three limbs run lengthwise in the body. Most of the posterior limb comprises the **spiral intestine,** or **spiral valve.** This part of the gut is large in diameter and tapers at each end. The mucosa is thrown into a single prominent fold as in the lamprey, but here the attachment of the fold to the wall of the intestine spirals more tightly as it runs down the tract. The structure of the organ resembles a spiral staircase in a circular tower. The number of turns ranges among the species from 5½ to 50. The functional advantage is great increase in the surface area of the epithelium. An equivalent increase could be achieved by lengthening and coiling the entire intestine, but spiral structure of the mucosa is better adapted to the long slender body cavity of the fish. The structure just described is modified in one family of sharks: The attachment of the mucosal fold to the intestinal wall does not spiral but instead runs lengthwise. The height of the fold, however, exceeds the diameter of the intestine, so the fold rolls up into a scroll having about 2½ turns.

Some selachians have a short segment of intestine of relatively small diameter between the stomach and spiral intestine. A short rectum joins the cloaca. A dorsal appendage from the rectum is called the **rectal gland.** Its function is salt excretion.

The gut of CHIMAERAS is somewhat angled anteriorly but in the absence of a stomach is not N-shaped. A cloaca is absent. There is no rectal gland, but similar glandular tissue is found in the wall of the rectum.

Turning back to PLACODERMS, it is of importance that one fossil of an antiarch shows the imprint of a spiral intestine. Further, some "coprolites" are in fact fossilized, spiral intestines of extinct fishes. This complicated organ is clearly primitive among vertebrates.

The intestine of BONY FISHES is more variable than that of cartilaginous fishes. It is rarely straight, commonly thrown into one or two S curves, and occasionally coiled. Its length may be less than that of the body but is usually somewhat longer than in cartilaginous fishes and reaches twelve body lengths in some species. A spiral intestine is present in all modern bony fishes except teleosts (and several of these have remnants of the valve). Ray-finned fishes are distinctive for another structure that increases the surface of the intestine; adjacent to the stomach the intestine develops diverticula called **pyloric caeca.** Most fishes have scores or hundreds of caeca, but some have few and several have none. The tubular caeca may open into the intestine individually or may cluster to form a compound organ. Histologically they resemble the adjacent intestine. Among bony fishes, only dipnoans and the surviving crossopterygian have a cloaca. There is no rectal gland.

Tadpoles have long coiled intestines, but adult AMPHIBIANS have relatively short and simple digestive tracts ranging in length from $1/2$ to $3^{1}/4$ times the length of the body. As in other tetrapods, a coiled small intestine is set off from a shorter large intestine. At the boundary between the two there may be a single small colic caecum. This structure is apparently vestigial in many surviving amphibians, yet its presence in the class must be considered primitive. A cloaca is present.

The intestine is straight in most snakes and amphisbaenians but is otherwise moderately coiled in most REPTILES. Its length usually ranges from $1/2$ to 2 times the body length but tends to be longer in turtles. Small and large intestines are distinct. A dorsal colic caecum of small or moderate size is present in many species but has secondarily been lost by others.

The duodenum of BIRDS always forms a long narrow loop which lies ventral in the body cavity and is tightly joined to the pancreas. The remainder of the small intestine is relatively long and forms various complicated patterns of folds and coils which are constant within families. The large intestine is short, nearly straight, and villous. It is in a dorsal position. Two colic caeca are usual—rarely one or more than two are present. These are finger-like in form, often of considerable length, and

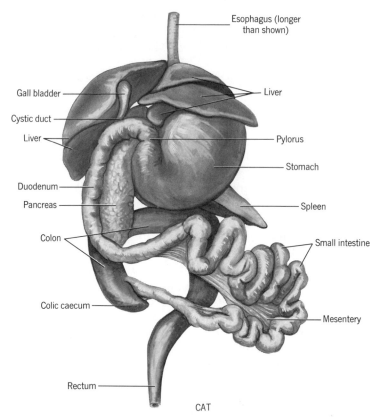

Esophagus (longer than shown)

Gall bladder

Cystic duct

Liver

Duodenum

Pancreas

Colon

Colic caecum

Rectum

Liver

Pylorus

Stomach

Spleen

Small intestine

Mesentery

CAT

FIGURE 10-9
DIGESTIVE TRACT OF A CARNIVOROUS MAMMAL.
Ventral view.

join the gut laterally or ventrally. A cloaca is present and has a dorsal diverticulum called the **cloacal bursa** (or bursa of Fabricius) which functions in the formation of antibodies.

The gross morphology of the intestine of MAMMALS is highly variable and correlates with diet and systematics only in a general way. The intestine may be as short as 2–6 body lengths (many insectivores and carnivores) or as long as 20–25 body lengths (some artiodactyls and marine mammals). A duodenal loop is usually present but is less tightly bound than in birds. Further, compared to birds, the pattern of folding of the small intestine is less regular and the large intestine tends to be longer and more bulky—particularly in herbivores. Several unrelated mammals have paired colic caeca. This may be the primitive condition, but a single ventral caecum is the rule. It may be small (even secondarily absent) but can be two or more times the length of the body. Caecum and adjacent intestine form tightly spiraled coils in some rodents, lemurs, and certain other mammals (see Figure 25-21). The lining of the caecum and large intestine of some mammals (and more rarely of reptiles and birds) forms a series of transverse folds or a single spiral fold which is reminiscent of the intestinal valve of fishes but here is less developed internally and more evident externally as a series of shallow bulges along the gut. Only monotremes among mammals have a cloaca.

Liver and Gall Bladder.   The vertebrate liver is unique to the subphylum and varies little among the classes. It is the largest organ of the body. The functions of the liver are many and diverse: It is a storage depot for carbohydrate and (particularly in cyclostomes and fishes) for fats. It converts protein to carbohydrate or fat with the release of nitrogenous waste which is transported to the gills or kidneys for elimination. It elaborates much of the yolk which the maternal body transfers to the growing eggs. The embryonic liver (also the adult organ of fishes) produces blood cells, and the adult liver removes "old" red cells from the blood stream. Various toxicants can be removed from the blood by the liver, and substances needed for clotting are released. Several vitamins are manufactured or stored. Finally, the function that relates the liver to digestion—bile is secreted into a duct system and delivered to the duodenum where it emulsifies fats, thus making them digestible by pancreatic enzymes.

Liver and gall bladder develop from one or (usually) two ventral **hepatic diverticula** from the gut tube just posterior to the stomach. With few exceptions (birds), the more posterior diverticulum forms the gall bladder, and the anterior diverticulum (which may be somewhat paired in fishes) branches and expands to become the liver. However, liver tissue may also originate in adjacent tissues and migrate *toward* the gut, in which case the secretory tissue will be of mesodermal as well as entodermal origin.

Blood coming to the liver from the viscera in the hepatic portal system seeps through specialized capillaries called **sinusoids** before collecting in the hepatic veins to exit from the organ. The sinusoids are suspended in a labyrinth of tunnels which weave within a meshwork of interconnecting cellular walls. Sphincters control blood flow in the sinusoids. The walls between sinusoids are one cell thick in mammals and some birds; they are two cells thick in other vertebrates. Adjacent cells are held together by the tiny bile channels which follow their borders and also by projections which fit like snap fasteners into holes in adjacent cells. In higher vertebrates the vessels and ducts of the liver are arranged into polyhedral units called **lobules.**

The hepatic diverticulum of amphioxus is a single hollow organ which lies against the right side of the gut. Its fine struc-

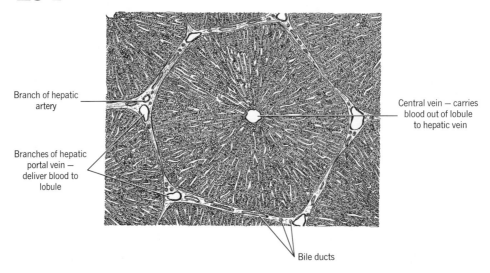

Branch of hepatic artery

Central vein — carries blood out of lobule to hepatic vein

Branches of hepatic portal vein — deliver blood to lobule

Bile ducts

FIGURE 10-10
ONE LIVER LOBULE AND PARTS OF ADJACENT LOBULES as seen in cross-section.

ture and functions are unlike those of the vertebrate liver with which it is homologous, if at all, only in a general way.

The liver tends to be lobed, and two principal lobes are frequent. However, the lobes may be side by side or one in front of the other, and the organ can have no lobes or several arranged in a variety of patterns which are without known functional or systematic significance.

Bile is usually prevented from entering the gut unless chyme is leaving the stomach. At other times it backs up into the gall bladder which stores it temporarily and concentrates it as much as ten times. The **cystic duct** from the gall bladder joins the **hepatic duct** from the liver to continue to the duodenum as the **common bile duct.** The amount of bile secreted in relation to body weight varies widely among vertebrates. The presence and size of the bladder correlates with rate of bile secretion, intermittence of feeding, and fat content of the diet. The organ is always present in carnivores. It is lacking in the adult lamprey, several teleosts, and in certain herbivores distributed in five families of birds and six orders of mammals.

Pancreas. The pancreas is a pale-colored organ that lies adjacent to the duodenum. It is found only in vertebrates, and all vertebrates have a pancreas. It is always a compound organ having both **exocrine** (secreting into a duct system) and **endocrine** (secreting into the blood) functions. About half a dozen enzymes are present in the pancreatic juice. These are able to digest nearly all foodstuffs. The endocrine secretions, insulin and glucagon, are essential for control of the intermediate metabolism of carbohydrates.

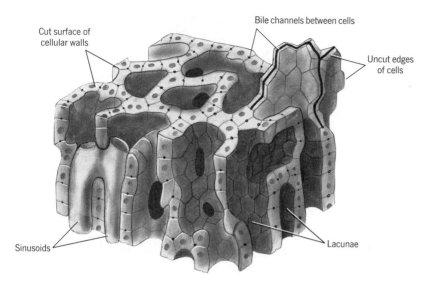

Cut surface of cellular walls

Bile channels between cells

Uncut edges of cells

Sinusoids

Lacunae

FIGURE 10-11
STEREOGRAM OF A FRAGMENT OF MAMMALIAN LIVER.

The pancreas has its origin in aggregations of cells in the wall of the embryonic foregut at the level of the hepatic diverticula. These form a dorsal pancreatic diverticulum and a pair of ventral diverticula. Growth brings the branching diverticula together and they contribute in various combinations to the adult organ; ventral derivatives are stressed in amphibians and reptiles, whereas one or both may degenerate in fishes. The left ventral diverticulum tends to degenerate in mammals. The dorsal diverticulum contributes part of the exocrine tissue and seemingly all of the endocrine tissue. It is apparent from this complicated ontogeny why the pancreas may have one, two, or three ducts with considerable individual as well as systematic variation.

The exocrine tissue of the pancreas is tubular, branched, and alveolar. Individual cells are pyramidal. Those at the ends of the alveoli contain granules which are converted to enzymes when released. The endocrine tissue consists of several kinds of cells which form small scattered aggregations called **islets of Langerhans.**

Amphioxus has no pancreas, but isolated cells of the exocrine type are found in the wall of the gut around the opening of the hepatic diverticulum. In cyclostomes the pancreatic cells form aggregates under the duodenal epithelium. Dipnoans do not have a discrete pancreas but the organ is present in the intestinal wall. The organ is free and compact in selachians. It is diffuse in teleosts, being scattered as flecks of tissue in the mesenteries or along the blood vessels that penetrate liver and spleen. The pancreas of birds is somewhat three-lobed and usually has three ducts. In other tetrapods the organ is more

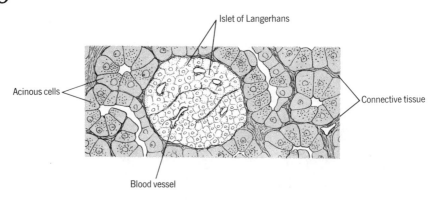

Islet of Langerhans

Acinous cells

Connective tissue

Blood vessel

FIGURE 10-12
SECTION OF THE PANCREAS OF A MAMMAL.

compact and usually has two ducts or one, which may join the common bile duct.

Comparative Anatomy of Coelom and Mesenteries

Transitory coelomic cavities occur in the myotomes and sclerotomes (Figures 5-3 and 8-3). Similar cavities may occur in the ventral portions of the visceral arches and in the developing tail musculature of embryos (and adults of several fishes). From these clues it may be inferred that the coelom was once more extensive. However that may be, the functional coelom of surviving vertebrates occurs only within derivatives of the hypomere and extends at most from the level of the posterior part of the pharynx to the cloaca. Initially, in both phylogeny and ontogeny, the gut tube is straight and is supported dorsally and ventrally by continuous mesenteries established from splanchnic mesoderm as right and left coelomic cavities expand (Figure 5-4). This simple structure becomes much complicated in adults of most vertebrates by partitioning of the coelom and by deletions, folding, and fusions of the mesenteries.

The mesenteries of AGNATHS and FISHES remain virtually straight and complete (dipnoans), are nearly absent except for portions related to stomach and liver (lampreys, selachians), or range between these extremes. Fusions and moderate folding are usual. Mesenteries of fishes are commonly pigmented—supposedly to protect light-sensitive gonads.

The heart of fishes lies anterior to the pectoral girdle and ventral to the posterior gill chambers. In hagfishes a tough **transverse septum** extends vertically upward from the ventral body wall posterior to the heart, partly separating an anterior **pericardial cavity** from a larger **peritoneal cavity.** In selachians the developing transverse septum temporarily separates the two cavities and then secondarily develops small orifices. Lampreys and other fishes retain a complete septum. (Figure 10-3).

These basic relationships have not been modified by URO-DELES. The dorsal mesentery is nearly complete. The small pericardial cavity remains far forward where it is separated by a transverse septum from the principal coelom which may now be called a **pleuroperitoneal cavity** because slender lungs are present. These are anchored to the lateral body wall by **lateral mesenteries.**

The heart of OTHER TETRAPODS lies at the level of the pectoral girdle or posterior to the girdle. The lungs are dorsal to the heart. (Parts of the liver intervene between heart and lungs in birds and some reptiles but not in mammals.) The heart is separated from the lungs (and liver if present) by more or less horizontal partitions which have their origin in the embryo as folds in the serous membrane of the right and left lateral body walls. These grow out to join in the midline of the body. They are called lateral mesocardia (birds) or **pleuropericardial membranes.** Posteriorly they join the transverse septum to form the adult pericardial membrane, or **pericardium.**

CROCODILIANS, SOME LIZARDS, AND BIRDS partition the pleuroperitoneal coelom into additional cavities by complex outgrowths and fusions of the mesenteries. The thoracic air sacs of birds separate a ventral oblique septum from a dorsal pulmonary diaphragm which is supplied with striated muscle. Their lungs grow up against the dorsolateral body walls, thus obliterating the pleural cavities in the adult.

MAMMALS separate paired **pleural cavities** from the peritoneal cavity by a **diaphragm.** The ventral portion of this organ is derived from the transverse septum. The dorsal portion is derived from the dorsal mesentery and from still another pair of outgrowths from the lateral body wall, these being called **pleuroperitoneal membranes.** The striated muscle of the diaphragm is derived from cervical myotomes (and is innervated by cervical nerves) because it develops at that level in the embryo and then migrates posteriorly as the neck lengthens and the lungs expand.

The embryonic liver of tetrapods starts to grow within the transverse septum. As it enlarges, it bulges out of the septum posteriorly and finally separates more or less completely from the developing diaphragm, trailing the **coronary ligament** behind. As the liver grows out of the septum, it grows into the ventral mesentery. The part of the ventral mesentery extending from liver to ventral body wall is the **falciform** (= sickle-shaped) **ligament;** the part between liver and gut tube is the **lesser omentum.** A small portion of another part of the ventral mesentery may anchor the bladder to the body wall. Between liver and bladder the ventral mesentery of tetrapods is missing. The dorsal mesentery is more complete and much complicated by folding and fusions. Between the stomach and body wall the dorsal mesentery of mammals becomes extended into a saclike **omental bursa** (= membrane + purse).

The pericardial cavity of anurans and reptiles is bordered dorsally by the pericardial membrane and ventrally by the body wall. In birds, the growing liver forces its way between body wall and membrane, thus wrapping the pericardial membrane nearly around the heart. In mammals, growth of the lungs does the same thing, this time removing the pericardial cavity entirely from the body wall (Figure 10-13). Right and left pleural cavities are then separated by dorsal and ventral mesenteries and by the pericardial membrane. The combined partition is called the **mediastinum** and is unique to mammals.

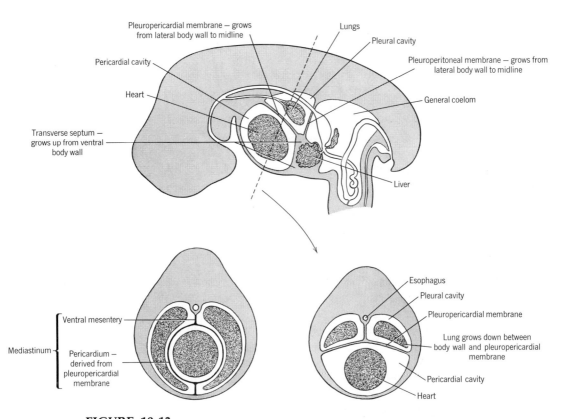

FIGURE 10-13
PARTITIONING OF THE COELOM IN A MAMMAL seen in sagittal section (above) and cross-sections (below). These developmental stages follow those shown in Figure 10-1.

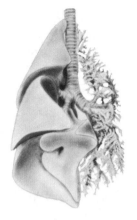

Respiratory Systems and Gas Bladder

The basic requirement of a respiratory organ is a surface through which gas exchange between blood and the external water or air can take place by diffusion. The surface must remain moist to let the cells live, must be large enough to permit sufficient gas flow, and must be thin enough to allow adequately rapid diffusion. Since the rate of diffusion also depends on the difference in oxygen and carbon dioxide concentration across the membrane, there is great advantage to perfusing the respiratory organ with blood that is relatively high in carbon dioxide content and low in oxygen. Similarly, there should be a rapid exchange of water or air on the other side of the membrane. The circulatory system controls the flow of

239

blood, and the muscular and skeletal systems control the flow of water or air, often in conjunction with respiratory ducts.

The principal respiratory organs of postembryonic vertebrates are gills and lungs, although the skin and several other structures are sometimes also adapted for respiration. Gas bladders are probably homologous with lungs. Gills and lungs are related only by the serial homology of embryonic primordia; they relate to virtually different systems.

The rather complicated structure of gills and lungs is distinctive at class and sometimes at subclass levels. At lesser taxonomic levels there is considerable variation, some of which is adaptive and some apparently not. Particular interest (and some remaining problems) center on the evolution of respiratory organs during the transition to terrestrial life.

Development and Structure of the Pharynx

Certain larval vertebrates have gills that develop from surface ectoderm and extend beyond the contour of the head. These are called external gills and are discussed in a subsequent section. Most gills lie within the contour of the head and are called internal gills. Internal gills develop in relation to the pharynx.

The pharynx forms from the portion of the foregut that lies between the developing oral cavity and the esophagus. It is lined by entoderm. The lateral walls of the embryonic pharynx develop six or more pairs of evaginations termed **pharyngeal pouches.** Opposite to the pouches, on the outside of the body, are shallow indentations called **visceral grooves.** Pouches and grooves are separated by thin partitions termed **closing plates.** Adjacent pouches are separated by **visceral arches,** each pair of which contains its respective pair of arteries or **aortic arches.**

Subsequent development depends upon the presence or absence of jaws, gills, and related structures in the adult. The first ancestral visceral arch and pharyngeal pouch are apparently retained by cephalaspids and perhaps by pteraspids. They are questionably present in other agnatha and are absent from gnathostomes. The true second arch and pouch are therefore universally numbered the first. The "first" pharyngeal pouch becomes a typical **gill chamber** in agnatha. In other fishes it is lost or is reduced in size and modified in function to become the cavity of the **spiracle.** Succeeding pouches of agnatha and fishes form gill chambers. The first pouch of tetrapods becomes the cavity of the middle ear; more posterior pouches form gill chambers in larval amphibians but otherwise lose their pouch-like character after contributing glandular and lymphatic tissues from their epithelial linings. At the placoderm level of evolution, however, a posterior pair of pouches probably became the primordia of air bladders and lungs.

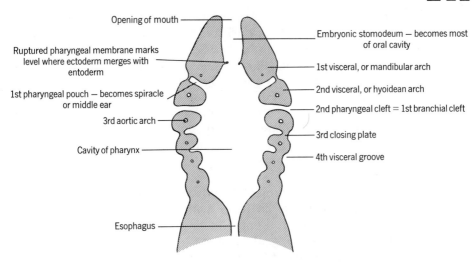

Opening of mouth

Embryonic stomodeum — becomes most of oral cavity

Ruptured pharyngeal membrane marks level where ectoderm merges with entoderm

1st visceral, or mandibular arch

2nd visceral, or hyoidean arch

1st pharyngeal pouch — becomes spiracle or middle ear

2nd pharyngeal cleft = 1st branchial cleft

3rd aortic arch

3rd closing plate

Cavity of pharynx

4th visceral groove

Esophagus

FIGURE 11-1
FRONTAL SECTION OF THE HEAD AND PHARYNX OF THE LARVA OR EMBRYO OF A JAW-BEARING VERTEBRATE at a developmental stage corresponding to that of the lower drawing in Figure 10-1. Compare also with Figure 7-4.

The closing plates of gill-bearing vertebrates rupture in the embryo to establish communication between the gill chambers and the outside of the body. Tetrapods retain the first closing plate as the eardrum; succeeding closing plates and visceral grooves of amniotes and of most adult amphibians lose their identity and leave no derivatives.

The first visceral arch (the mandibular arch) becomes the jaws of gnathostomes. The second (or hyoid) arch usually supports the jaws in fishes and contributes to the middle ear in tetrapods. As was explained in Chapter 7, the more posterior visceral arches support the **gill bars** of fishes and have various derivatives in tetrapods. The respiratory epithelium of gills develops from the margins of visceral arches near the positions of the embryonic closing plates; it is considered to be entodermal in cyclostomes but probably is more often ectodermal in gnathostomes. As explained in Chapter 9, musculature related to derivatives of visceral arches develops from the branchial portion of the hypomere; that of the floor of the pharynx is hypobranchial in origin.

General Structure and Function. The basic structure of internal gills is remarkably similar for all fishes. Each gill bar consists of a part of the visceral skeleton, the branchial derivatives of the corresponding aortic arch, the associated cranial or cervical nerve, intrinsic branchial muscles, and related epithelium. The

Gills and Gill Chambers

bars of some fishes are continued by **gill septa** of supportive tissue which extend toward the surface of the body. On their pharyngeal (inner) margins most bars carry one or more rows of stiff strainers called **gill rakers.** A typical gill bar bears two rows of **gill filaments** which are on opposite sides of the septum if a septum is present. Each row may be likened to a comb, with the long delicate filaments corresponding to the teeth of the comb. A bar with filaments (and septum if present) forms a partition between adjacent gill chambers. It follows that the anterior row of filaments faces into one chamber whereas the other row faces into a succeeding chamber (see Figure 11-3). The filaments are subject to muscular control.

The gill filaments, in turn, bear, on opposite surfaces, tiny parallel **gill lamellae** numbering in each series about 20/mm of filament but ranging at least from 11 to 31/mm of filament. Lamellae of adjacent filaments touch or even interdigitate, and water passing over the gills must seep through the minute crevices of the mesh thus formed. The total respiratory surface per unit weight of fish varies about tenfold according to the activity of the species. For a sea bass, which has an intermediate value, an individual weighing 20 kg has a respiratory surface of about 91,600 cm² (60 ft² for a fish of 44 lb).

An **afferent branchial artery** (afferent = toward + to bear) enters each gill bar from below. As it extends upward it gives off **filamental vessels,** one of which loops to the apex of each filament and returns to drain into an **efferent branchial artery** (efferent = away + to bear) which continues upward and out of the gill. Capillary beds arch across the loops of the filamental

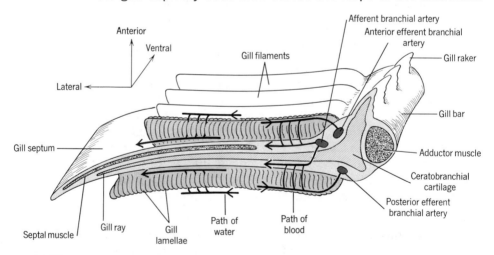

FIGURE 11-2

THE STRUCTURE OF GILLS illustrated by an elasmobranch fish. Somewhat stylized: gill filaments are longer, and gill lamellae are more numerous than shown; blood capillaries within lamellae are not shown.

vessels and extend into the lamellae.The epithelium covering the lamellae is a single cell layer in thickness.

Lamellae are oriented parallel to the stream of water which always passes between them from the inside of the gill apparatus to the outside. Blood always flows in the lamellae from the outside of the apparatus inward. Consequently, when blood first enters the area of respiratory exchange it contacts water that is already depleted in oxygen. This blood, however, has minimum oxygen content, so a gradient exists and oxygen immediately diffuses into the blood. As the blood continues through the lamellae its oxygen content increases, but so does the oxygen content of the ever "fresher" water with which it comes in contact. Thus, a diffusion gradient is maintained and gaseous exchange continues until the blood leaves the lamellae. The mechanism is so efficient that up to 80% of the oxygen in the water is removed. The diffusion of carbon dioxide from blood to water is simultaneously favored by the same mechanism. The transfer of heat or diffusible materials between currents of gas or liquid passing one another in opposite directions is called **countercurrent exchange.** It was perfected several hundred million years ago in the gills and gas bladders of fishes and, as will be explained elsewhere in this book, is the basis for other important adaptations of vertebrates. Human engineers started to use countercurrent exchange somewhat more recently, e.g., for industrial purposes involving heat exchange.

A single gill bar with its anterior and posterior rows of respiratory filaments is called a **holobranch** (= entire + gill). If a bar bears filaments on one surface only, it is a **hemibranch** (= half + gill). The filaments on the posterior surface of the mandibular arch (facing into the first or spiracular gill chamber) are often modified to serve a nonrespiratory function and then comprise a **pseudobranch** (= false + gill).

The universal respiratory function of gills is exchange of oxygen and carbon dioxide between blood and environment. Gills also function importantly in excretion and osmoregulation. Bony fishes excrete nearly all their nitrogenous waste from the gills. It is transported in the blood as urea but is mostly converted in the gills to ammonia which is lost by diffusion. Gills of freshwater bony fishes passively admit water and actively absorb salts; gills of marine bony fishes pass little water and actively excrete salts. Cartilaginous fishes are unique in that their gills are nearly impervious to urea which is retained in high concentration in the blood and tissues. These differences among the fishes will be correlated with kidney function and osmoregulation in Chapter 13.

Comparative Anatomy of Internal Gills. The ancestors of vertebrates were probably filter-feeders with visceral arches specialized for trapping food particles in sticky mucus. To judge by cephalochordates and urochordates, such arches were numerous, delicate, and surrounded by a single large chamber through which water was circulated. Certain cephalaspids and pteraspids also had large gills and may have filtered or strained some of their food. However, the visceral arches of all vertebrates number fewer than twenty, and their gills, when present, function primarily for respiration.

Gills are commonly classified in three categories according to arrangement and relation to supportive and protective tissues. Some gills, however, are intermediate in nature.

Pouched gills are characteristic of agnatha. The gill filaments are arranged over the surface of discrete, spherical or lenticular, pouch-like gill chambers (Figures 11-3 and 11-4). Each pouch may have its own external pore (never a large slit) to the surface of the body (cephalaspids, anaspids, lampreys), or the pouches on each side of the body may communicate with the outside by a common duct and pore (pteraspids, some hagfishes). Similarly, each pouch characteristically has its own internal pore to communicate with a typical pharynx (hagfishes, probably cephalaspids) or with a division of the pharynx that is separate from the path of food (lampreys).

Pteraspids had 5 to at least 9 pairs of pouched gill chambers, cephalaspids had 10 or 11 pairs, and anaspids had 8–15. Lampreys have 7 pairs of pouches, whereas hagfishes have 6–15; the number is somewhat variable even within species. The point to note is the range and variability, not the exact numbers.

Ostracoderms probably took water into the pharynx by way of the mouth, passed it through the gill pouches, and expelled it from the head through the branchial pores. Cyclostomes also expel water through the branchial pores, but are unique in that the specialized use of the mouth for feeding and prehension precludes its constant use in respiration. Instead, water may enter and leave the gill pouches from the external branchial pores (flow is thus tidal) though water probably passes the lamellae in only one direction. The visceral skeleton is a continuous, unjointed lattice of cartilage. Water is actively expelled from the pouches by muscular contraction; it reenters the pouches as the result of the inherent resilience of the visceral skeleton.

Virtually nothing is known of the visceral arches and gill chambers of placoderms; their gills cannot yet be classified.

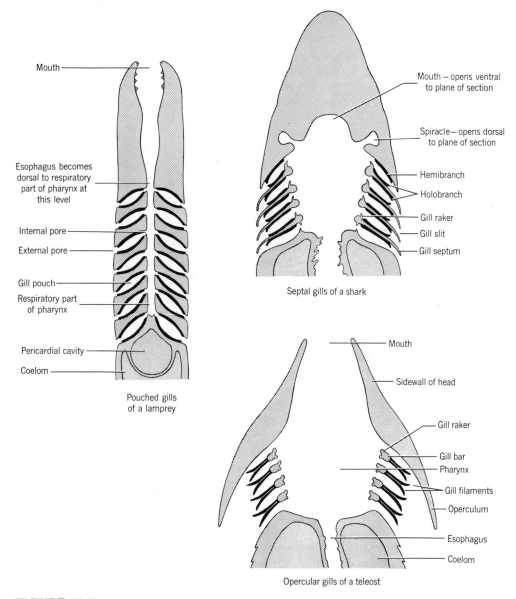

Mouth

Esophagus becomes
dorsal to respiratory
part of pharynx at
this level

Internal pore

External pore

Gill pouch

Respiratory part
of pharynx

Pericardial cavity

Coelom

Pouched gills
of a lamprey

Mouth — opens ventral
to plane of section

Spiracle — opens dorsal
to plane of section

Hemibranch

Holobranch

Gill raker

Gill slit

Gill septum

Septal gills of a shark

Mouth

Sidewall of head

Gill raker

Gill bar

Pharynx

Gill filaments

Operculum

Esophagus

Coelom

Opercular gills of a teleost

FIGURE 11-3
TYPES OF VERTEBRATE GILLS shown by frontal sections of the head and pharynx.

Septal gills differ from pouched gills in that the gill chambers
tend to be larger, to communicate more widely with the
pharynx internally, and to communicate with the outside of the
body through vertical **gill slits** instead of by pores. The serial
slits cause the supportive tissue that joins the gill bars to the
surface of the body to be in the form of plate-like **gill septa**

which share in the support of the gill filaments. This is the basis for the term **elasmobranch** (= plate + gill) which is applied to fishes with septal gills. These are Selachii, Cladoselachii, and Pleuracanthodii.

The first gill chamber and cleft of elasmobranchs is reduced to a spiracle. On the anterior face of the spiracular chamber is a vascular pseudobranch which receives oxygenated blood and apparently monitors it in some way before delivering it to the eye.

Elasmobranchs nearly always have a hemibranch on the anterior face of the first branchial chamber and then four holobranchs. There are four posthyoidean gill-bearing arches and five gill clefts. Three genera of sharks have six clefts and one has seven. Perhaps the greater number is the ancestral condition, but this is not known. Stability, not variability, in number of clefts should be stressed for elasmobranchs.

Sharks usually draw water into the mouth and pharynx by action of hypobranchial muscles. (Alternatively, actively swimming sharks can merely open the mouth and allow water to flow in.) The same action passively closes the valve-like external gill slits. The gill chambers are next enlarged by branchial muscles as the oral cavity is compressed. Either the jaws are closed or the mouth is functionally sealed by oral valves. Thus, the oral cavity changes from suction pump to force pump and moves the water through the gill filaments to the portions of the gill chambers that are external to the gills. When the pressure there surpasses the outside pressure, the gill slits open, water is forced out, and the cycle is completed. The entire sequence can be reversed, thus forcing jets of water out of spiracles or mouth to clean the gills or reject food. When rays rest on the bottom, their ventral mouths are against the substrate. They may then suck water into the pharynx through their large, dorsally positioned spiracles.

Holocephalians have gills that are intermediate between septal gills and those of the next category. Septa are present but are much reduced so that effective serial gill slits are absent. Instead, a unique fleshy operculum covers the gills. These fishes are also distinctive for having an anterior hemibranch, three instead of four holobranchs, and then a posterior hemibranch. The fifth cleft is closed.

Opercular gills are characteristic of Osteichthyes. Septa are very short in Dipnoi and otherwise are virtually absent in opercular gills; only the gill bars remain to anchor the gill filaments. A bony operculum is present to protect the otherwise exposed filaments and to contribute to the pump and valve system of the gill apparatus.

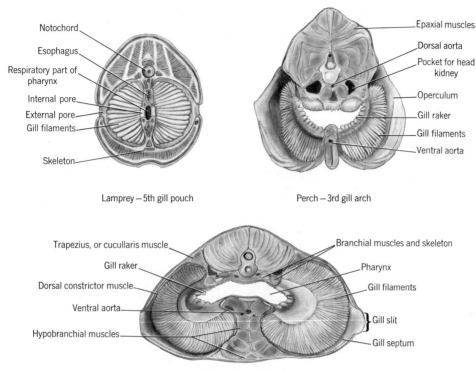

Lamprey — 5th gill pouch

Perch — 3rd gill arch

Shark — 3rd gill chamber

FIGURE 11-4
POUCHED, SEPTAL, AND OPERCULAR GILLS as seen in posterior views of cross-sections of the head made through the pharynx.

A spiracle is usually present in Chondrostei, and these fishes have a pseudobranch on the anterior face of the spiracular chamber. Other bony fishes lose the spiracle but may retain the pseudobranch which then fills a depression on the wall of the pharynx or moves back to the anterior face of the first branchial chamber. The structure of pseudobranchs is various and their function is not adequately known: They may have a reduced number of free filaments, may be covered by epithelium, or may assume a glandular character and sink well below the surface. They are thought to contribute to the special metabolic requirements of the eye and other organs in ways that include enzymatic control of the dissociation of carbonic acid in the blood.

A hemibranch is present on the anterior face of the first branchial chamber (posterior wall of hyoid arch) of Dipnoi and some Chondrostei, as it is also in elasmobranchs. Otherwise this hemibranch is missing in bony fishes, although, as just noted, the pseudobranch of the spiracular chamber may move to a corresponding position when its "own" chamber is lost.

The functional gills of bony fishes usually consist of four holobranchs (starting with the third visceral, or first branchial arch). The complement is reduced to three, two, and even one holobranch (plus a posterior hemibranch in some instances) by some Dipnoi and some Teleostei for which the gas bladder has come to function in respiration.

We have seen that ventilation of septal gills tends to be in three phases: Water enters the oral cavity, then passes over the gills to the lateral portions of the gill chambers, and then is expelled. Ventilation of opercular gills is similar except that the second and third phases tend to be combined. The first phase is relatively short, so the passage of water over the gills is nearly continuous.

Few trends can be noted in the evolution of gills—particularly when it is remembered that septal gills as exemplified by elasmobranchs are not in the ancestry of opercular gills (although similar gills of unknown placoderms may have been ancestral). There has been a general tendency to reduce and to stabilize the number of gill bars and chambers. Within the Osteichthyes, but not the Chondrichthyes, there was an early trend to loss of the spiracle and loss of septa (but among amphibians, some larval caecilians and several salamanders still retain a spiracle!) Other variations relate to habit or are seemingly nonadaptive (for instance, the determination of which holobranchs will be retained as the series is reduced).

External Gills. External gills develop from skin ectoderm of the branchial area but are not directly related to the visceral skeleton or branchial chambers. They are filamentous or feather-like, and their epithelium may be ciliated. Blood supply is indirectly from the second aortic arch if there is a single pair of gills and from several aortic arches if there are more. Muscles are present to move the gills.

External gills occur in the larvae of two of the three Recent genera of dipnoans, the chondrostean *Polypterus,* one teleost, and amphibians. They occur also in adults of **perennibranchiate** (= through the year + gill-bearing) urodeles, which, however, are neotenous in this respect (review p. 23). The principal function of external gills is respiratory, but they develop early in ontogeny—before the pharyngeal slits are formed—and in some instances absorb yolk or other nutrients before the fish hatches. It is clear that external gills are a larval adaptation which evolved early in the phylogeny of bony fishes; they probably occurred in crossopterygians. There is no evidence that external gills ever occurred in adult fishes.

The terms internal and external gills are not always apt. Some larval rays have extensions from their "internal" gills

African lungfish, *Lepidosiren*

Salamander, *Siren*

FIGURE 11-5
EXTERNAL GILLS in larvae of a fish and amphibian seen, respectively, in lateral and dorsal views.

which stream far out of the gill slits. Young tadpoles of anurans have "external" gills which are hidden under a fleshy operculum. Some older tadpoles have gills of uncertain homology.

Development and Origin. Internal organs that are filled with air and function primarily in respiration are here called lungs whether they are the familiar paired structures of tetrapods or the unpaired structures of certain fishes. Internal organs that are filled with gas but are not respiratory are called gas bladders. These occur only in bony fishes.

The gas bladder develops from a single evagination of the gut tube near the level of the stomach. The evagination is usually middorsal to the gut but may be to one side of the midline. As it grows, it is invested with mesoderm which will contribute muscles and supportive tissue to the adult organ. The respiratory system of amniotes also develops from a single evagination of the gut tube. In this instance, however, the evagination is midventral and close behind the pharynx. The initial primordium quickly bifurcates to form two primary lung buds which, in turn, divide further to produce whatever respiratory tree is to be present. Again, mesoderm contributes supportive tissues.

Lungs of air-breathing Actinopterygii (except *Polypterus*) are single and develop dorsal to the gut. Lungs of *Polypterus* and Dipnoi may be single or paired; they develop ventral to the gut.

These facts have long been known, and early in the twentieth century provided E. S. Goodrich and other anatomists with material for analysis and speculation about the origin and phylogeny of lungs and gas bladders. Several more recent observations are pertinent: (*1*) A fossil of an antiarch shows imprints of paired, ducted bladders or lungs. (2) The tissue that evaginates to form the respiratory system of amniotes takes its origin from paired areas of entoderm located just posterior to the

Gas Bladders and Lungs

pharyngeal pouches. (3) In anurans and caecilians among amphibians, the lungs do in fact develop from paired lateral evaginations of the gut tube.

From these clues it appears that: (*1*) The evolution of gas bladders or (more likely) lungs extends back to the placoderm ancestors of bony fishes, though probably not to the common ancestors of bony and cartilaginous fishes. (*2*) Lungs were initially paired lateral organs which developed in series with the pharynageal pouches. (*3*) These ancient organs shifted ventrally to form the respiratory structures of tetrapods, dipnoans, and some primitive ray-finned fishes. (*4*) In most ray-finned fishes the ancient paired organs either shifted dorsally to merge over the gut and become a single bladder, or one of the initial lungs was lost leaving the other to shift around the gut to right or left. Embryology and the morphology of related blood vessels indicate that different groups of fishes made the shift in different ways.

Structure and Function of Gas Bladders.   Because of its mode of origin, the embryonic gas bladder is always joined to the gut by a **pneumatic duct.** This duct is lost early in development, is lost late in development, or is retained throughout life, according to species. A bladder having a duct (and also a fish having such a bladder) is said to be **physostomous** (= bladder + mouth). A bladder (or fish) having no duct is **physoclistous** (= bladder + closed). Physostomes are usually fishes that are relatively primitive or unspecialized and live in freshwater. Conversely, physoclists tend to be relatively specialized and to be marine.

Gas bladders and lungs of fishes may be located high in the body cavity above the animal's center of mass. This arrangement enables the fish to remain upright without expending muscular effort. Of fishes having the gas bladder below the center of mass, most must use their fins to counter a tendency to roll, but one catfish has solved the problem another way—it habitually swims upside down.

Gas bladders comprise about 4–11% of the body by volume. They may be long or short, straight or curved, simple or partitioned into two or three more or less distinct parts. Gas is secreted into the bladder from the blood. Rarely (and primitively?) an extensive part of the epithelium of the bladder is vascularized for this purpose. Typically, the secretory area is limited to one or several anterior **gas glands** where the epithelium is columnar and tightly folded on itself. Underlying each gas gland is a **rete mirabile** (= net + marvelous) consisting of as many as tens of thousands of uniquely long afferent capillaries, all oriented in the same direction, among which pass a like number of efferent capillaries. Gas gland and rete mirabile together are called a **red body.** In the rete the surface contact between blood approaching and leaving the bladder may be more than a square meter. Gasses pass by countercurrent exchange from one set of capillaries to the other, thus maintaining maximum gas content in the gas gland. However, the amazing physiology of the gland cannot be explained by diffusion gradients alone. The oxygen content within the bladder of certain deep-sea fishes reaches nearly 1000 times the oxygen content of the surrounding seawater, and the partial pressure of oxygen in the bladder ranges as high as 200 atm.

Gas is resorbed from the epithelium of the posterior part of the bladder or from a portion thereof which may be set off and then takes the name **oval body.** Arterial blood reaches the red body from the celiacomesenteric artery (which delivers to the anterior viscera) and reaches the ovale from the dorsal aorta. It leaves these structures, respectively, in the hepatic portal and postcardinal veins. This is in sharp contrast to the blood circulation of lungs. The red body is innervated by the vagus nerve and the oval body by the sympathetic nervous system. The lining of the gas bladder may be thin or fibrous. In some fishes the organ is more or less covered by striated muscle.

The most important function of virtually all physoclistous bladders, and also of some physotomous bladders, is that of **hydrostasis.** When this is true, the common term "swim bladder" is an appropriate substitute for gas bladder. The term "air bladder" is less apt; the mixture of oxygen, nitrogen, and carbon dioxide in the bladder approximates that of air when the fish is near the surface but changes as the fish descends, reaching nearly 90% oxygen in some instances.

By secreting or resorbing gas the fish can alter the volume of its bladder. This changes the mass per unit volume of the fish

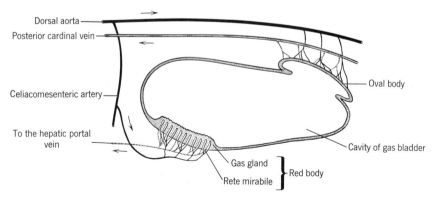

FIGURE 11-6

GENERAL STRUCTURE AND BLOOD SUPPLY OF A PHYSOCLISTOUS GAS BLADDER.

and allows it to adjust its density to that of the environment. Adjustment is under reflex control. It is not rapid enough to be used as a means of moving from less dense surface water to more dense deep water (or the reverse), but once the fish has changed its depth by swimming, it can maintain its level with the gas bladder. Some fishes make daily vertical excursions of as much as 1000 m. Fishes that inhabit shallow rapid streams and marine fishes that remain at constant depth tend to reduce or lose the gas bladder. Some deep-sea fishes have a bladder, and some do not. Bottom fishes such as flounders and halibuts have no bladder.

A second function of the gas bladder is **sound production.** The layman's conception of fishes as being silent is not generally correct; many fishes produce buzzing, rasping, squeaking, clicking, or other sounds. The bladder may serve as resonator for sounds produced by rubbing the pharyngeal teeth together or scraping certain bones together. Vibrations may be produced by action of the intrinsic or extrinsic muscles of the bladder itself, and some physostomes produce sound by the controlled passage of air between the bladder and gut.

Another function of the bladder is **sound** and **pressure reception.** The soft tissues of the fish cannot receive sound waves because their density is nearly the same as that of water. Compact bone may serve, but the compressible gas bladder is far superior and has been adapted for that purpose by several orders and many families of fishes. Sound vibrations received by the bladder must be transmitted to the inner ear in order to be sensed. This is accomplished either by paired long extensions of the bladder into the back of the skull (cods, herrings) or by paired chains of four (sometimes three) ossicles which impinge on the bladder posteriorly and on the perilymph of the internal ear anteriorly (minnows, catfishes, carps). The ossicles are derived from processes of anterior vertebrae and are called the **Weberian apparatus.**

Comparative Anatomy of Lungs. Typical lungs of tetrapods are readily distinguished from gas bladders of fishes: (1) Lungs are respiratory in function. (2) They are paired, their epithelium is folded and branched, and they join the ventral side of the gut tube by a duct called the **trachea.** (3) Arteries to typical lungs are related to the sixth aortic arch and carry blood that is low in oxygen content. Venous blood from lungs returns directly to the heart without prior mixing with blood of low oxygen content. Effective lungs are associated with an effective double circulation through the heart.

Most of these differences between gas bladders and "typical" lungs also distinguish the lungs of DIPNOI from gas bladders—but arterial blood comes from the third or fourth branchial arch, and the lung may be bilobed but not double. The lungs of *Polypterus* are paired (though one is anterior to the other) and ventral, but they are smooth within, they have a unique lining of striated muscle, their venous blood becomes mixed before reaching the heart, and the circulation through the heart is not double. The lungs of the bowfin and garpike, among Holostei, are dissected within but otherwise agree with those of other ACTINOPTERYGII in being similar in structure and relationships to those physostomous bladders that function in hydrostasis. Thus, the distinction between lungs and gas bladders is somewhat arbitrary.

Air breathing in fishes is an adaptation either to occasional survival in warm and stagnant water which is low in dissolved oxygen or (in dipnoans) to periodic draught. These fishes probably continue to rely largely on the gills for ridding the body of carbon dioxide. At least 23 genera of bony fishes are habitual air breathers. Many of these have reduced gill area, and some suffocate if prevented from breathing at the surface. Internal nares, even when present, are not used in respiration; the fish gulps air and using the pharynx as a pump, transfers it to the lung via the pneumatic duct. The entrance to the duct may be controlled only by a muscular sphincter or by both muscles and cartilages.

Lungs of amphibians probably have never been much advanced over those of dipnoans. The more terrestrial LABYRINTHODONTIA possibly had the most advanced lungs of the class. Some had well-developed ribs and probably used the thorax in ventilating their lungs.

ANURA have large but short lungs. The interior of the lung is an open sac, but the walls have partitions of the first, second, and third order providing a total respiratory surface of about 1 cm²/g body weight (15–20 cm² for a medium-sized frog). The very short trachea divides into two short **bronchi,** one leading to the apex of each lung. The epithelium of these ducts is ciliated and thus is able to clean the respiratory system. Cartilage may support the walls of trachea and bronchi against collapse. The opening from trachea to pharynx is called the **glottis.** It is a longitudinal slit which is usually flanked by a dorsal pair of **arytenoid cartilages** (arytenoid = cup-shaped), which support vocal cords, and a ventral pair (often fused) of **cricoid cartilages** (cricoid = ring-shaped). These cartilages are regarded as derivatives of posterior visceral arches of ancestors. Together they constitute the **larynx,** a structure characteristic of tetrapods. The larynx is joined by ligaments to the hyoid apparatus.

Anurans ventilate their lungs with the buccopharyngeal cavity much as fishes ventilate gills—but with some differences. The internal nares are functional for the first time in vertebrate history. With glottis closed, and nares open, air is sucked into the large buccopharyngeal space by lowering the throat. Then

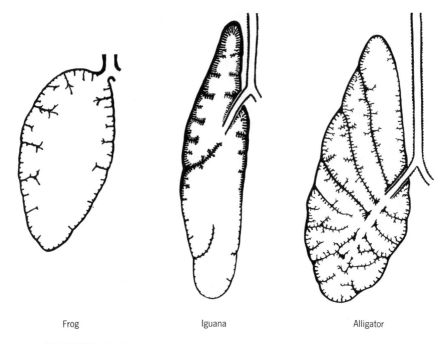

Frog   Iguana   Alligator

FIGURE 11-7

LUNG STRUCTURE OF SOME LOWER TETRAPODS as seen in frontal section. Somewhat stylized and two-dimensional. A finer order of branching is not visible at this scale.

with glottis and nares both open, air escapes from the body under pressure from the resilient lungs. In the process the air passes through the top of the buccopharyngeal space but does not mix with the "fresh" air already stored in the ventral part of the space. The nares then close and raising of the throat forces the fresh air into the lungs via the open glottis. Thus, the tidal flow of respired air causes mixing of fresh air with some residual used air in the lung. The greater diffusivity of carbon dioxide allows much of this gas to escape through the moist skin, independent of the breathing apparatus. The buccopharyngeal pulse pump of the anuran is inefficient in several respects but favors the vocalization of these animals. Further, their manner of ventilating the lungs resembles that of some air-breathing fishes and may be as much inheritance as accommodation.

APODA, having long slender bodies, usually retain only the right lung. (Their larvae have both lungs.) Otherwise, their respiratory system resembles that of Anura. Lungs of URODELA have regressed; most species have lost their lungs entirely. Where present, they are long slender sacs with smooth walls. Their respiratory function is supplemented by the skin (and sometimes also by gills), and they may serve also as crude hydrostatic organs. The trachea is longer than in anurans. A larynx is present but relatively simple. Vocal cords are absent.

Lungs of REPTILES are large and varied. The degree of partitioning may be similar to that of the most advanced amphibians, but the approximate surface area per unit of body weight seems to be larger, and the architecture is different. Reptilian lungs are usually more partitioned anteriorly than posteriorly. This is particularly true of snakes where one long lung (the other being much reduced or lost) is completely smooth posteriorly. Trachea and bronchi are longer than for amphibians (much longer in snakes) and are supported by cartilaginous rings which may be closed or open dorsally. Each bronchus enters its lung near the middle or near the anterior end but not at the apex. Without further division, the bronchus continues within the lung, or opens into more or less through airways. The larynx again consists of cricoid and arytenoid cartilages (variously shaped and fused) which (except in snakes) are joined to the hyoid apparatus. Vocal cords are present only in some lizards. Many reptiles hiss by passing air slowly through a partly closed glottis, at the edge of which there may be an erectile, sound-producing flap. (Certain snakes hiss by rubbing the scales of adjacent body regions together and amplifying this sound via their inflated body.)

When the mouth is closed, the glottis may be functionally joined to the internal nares by juxtaposition (snakes) or by grooves in the buccal epithelium. The throat no longer serves as a pulse pump. Instead, intercostal and other muscles alter the size of the thorax to ventilate the lungs (except in turtles, where special muscles within the shell are effective and motion of the limb pockets compensates for volume changes of the lungs).

Skipping birds for the moment, lungs of MAMMALS have the same basic plan as those of reptiles but are much more complicated and more efficient. Lobulation of the lungs is variable and without systematic or evident adaptive import. Lobes may be absent (horse, whales, sea cows, some bats) but usually there are at least two lobes on the left and at least three on the right. The lobes may or may not be divided into lobules which can be conspicuous or faint.

Of more functional importance is the nature of the respiratory tree. The trachea (like that of reptiles) is supported by rings of hyaline or fibrous cartilage which are here incomplete dorsally. Elastic connective tissue joins ring to ring and completes the tube where cartilage is absent. The resultant structure is ideal for holding the airway open yet allowing the tube to twist as the neck is turned, change length with swallowing, and change diameter if there is marked alteration of internal pressure, as in coughing. The trachea is lined by ciliated epithelium. Smooth muscle and mucous glands are present in the walls.

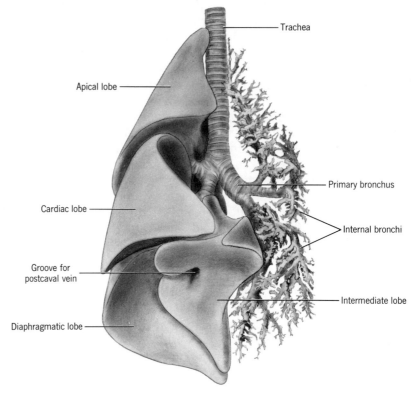

Trachea

Apical lobe

Primary bronchus

Cardiac lobe

Internal bronchi

Groove for
postcaval vein

Intermediate lobe

Diaphragmatic lobe

FIGURE 11-8

**MAMMALIAN LUNG illustrated by the dog. Ventral view. Dissected
on the left to show the bronchial tree.**

The trachea divides into right and left primary bronchi, each
of which enters its lung somewhat anterior and dorsal to the
center. Within the lungs the primary bronchi divide into smaller
bronchi which divide again through numerous generations of
branching. All internal bronchi are supported by cartilage
which, however, may be in plates rather than rings. The airways
are continued by branching **bronchioles** which are membranous, not cartilaginous. These, in turn, open into nonciliated
respiratory bronchioles, where gaseous exchange begins, and
finally into **alveolar duct systems** which are clusters of about
20 hemispherical **alveoli** all opening into a common terminal
chamber (Figure 11-10).

Lung volume is nearly proportional to body size, but alveolar
size (and hence respiratory surface) varies with metabolic rate.
Total surface ranges from about 8 cm^2/g body weight (some
primates) through 50 cm^2/g (mouse) to 100 cm^2/g (bat). In
other terms, the respiratory surface of the human lung is about
70 m^2, or 40 times the surface area of the body, and a 20 kg

dog has a respiratory surface of about 50 m², or 538 ft² (compare with fish of same weight, p. 242).

Pulmonary vessels may follow the airways, or the arteries only may follow according to species. Alveoli are richly supplied with capillaries where blood is separated from air only by the endothelium of the capillary, a basement membrane, and an exceedingly thin alveolar epithelium.

The large larynx is attached to the hyoid apparatus. Paired arytenoid cartilages help support and control the vocal cords. The cricoid cartilage is single. Two additional cartilages are present that are lacking in other vertebrates: a large ventral **thyroid cartilage** (thyroid = shield-shaped) and a cartilage in the **epiglottis.** The epiglottis is a stiff valve-like flap which guides air between posterior nares and glottis during respiration and, to keep food out of the respiratory system, closes the glottis during swallowing.

Mammalian lungs are inflated by the negative pressure that results from contraction of the external intercostal muscles and the dome-shaped diaphragm. (Additional muscles function in forced breathing.) Lungs deflate largely by their inherent elasticity, but the internal intercostals and abdominal muscles also serve in forced breathing. Surface tension of the liquid film within the small alveoli would cause the lungs to collapse were a lipoprotein not secreted to reduce the tension.

The lungs and related air sacs of BIRDS are relatively uniform within the class, and are unique among animals. They are often considered to constitute the most efficient of vertebrate respiratory systems. Such judgements are somewhat speculative, but at least some birds can survive at higher altitudes than can mammals. The trachea, like the neck, is long. Indeed, in certain large birds of several orders, it may form one or more loops near or within the sternum, thus becoming even longer than the body. Its rings are usually complete at its posterior end but may be incomplete on the dorsal side anteriorly. The trachea may have a different resting diameter at different levels. Its volume probably relates to voice quality and to the concentration of carbon dioxide in the air reaching the lungs.

The lungs are relatively small and compact. They are located in the dorsal part of the thorax where they are partly dissected by the heads of several pairs of ribs. Each **primary bronchus** enters its respective lung ventrally somewhat anterior to the center of the organ. It continues for a short distance within the lung, then in some species expands into a vestibule which, in turn, narrows to become a **mesobronchus.** About four secondary bronchi, or **ventrobronchi** join the primary bronchus. These turn medially and all but the first branch in the ventral part of the lung. Six to eight **dorsobronchi** join the dorsal side of the mesobronchus and branch in the dorsal part of the lung. Ven-

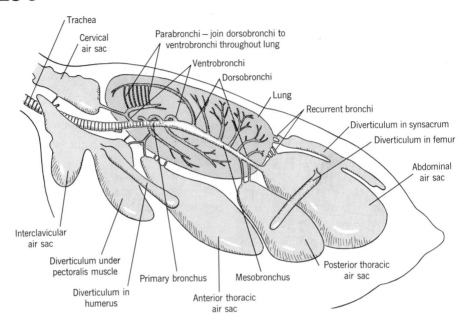

FIGURE 11-9

AVIAN RESPIRATORY SYSTEM in left lateral view. Size of lung is exaggerated. Pattern of ducts is somewhat distorted because of two-dimensional representation; dorsobronchi and ventrobronchi loop close to the surface of the lung.

trobronchi and dorsobronchi are joined by numerous **parabronchi** which are about 1 mm in diameter. Many thousands of respiratory pockets, measuring about 0.3 mm in length, branch at right angles from each parabronchus and it is in these spaces that respiratory exchange takes place.

Lungs of birds function in relation to air sacs which are devoid of respiratory epithelium but contribute to ventilation of the system. The mesobronchi terminate in large paired **abdominal air sacs.** Extensions from these sacs penetrate the synsacrum, femur, and thigh muscles. Paired **posterior thoracic air sacs** join the mesobronchi; paired **anterior thoracic air sacs** usually join the third ventrobronchus. These sacs also have secondary connections to the duct system by small recurrent bronchi. Smaller paired **cervical air sacs** pneumatize cervical vertebrae and muscles; they join the first ventrobronchus. Finally, an unpaired **interclavicular air sac** sends diverticula into the humeri and among muscles of the axillas and shoulders. The entire system is somewhat variable and exceedingly complex.

When the bird is at rest, the sternum is rocked downward and forward for inspiration and is raised for expiration. Each phase is active. The volume of the lungs changes little, but contraction of a thin muscle under the lungs causes some

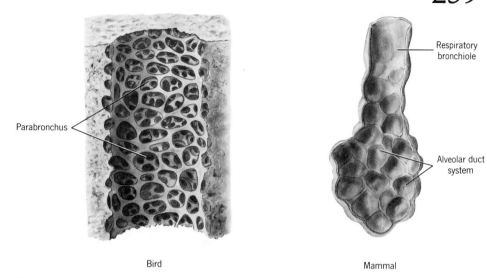

Parabronchus

Respiratory
bronchiole

Alveolar duct
system

Bird

Mammal

FIGURE 11-10

**CONTRAST BETWEEN THE SITES OF GASEOUS EXCHANGE OF THE AVIAN
AND MAMMALIAN LUNG.**

enlargement during expiration. The path of air has been much
studied. Some aspects remain uncertain and are perhaps variable. It now seems probable that most inspired air flows
directly to the posterior air sacs and then through the lung
tissue to the anterior sacs where it is held briefly prior to expiration. Since airflow in the parabronchi is largely in one direction rather than tidal, it is likely that air and blood enter into
countercurrent exchange, thus increasing efficiency. The function of the system during flight is virtually unknown. It must be
assumed that ventilation is increased enormously—perhaps
due in part to direct action of flight muscles on air sacs. The air
sacs serve as evaporative coolers during periods of activity.

Birds have a larynx of characteristic form. Its cartilages are
the same as those of reptiles. Vocal cords are absent. Sound is
produced instead by a unique **syrinx** located at or near the
bifurcation of the trachea. A portion of the outer or (more
often) mesial wall of each bronchus (or less frequently the wall
of the trachea, or of trachea and bronchi in combination) is a
thin membrane supported at its margins by modified carti-

FIGURE 11-11

**SKELETAL SUPPORT OF THE SYRINX OF A
MERGANSER,** *Mergus*, **seen in ventral view.**

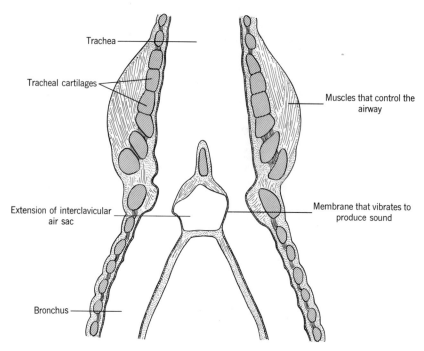

Trachea

Tracheal cartilages

Muscles that control the airway

Extension of interclavicular air sac

Membrane that vibrates to produce sound

Bronchus

FIGURE 11-12
STRUCTURE OF THE SYRINX OF A EUROPEAN BLACKBIRD as seen in longitudinal section.

laginous rings. This membrane impinges on air sacs and thus is free to vibrate in the stream of air passing through the system. The organ is sometimes asymmetrical and can be conspicuous in birds with resonant voices. Complicated intrinsic and extrinsic muscles control tension of the membranes and also the characteristics of the resonating system. The muscles are variable among species and often are used in systematics.

Miscellaneous
Respiratory
Structures

Many structures other than gills and lungs have been adapted by one or another species to serve in respiration. The yolk sac respires for the larvae of many fishes and the allantois serves for the fetuses of amniotes. Some fishes supplement gill respiration with opercular pouches which branch from the gill chambers and have vascular linings. Other fishes respire in part through the lining of the hindgut, lining of the stomach, or bases of the fins. The skin and the buccopharyngeal cavity are supplementary or principal respiratory organs for various Lissamphibia.

12

Circulatory System

Animals must have a system of internal transport unless their size and structure place all their cells in close proximity to an aqueous environment. Vertebrates, being large and solid, have the most highly evolved circulatory system in the animal kingdom. It functions to transport respiratory gases, nutrients, metabolic wastes, hormones, and antibodies. It serves (in conjunction with the kidneys and some other organs) in maintaining the internal environment. It removes toxic and pathogenic materials from the body, and may function (with muscles and the integument) in temperature regulation. Further, it has the capacity to repair leaks, compensate for damage, and respond with amazing versatility to the varying requirements of the moment.

General Function and Nature of the System

FIGURE 12-1
ARTERIOLE, VENULE,
AND NETWORK
OF CAPILLARIES

The vertebrate circulatory system has two components, a **blood-vascular system** and a **lymphatic system.** The former consists of the heart, blood vessels, and blood. **Arteries** distribute blood from heart to tissues, and **veins** return blood from tissues to heart. Small arteries (arterioles) are joined to small veins (venules) by **capillaries** which form a network within the tissues. Capillaries are usually about 1 mm long, scarcely larger in diameter than a single red blood cell, and only a fraction of a millimeter distant from one another. Physiological exchange between blood and tissues takes place through their thin walls. In several places in the body (digestive organs, kidneys, hypophysis) blood that has passed a capillary bed elsewhere enters a second capillary bed before reaching the heart. The veins between two capillary networks constitute a **portal system.**

The blood-vascular system of vertebrates (unlike that of invertebrates, except annelids) is a continuum of ducts and is therefore said to be a **closed system.** However, fluid constituents of the blood leak out of the capillaries driven by diffusion, osmosis, and the hydrostatic pressure produced by the heart. Fluids would accumulate in the tissues and cause swelling were they not drained away by the second component of the circulatory system, the lymphatic system. Tissue fluids enter net-like or blindly ending **lymphatic capillaries** where they constitute **lymph.** Lymph passes slowly into larger and larger **lymphatic vessels** until it is discharged into the venous system at several points. Lymphatic capillaries in the gut are called **lacteals.** They absorb the digested long-chain fats; their lymph is whitish after a fatty meal (lacti = milky).

The heart provides little pressure to drive venous blood and none to drive lymph. Their passage is assisted by pressures on their vessels resulting from respiratory and other motions of the body and probably by their own intrinsic musculature. Most veins and lymphatic vessels of tetrapods have valves to prevent backflow. Many lower vertebrates also have lymph sinuses, some of which beat weakly as **lymph hearts.**

Ontogeny and phylogeny are more strikingly related in the circulatory system than in any other. Embryonic hearts, arteries, and veins of the higher vertebrates closely resemble the corresponding organs of remote ancestors. The pattern of circulation changes in development much as it must have changed in evolution, and it is fascinating to observe the sometimes marked changes in progress because the system functions continuously throughout.

The circulatory system also has more individual variation than any other. This is particularly true of the veins and is the combined result of a complicated ontogeny (which provides many opportunities for deviations) and the near functional equivalence of various departures from the norm.

Further, the system is exceedingly adaptable. A piece of a vein grafted into an artery transforms structurally to become an artery. Nearly any vessel of the body can be tied off without serious inconvenience to the animal if the obstruction is made slowly enough for the system to compensate by enlarging alternate routes. Blood can, and does, flow in either direction in several vessels. Blood can be diverted toward or away from any given part of the body, the volume of circulating blood can be changed, and the rate of circulation can be varied about fivefold.

In spite of marked individual variation, parts of the system are useful in systematics. This is particularly true of the heart and major circuits at levels of the class and subclass, and of patterns of arteries in the limbs and basicranial area at levels of the order and family. Some variations relate to posture or habit, but the mechanisms involved are often poorly understood.

Development and Histology

The circulatory system is the first of all organ systems to become functional during development. This is hardly surprising since differentiation and growth of all parts of the body are soon dependent on internal transport. The chick heart starts to pulsate at about 30 h of incubation when the body is not yet large enough to enclose the organ; the human heart becomes functional at 4 weeks when the embryo is scarcely 5 mm long.

The first visual indication of the formation of the system is the appearance on the yolk sac of solid isolated masses of mesoderm called **blood islands.** The peripheral cells of adjacent islands gradually become contiguous and form a network of tiny vessels. The deeper cells separate from one another and become blood cells.

Vessels soon form also from mesenchyme within tissues of the body. They are initially all of about the same size and together comprise a continuous network in relation to organs of rapid growth such as the central nervous system and eyes. Gradually certain channels enlarge or merge to become the first arteries and veins. All parts of the system are initially paired and symmetrical. However, the primordia of the heart and certain arteries fuse in the midline very soon after their appearance, and asymmetries are established early in certain veins.

Some major lymphatic vessels develop in relation to veins and subsequently become independent. More peripheral parts of the system probably have an independent origin.

Even before blood islands develop, the part of the splanchnic layer of the hypomere that is just posterior to the pharynx and ventral to the gut (or will soon come into that position—there are complicating changes of position in mammals) becomes markedly thicker on both sides of the body. These mesodermal folds approach one another in the midline and fuse to form a longitudinal tube. The tube is fixed to surrounding tissues at each end but otherwise becomes free as it passes through an expanded portion of the coelom. The free section establishes four chambers which begin to contract in sequence and thus become the heart.

Arteries, veins, and capillaries are at first indistinguishable histologically. Each is a tube formed from thin, flat endothelial cells which are loosely wrapped on the outside in a meshwork of connective tissue. This is the definitive structure of capillaries. Arteries and veins each retain these tissues as a **tunica interna** but add more peripheral tissues as they mature. Arteries develop a thick **tunica media** which usually consists of circularly arranged smooth muscle fibers but in the largest arteries consists instead of yellow elastic fibers. Arteries are completed by a thinner **tunica externa** of longitudinally oriented connective tissue. The scanning microscope reveals that at least some arteries of at least some mammals are lined by a meshwork of irregular projections which vastly increase their surface.

Veins are usually larger in diameter and thinner-walled than corresponding arteries. They are also more variable in structure, however, and in some instances are much like arteries. The tunica media is typically thin and may be indistinct. The tunica externa, by contrast, is as thick or thicker than that of

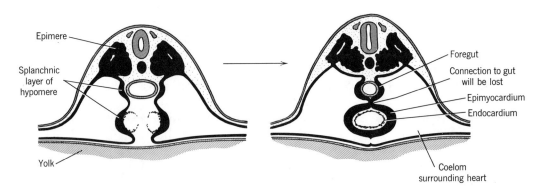

FIGURE 12-2
FORMATION OF THE HEART IN AN EMBRYO HAVING MUCH YOLK. For level of sectioning, see upper drawing of Figure 5-3. Compare also with Figure 5-4.

arteries. At death, the more muscular arteries squeeze most of the blood into the veins. When cut, the arteries stand open but empty, whereas the veins usually are either gorged with blood or collapsed.

Lymphatic capillaries resemble blood capillaries but are larger, more irregular in shape, and more permeable. Their endothelial cells are relatively large. Lymphatic vessels are constructed somewhat like veins, although the three layers tend to be less distinct.

When the embryonic heart is first established it has two layers, an internal endocardium and a thicker epimyocardium. The **endocardium** corresponds to the tunica interna of the vessels, but in the mature heart differs in having a thick layer of elastic connective tissue under the endothelial lining. The epimyocardium slowly matures—while functioning—into a myocardium and an epicardium. The **myocardium,** or heart muscle, represents an enlargement of the tunica media of arteries. The **epicardium** corresponds to the tunica externa, but differs in being a serous membrane underlain by connective tissue.

Heart muscle, or **cardiac muscle,** resembles skeletal muscle in being organized into fibers which have striated myofibrils. It differs from skeletal muscle in that the fibers branch and recombine, nuclei are central rather than peripheral, and **intercalated disks** are spaced along the mature fibers at right angles to their long axes. These disks interrupt the myofibrils and apparently divide the fibers into cell areas. The fiber bundles are arranged in layers which spiral around the individual heart chambers; contraction usually constricts a chamber in all dimensions (though there are exceptions—particularly among fishes). Further characteristics of cardiac muscle are its

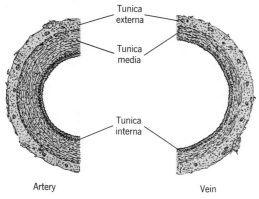

Tunica externa

Tunica media

Tunica interna

Artery   Vein

FIGURE 12-3
STRUCTURE OF BLOOD VESSELS AS SEEN IN CROSS-SECTION.

inherent capacity to beat and its ability to conduct the stimulus to beat throughout its own substance. In the adult heart, the beat is initiated at a locus called the **pacemaker** (usually located near the right atrium) and is conducted to all the chambers in coordinated sequence by somewhat modified muscle fibers called **stimulation fibers** (another name is Purkinje fibers). The beat of the adult heart is also influenced by the autonomic nervous system and associated chemical compounds.

Blood and Blood-Forming Tissues

Vertebrate blood consists of blood cells of various types suspended in a fluid called **plasma.** Plasma is an aqueous solution of nutrients, metabolic wastes, salts, hormones, and proteins. Blood proteins are probably formed by the liver. They include albumens, the most abundant; fibrinogen, which contributes to the complicated clotting reaction; and several globulins, which respond to the entry of certain foreign materials into the body. Lymph is essentially a plasma with reduced protein content. Serum is the fluid remaining after blood has clotted.

Blood cells are of three principal kinds: red cells or **erythrocytes,** white cells or **leucocytes,** and **thrombocytes.** Erythrocytes occur only in blood vessels. They tend to be smaller than leucocytes, yet vary considerably in size, being relatively large in amphibians and small in mammals. They are usually flat, oval or round, and nucleated, but in nearly all mature mammals are round and enucleate. Erythrocytes are rich in the red protein hemoglobin which combines readily with oxygen and is responsible for the efficiency of these cells in oxygen transport. They also transport carbon dioxide.

There are only 1% (mammals) to 10% (fishes) as many leucocytes as erythrocytes in the blood stream, but only leucocytes occur in the lymphatic system. These cells actively move through capillary walls and quickly aggregate at sites of local infection. Some leucocytes destroy foreign bodies by engulfing them—a process called **phagocytosis** (= to eat + cell + process). These cells are also involved in the immune response. It appears that leucocytes that enter the tissues may change function and become typical connective tissue cells; blood is usually classified as a special type of connective tissue.

Many varieties of leucocytes have been identified among the different vertebrates, and intergrades make classification difficult. Two main kinds are granulocytes and lymphoid cells. **Granulocytes** (also called polymorphonuclear leucocytes) are large cells. The nucleus is subdivided into two or more lobes and the cytoplasm is highly granular. Three subtypes of granulocyte are named on the basis of staining properties: neutrophils (by far the most abundant in all vertebrates except reptiles), acidophils (present in all vertebrates), and basophils (least abundant and absent in some fishes). **Lymphoid cells** have a central, unlobed nucleus and lack cytoplasmic granules. Large lymphoid cells are called monocytes; small cells (which are much more abundant in all vertebrates) are lymphocytes.

Thrombocytes are small, nucleated, spindle-shaped cells of the blood stream. They occur in all vertebrates except mammals, which have instead enucleate cell fragments called platelets. When thrombocytes or platelets escape through a cut in a blood vessel they adhere to other tissues and disintegrate, thereby releasing a material that initiates the clotting process (thrombocyte = clot + cell).

Unlike other cells, blood cells are short-lived; they survive for days, weeks, or several months and then for some reason are destroyed. Consequently, they are produced and removed continuously from early embryonic life to death. For the most part, cell formation (**hemopoiesis**) and destruction occur in the same tissues. Many of these tissues also contribute to antibody production and clean the blood by filtration and the removal of many kinds of pathogens.

Blood islands of the yolk sac produce the first blood of most vertebrates. Blood also forms in conjunction with vessels of the body generally, and vessels continue to produce red cells in adult fishes. The digestive tract, thymus, kidney, liver, and parts of the nasopharynx produce blood in the embryos only of some vertebrates and in embryos and adults of others. The gonads and pericardial membrane are active in restricted groups of adult fishes. **Lymph nodes** occur sparingly in certain

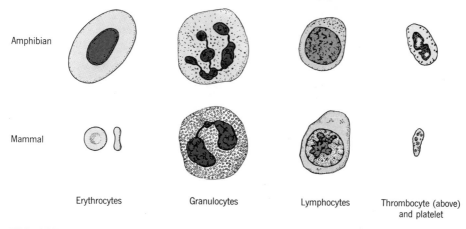

Amphibian				
Mammal				
Erythrocytes	Granulocytes	Lymphocytes	Thrombocyte (above) and platelet	

FIGURE 12-4
REPRESENTATIVE VERTEBRATE BLOOD CELLS.

water birds and are otherwise limited to mammals where they are abundant along the lymphatic vessels in certain areas. They are whitish lumps of variable size and shape which produce quantities of lymphocytes.

Islets and cords of splenic tissue occur under the submucosa of the gut of hagfishes and within the lengthwise intestinal fold of lampreys. In other vertebrates the **spleen** is a discrete reddish organ located in the dorsal mesentery. It may be elongate or compact and relatively small (birds) or large (mammals). The spleen of mammals lies to the left of the stomach. In this class it produces lymphocytes, destroys erythrocytes, and also stores quantities of erythrocytes for release when the body has need of them. In all other classes the spleen produces all kinds of blood cells. Ungulates and some other mammals also have **hemal nodes** which are small nodules of splenic tissue distributed along blood vessels of the gut, kidney, and liver.

There are two kinds of bone marrow: Yellow marrow occurs in the larger cavities of the long bones and is fatty; red marrow occurs in the ribs, vertebral centra, and epiphyses of long bones. The types are not always distinct. Red marrow is hemopoietic in tetrapods. In mammals it produces all types of blood cells except lymphoid cells; in the other classes it produces all types of blood cells.

In general, hemopoietic tissue consists of a matrix of connective tissue intimately related to a rich blood supply. The supportive framework often includes **reticular tissue,** the stellate cells of which are phagocytic and can differentiate into various kinds of blood cells. Commonly, the blood is carried in **sinusoids** which differ from capillaries in being larger and more irregular. They anastomose freely and carry venous blood. (An **anastomosis** is a net-like intercommunication of vessels.) Similar lymphatic sinusoids carry lymph in the lymph nodes.

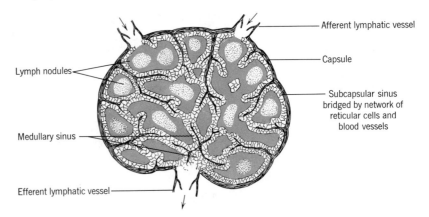

Afferent lymphatic vessel

Capsule

Subcapsular sinus bridged by network of reticular cells and blood vessels

Lymph nodules

Medullary sinus

Efferent lymphatic vessel

FIGURE 12-5
DIAGRAM OF A SECTIONED LYMPH NODE.

The spleen and lymph nodes are encapsulated by dense connective tissue which also sends strands into the substance of the organs. (The capsule of the spleen may also contain muscle.) Interstices of the supportive matrix are packed with formative cells of various types and by large phagocytic cells called macrophages. The spleen is further characterized by having areas of white pulp, which is supplied by arteries and has only leucocytes in its tissues, and areas of red pulp, which is supplied by venous sinusoids and includes erythrocytes in its tissues.

Evolution of the Heart

Primitive Single-Circuit Pump.   Amphioxus has no heart. It has instead a pulsating vessel in the position where the heart evolved in vertebrates. This vessel approximates the embryonic primordium of the vertebrate heart and is apparently homologous.

The structure of the adult ancestral vertebrate heart can be inferred from the structure of the embryonic heart of descendants. It is a nearly straight tube having four chambers which contract in sequence to pump a single stream of unoxygenated blood forward in the body. A thin-walled **sinus venosus** receives blood from the great veins and empties it through a simple **sinuatrial valve** into the **atrium.** The atrium, also thin-walled but muscular, expels blood through one or more rows of **atrioventricular valves** into a large thick-walled **ventricle.** The ventricle

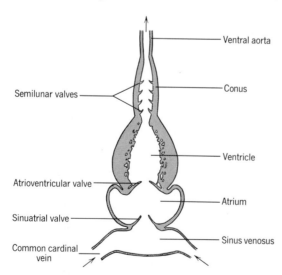

FIGURE 12-6

HYPOTHETICAL ANCESTRAL VERTEBRATE HEART as seen in frontal section. This structure is closely approximated in the ontogeny of all vertebrate hearts.

pumps into the **conus** (also called conus arteriosis) which looks like an enlarged artery and is lined with several rows of cup-shaped **semilunar valves.**

Hearts of CYCLOSTOMES and FISHES (except dipnoans) vary widely in detailed structure but depart little from the general ancestral plan. In fishes the heart is relatively far forward in front of the pectoral girdle and under the posterior gills. The sinus venosus ranges from large (most sharks) to small (cyclostomes) and is often of irregular shape. It may have smooth rather than cardiac muscle. The atrium is relatively large and usually shifts to a position dorsal to the ventricle. The inner layer of the ventricle is exceedingly spongy. In teleosts this chamber is conical, with apex pointing posteriorly. The conus varies from long and active as a pumping organ (cartilaginous fishes, holosteans, *Polypterus*, bowfin, dipnoans) to virtually absent (cyclostomes, teleosts). When large it contains two to eight rows of semilunar valves (some of which, however, may not span the channel of the organ). The conus prevents backflow of blood as the ventricle fills. Teleosts have a **bulbus arteriosus** within the pericardial cavity in the position of the conus of other fishes. This organ is highly elastic and passively evens the flow of blood into the af-

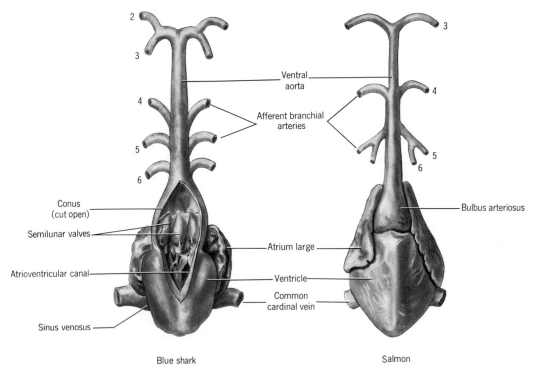

Blue shark

Salmon

FIGURE 12-7

HEARTS AND RELATED VESSELS OF REPRESENTATIVE FISHES. Ventral views.

ferent branchial arteries. (The short divided ventral aorta of some urodeles is also called a bulbus arteriosus.)

The pericardial cavity of elasmobranchs, being bordered in part by the skeleton, is semirigid. Consequently, as the ventricle contracts, blood enters the sinus venosus and atrium partly by suction. The location of a pacemaker is obscure in fishes. Hearts of fishes are relatively small because fishes have relatively small blood volume. As expected, active fishes have larger hearts than their sluggish relatives. Cyclostomes and many fishes have accessory hearts or pumping mechanisms elsewhere in the body. Some of these are mentioned later in this chapter.

Transition to Double-Circuit Pump. For reasons explained later in this chapter, vertebrates that respire exclusively with gills can get along with a heart that pumps one stream of blood, whereas active vertebrates that respire exclusively with lungs must have hearts that pump one stream of oxygenated blood and another completely independent stream of unoxygenated blood. The evolution of a double-circuit pump required extensive changes and was slow to be completed. The hearts of dipnoans, amphibians, and reptiles are transitional between single- and double-circuit pumps. These animals can survive with such pumps because they are less active than birds or mammals and because for most the lungs are not the only respiratory organ. Some of these animals can benefit by adjusting the circulation, according to circumstance, to approach either the single- or double-circuit condition.

The hearts of DIPNOI and AMPHIBIA are similar. The atrium is partly divided (dipnoans and urodeles) or completely divided (anurans). Pulmonary blood from the lungs enters the left chamber; systemic blood from the remainder of the body enters the sinus venosus which joins the right chamber. (The sinus venosus is usually large but is reduced in anurans.) The ventricle is partly divided in dipnoans and undivided in amphibians. Even though the ventricle is single, the very spongy nature of its walls reduces mixing of blood. The conus is large and no longer contractile. It is partly divided within by various configurations of semilunar valves (anurans) and by a longitudinal partition, or **spiral valve,** which reduces mixing and deflects the two streams of blood into different vessels. The mechanism has been much studied in frogs using a variety of ingenious techniques. The evidence indicates that there is remarkably little mixing of the two streams of blood. Sequence of beat and differential pressures supplement structure in the control of function.

It is not surprising that urodeles, being less dependent than anurans on a pulmonary circulation, have a less effective double pump. Regression has gone so far in lungless salamanders that even the interatrial septum is reduced or lost.

Extinct rhipidistian fishes of the infraclass Crossopterygii and Labyrinthodontia may have had more completely double-circuit hearts than the somewhat specialized (Anura) and regressed (Urodela) hearts of modern Lissamphibia.

REPTILES have a more nearly complete double circulation. The sinus venosus is large (turtles), small, or vestigial, and relates only to the right atrium. The atrium is completely divided, and the pulmonary and systemic veins have the same relationships as in amphibians. The ventricle is completely divided in crocodilians and partly divided in other reptiles. Reptiles are unique among vertebrates in that the conus has become divided into three channels, a **pulmonary trunk** leading to the lungs, and right and left **systemic trunks** to the general circulation.

This heart appears to have all the "parts" needed to make a complete double pump, yet even when the ventricle is fully divided there may be some mixing because of the assembly of the parts. The right systemic trunk tends to join the left ventricle in lizards but the right ventricle in turtles and crocodilians; the right atrium joins the right ventricle in crocodilians

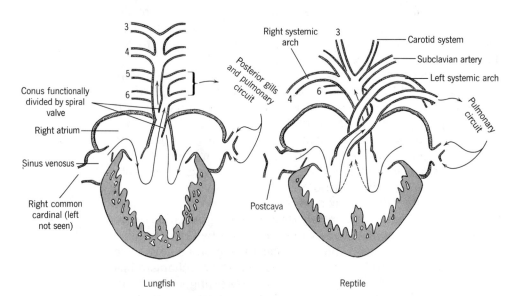

FIGURE 12-8

DIAGRAMS OF HEARTS THAT ARE TRANSITIONAL TO DOUBLE-CIRCUIT PUMPS. Ventral views of frontal sections. Somewhat distorted by two-dimensional representation—sinus venosus and pulmonary veins enter atria more dorsally; divisions of conus spiral in space. There is considerable variation within the taxa represented. Aortic arch derivatives are numbered.

but the left ventricle in lizards and turtles. The consequence is that mixed blood tends to enter the left systemic arch, whereas oxygenated blood enters the right arch. The origin of the "problem" is the lack of alignment among interatrial septum, interventricular septum, and septa of the conus. The seeming defect may, in fact, be advantageous in that it provides versatility. Thus, when a turtle remains submerged it may bypass the virtually nonfunctioning pulmonary circuit. In normal circumstances there is little mixing of pulmonary and systemic streams of blood. It is probable, however, that mammals evolved from reptiles that divided the conus into two instead of three channels.

Double-Circuit Pump. Adult BIRDS and MAMMALS have completely double circulations: a low-pressure pulmonary circuit using the right side of the heart and a high-pressure systemic circuit using the left side. The sinus venosus is vestigial in birds and absent in adult mammals, the embryonic sinus having merged with the right atrium. The atrium is divided and relatively smaller than in many fishes. The ventricle is divided and disproportionately strong on the left side. The embryonic conus divides into a pulmonary trunk joining the right ventricle and a systemic trunk joining the left. Unlike the condition in lower vertebrates, the systemic arch is single. It loops to the right in birds and to the left in mammals.

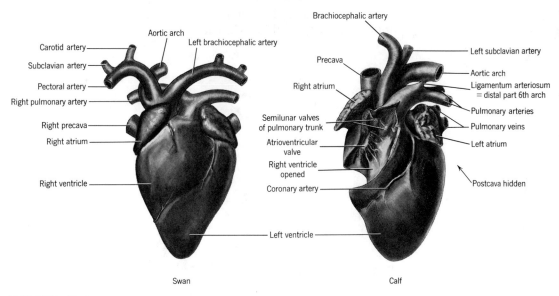

Swan   Calf

FIGURE 12-9

HEARTS AND RELATED VESSELS OF A BIRD AND A MAMMAL. Ventral views. Fat dissected away. The left subclavian and left common carotid arteries branch independently from the aorta in some other mammals.

In spite of similarities, the avian and mammalian hearts clearly evolved double circulations independently. Embryonic development indicates that the interatrial and interventricular septa are not homologous and that the valves are dissimilar.

The Initial Pattern of Blood Vessels

The pattern of vessels that comprises the initial functioning system of the embryo is basically the same for all vertebrates. Virtually the same pattern is also found in adult amphioxus. It can, therefore, confidently be considered primitive for vertebrate phylogeny as well as ontogeny (Figure 12-11). It should be learned as our point of departure.

The heart pumps blood forward under the pharynx in the **ventral aorta** (also called truncus arteriosus)—a vessel which is single where it leaves the heart but sometimes paired under the anterior branchial arches. In embryos, the ventral aorta distributes its blood to paired **aortic arches** which run upward through the visceral arches. In adults of lower vertebrates, the aortic arches have differentiated into proximal (toward the heart) **afferent branchial arteries,** gill capillaries, and distal **efferent branchial arteries.**

The principal distributing vessel of the body is the **dorsal aorta.** This vessel is paired throughout its length when it first differentiates. It may remain paired dorsal to the pharynx, but at more posterior levels immediately fuses in the midline. Blood entering the paired dorsal aortae from the anterior aortic arches (or anterior branchial efferents) runs forward and continues into the head in extensions of the aortae called **internal**

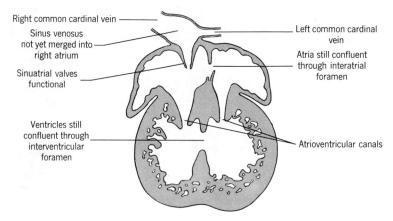

Right common cardinal vein

Sinus venosus not yet merged into right atrium

Sinuatrial valves functional

Ventricles still confluent through interventricular foramen

Left common cardinal vein

Atria still confluent through interatrial foramen

Atrioventricular canals

FIGURE 12-10

RECAPITULATION IN THE HEART OF A MAMMALIAN FETUS as seen in a ventral view of a nearly frontal section. Not seen in this plane of sectioning are the small pulmonary veins and the conus, which is starting to divide into pulmonary and systemic trunks.

carotid arteries. Blood entering the aortae from the posterior arches (or efferents) runs posteriorly into the unpaired dorsal aorta whence it is distributed by three sets of branches: Paired and segmental **dorsal branches** deliver blood chiefly to the body musculature, the largest branches serving the paired appendages. Paired **lateral branches** deliver to urogenital organs. Prim-

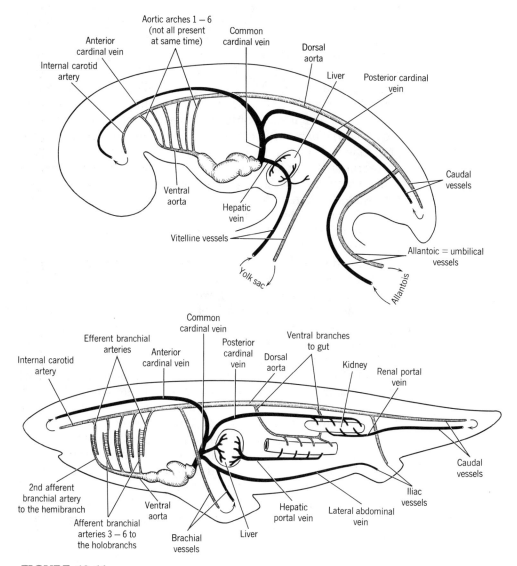

FIGURE 12-11

BASIC PATTERN OF THE VERTEBRATE CIRCULATORY SYSTEM as seen in an amniote embryo (above) and an adult fish at the shark level of evolution (below). Not shown: subcardinal and supracardinal veins, dorsal and lateral branches of dorsal aorta. All vessels paired except dorsal and ventral aortae, caudal vein, and vessels of gut.

itively these are numerous and segmental. **Ventral branches** deliver to the yolk sac and viscera. These branches are initially numerous but only vaguely segmental. They are unpaired with two exceptions: The **vitelline arteries,** which serve the yolk sac, and **allantoic arteries,** which serve the allantois, represent pairs of branches that are established before the embryonic primordia of the dorsal aorta have fused.

The initial ontogenetic and phylogenetic pattern of veins is more complicated and comprises three somewhat independent systems. The **subintestinal–vitelline system** drains the tail, digestive tract, and yolk sac. A **caudal vein** runs forward to an **anal loop** which skirts the anal area. **Subintestinal veins** (single posteriorly and paired anteriorly) continue the drainage forward, receive tributaries from the digestive tract, and, after penetrating the liver, empty into the common cardinal veins (see below) close to the heart. This completes the subintestinal component of the system as it may have occurred in ancestral vertebrates. A vitelline component consists of large **vitelline veins** which come from the yolk sac and enter the subintestinals. Morphologists find it convenient to think of the vitelline veins as large precocious branches of the subintestinals. Embryologists are more likely to regard the subintestinals as belated branches of the vitelline veins.

A second system of veins is the **cardinal system** which drains the head, dorsal body wall, and kidneys. Its principal vessels are the **anterior cardinal veins,** which are lateral to the internal carotid arteries; **posterior cardinal veins,** which are located in or adjacent to the dorsal portion of the kidneys; and short **common cardinal veins,** which are formed by the confluence of the preceding vessels. After being joined by veins of other systems these return all blood to the heart. Several additional pairs of longitudinal veins are related to the embryonic kidneys. These anastomose freely with one another and drain into the posterior cardinal veins. Two of these that contribute to the story to follow are the **subcardinal** and **supracardinal veins.**

The third system of veins is the **abdominal system** which drains the ventral body wall and paired appendages. It consists of paired **lateral abdominal veins** which receive **iliac** and **subclavian branches** from the posterior and anterior appendages and drain into the common cardinal veins near the entrances of the subintestinal veins.

Evolution of Blood Vessels   **Derivatives of Aortic Arches and Related Vessels.** Embryos of vertebrates nearly always have six pairs of aortic arches, but the first (located in the mandibular visceral arch) is always lost or reduced to remnants in the adult.

CYCLOSTOMES have a long ventral aorta which is paired anteriorly. There are eight or more aortic arches. Afferent and efferent branchial arteries have different and distinctive patterns of branching in lampreys and hagfishes. The dorsal aorta is single over the gills (lampreys) or paired (hagfishes).

ACTINOPTERYGII except Chondrostei are advanced in that the second aortic arch, like the first, has been modified or lost. They are relatively primitive in that their afferent and efferent branchial arteries are simple and free of loops and branches. Arches 3–6 characteristically serve the four holobranchs. The efferent branchials tend to lie at the levels of the respective visceral arches. Distally, the afferent branchials branch in various patterns from the moderately long ventral aorta. Over the pharynx are paired dorsal aortae which extend forward under the brain as internal carotid arteries. Jaws, orbits, mouth, and pseudobranchs are supplied by arteries of varied patterns, doubtful homologies, and many names.

DIPNOI and CHONDROSTEI vary among themselves but tend to differ from typical ray-finned fishes in that (1) the air-breathers among them have true pulmonary arteries which branch from the efferent segments of the sixth arches, (2) the second aortic arch is retained (some dipnoans excepted) to serve the hemibranch on the hyoidean visceral arch, (3) the fifth and sixth afferents of the air breathers tend to branch from the ventral aorta just anterior to the partly divided conus which diverts into them the relatively unoxygenated stream of blood, and (4) the efferent branchial arteries tend to fork within the gills and to bunch over the gills.

The phylogenetically important extinct rhipidistian fishes probably had simple efferent drainages as in ray-finned fishes. They certainly also had pulmonary arteries joining the sixth arch.

CHONDRICHTHYES also retain the second aortic arch to serve the hyoidean hemibranch. Thus, there are five (rarely more) functional afferent branchial arteries. There are typically four efferent branchials, each collecting from an efferent loop which drains the two hemibranchs facing into one gill chamber. Numerous shunts join adjacent loops. The efferents lie opposite the gill chambers, not the visceral arches. The dorsal aorta is unpaired over the pharynx.

Adult TETRAPODS lack both first and second aortic arches. A characteristic carotid system delivers blood to the head. It consists of (1) common carotid arteries derived from segments of the ancestral (and embryonic) paired ventral aortae between the ventral roots of the third and fourth arches, (2) external carotid arteries which supply the throat and ventral part of the head and possibly represent the paired ventral aortae anterior

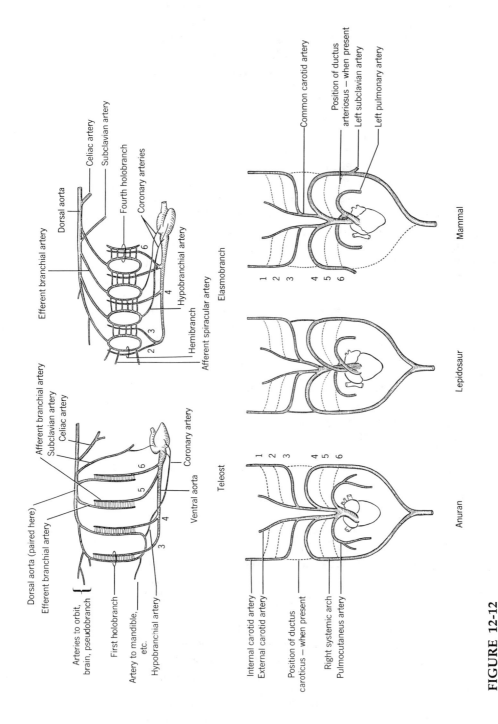

FIGURE 12-12

DERIVATIVES OF AORTIC ARCHES AND RELATED VESSELS OF REPRESENTATIVE VERTEBRATES. Semidiagrammatic. Left lateral views (above) and ventral views (below). Aortic arches are numbered.

to the ventral roots of the third arch, and (3) internal carotid arteries which supply the brain and much of the head and are derived from the ancestral internal carotids, plus the paired dorsal aortae between the dorsal roots of the first and third arches, plus the third arch distal to its junction with the external carotid.

URODELA retain paired third, fourth, and sixth arches and usually also a pair of fifth arches. Afferent and efferent branchial arteries are, of course, present in perennibranchiate species, but are joined by shunts which allow some blood to bypass the gill capillaries. Pulmonary arteries branch from the sixth arches. (There are variations on the typical pattern, and, unfortunately for students of morphology, the familiar *Necturus* is unusual in that common carotids are not identified and the proximal parts of the sixth arches are missing.)

ANURA have a "standard" carotid system to transmit blood to the head. The fourth arches are systemic arches transmitting blood to the posterior part of the body. The short segment of the paired dorsal aortae between the dorsal roots of the third and fourth arches (called the **ductus caroticus**) is thus a useless shunt between diverging streams of blood; although retained by most urodeles, it is lost in anurans. The fifth arch is always absent in the adult. The useless (indeed detrimental) part of the sixth arch that is distal to the pulmonary artery (called the **ductus arteriosus**) is lost. The proximal part of the sixth arch contributes to the base of the pulmonary artery.

The basic pattern of aortic arch derivatives in modern REPTILES is the same as for anurans, with the important difference that the conus has divided into three instead of two channels. Instead of deflecting most unoxygenated blood into the pulmonary artery and most oxygenated blood into the combined carotid and systemic systems (as in anurans), reptiles deflect most unoxygenated blood into the pulmonary system, most oxygenated blood into the right systemic arch, and (according to species) mixed blood into the left systemic arch (Figure 12-8). The carotid system (which is variously modified in the different groups) relates only to the right arch. It is curious that the ductus caroticus still persists in some lepidosaurs, and the ductus arteriosus in *Sphenondon* and several other species.

BIRDS modify the carotid system so that internal carotids substitute for common carotids in the long neck. Right and left vessels lie side by side in the chick, and it is of interest to systematists that both or either may be retained in the different orders. The right, fourth aortic arch becomes the systemic arch. Characteristic brachiocephalic arteries, which branch from the systemic arch, give rise to carotid and subclavian arteries, the pattern of branching being various.

MAMMALS retain only the left systemic arch. The carotid system is less modified than in birds, but the proximal relations of the common carotids to the arch and the subclavian arteries are often asymmetrical.

Other Arteries. Posterior to the pharynx the **dorsal aorta** is the most conservative vessel of the body. It is always a large, median longitudinal artery lying ventral to the notochord or spine. It extends into the tail as the **caudal artery.** Further, there is little evolutionary significance to the variations of the branches of the aorta. **Ventral visceral branches** may be numerous (amphibians) but typically include only a **celiac artery** to stomach, duodenum, liver, and pancreas and one or several **mesenteric arteries** to the remainder of the gut.

Lateral visceral branches serve the urogenital organs. They are numerous if these organs are long (most anamniotes) and otherwise few. They take their names from the organs served — renal, ovarian, spermatic.

Dorsal somatic branches of the dorsal aorta serve spinal cord, muscles, and skin. They are segmental in fishes, but in tetrapods tend to be modified for functional reasons noted below. The large **subclavian arteries** extend into the pectoral appendage as **brachial arteries** (do not confuse with branchial arteries); the corresponding **iliac arteries** extend into the pelvic appendages as **femoral arteries.** The subclavians join the unpaired dorsal aorta in urodeles and most fishes, the paired systemic arches of anurans, the left systemic arch of mammals, the right systemic arch in birds and some reptiles, and the carotid system of other reptiles. To explain these differences would provide a useful review of earlier sections of this chapter.

Derivatives of Subintestinal System. The subintestinal veins of embryos are paired and continue to the yolk sac as vitelline veins. Adults of all vertebrates lose the vitelline veins and establish a single large **hepatic portal vein** (hepatic = of the liver) by the selective retention of parts of the left and right subintestinals and of several anastomoses which occur between them within and just posterior to the liver. Inside the liver the system breaks up into sinusoids, and anterior to the liver it continues toward the heart as one or more **hepatic veins** which drain the liver.

Primitively, the hepatic portal vein drains the gut and also the tail via the caudal vein and anal loop. This condition is re-

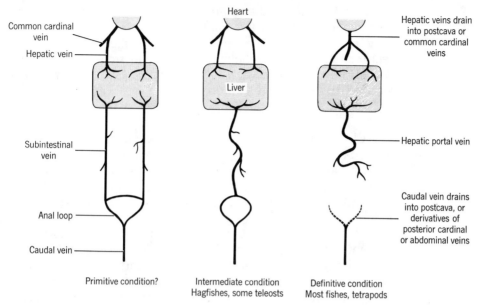

Heart

Common cardinal vein

Hepatic vein

Hepatic veins drain into postcava or common cardinal veins

Liver

Subintestinal vein

Hepatic portal vein

Anal loop

Caudal vein drains into postcava, or derivatives of posterior cardinal or abdominal veins

Caudal vein

Primitive condition?

Intermediate condition Hagfishes, some teleosts

Definitive condition Most fishes, tetrapods

FIGURE 12-13
EVOLUTION OF THE SUBINTESTINAL SYSTEM OF VEINS. Ventral views.

tained by hagfishes and some teleosts. (Birds also join the hepatic portal system to the caudal drainage, but the homology of the connecting vessel is disputed. Lampreys are anomalous in that blood from the gut enters the cardinal system.) In all other vertebrates the hepatic portal loses its connection to the anal loop. Consequently, the caudal blood must enter another system of veins (see below).

Derivatives of Cardinal System. Anterior cardinal veins drain the brain and part of the head in cyclostomes and fishes. In tetrapods the same vessels are called **internal jugular veins** and are somewhat modified peripherally by various combinations of the parent vessels with their embryonic tributaries. The more ventral and external parts of the head are drained in fishes by veins of uncertain homologies usually called **inferior jugular veins.** The same regions are drained in tetrapods by **external jugular veins** which are apparently enlarged tributaries of the primitive anterior cardinals. They join the internal jugulars in the neck. The common cardinal veins are paired in fishes; one or the other is lost in adult cyclostomes. Venous sinuses are common in the cardinal and jugular drainages—particularly in cyclostomes, cartilaginous fishes, and (adjacent to the brain) in mammals.

Between the heart and the confluence of internal and external jugulars the derivatives of the cardinals of tetrapods are called **precavae** (also anterior vena cavae). These vessels are paired in tetrapods other than mammals and also in some primitive mammals (most marsupials and rodents, insectivores, and some others). In other mammals a shunt called the **innominate vein** (or left brachiocephalic) carries blood from the left jugulars to the right precava. The left precava is then lost except for a contribution to the coronary sinus.

CYCLOSTOMES preserve the primitive condition of the posterior cardinal veins—posteriorly they join the anal loop of the subintestinal system and anteriorly they enter the remaining common cardinal. They drain the urogenital organs and body wall and share the drainage of the tail with the hepatic portal vein.

In FISHES an interruption develops in the posterior cardinals just anterior to the kidneys. Blood from the anal loop (which is now usually detached from the hepatic portal system) usually flows into the posterior segments of the divided posterior cardinals—now called **renal portal veins**—passes into the tissues of the kidneys, and emerges into subcardinal veins which transmit it to the anterior segments of the posterior cardinals. Certain segmental veins enter the kidney directly. The **renal portal system** is thus established (which, as we shall see later, is of great physiological importance). In some species of teleosts the renal circulation is asymmetrical, and in some, all or part of the caudal blood can bypass the kidneys. The posterior cardinals of most cartilaginous fishes expand into sinuses near the heart.

DIPNOANS and URODELES retain the same pattern of vessels but add a new vessel, the postcava (see below), which receives most of the blood from the kidney. The reduced anterior segments of the posterior cardinal veins drain the body wall and perhaps receive a trickle of blood from the anal loop if the interruptions in their course are incomplete.

ANURANS and REPTILES are further advanced in that the anterior segments of the posterior cardinals are usually represented only by variable **vertebral veins** which drain the anterior part of the thorax. (Again, *Sphenodon* retains the primitive condition.) All blood in the renal portal veins now enters the kidneys, but in some species part of the blood is shunted through the organ to the postcava without entering a capillary bed.

The situation is the same in BIRDS except that nearly all blood passes through the kidney to the postcava. It enters kidney tissues only if a valve closes and occludes the direct route. MAMMALS have no renal portal system at all. An **azygous vein** and, in some species, a **hemiazygous vein,** which drain part of the thorax, are the only derivatives of the anterior

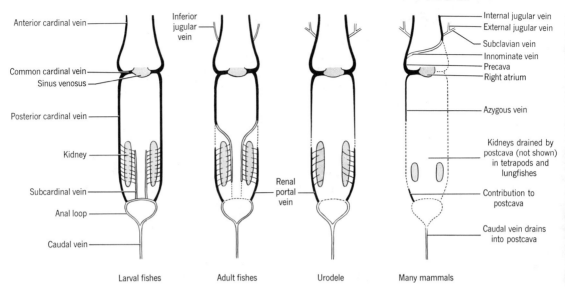

FIGURE 12-14
**EVOLUTION OF THE CARDINAL SYSTEM OF VEINS (solid black) AND SOME
RELATED VESSELS (open). Ventral views.**

segments of the posterior cardinals. A derivative of one of the
posterior segments will be mentioned below.

Derivatives of Abdominal System.   The primitive pattern of the
abdominal system is shown by CHONDRICHTHYES. Posteriorly
the paired **lateral abdominal veins** receive small **iliac veins** from
the pelvic fins but do not connect with other venous systems.
Anteriorly they join the **subclavian veins** from the pectoral fins
and then empty into the common cardinals.

AMPHIOXUS, CYCLOSTOMES, and RAY-FINNED FISHES
lack this system altogether. The iliac veins of these fishes join
the posterior cardinal veins (or their derivatives, the renal
portal veins), and the subclavian veins join the posterior or
common cardinals.

DIPNOANS, AMPHIBIANS, and REPTILES have modified the
ancestral pattern in similar ways. The abdominal veins may
remain paired (turtles and crocodilians) but usually fuse in the
midline of the body wall to form a **ventral abdominal vein.** Posteriorly, the abdominal system is joined to the renal portal
veins by paired **pelvic veins.** Anteriorly, it enters the right
common cardinal vein (dipnoans) or shifts to join the hepatic
portal system within, or just posterior to, the liver. The subclavian veins do not take part in this shift and thus are left to
enter cardinal vessels or their derivatives, the precavae. These
changes bring the appendicular veins into the same relationships they have in fishes that lack the abdominal system.
There is the difference, however, that when present, the ab-

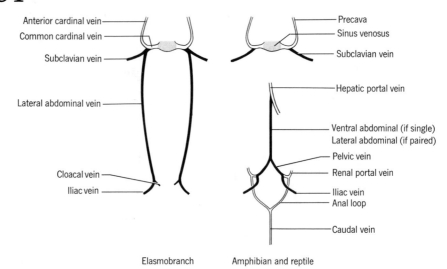

Elasmobranch   Amphibian and reptile

FIGURE 12-15

EVOLUTION OF THE ABDOMINAL SYSTEM OF VEINS (solid black) AND SOME RELATED VESSELS (open). Ventral views.

dominal system of these animals joins the two great portal systems of the body.

Embryonic amniotes have paired abdominal veins—the primitive condition. Fetuses retain both (most reptiles) or the left only (birds and mammals) as **allantoic veins** (reptiles and birds) or the **umbilical vein** (mammals). These vessels enter the liver as their antecedents did in ancestral amphibians. Within the liver a channel called the **ductus venosus** usually connects them directly with the hepatic veins so the fetal circulation is not functionally portal.

Adult BIRDS are usually considered to lack abdominal veins. There is a vein (variously named) that may possibly be a derivative. The system is absent in all adult MAMMALS except the monotreme *Echidna.*

Postcava: A New Vessel from Leftovers. In fishes, veins that relate to the kidneys are often asymmetrical. Among these are the anterior segments of the posterior cardinal veins and their continuations derived from the embryonic subcardinals. In DIPNOANS, these vessels persist on the left but are so much larger on the right that the resultant channel is given a new name, the **postcava** (also posterior vena cava). This vessel drains the tail and both kidneys and shares with the abdominal system the drainage of the pelvic appendages.

In AMPHIBIANS, REPTILES, and BIRDS the postcava does not continue into the caudal vein and is, therefore, a short vessel extending only to the posterior end of the kidneys. The proximal end of the vein is now derived from a right hepatic vein instead of from the anterior segment of the right posterior cardinal. This releases the cardinal which persists as such in urodeles and (with the left postcardinal vein) may form small vessels from the body wall in the other vertebrates of this group.

In MAMMALS the right supracardinal vein and right posterior segment of the posterior cardinal vein are added (in various proportions) to the contribution of the right subcardinal and hepatic veins to complete the postcava. This lengthens the vessel and again joins it to the caudal vein. Portions of the corresponding veins on the left and anastomoses between left and right may also be incorporated, and the adult vessel is normally paired in various species in several orders. The urogenital veins form in conjunction with the postcava, chiefly from the subcardinal veins. Considering this complicated ontogeny it is not surprising that there is much individual variation in these vessels.

It is desirable to relate the structure of the circulatory system to its various functions. One way to do this is to trace the path of blood from one point to another in a given animal. Thus, a blood cell traveling from the left ventricle of an adult mammal to the intestine and back to the starting point by the most

Functional Units of the Blood-Vascular System

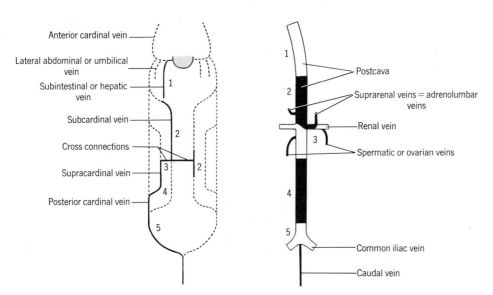

FIGURE 12-16

FORMATION OF THE POSTCAVA IN A MAMMAL. Left, embryonic and evolutionary primordia; right, adult derivatives. Ventral views. Contributions of the different primordia vary among tetrapods.

direct route would traverse—in order—the systemic trunk and arch, dorsal aorta, superior mesenteric artery, capillary bed of intestine, hepatic portal vein, liver sinusoids, hepatic veins, postcava, right atrium, right ventricle, pulmonary artery, capillary bed of lungs, pulmonary vein, left atrium, and left ventricle.

A similar method of study is to compare like circuits of different animals. For instance, to travel from a pelvic appendage to the heart (the most complex example), blood of a cartilaginous fish could only traverse lateral abdominal and common cardinal veins. In ray-finned fishes it would traverse a renal portal vein; probably, but not certainly, the tissues of the kidney; and the posterior and common cardinals. Dipnoans would offer the choice of renal portal, kidney, and postcava; or pelvic vein, ventral abdominal, and common cardinal. In urodeles the blood could course through (1) renal portal, kidney, and postcava; (2) pelvic vein, ventral abdominal, hepatic portal, liver sinusoids, hepatic vein, and postcava; or (3) renal portal, posterior cardinal, and precava. Anurans offer choices (1) and (2) above. Reptiles also offer routes (1) and (2), but kidney tissues might be eliminated from route (1). Further, a small quantity of blood might follow route (3) in certain reptiles although instead of the posterior cardinal the equivalent vertebral veins would be followed. In birds, nearly all blood would shunt through the kidney to the postcava. The blood could enter only the postcava of mammals. (The above discussion disregards small collateral systems.)

Students can practice on circuits of their own selection. It would be a mistake to identify the channels and learn the evolution of the elements of the system without attention to their functional relationships.

Another way to relate structure to function is to consider the qualitative changes that occur in the blood in its travels. In this context one can consider a circuit to be one capillary bed where the blood is modified in a given way, another where the original quality is restored, and the intervening channels. To simplify the analysis I shall continue by commenting on the functional requirements and correlated evolutionary changes in units of the system that might be considered half circuits—capillary beds that modify the blood in specific ways together with the related afferent and efferent vessels. Quality of the blood is also modified by the mixing of different streams of blood in the heart and vessels.

Metabolic Units. The term "metabolic units" will be applied to the parts of the circulatory system that relate directly to tissue metabolism. All living tissues take oxygen and nutrients from the blood and release carbon dioxide and other wastes. However, the requirements of the various tissues differ widely: Cartilage, fat, and other supportive tissues have low metabolism—a meager blood supply meets their needs. Liver and kidneys perform their complicated tasks with surprisingly little oxygen: Kidneys of some vertebrates receive only portal blood, and the livers of other vertebrates receive nearly exclusively

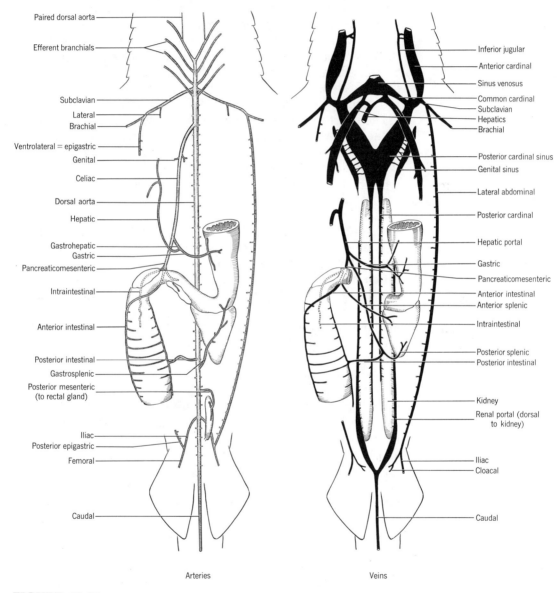

Arteries   Veins

FIGURE 12-17

CIRCULATORY SYSTEM OF THE SHARK *Squalus*. **Ventral views. For the ventral aorta and branchial vessels see Figures 12-7 and 12-12.**

portal blood. Muscles and glands, in contrast, usually require much oxygen and nutrients when active (particulary for mammals). Heart muscle and the central nervous system are the most stringent in their needs.

Vessels that convey blood to the muscles and nervous system of FISHES (except Dipnoi) are relatively primitive; the only modifications of functional consequence are in certain aortic arches. Since the tissues need oxygenated blood, their supply is always from branchial efferent arteries. The first aortic arch, however, appears in the embryo and is subsequently lost because the first visceral arch is converted to jaws and forms no gills. If retained, the oxygen-poor blood of this aortic arch would be detrimental. Similarly, the second aortic arch appears and is then lost (or much modified) unless the second visceral arch forms a hemibranch.

The heart muscle of fishes cannot be supplied by blood from adjacent vessels because that blood is poor in oxygen. It is supplied instead by paired **coronary arteries** which branch from the dorsal aorta, subclavian arteries, one of the more posterior pairs of branchial efferents, or from a combination of these.

Vessels that convey blood to the actively metabolizing tissues of TETRAPODS must adapt to several altered conditions. The heart is now pumping two streams of blood. If separation of the two streams is complete, all systemic blood is equivalent and quality is not a factor in its distribution. If separation is incomplete, the configuration of the heart and proximal arteries is such that the best oxygenated blood goes to the brain and head where it is most needed. Some interesting variations of the subclavian and other proximal arteries — particularly of reptiles — represent "successful experiments" in the symmetrical distribution of blood relative to its quality.

Coronary arteries no longer need return to the heart by devious vessels from the efferent side of gills; instead they branch from the systemic trunk where it leaves the heart. (Some amphibians and fishes lack coronary vessels. Their cardiac tissue utilizes the blood in the heart chambers.)

The strong axial muscles of fishes are segmental; they are efficiently served by segmental vessels. Tetrapods, having lost the segmental arrangement of most of their muscles, supplement segmental vessels with longitudinal vessels. These form in the embryo by the anastomosis of the distal ends of adjacent segmental arteries and the subsequent loss of the proximal ends of all except the one or two that will deliver blood to the resultant longitudinal vessel. The vertebral, ventral thoracic, and epigastric arteries are of this nature.

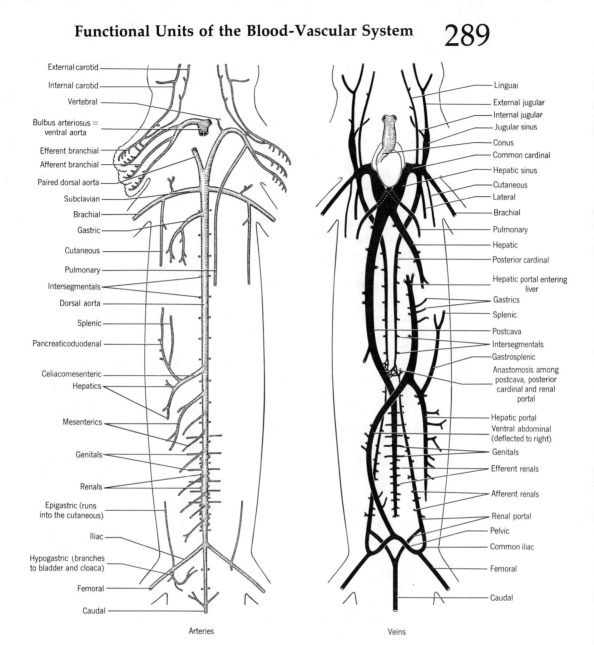

FIGURE 12-18
CIRCULATORY SYSTEM OF THE URODELE *Necturus.* **Ventral views.**

Fishes maintain a nearly constant body configuration; vessels may be interrupted by injury but not occluded by accidents of posture. Further, gravity does not affect blood circulation of fishes because of the buoyancy of the surrounding water. Tetrapods temporarily restrict or occlude vessels as posture is changed, and gravity does affect the distribution of blood. Several provisions are made to ensure that tissues will not be deprived. Arterial blood pressure tends to be higher than in fishes, and blood volume is relatively greater. Most parts of the body have a double blood supply by vessels that are parallel (e.g., ulnar, radial, and interosseus arteries of forearm) or that form distribution loops having inflow at each end (e.g., volar arches which deliver blood to digits, arches along curvatures of stomach and intestine, confluence of ventral thoracic and epigastric arteries in body wall, confluence of internal carotid and vertebral arteries in a common distribution loop ventral to the brain). Further, throughout the body, adjacent major arteries are joined by frequent small anastomosing vessels, and isolated major arteries have frequent small branches that form somewhat meandering and criss-crossing adjacent channels. These small vessels comprise the **collateral circulation.** They are standby channels in case of occlusion or injury to the major channels. Fishes also have a collateral circulation, but it is enhanced in tetrapods.

The venous drainage of the more active tissues is similarly primitive in FISHES, and such variations as are found (as in the lateral abdominal system, veins from second aortic arch, and veins of head) seem not to relate to the function of metabolic units. The veins of fishes have valves only at the junctions of certain tributaries with major channels. This is apparently because the channels are relatively direct (there being no long neck or appendages) and the flow of blood is not complicated by gravity. Nevertheless, venous pressure is very low in fishes, and some of them have evolved accessory venous pumping mechanisms. Caudal musculature of elasmobranchs squeezes blood from segmental veins through valves into the caudal vein which, being in the rigid hemal canal, is not constricted by the muscles. In hagfishes (and perhaps some other fishes as well) respiratory movements hasten venous return in nearby vessels.

The patterns of veins that drain the more active tissues of TETRAPODS vary markedly from those of fishes (except dipnoans) and also vary among the tetrapod taxa. Again, however, the functional interpretation of the variations is not evidently related to the requirements of metabolic units. The veins, like the arteries, tend to develop rich collateral circulations. Valves are usually present; the pull of gravity contributes to their necessity.

Respiratory Units. The function of respiratory units is to restore the concentrations of respiratory gases from the levels occurring in efferent vessels of metabolic units to the levels required in afferent vessels of metabolic units. Blood delivered to the respiratory organ should be low in oxygen content and high in carbon dioxide content, and blood collected from the organ should reach the metabolizing tissues with little or no prior mixing with less oxygenated blood.

We have learned that blood oxygenated in the gills of fishes and branchiate amphibians continues directly to the metabolizing tissues without returning to the heart. This plan has the advantage that there can be no detrimental mixing of blood on either side of the respiratory organ. It has the disadvantage that blood pressure in the dorsal aorta can be only one-quarter or one-half of the pressure in the ventral aorta. This accounts for the accessory venous pumping systems mentioned above, and doubtless also accounts in part for the shunts that are common from afferent to efferent branchials: By bypassing the gill capillaries with part of the blood the drop in pressure can be minimized at the expense of respiratory efficiency.

Air-breathing vertebrates ultimately solved the problem of maintaining blood pressure in vessels leading to the metabolizing tissues by returning all blood from the respiratory organ to the heart whence it could be pumped to the tissues. Solving one problem created others, however, and as we have seen, the evolution of a complete double circulation was very slow.

It would be reasonable for air-breathing fishes to supply blood to their lungs from the ventral aorta or branchial afferent arteries—those being vessels having both oxygen-poor blood and adequate blood pressure. It is surprising that none did either (insofar as we can judge from living fishes). Instead, these fishes supply their lungs with blood coming either from the dorsal aorta (or paired dorsal aortae or branch of the dorsal aorta) or from the efferents of the fifth and sixth aortic arches. Either arrangement can be effective only if the more posterior gills do not accomplish a significant degree of oxygenation of the blood before it is sent on to the lungs. This circumstance would pertain in stagnant water or if much blood were shunted past the gill capillaries.

In any event, to take blood to the lungs from dorsal aorta or posterior branchial efferent arteries has what appears to be a serious weakness: Blood flowing to the posterior part of the body can be no richer in oxygen than blood that will go *to* the lungs if the latter comes from the aorta, and scarcely richer if it comes from efferents of the fifth and sixth arches. (It can be a little richer because of mixing of fifth and sixth arch blood with blood from the fourth arch which has received the more oxygenated stream from the heart.) It is not efficient to mix oxygen-rich blood from the lungs with oxygen-poor blood from the tissues and then send the mixture to both lungs and tissues. It is not efficient, but it must be acknowledged that it is adequate because it has met the needs of air-breathing

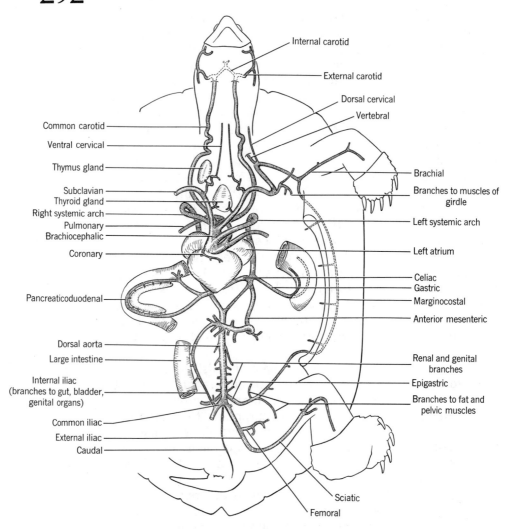

Internal carotid

External carotid

Dorsal cervical

Vertebral

Common carotid

Ventral cervical

Thymus gland

Brachial

Subclavian
Thyroid gland
Right systemic arch
Pulmonary
Brachiocephalic

Coronary

Branches to muscles of girdle

Left systemic arch

Left atrium

Celiac
Gastric
Marginocostal

Anterior mesenteric

Pancreaticoduodenal

Dorsal aorta
Large intestine

Internal iliac
(branches to gut, bladder,
genital organs)

Common iliac
External iliac
Caudal

Renal and genital branches

Epigastric

Branches to fat and pelvic muscles

Sciatic
Femoral

FIGURE 12-19
ARTERIES OF A TURTLE, family Testudinidae.

fishes. (Air-breathing fishes eliminate most carbon dioxide by way of the gills, not the lung, and this may have contributed to the evolutionary dilemma. Their need for gills is continuous; their need for lungs is intermittent.)

This general plan also serves the needs of the embryos and fetuses of amniotes. The respiratory organ is a yolk sac or allantois (or placenta which is derived in part from the yolk sac or allantois). The vitelline arteries to yolk sac, and allantoic or umbilical arteries to allantois or placenta, are like the pulmonary arteries of some fishes in being branches of the dorsal aorta. For fetuses also, therefore, the posterior part of the body must get along with blood having the same oxygen content as that going *to* the respiratory organ.

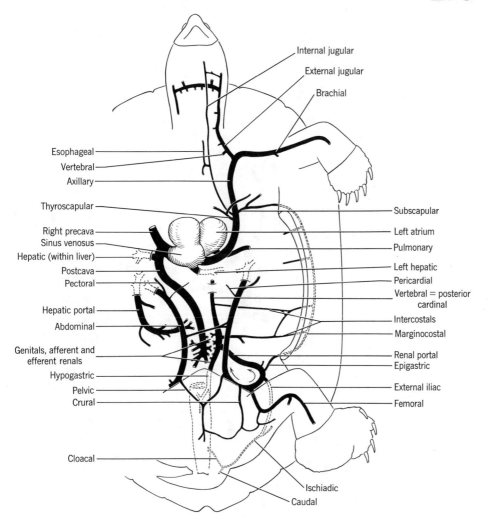

FIGURE 12-20
VEINS OF A TURTLE, family Testudinidae.

Fetuses can alter this situation only by abandoning their fetal respiratory organs at birth or hatching. Adult fishes that supply blood to the lungs from the dorsal aorta cannot alter the situation and seemingly have no need to do so. Urodeles, most dipnoans, and several teleosts (and certainly also the ancestral crossopterygians) have the potential to "improve" the system, and they have succeeded in varying degrees. Adult birds and mammals deliver only oxygen-rich blood to all metabolizing tissues. The first requirement was to separate more and more successfully the systemic and pulmonary streams of blood in the heart, and to deliver the systemic stream to the anterior arches and the pulmonary stream to the posterior arches. The second requirement was to keep the streams separate beyond

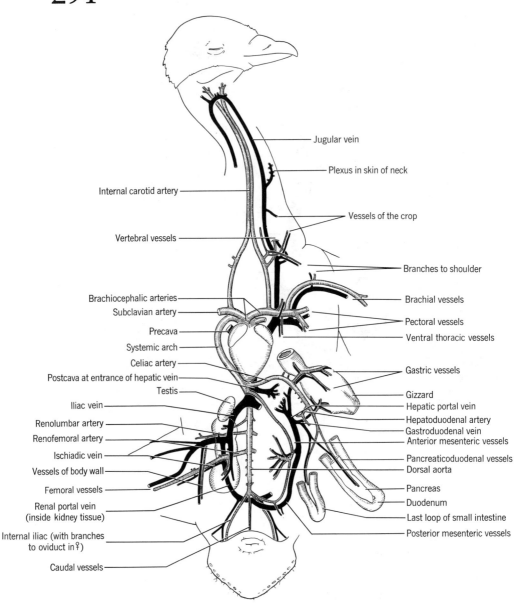

Jugular vein

Plexus in skin of neck

Internal carotid artery

Vessels of the crop

Vertebral vessels

Branches to shoulder

Brachiocephalic arteries

Brachial vessels

Subclavian artery

Pectoral vessels

Precava

Ventral thoracic vessels

Systemic arch

Celiac artery

Gastric vessels

Postcava at entrance of hepatic vein

Gizzard

Testis

Hepatic portal vein

Iliac vein

Hepatoduodenal artery

Renolumbar artery

Gastroduodenal vein

Renofemoral artery

Anterior mesenteric vessels

Ischiadic vein

Pancreaticoduodenal vessels

Vessels of body wall

Dorsal aorta

Femoral vessels

Pancreas

Renal portal vein
(inside kidney tissue)

Duodenum

Last loop of small intestine

Internal iliac (with branches
to oviduct in ♀)

Posterior mesenteric vessels

Caudal vessels

FIGURE 12-21
CIRCULATORY SYSTEM OF THE PIGEON.

the roots of the arches. This was accomplished by (*1*) joining the pulmonary artery to the sixth aortic arch only (tetrapods, most dipnoans, and the chondrostean *Polypterus),* (*2*) deleting the fifth arch (amniotes and most amphibians), and (*3*) deleting the portion of the sixth arch extending from the root of the pulmonary artery to the dorsal aorta (amniotes except the lepidosaur *Sphenodon,* and anurans). These changes are recapitulated by the embryos of amniotes.

The evolution of an effective respiratory circulation required one further change for air-breathing vertebrates: Blood returning to the heart from the lungs had to be kept separate from systemic blood. Most fishes drain blood from the gas bladder into the postcardinal or hepatic portal veins, which are far from the heart. *Polypterus* drains blood from the lung into hepatic veins. These are close to the heart, but nevertheless, mixing with systemic blood is assured. The bowfin, a holostean, drains blood from the lung into the left common cardinal vein where it mixes with only part of the systemic blood. The pulmonary veins of dipnoans and tetrapods, at last, enter directly into the left, or "oxygenated side" of the heart.

As noted at the end of Chapter 11, various fishes and amphibians supplement branchial or pulmonary respiration by respiring also through the skin or lining of some part of the alimentary canal. These respiratory units all appear "imperfect" because of shortcomings of the associated circulation. Thus, although the cutaneous artery to the skin of amphibians is a branch of the pulmonary artery, and therefore carries only blood that is rich in carbon dioxide, blood in the cutaneous vein mixes with anterior systemic blood before reaching the heart.

Nutritive Units. Nutrients enter the blood from the gut of the adult vertebrate and from the yolk sac of most embryos. The allantois has only a subsidiary nutritive function in many amniotes (absorbing albumen or salts from the eggshell), but in most eutherian mammals it becomes the fetal contribution to the placenta and thus is the nutritive organ of the fetus. (The yolk sac instead contributes to the placenta in many marsupials and rodents.)

Blood flowing to the nutritive organ of all vertebrates, except larvae of bony fishes and amphibians, traverses ventral branches of the dorsal aorta. This blood may be oxygenated (if flowing to the adult gut or to the yolk sac of larval cartilaginous fishes) or mixed (if flowing to the yolk sac or allantois of fetal amniotes). Blood flowing to the distinctive yolk sacs of larval bony fishes and to the gut of larval amphibians traverses any of several major venous drainages and thus is portal blood and oxygen-poor. It has been thought that there must be unknown significance in these differences. A more probable interpretation is that a nutritive organ is little dependent on oxygen and can, therefore, receive oxygenated or unoxygenated blood at the "convenience" of the evolutionary process.

Blood from the nutritive organ is usually processed by the liver before it reaches the heart. It always mixes with posterior systemic blood before it reaches the heart and usually mixes with anterior systemic blood within the heart (this mixing is only partial in fetuses of some amniotes). It follows that no

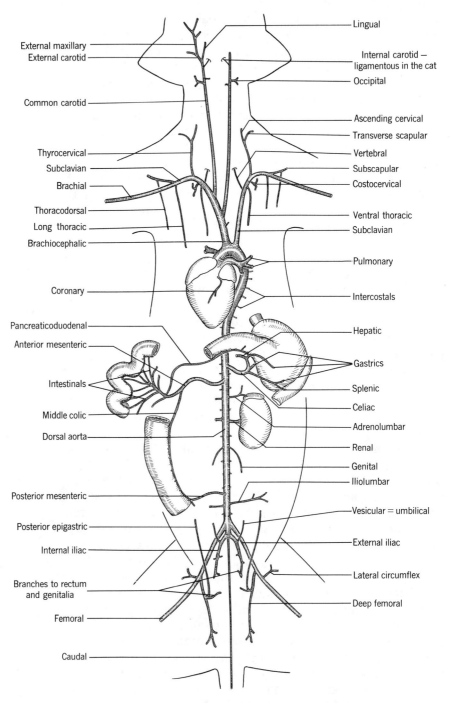

External maxillary
External carotid
Common carotid
Thyrocervical
Subclavian
Brachial
Thoracodorsal
Long thoracic
Brachiocephalic
Coronary
Pancreaticoduodenal
Anterior mesenteric
Intestinals
Middle colic
Dorsal aorta
Posterior mesenteric
Posterior epigastric
Internal iliac
Branches to rectum
and genitalia
Femoral
Caudal

Lingual
Internal carotid —
ligamentous in the cat
Occipital
Ascending cervical
Transverse scapular
Vertebral
Subscapular
Costocervical
Ventral thoracic
Subclavian
Pulmonary
Intercostals
Hepatic
Gastrics
Splenic
Celiac
Adrenolumbar
Renal
Genital
Iliolumbar
Vesicular = umbilical
External iliac
Lateral circumflex
Deep femoral

FIGURE 12-22
ARTERIES OF THE CAT. Ventral view.

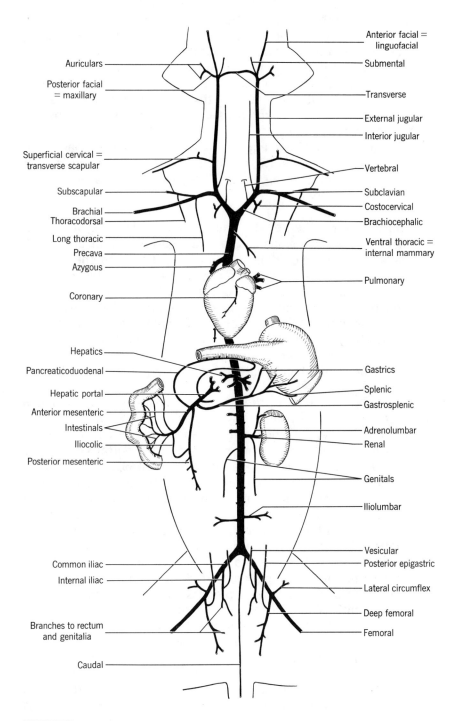

FIGURE 12-23
VEINS OF THE CAT. Ventral view.

Auriculars

Posterior facial = maxillary

Superficial cervical = transverse scapular

Subscapular

Brachial
Thoracodorsal

Long thoracic
Precava
Azygous

Coronary

Hepatics
Pancreaticoduodenal

Hepatic portal
Anterior mesenteric
Intestinals
Iliocolic
Posterior mesenteric

Common iliac
Internal iliac

Branches to rectum and genitalia

Caudal

Anterior facial = linguofacial
Submental

Transverse

External jugular
Interior jugular

Vertebral

Subclavian
Costocervical
Brachiocephalic

Ventral thoracic = internal mammary

Pulmonary

Gastrics
Splenic
Gastrosplenic

Adrenolumbar
Renal

Genitals

Iliolumbar

Vesicular
Posterior epigastric

Lateral circumflex

Deep femoral
Femoral

organ other than the liver receives blood having maximum concentration of nutrients. Further, the yolk sacs of larval bony fishes and the gut of larval amphibians are the only nutritive organs to receive blood that is low in nutrients (because it comes in a portal system). All other nutritive organs receive mixed blood that is as rich in nutrients as the blood going to other organs. That the body has not provided a maximum gradient of nutrients across the absorbing membranes is perhaps because absorption is largely by active cell transport rather than by diffusion.

In general, nutritive units of the circulatory system have relatively wide latitude in the quality of blood received and in their relation to metabolizing tissues. Their evolution has, accordingly, been conservative.

Hepatic Unit. The hepatic unit of the circulatory system consists of the liver sinusoids, a large portal system and small hepatic artery which deliver blood to the liver, and one or more hepatic veins which drain the organ. The afferent blood always includes all the blood draining the nutritive organ and in addition may (in amphibians and reptiles) include venous blood from the hind limbs, tail, and abdominal wall. The efferent blood always mixes with other blood before reaching the heart. The liver alters the quality of blood in numerous ways (see p. 233), many of which relate to conversion or storage of foodstuffs.

The hepatic unit is the most conservative of all principal units of the circulatory system. Functional interpretation requires the explanation of constancy rather than of change. Why is the adult liver always provided with a portal system and with only a relative trickle of arterial blood? It has been suggested that slow-moving blood of low oxygen content is necessary to liver function. This seems doubtful, however, since, in fetuses of amniotes, it receives quantities of oxygenated blood from the allantois or placenta. A more probable explanation is that the liver *can* function on venous blood and that it would, therefore, be extravagant to "waste" on the organ the necessary large volume of oxygenated blood under high pressure.

Why does the liver always process all the blood from the nutritive unit? (The placental blood of some mammals shunts through the fetal liver without change, but the nutrients in this blood have been processed by the maternal liver.) One suggestion is that the unprocessed nutrients would be detrimental in the general circulation or unavailable to metabolizing units. This is probable, but has not been adequately substantiated. Another suggestion is that the liver can best process nutrients if it receives them in high concentration. This could hardly be so, because amphibians and reptiles then would not dilute the nutrients by mixing in blood from the posterior part of the general circulation. A possible explanation is that the relatively large liver develops in the path of both the subintestinal and abdominal systems, and it is these systems that ultimately contribute all the portal blood. A difficulty with this explanation is that the postcaval vein *also* courses near or through the liver, yet never releases its blood to that organ. This problem seems not yet to have a full answer.

Renal Unit. The vertebrate kidney usually has two capillary beds. One is organized as tiny knots called glomeruli which lie within the renal capsules and are associated with the filtration phase of kidney function. The other is organized as nets which surround the nephric tubules and are associated with excretion and selective resorption.

Glomeruli are always supplied by renal arteries, which are lateral branches of the dorsal aorta. Capillaries surrounding the tubules may take "seconds" on the blood from the glomeruli, but usually are supplied by a renal portal system. The two capillary beds merge, so blood from the two sources mixes within the kidney.

Animals that produce a large volume of urine (freshwater fishes, amphibians, mammals) must have a large high-pressure (arterial) flow of blood to the glomeruli, because it is this pressure that is the driving force of the filtration process. These animals can either reuse the same blood for the tubules (mammals) or use portal blood instead. The oxygen content of this blood is unimportant, it is low pressure that is required so reabsorption can take place. Animals that conserve water and salts (marine fishes, reptiles, birds) must have a small high-pressure flow to the glomeruli (or none in some fishes) so filtration will be limited. In order to supply enough low-pressure blood to the tubules, most marine fishes *must* have a renal portal system.

We have learned that blood drains out of the kidneys into either subcardinal and posterior cardinal veins or into a postcava. It makes no functional difference which, because in either circumstance this blood is thoroughly mixed with other blood before reaching the heart.

Evolution of the Lymphatic System

Amphioxus has no lymphatic system. CYCLOSTOMATA, CHONDRICHTHYES, and CHONDROSTEI have networks of fine vessels that resemble lymphatic vessels but differ in containing some red blood cells and in joining the veins and venous sinuses in many places. The term **hemolymphatic system** is appropriate for these vessels. The more superficial channels tend to be more like veins and the visceral channels more like lymphatics. **Hemolymph propulsors** are present in hagfishes but absent in these other fishes. These valved reservoirs passively propel hemolymph as extrinsic muscles impinge on them. The hemolymphatic system of these vertebrates almost certainly represents an early stage in the evolution of the true lymphatic system.

The lymphatic system is incomplete in Dipnoi. It is fully developed in HOLOSTEI and TELEOSTEI. Four **subcutaneous ducts** are typical: one dorsal, one ventral, and two lateral. One or usually two **subvertebral ducts** (also called cardinal ducts) extend the length of the body cavity. Cranial and visceral networks complete the system. Small pairs of **lymph propulsors** are usually present in the tail and near the pharynx. There are no valves in the vessels. Lymph usually enters the venous system by a pair of openings in the anterior cardinal veins and another pair near the iliac veins.

AMPHIBIANS and REPTILES agree with teleosts in having well-developed lymphatic systems with subcutaneous, subvertebral, and visceral vessels. Valves are absent. Urodeles usually have numerous small pairs of segmentally arranged **lymph hearts** (with their own intrinsic musculature) and also a larger pair near the bases of each set of limbs. Anurans have fewer small hearts but retain the two larger pairs. They have large **lymph sinuses** under the skin. Reptiles have the posterior pair of hearts which drain into the renal portal veins. They are located at the ends of transverse processes of basal caudal vertebrae.

The subvertebral ducts of BIRDS are also called **thoracic ducts.** They drain into the precavae. Smaller ducts may enter the iliac veins. Valves may be present but are not numerous. Small lymph hearts appear in the fetus but are rarely present in the adult.

Most MAMMALS have a large single thoracic duct. It is derived from the subvertebral ducts and drains most of the body and enters the left jugular or subclavian vein. A small duct on the right drains the forelimb and shoulder on that side and enters the right subclavian vein. Numerous bicuspid valves are present in the vessels. Lymph hearts are absent. **Lymph nodes,** sparcely represented in the other tetrapod classes, are numerous in mammals—particularly among the viscera, in the neck, and at the bases of the limbs. The system has many anastomoses and parallel channels. Lymphatic vessels penetrate most tissues but are absent from the central nervous

FIGURE 12-24

LYMPHATIC SYSTEM. Ventral views. The teleost and lizard are composite drawings. Modified from Kampmeier.

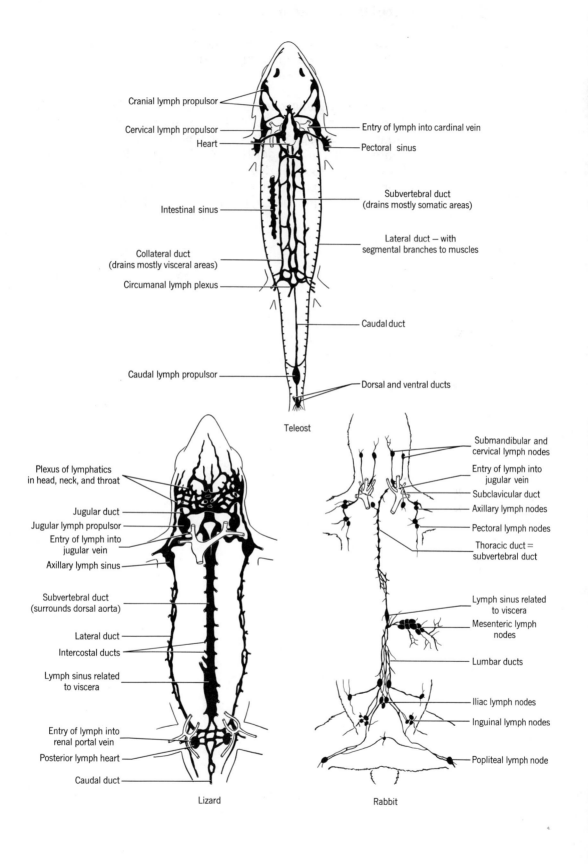

Cranial lymph propulsor

Cervical lymph propulsor

Heart

Entry of lymph into cardinal vein

Pectoral sinus

Intestinal sinus

Subvertebral duct
(drains mostly somatic areas)

Lateral duct — with
segmental branches to muscles

Collateral duct
(drains mostly visceral areas)

Circumanal lymph plexus

Caudal duct

Caudal lymph propulsor

Dorsal and ventral ducts

Teleost

Plexus of lymphatics
in head, neck, and throat

Jugular duct

Jugular lymph propulsor

Entry of lymph into
jugular vein

Axillary lymph sinus

Subvertebral duct
(surrounds dorsal aorta)

Lateral duct

Intercostal ducts

Lymph sinus related
to viscera

Entry of lymph into
renal portal vein

Posterior lymph heart

Caudal duct

Lizard

Submandibular and
cervical lymph nodes

Entry of lymph into
jugular vein

Subclavicular duct

Axillary lymph nodes

Pectoral lymph nodes

Thoracic duct =
subvertebral duct

Lymph sinus related
to viscera

Mesenteric lymph
nodes

Lumbar ducts

Iliac lymph nodes

Inguinal lymph nodes

Popliteal lymph node

Rabbit

system, bone marrow, deeper parts of the liver and spleen, epithelium of the skin, cartilage, and placenta.

The **thymus gland** occurs in all vertebrates except possibly the cyclostomes. It is derived from the epithelium of one or (usually) several pharyngeal pouches and comes to lie in the gill area, neck, or anterior thorax. It may be diffuse or discrete. The thymus produces lymphocytes and also establishes the immunological potential of the young animal. It relates as closely to the lymphatic system as to any other.

13

Urogenital System

Function and Characteristics. Many organs contribute to the elimination of waste products from the body. The excretory functions of lungs, gills, skin, and parts of the digestive system have been mentioned, and the role of salt glands will be noted later in this chapter. Kidneys, however, are nearly always the primary adult excretory organs and as such perform two related and essential functions: (1) They remove the nitrogenous waste products of protein metabolism and also many other harmful substances, and (2) they eliminate controlled amounts of water and salts, thus maintaining the internal environment within the narrow limits necessary for life.

General Nature and Development of Kidneys

303

Embryologists and morphologists have studied the urinary system for various reasons. Like the pharynx and circulatory system, it is outstanding for the recapitulation which takes place during its development: The system was much studied several generations ago in an effort to elucidate vertebrate origins. Conclusions were tentative, however, and most current research relates structure to the function of the system instead of to its evolution. Because the fine structure of kidneys correlates with physiological needs, it varies within certain large taxa (teleosts, rodents) having diverse adaptations and is usually not a key to close systematic relationships. Further, the urinary system is of interest because it provides striking examples of gradients. Embryos and ancestors each form a series of excretory structures of increasing complexity which replace one another in orderly sequence in time and space.

The kidneys of vertebrates are compact organs derived from mesoderm and consisting of numerous **nephric tubules** (also called renal tubules—"nephros" and "ren" both mean kidney) that open either into the general coelom or into cup-shaped spaces called **renal capsules** which are coelom derivatives. Such kidneys are called **metanephridia.** In the metanephridia of vertebrates, quantities of water and other constituents of the plasma leave the blood by filtration from small knots of capillaries called **glomeruli** (= small balls). If the capillary bed is surrounded by a renal capsule, it is said to be an **internal glomerulus.** Glomerulus and capsule together form a **renal corpuscle.** If the capillary bed discharges instead into the general coelom, it is then called an **external glomerulus** and the filtrate reaches the tubules indirectly. A second capillary bed meshes around each tubule where many constituents of the filtrate are selectively returned to the bloodstream. A tubule with related corpuscle is the functional unit of the vertebrate kidney and is called a **nephron.** There are thousands or even millions of nephrons in adult kidneys. Urine, the product of excretion by the kidneys, moves out of all the tubules into a long pair of **nephric ducts** which open into the cloaca or a derivative of the cloaca.

Amphioxus has short excretory tubules derived from ectoderm that do not open into coelom derivatives but instead terminate blindly at their inner ends in flask-shaped cells which drain wastes from the body fluids surrounding them. At its other end each tubule drains into the atrium, which in a sense is outside the body. Such kidneys are called **protonephridia** and are found also in many invertebrates. According to taxon, each flask-like cell has either cilia (making it a flame cell) or a

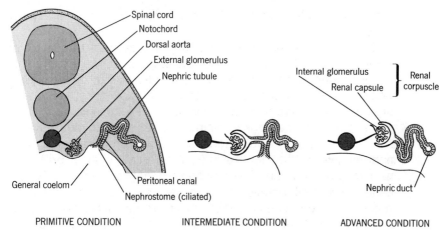

FIGURE 13-1
RELATIONS OF NEPHRIC TUBULE TO THE GLOMERULUS AND GENERAL COELOM. Cross-sections.

flagellum (making it a solenocyte) which drives fluid along in the tubules.

Protonephridia have been considered ancestral to meta-nephridia, but if so, the changes that brought the tubules into relation with the coelom or its derivatives, that caused the tubules to share a duct system, that made the common ducts discharge into the cloaca instead of outside the body, and that caused them to develop from mesoderm instead of ectoderm are all lost in antiquity. In truth, the kidneys of amphioxus do not support close affinity of Cephalochordata and Vertebrata.

A somewhat more attractive hypothesis is that ancient gen-ital ducts (called gonoducts or coelomoducts) became meta-nephridia. Still another theory postulates joint origin from protonephridia and genital ducts.

Development.   Recall that the embryonic mesoderm differen-tiates on each side of the body into a segmented dorsal epi-mere, a small lateral mesomere, and an unsegmented ventrolat-eral hypomere (Figure 5-4). Earlier chapters have presented the derivatives of the epimere and hypomere; we now come to the relatively inconspicuous but important mesomere.

The mesomere is as long as the general coelom, or even a little longer. Gradually it pinches off from the overlying epi-mere. Anteriorly it remains relatively thin and becomes seg-mented into units called **nephrotomes.** Posteriorly it does not become segmented and is called the **nephrogenic cord.** Nephro-tomes and nephrogenic cord merge into one another at a level which varies according to species and is unimportant to sub-sequent development. At mid and posterior levels the developing kidney bulges somewhat into the coelom to form a **nephric ridge.**

The differentiation of the mesomere is subject to control by induction: It will not form normal organs in the absence of either the adjacent epimere or hypomere, and undifferentiated mesoderm from elsewhere in the body will form urogenital organs if transplanted to the position of the mesomere.

The general coelom or splanchnocoel occurs in the hypomere. It is initially continuous with small coelomic spaces or **nephrocoels** in the nephrotomes (see again Figure 5-4). The narrow channels that join splanchnocoel to nephrocoels are called **peritoneal canals,** and their openings into the splanchnocoel are **nephrostomes.** (Early authors used the term "peritoneal funnel" instead and reserved "nephrostome" for the opening from nephrocoel to nephric tubule.) Functional peritoneal canals and nephrostomes are ciliated (Figure 13-1). Posteriorly, at levels of the nephrogenic cord, peritoneal canals may be present, but in most amniotes they fail to form. In their absence, spaces regarded as nephrocoels form within the cord by a process of cavitation. Nephrocoels are important because they become the renal capsules of the mature kidney.

Nephric tubules develop either as outgrowths of the walls of nephrocoels (probably the primitive method) or from small solid masses of mesenchyme. Gradually they lengthen and become more or less convoluted. The anterior tubules are the first to mature, and their outer ends join to form a nephric duct on each side of the body which then extends itself caudad by terminal proliferation until it has grown all the way to the cloaca (Figure 13-2). As the more posterior tubules mature they

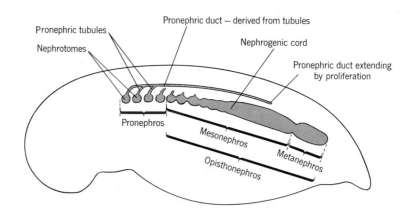

FIGURE 13-2
DEVELOPMENTAL ORIGIN OF THE ELEMENTS OF THE EXCRETORY SYSTEM. Left lateral view of a generalized vertebrate embryo. (Pronephric duct is actually lateral, not dorsal to the nephrogenic cord.)

merely join the preexisting duct. Indeed, the duct induces their development.

(The developmental steps outlined above are typical, but it would be misleading to imply that they are constant. Anterior renal corpuscles may approach the splanchnocoel and open out into it in such a way that the renal capsules disappear and the formerly internal glomeruli become external. Conversely, external glomeruli may become enveloped in folds of the peritoneal lining so that they become secondarily internal and discharge into renal capsules that are derived from the splanchnocoel instead of from nephrocoels. Adjacent nephrocoels and glomeruli may fuse, thus forming a giant corpuscle called a **glomus.** Another departure from the "standard" developmental process that has reference to the nephric duct will be noted below).

Archinephros: Ancestral Kidney. Although the mode of origin of vertebrate kidneys from invertebrate kidneys is highly speculative, the general nature of the early vertebrate kidney can be approximated from embryological and comparative morphological evidence. This somewhat hypothetical organ is called an **archinephros,** which means ancient kidney. It was derived from the entire mesomere and hence was long. (Another name for archinephros is holonephros which means entire kidney.) It was segmented throughout most or all of its length and had one pair of nephrocoels and one pair of tubules per pair of nephrotomes. These particular nephric tubules are identified as **archinephric tubules** and their long duct as the **archinephric duct.** Glomeruli were large. It is not certain if they were initially internal or external. In either event, nephrostomes were present.

An archinephros is found only in larvae of hagfishes and caecilians. Ostracoderms and placoderms may have had such organs. Usually, the developing mesomere has a marked tendency to become subdivided into regions. Development always sweeps along the mesomere from anterior to posterior, but the wave is interrupted at two levels so that anterior, middle, and posterior kidneys form in sequence. The more anterior regions usually degenerate as posterior regions become functional. Further, there is a tendency for the tissues to fail to mature between adjacent regions, thus establishing gaps. The more posterior break may be omitted, however, so two instead of three pairs of kidneys are formed in sequence. The various resultant kidneys will now be reviewed in order.

Evolution of Kidneys

Pronephros: Larval Kidney or Specialized Remnant. The most anterior kidney and the first to form is the **pronephros.** Relatively few nephrotomes are incorporated—one to twelve pairs according to species, but rarely more than four. All participating segments form discrete units in the early larvae of some species (notably in hagfishes and caecilians) but before becoming functional the pronephros is usually modified by the degeneration of the first and last pairs or so of tubules and by the partial or complete fusion of the remaining segments to form a glomus on each side of the body. Right and left primordia even fuse in the midline to form a single glomus in many Actinopterygii. Pronephric tubules are relatively simple unless related to a glomus, in which case they may become long and coiled. Glomeruli are internal (caecilians, most bony fishes), external (amphibians except caecilians), or even intermediate (some birds and reptiles). Ciliated nephrostomes are often present and may open into that portion of the anterior splanchnocoel that will become pericardial cavity in the adult.

The pronephros appears in at least rudimentary form in all vertebrates. It is functional in the free-living larvae of bony fishes and amphibians and possibly briefly in embryos of some reptiles. Renal corpuscles fail to form in cartilaginous fishes. A somewhat modified pronephros—usually compacted into a glomus—remains functional in adults of hagfishes and several bony fishes, where it may be called the head kidney because it is usually not the only functional kidney of these animals. Where not functional in the adult, pronephroi may vanish as development proceeds (cartilaginous fishes, most amphibians, birds) or may leave derivatives which are lymphatic or glandular in nature. The duct of the pronephros, which is identified as the **pronephric duct,** is the most constant part of the organ and is not lost even though the tubules degenerate.

It is seen that the pronephros is highly variable in development and structure. This is the result of ancient origin and reduced function. Few phylogenetic conclusions can be drawn, and the nonspecialist will not find it profitable to memorize individual peculiarities.

Opisthonephros: Kidney of Anamniotes. It was explained above that there is an anterior break in both time and space in the development of each mesomere and that the kidney that forms anterior to the break is the pronephros. If a second break forms in the mesomere, then middle and posterior kidneys develop in sequence. The middle kidney is called the **mesonephros** and is present only in fetuses of animals that retain the posterior kidney, or **metanephros,** and none other as adults. These are the amniotes. If all or most of the mesomere posterior to the pronephros instead forms one kidney, then that kidney is called an **opisthonephros.** Such a kidney is typical of late larval and adult anamniotes.

[In spite of the distinction made above, the terms meso-nephros and opisthonephros are used somewhat in-terchangeably and not entirely without reason. Some ver-tebrates complicate the concept of multiple kidneys by having an opisthonephros that excludes most of the posterior segment of the mesomere (anurans) or that is incompletely divided into anterior and posterior portions (dipnoans, many urodeles, some elasmobranchs, some teleosts). Some other teleosts have com-pletely separate middle and posterior kidneys and retain each as adults. These organs are not identified as mesonephros and metanephros (as would be logical) but instead as an opistho-nephros divided into a trunk kidney and a tail kidney. The more posterior kidneys tend to be larger and more complex than an-terior kidneys, but there are no single structural distinctions between adjacent organs. Nephrons at opposite ends of an opisthonephros may be more dissimilar than one of the anterior end of the opisthonephros and one of the pronephros. The student should be cognizant of the concept of variable yet graded development along the length of the mesomere and not be unduly dismayed by the multiplicity of terms or difficulties in their definitions.]

The opisthonephros, then, develops later in time than the pronephros and forms from much or all of the long middle and posterior portions of the mesomere. It is initially segmental, at least anteriorly, and remains so in hagfishes. In other ver-tebrates, one to twenty generations of secondary tubules develop from the primary tubules, thus causing the organ to lose any segmental character. Opisthonephric tubules tend to be more coiled than unfused pronephric tubules. The entire organ bulges more into the coelom. Anteriorly, the glomeruli may be external but usually all are internal and relate to renal capsules. Nephrostomes are usually present in early develop-ment. Some of them may be retained (some sharks, several primitive bony fishes, some amphibians), but usually all are lost.

Opisthonephric tubules usually tap into the preexisting pro-nephric duct which then can be called the **opisthonephric** (or in amniotes the mesonephric) **duct.** (Even more terms are used — few anatomical structures have so many names.) In cy-clostomes, some amphibians, and some bony fishes, however, the opisthonephric tubules contribute to the formation of the posterior portion of the duct, and in caecilians the duct sends buds to meet the growing tubules. Which mode of origin is the most primitive is somewhat in doubt.

The opisthonephros of teleosts is particularly variable as to position, length, and subdivision. The paired organs of embryos sometimes fuse to become a single median kidney. The anterior end of the organ is often lymphoid and not excretory.

The shorter but otherwise equivalent mesonephros is functional in fetuses of amniotes and even briefly after hatching or birth in some reptiles, monotremes, marsupials, and some ungulates. The organ varies from small (some rodents) to medium (carnivores, primates) to large (some ungulates), its development being inversely proportional to the excretory efficiency of the placenta.

Metanephros: Kidney of Amniotes. The metanephros is the most posterior of the kidneys and is the last to develop in both ontogeny and phylogeny ("meta" = later in time). The part of the nephrogenic cord from which it develops lies opposite to only one or two body segments. The organ itself is never segmental, and hundreds of thousands of nephrons are formed by successive generations of budding. Glomeruli are always internal, and nephrostomes are absent. Metanephroi differ from other kidneys in being of dual origin. Early in development a **ureteric bud** grows out from each mesonephric duct near its entrance into the cloaca and approaches the differentiating nephrogenic cord. The lengthening neck of the bud becomes the duct of the metanephros, or **ureter,** and the end branches within the kidney to form the **collecting tubules** which become continuous with the nephric tubules derived from the nephrogenic cord. The entire organ migrates forward somewhat as it matures, particularly in mammals. It is of interest that the

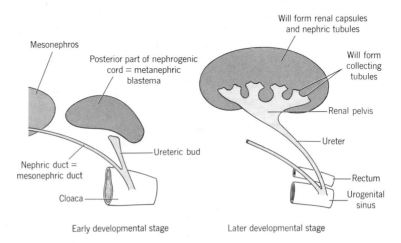

Early developmental stage   Later developmental stage

FIGURE 13-3

STAGES IN THE DEVELOPMENT OF THE METANEPHROS AND URETER.

tail kidney of certain teleosts, and also the posterior part of the opisthonephros of some other anamniotes, are drained by independent ducts which form about as the ureter forms. These are called **accessory ducts** rather than ureters, but they are nonetheless probably homologous with the amniote ureter.

The metanephroi of reptiles are of varying irregular shapes and may be crenulated by lobulation. The kidneys of birds fit into hollows in the synsacrum and are usually three-lobed. Kidneys of mammals have a collecting basin called the **renal pelvis** into which urine oozes from one or more **renal papillae** which project into the pelvis. The organs are bean-shaped, simple, and smooth in primitive and small mammals and in primates and most carnivores. In some other species they are smooth externally but have numerous lobes internally and multiple or otherwise complicated papillae. In marine mammals and some ungulates and other large species they are compound organs resembling clusters of rat kidneys. The pattern of branching of the collecting tubules varies, but the development of compound kidneys in large animals seems to be a means of avoiding very long collecting tubules which would require excessive pressure for driving the contained fluid.

A more detailed account of kidney structure is now desirable to provide a basis for interpretation of anatomy in relation to function. The glomerulus is a tuft of capillary loops and anastomoses which hangs into the renal capsule. The capsule is thus cup-shaped, with an outer, or **parietal, wall** and an inner, or **visceral, wall.** The parietal wall is of squamous or cuboidal epithelium which rarely is secretory but usually is not. In higher vertebrates the visceral wall enfolds the capillaries of the glomerulus. Its cells (podocytes) have numerous finger-like projections called **pedicels** which reach to the basement membrane immediately under the endothelium of the capillaries. The electron microscope reveals apparent pores in this endothelium which thus create a filter. The renal corpuscle is, indeed, a device for filtering from the blood quantities of a fluid which is essentially a protein- and fat-free plasma. The driving force of the filter is the relatively high pressure of arterial blood (review p. 299). In mammals the total filtration surface is roughly half the body surface, and the daily production of filtrate is equal to many times the volume of the extracellular fluids of the entire body.

Renal capsules of poikilothermous vertebrates are usually separated from their convoluted nephric tubules by narrow ciliated **neck segments.** Pronephric tubules consist of ciliated columnar or cuboidal epithelium. Other nephric tubules are characteristically divided into a **proximal tubule** and a **distal**

Kidney Structure in Relation to Function

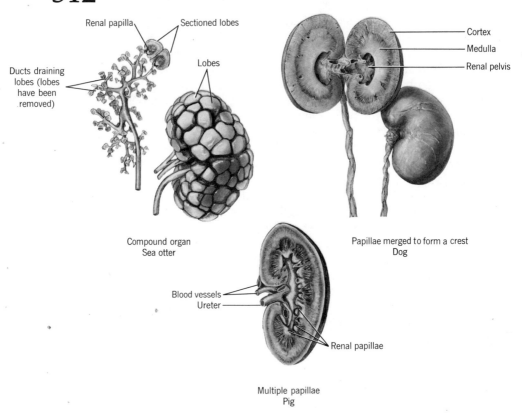

Renal papilla

Sectioned lobes

Ducts draining
lobes (lobes
have been
removed)

Lobes

Cortex

Medulla

Renal pelvis

Compound organ
Sea otter

Papillae merged to form a crest
Dog

Blood vessels
Ureter

Renal papillae

Multiple papillae
Pig

FIGURE 13-4

REPRESENTATIVE MAMMALIAN KIDNEYS.

tubule which are separated by a short yet sharply demarcated **intermediate segment.** The proximal tubule has cuboidal or low columnar cells with microvilli. The intermediate segment is thin and ciliated. Cells of the distal tubule are cuboidal and have few microvilli or none. The proximal tubule returns sugars, amino acids, vitamins, and various salts to the bloodstream and may secrete certain foreign materials into the filtrate. The distal tubule acidifies the filtrate and removes sodium and chloride ions. Water may be returned to the blood by both tubules. The details of tubule function are varied and complicated. Excretion is influenced by hormones in at least some classes.

The progenitors of Vertebrata were certainly marine, as are the related Cephalochordata, Urochordata, Hemichordata, and Echinodermata. Most marine invertebrates have body fluids that are isomotic (equal in osmotic concentration) to seawater. Little water enters or leaves their bodies, and their excretory organs lack filtering devices for removing water.

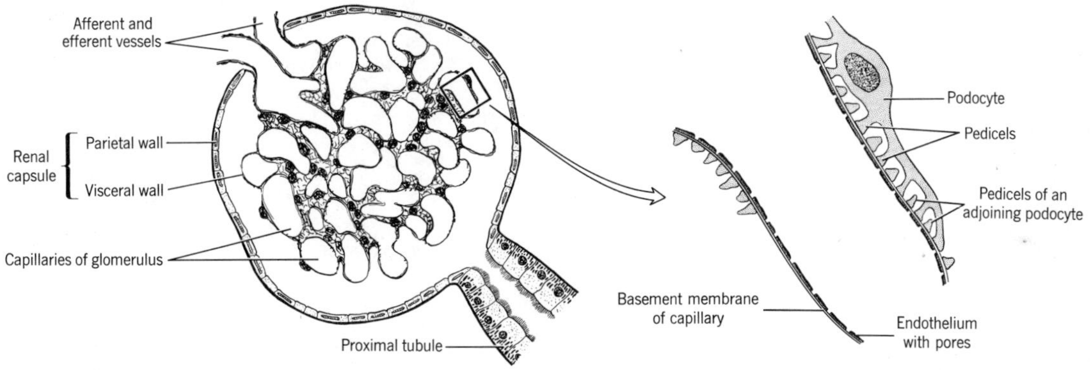

FIGURE 13-5

RENAL CORPUSCLE OF A MAMMAL (left) AND THE RELATION OF THE VISCERAL WALL OF THE RENAL CAPSULE TO A CAPILLARY (right).

Most vertebrates, regardless of habitat, have body fluids that are hypoosmotic (lower in osmotic concentration) to seawater but hyperosmotic (higher in concentration) to freshwater. Consider first the FRESHWATER FISHES and AMPHIBIANS. Water constantly enters their bodies from the environment because of the osmotic gradient. The skin may be moderately waterproof, but the gills and oral membranes admit much water. Consequently, even though these animals scarcely drink at all, they must eliminate copious quantities of urine to maintain their water balance. Accordingly, they have prominent renal corpuscles. Short proximal and distal tubules return solutes to the bloodstream, yet the urine is more salty than the environment, and salt intake is essential. Some salt is taken with the food. Many amphibians can selectively absorb salt through the skin, and most of the fishes can extract salt from freshwater with the gills.

MARINE BONY FISHES have the opposite problem with osmosis: Their body fluids constantly tend to leak away into the sea. In order to conserve water they must pass little urine and therefore have relatively small and poorly vascularized renal corpuscles. Some species have only rudimentary renal corpuscles or none at all. In order to replace water these fishes drink freely, which introduces an excess of salt. The gills excrete monovalent ions, and the proximal tubules of the kidneys excrete divalent ions. Distal tubules, which in other animals usually absorb salts, are usually absent.

CARTILAGINOUS FISHES (and the modern crossopterygian, *Latimeria*) are also marine but have solved the problem of water balance in a different and unique way. Nitrogenous wastes of vertebrates are produced and excreted in many

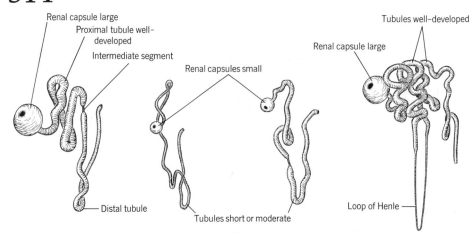

Renal capsule large
Proximal tubule well-developed
Intermediate segment

Renal capsules small

Tubules well-developed
Renal capsule large

Distal tubule

Tubules short or moderate

Loop of Henle

Fresh water fishes, amphibians
Much water eliminated

Marine bony fishes
Water retained

Reptiles
Water retained

Mammals
Water filtered and resorbed

FIGURE 13-6

SOME RELATIONS BETWEEN STRUCTURE AND FUNCTION OF THE NEPHRON.
(There is considerable variation within the groups illustrated.)

forms but in the bloodstream occur primarily as urea. Urea is largely converted to toxic ammonia in the gills of bony fishes, and the ammonia is then immediately excreted by the gills. Urea is eliminated by the kidneys of amphibians, many turtles, and mammals. These animals excrete urea because they do not need it, but it is scarcely toxic and may to a degree be eliminated because it diffuses readily and cannot be retained. The only vertebrates that can retain urea are the cartilaginous fishes. An unknown mechanism makes their gills impervious to the substance, and their large and distinctive nephric tubules are able to return urea to the blood from the filtrate. Consequently, the blood contains enough urea to be slightly hyperosmotic to seawater. Some water enters the gills. Little water is drunk, and urine volume is moderate. Renal corpuscles are unexpectedly large. Excess urea is excreted by the kidneys. Excess salt is also excreted by the kidneys which, however, may fall behind in the elimination of the monovalent ions. These are then excreted by the rectal gland which discharges into the hind gut.

CYCLOSTOMES are largely marine, but some lampreys migrate into freshwater. The structure and function of their kidneys vary among species and are not adequately known. Renal corpuscles are unaccountably the largest of all vertebrates. Neck segments are long, and distal tubules are lacking. Hagfishes control water intake by having blood of uniquely high salt concentration. In freshwater, lampreys absorb some salt through the skin and excrete much urine.

Amniotes use water to carry out excretory wastes and to condition the surfaces of the lungs. They also lose water through the skin. Since drinking is the only source of water in a usually dry environment, it is desirable for the animal to conserve water. This is done in various ways, but most importantly by the kidneys. The kidneys of mammals do the job in one way whereas those of reptiles and birds do it in another.

Urea is highly soluble. It flushes from the body in water and escapes from aquatic embryos into the environment. (Some sharks must provide their embryos with waterproof cases to conserve urea for their special needs.) Most BIRDS and REPTILES have insufficient water available to pass dilute urine, and their terrestrial eggs have no way to dispose of soluble urea. They excrete, instead, mostly uric acid which is insoluble and can be passed from the body in a semisolid state with very little water. Nitrogen in uric acid is relatively concentrated and is not toxic; embryos can allow it to accumulate in fetal membranes or bladders for discarding at hatching. The development of this excretory product was a major requirement for the evolution of the terrestrial or **cleidoic egg** ("kleido" = locked up). Since the formation of quantities of filtrate would be detrimental, these animals have small, poorly vascularized renal corpuscles. Nephric tubules are short in reptiles and of moderate length in birds. Both proximal and distal tubules are present. Cilia are absent from the intermediate segments—apparently because greater filtration pressure renders them unnecessary. Most of these animals can eliminate excess salt in the urine. However, various marine species of both reptiles and birds take quantities of salt with the food and have only seawater to drink (if they drink at all). Many of these animals (sea snakes, marine iguanas, sea turtles, cormorants, albatrosses, petrels, gulls, terns, sea ducks, etc.) have salt glands which excrete a very concentrated brine. These glands may be derived from lacrimal glands or nasal or orbital glands and are variously located near the eyes, jaws, or tongue, or in the nasal chamber.

MAMMALS excrete urea which is removed from the blood by prominent renal corpuscles. Quantities of dilute filtrate are thus produced. The mammalian kidney, however, is more effective than any other at returning water from the filtrate to the blood. Only about one one-hundredth of the filtrate is passed as urine. Concentration is dependent on the distinctive **loops of Henle** (Figure 13-6) which are derived from parts of the proximal and distal tubules and extend toward the pelvis of the kidney. (There is no intermediate segment.) Renal corpuscles and convoluted parts of the tubules are located in the **cortex,** or outer zone of the organ ("cortic" = bark). The loops of Henle are located in the **medulla,** or inner zone ("medull" = pith). Some are long and some are of moderate length. The thin descending and thicker ascending limbs of a single loop are straight and immediately adjacent to one another. This is of functional significance because one of the means of modifying the urine is the cycling of sodium by countercurrent exchange between the limbs. Mammals are the only vertebrates to pass urine that is more concentrated than the blood. Birds also have loops of Henle (apparently independently evolved), but their loops are short and are borne by only a fraction of the nephrons.

Reproductive Strategies

The diverse reproductive strategies of vertebrates often relate directly or indirectly to structure.

Nearly all vertebrates are **dioecious** (having separate male and female individuals), but hagfishes and some teleosts are **hermaphroditic** (have both sexes functional in the same individual). Eggs and sperm are then produced in different parts of the gonads and usually at different times, thus assuring cross-fertilization. Several teleosts, however, are self-fertilizing. Some perches, darters, basses, and lizards are **parthenogenetic,** the eggs maturing without entry of sperm.

Most vertebrates are **oviparous;** that is, they lay eggs that mature outside the maternal body. Included are cyclostomes, most bony fishes, holocephalians, some elasmobranchs, most amphibians, most reptiles, birds, and monotremes. The number of eggs laid per season ranges from 1 to 20 (some birds), 100 (phythons), or even 28 million (several teleosts). They may be spherical, ovoid, teardrop-shaped, or tubular, and are variously provided with stalks, tendrils, filaments, floats, jelly coats, calcareous shells, or keratinized cases. Eggs may be broadcast singly, deposited in clumps, placed in nests, or carried about by a parent in the mouth (a catfish), in a pouch (sea horse, spiny anteater), on the back (some frogs), or elsewhere.

Brooding of eggs is common. It usually provides protection from predation, and according to species may also provide aeration (some fishes), humidity control (terrestrial amphibians, some squamates), protection from mold (some amphibians), temperature control (some reptiles, birds, platypus), and turning to prevent adhesions (birds). Most birds develop a naked, highly vascular **brood patch** on the breast.

Some species retain the eggs in the female reproductive tract until development is partly completed, thus reducing exposure to the vicissitudes of the environment. **Ovoviviparous** species retain their eggs in the body until hatching. Examples are found among fishes, amphibians of each order, and squamates. Since there is a tendency for such animals to reduce or eliminate the ancestral egg shells, they are not sharply demarcated from viviparous species, which give birth to "live" young.

Viviparity is found among teleosts (rockfishes, sea perches, blennies, cyprinodonts, etc.), elasmobranchs, amphibians (rarely), snakes, lizards, and all therian mammals. This reproductive strategy is a drain on the female, but provides maximum protection for larva or fetus. Viviparous reptiles include many alpine, arboreal, and aquatic species, particularly among forms with stout bodies and sluggish habits.

Viviparous teleosts either nourish the fetus within the hollow ovary or within the ovarian follicle. The elasmobranchs and reptiles retain the fetus in the oviducts, whereas mammals have evolved a uterus for the purpose. Respiratory gases and water are always exchanged between fetus and mother, and nourishment is usually provided from "milk" secreted by the female reproductive tract (some fishes), juices released by the lysis of maternal epithelia or blood (early stages in mammalian development), or physiological exchange between fetal and maternal blood streams. The latter mechanism is facilitated by a **placenta,** which is an intimate juxtaposition of fetal and maternal tissues in such a way as to assure a large area of contact. A wide variety of pleats, folds, filaments, and villi have evolved for the purpose. The yolk sac is commonly involved on the fetal side, though the allantois supplements or substitutes in most mammals.

Various fishes and mammals have evolved provisions for the storage of sperm cells in the male body long after the seasonal loss of testicular function, or within the female body for as long as 10 months after copulation. Some mammals in at least five orders postpone development instead by delayed implantation of the blastocyst (corresponding to the blastula of other vertebrates). These mechanisms time birth with seasonal food, dormancy, or return to breeding grounds.

Vertebrates with short brooding periods (11 days for several birds) or gestation periods (13 days for some marsupials) tend to have small, dependent hatchlings or young. Other animals have long brooding periods (about 79 days for one albatross) or gestation periods (22 months for the elephant) and usually have larger, more active young (ducks, grouse, ungulates, cetaceans). Sexual maturity is reached by several fishes by the time of hatching, and in humans and some large vertebrates only after many years.

Early Development and Ancestry of Gonads

Embryonic primordia of gonads come from two diverse sources. The first is the mesomere. After opisthonephroi, or mesonephroi, as the case may be, are well-established, **genital ridges** appear on their mesial surfaces. These ridges are shorter than the nephric ridges which preceded them, yet are longer than the definitive gonads—particularly in amniotes. Anterior and posterior parts of the genital ridges that do not form gonads form instead fat bodies and mesenteries that support the gonads. The latter are called **mesorchia** in males and **mesovaria** in females. Where reflected over the genital ridges, the lining of the coelom thickens to become the **embryonic germinal epithelium.** At the center of the developing organ is another type of tissue called **blastema.** It is derived from undifferentiated cells of the adjacent kidney. The germinal epithelium soon subtends sheets of tissue called **primary sex cords** which penetrate into the blastema.

The second source of gonadal tissue is unexpected and provides one of embryology's more fascinating stories. Several thousand cells of the entoderm of the yolk sac become distinctively large. Some of these stay behind, but many migrate individually and actively into the splanchnic mesoderm of the gut, up the body stalk and mesenteries, and laterally into the genital ridges. These are the **primitive sex cells.** They divide in route, yet some are lost along the way. The remainder space themselves in the germinal epithelium and sex cords where they have important derivatives, as noted below. They have been much studied, yet the cause of their migration and the developmental and evolutionary significance of their remote origin remain unknown.

In summary, the early gonad consists of a superficial germinal epithelium and a deep blastema. Sex cords (actually sheets—the term is not apt) penetrate the blastema, and primitive sex cells are distributed within the epithelium and sex cords. The outer portion of the organ is called the **cortex;** the inner portion, including blastema and sex cords, is the **medulla.** Although the sex of an individual is determined at fertilization, development is identical in the two sexes to the point de-

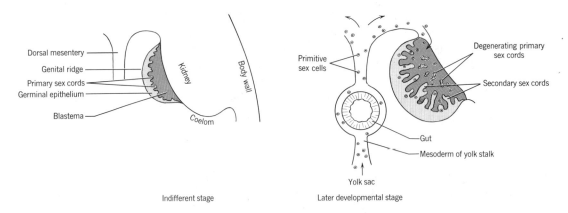

Dorsal mesentery
Genital ridge
Primary sex cords
Germinal epithelium
Blastema
Kidney
Body wall
Coelom

Indifferent stage

Primitive sex cells
Degenerating primary sex cords
Secondary sex cords
Gut
Mesoderm of yolk stalk
Yolk sac

Later developmental stage

FIGURE 13-7

STAGES IN THE DEVELOPMENT OF THE AMNIOTE OVARY. Cross-sections of the body.

scribed, and this phase of development is called the **indifferent stage.** The indifferent stage may be completed early in embryonic development (mammals) or delayed until adult body size is approached (hagfishes).

Gonads of the ancestral vertebrate probably extended the entire length of the coelom. They may have had gonocoels, though these are absent from the adult organs of known vertebrates. Further, ancestral gonads may have been at least partly segmental as they are in amphioxus and some embryonic urodeles and caecilians. Finally, the ancestral organs may have been bisexual. Dominance of the cortex, and of its hormone, cortexin, produces a female and dominance of the medulla, and of its hormone, medularin, produces a male. At the end of the indifferent stage, cortex and medulla are in potential physiological competition. The genetic sex usually masks the other, but under abnormal or experimental conditions the nongenetic sex may prevail.

The structure of gonads varies within some taxa (Teleostei, Reptilia), but in general is uniform for each class and too conservative to provide evidence for evolutionary lineages.

Structure of Gonads

Ovary. The germinal epithelium of the indifferent stage ovary becomes the thin peritoneal covering of the organ and the important **adult germinal epithelium.** The latter is one cell layer in

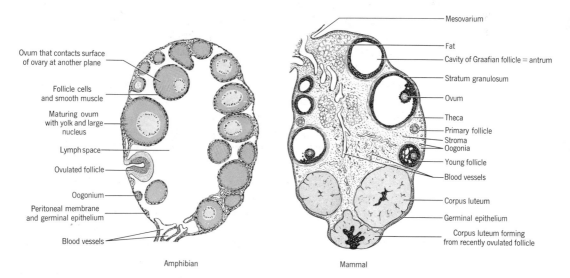

Ovum that contacts surface of ovary at another plane
Follicle cells and smooth muscle
Maturing ovum with yolk and large nucleus
Lymph space
Ovulated follicle
Oogonium
Peritoneal membrane and germinal epithelium
Blood vessels

Mesovarium
Fat
Cavity of Graafian follicle = antrum
Stratum granulosum
Ovum
Theca
Primary follicle
Stroma
Oogonia
Young follicle
Blood vessels
Corpus luteum
Germinal epithelium
Corpus luteum forming from recently ovulated follicle

Amphibian   Mammal

FIGURE 13-8

SECTIONS OF CONTRASTING TYPES OF OVARIES. (It is unlikely that all developmental stages shown would be present at the same time in the reproductive cycle.)

thickness and includes thousands of **oogonia,** some of which sink below the surface in each consecutive breeding cycle to enlarge and undergo maturation divisions and thus become **ova.** It is probable that the primitive sex cells within the epithelium form all ultimate germ cells in some species and only part of them in others. Even if they contribute only part of the germ cells, their presence seems to be essential for the maturation of all. Primary sex cords degenerate and are replaced by **secondary sex cords.** These form **follicle cells,** which surround, nourish, and support the ripening eggs, and **theca** or envelopes which enclose the follicles. Following ovulation, the follicles of some vertebrates are temporarily converted to **corpora lutea** which have an endocrine function. The blastema becomes only the **stroma,** or matrix of connective tissue and vessels within the ovary.

Ovaries vary widely in character, being long or short, compact or flat, smooth or lumpy, solid or flabby, according to species. Eggs differ in size in proportion to the amount of yolk present. Relatively few eggs ripen at one time if the eggs are large (cartilaginous fishes, reptiles, birds) or if the animal is ovoviviparous (many fishes, some caecilians, and some reptiles) or viviparous (mammals). The production of eggs is always cyclic, and the size of the ovaries of some vertebrates fluctuates markedly according to breeding condition. Long-bodied amphibians have long slender ovaries. Nevertheless, paired ovaries of large size are not easily accommodated in a streamlined body. Consequently, the two organs of some animals fuse more or less completely (lamprey, many teleosts) whereas other animals partially or completely suppress the left (many cartilaginous fishes) or right organ (hagfishes, most birds).

Ovaries of crocodilians, turtles, birds, and mammals are solid with relatively much stroma. Those of cyclostomes, cartilaginous fishes, dipnoans, and some primitive ray-finned fishes are also solid but less compact. In amphibians, regression of the blastema leaves one (urodeles) or several (caecilians and anurans) large lymph spaces within the ovary. Stroma is then virtually absent and the organ is soft and pleated. Ripening eggs hang into the central cavity but at ovulation rupture outward into the coelom. Smaller central lymph spaces occur also in ovaries of Lepidosauria.

Most teleosts have hollow ovaries, but the cavity is of different origin. A thin margin of the developing organ curls over and fuses either to the body wall or to the ovary itself, thus sealing off a bit of the general coelom. Ovulation is into the resultant cavity. In a restricted sense, the ovary only appears to be hollow. The coelomic space that is adapted to receive the eggs is continuous with the special oviducts of these fishes, so eggs never reach the unmodified coelom.

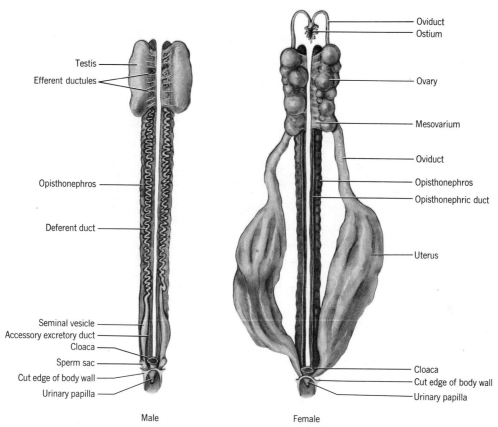

Male

Female

FIGURE 13-9

UROGENITAL SYSTEM OF THE ELASMOBRANCH *Squalus.* **Ventral views. In the male, sperm are transferred from the cloaca to grooves in the claspers (see Figure 8-14).**

Still another kind of cavity is present in ovaries of mammals. Other vertebrates have solid follicles, but in mammals a space appears within each maturing follicle which is then called a **Graafian follicle.** Corpora lutea are more prominent in mammalian than in other ovaries, and in this class there is a tendency for the ovaries to migrate posteriorly as they mature.

Testis. Testes tend to develop a little earlier than ovaries. Their development emphasizes derivatives of the medulla of the indifferent stage gonad; the germinal epithelium of the embryonic organ forms only the peritoneal covering of the adult testis. Primary sex cords do not degenerate, as they do in females, and secondary sex cords are not formed. The primary cords separate from the overlying germinal epithelium and become hollow, thus producing either sac-like **seminiferous ampullae** or slender coiled **seminiferous tubules.** When the organ is

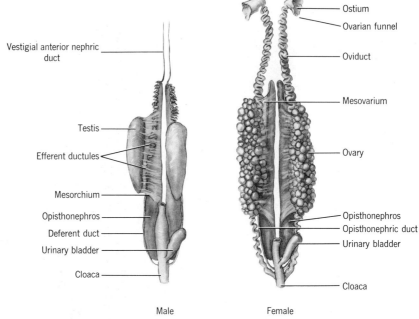

Male   Female

FIGURE 13-10

UROGENITAL SYSTEM OF THE AMPHIBIAN *Necturus.* **Ventral views.**

in breeding condition, the walls of ampullae or tubules consist of stratified epithelium including primarily germ cells at various stages of maturation. Undifferentiated **spermatogonia** are peripheral. As the **sperm cells** are maturing they are held for a time by supportive cells scattered among the germ cells. Finally the mature sperm are released into the duct system of the reproductive tract. Again, primary sex cells are considered to play a critical role in the formation of germ cells. The blastema of the indifferent stage gonad forms supportive tissue and in tetrapods also **rete cords** which hollow out to become **rete tubules.** These come to be continuous with the seminiferous tubules. They form an anastomosis within the testis and conduct sperm to the margin of the organ.

Testes are usually smoother, firmer, and smaller than the ovaries of the same species. Usually they are paired, but the two organs may be partially fused in elasmobranchs and are completely fused in adult cyclostomes. Testes are elongate in slender vertebrates (cyclostomes, most fishes, caecilians, urodeles) but compact and ovoid in some cartilaginous fishes, anurans, and amniotes. Some asymmetry in the size and position of the two members of a pair is not unusual. Seminiferous ampullae are usual for cyclostomes and fishes. Amphibians have ampullae or short tubules. Amniotes and some teleosts have tubules which usually branch and may end blindly (reptiles) or anastomose peripherally (mammals). There are dozens or even hundreds of seminiferous tubules in each testis.

Mammalian testes are distinctive for the degree to which several features are expressed: The organ is enclosed in a tough envelope of connective tissue called the **tunica albuginea** (= tunic + white). Internal lobulation is more pronounced than

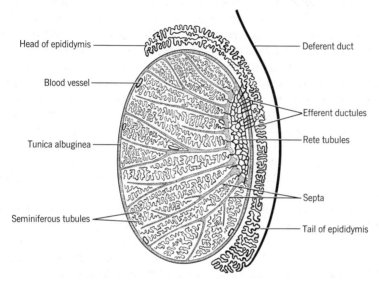

Head of epididymis

Blood vessel

Tunica albuginea

Seminiferous tubules

Deferent duct

Efferent ductules

Rete tubules

Septa

Tail of epididymis

FIGURE 13-11
STRUCTURE OF THE MAMMALIAN TESTIS. Diagrammatic longitudinal section.

Basement membrane

Primary spermatocyte

Supportive cell

Metamorphosing spermatids

Spermatogonia

Spermatids

Sperm

Lumen of seminiferous tubule

Interstitial cells

Blood vessel

FIGURE 13-12
CROSS-SECTION OF SEMINIFEROUS TUBULE OF A MAMMAL.

for most other vertebrates, and **septa** separate adjacent lobules. **Interstitial cells,** which produce male hormones, are packed in the interstices among the seminiferous tubules. Lower vertebrates also have male hormones and doubtless also interstitial tissue. However, in them the tissue is never so prominent as it is in mammals and for many it has not been identified at all.

Spermatogenesis ceases in testes that are warmer than about 36.5°C. Diurnal temperatures of birds usually exceed this figure by several degrees, but at night the body is cooler, and even during the day the testes are probably somewhat cooled by the abdominal air sacs. The testes of mammals are abdominal or pelvic in position if body temperature is relatively low (monotremes, edentates, whales, elephants). In many mammals (primates, most carnivores, most ungulates, and others) the testes descend at maturity out of the abdominal cavity into a cooler pouch of skin in the inguinal area called the **scrotum.** Small pockets of coelom extend into the scrotum to fascilitate the descent of the testes into the organ and to give them some freedom of motion within their sac. The paired canals joining scrotal coelom to general coelom usually close, but in such seasonal breeders as rodents may remain open so the testes can descend when active and return to the abdominal cavity at other times. In females the **large labia,** which flank the genital area in some species, are sexual homologs of the male scrotum.

Urogenital Ducts

Cyclostomes are unique in having nephric ducts but no genital ducts. Eggs and sperm are both released into the coelom whence they exit into the cloaca or urogenital sinus by way of a pair of **genital pores.** Similar openings, also called **abdominal pores,** are said to be present but not functional in various higher vertebrates: some sharks, dipnoans, and several primitive ray-finned fishes. They were probably present and possibly functional in ostracoderms and placoderms. The evolutionary origin of abdominal pores is obscure. Some anatomists believe they represent a posterior pair of ancient, segmentally arranged genital openings called **coelomoducts.** If so, their presence is a primitive rather than a specialized condition, a view which is supported by their wide distribution.

All other vertebrates have two pairs of urogenital ducts which remain identical in the two sexes to the end of the indifferent stage of development and then are modified according to sex and taxon. One pair is the **nephric ducts** which have already been described as the ducts of fetal kidneys other than metanephroi. The other pair is the **paramesonephric** ducts ("para" = beside). The phylogenetic origin of paramesonephric

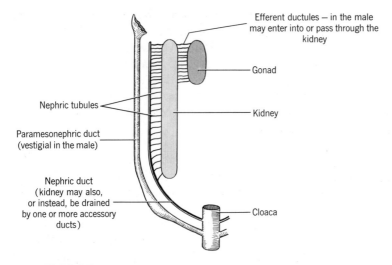

Efferent ductules — in the male may enter into or pass through the kidney

Gonad

Nephric tubules

Kidney

Paramesonephric duct (vestigial in the male)

Nephric duct (kidney may also, or instead, be drained by one or more accessory ducts)

Cloaca

FIGURE 13-13

THE PRINCIPAL UROGENITAL DUCTS AS THEY OCCUR EARLY IN BOTH PHYLOGENY AND ONTOGENY. Ventral view of right side only.

ducts is again obscure. In sharks and urodeles they form by a lengthwise splitting of the more precocious nephric ducts. In other vertebrates they either develop from long solid cords at the surface of the nephric ridges or from shorter anterior primordia which extend themselves back to the cloacal area. It is not possible to say what method of origin is primitive, but it has been shown experimentally that early association of the paramesonephric ducts with the nephric ducts is essential for the normal development of the former.

Other structures that contribute to the urogenital duct system of adults are the more anterior opisthonephric or mesonephric tubules, the ureter of amniotes, the accessory ducts of certain anamniotes, and, in teleosts, still other structures to be identified below.

The retention, loss, or modification of these fetal structures to produce the adult male and female patterns of ducts in the different taxa will now be summarized. To memorize the varied arrangements can be a bewildering and useless exercise, but a review of the kinds of arrangements with attention to several principles that are illustrated is worthwhile. First, one must be impressed by nature's resourcefulness in adapting these "raw materials" to her needs. Evolution is opportunistic and in this instance has selected a variety of patterns of near functional equivalence. Versatility and adaptability are assets in the game of evolution. Second, the developmental mechanism of hormonal influence is well-illustrated by the maturation of the urogenital ducts. The indifferent stage ducts are modified under the influence of sex hormones elaborated in the fetal gonads. Male and female patterns can be altered or even reversed experimentally, and variation between the two norms is common. Finally, the concept of **sexual homology** is exemplified. Organs of the two sexes that are derived from identical primordia of the indifferent stage, yet become different in structure and function, have an equivalence which differs from phylogenetic homology.

Male Ducts. The fetal paramesonephric ducts of males always regress. Vestiges are left which may be prominent (elasmobranchs and amphibians—particularly caecilians) but usually are not.

Except in cyclostomes, sperm are not released into the coelom. Instead they are conveyed in a closed system of ducts which usually is appropriated, at least in part, from the urinary system. If so, sperm cells leaving each testis enter a dozen or so **efferent ductules** derived from anterior nephric tubules. These ciliated ducts are small and cross the mesorchium to enter the nephric duct. (In some fishes the sperm enters the substance of the anterior part of the kidney before reaching the duct.) The nephric duct takes the new name **deferent duct** when it carries only sperm or both sperm and urine. It has ciliated cuboidal or columnar epithelium and walls of smooth muscle. In addition to conveying sperm it provides temporary storage and contracts during mating to ejaculate its contents. Various accessory organs are established by modification of parts of the deferent duct.

Typical CHONDRICHTHYES have paired deferent ducts which convey only sperm. The anterior end of each duct is convoluted to form an **epididymis** (= upon + testes) where sperm are stored and fluids augmented. The posterior end of each duct expands to become a **seminal vesicle.** The two vesicles may fuse before emptying into the cloaca. Urine is transported in several accessory ducts.

Various patterns serve the OSTEICHTHYES. Several primitive ray-finned fishes and one dipnoan pass sperm into the anterior end of the deferent duct which conveys both sperm and urine. Other primitive ray-fins and dipnoans have short paired sperm ducts, without apparent counterpart in other vertebrates, which run to the posterior ends of the opisthonephroi. There, efferent ductules cross into the nephric ducts which carry both urine and sperm for part of their length only. In teleosts the same distinctive sperm ducts, formed by folds of the peritoneum, extend posterior to the kidneys and either enter the nephric ducts just before their exit from the body or else exit independently.

All AMPHIBIA have deferent ducts, but in some species of each major group they only convey sperm and in others they

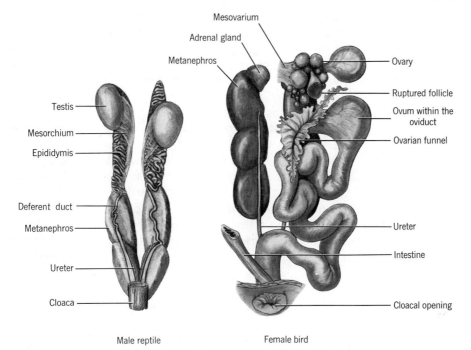

Male reptile   Female bird

FIGURE 13-14

UROGENITAL SYSTEM OF THE LEPIDOSAUR *Crotophytus* **AND THE BIRD**
Gallus. **Ventral views.**

also convey urine. Accessory ducts are present or absent as
needed. When present they drain principally the posterior parts
of the kidneys. Seminal vesicles (as defined above) are usually
present; epididymides (plural of epididymis) may be present.

Deferent ducts of AMNIOTES carry only sperm. Each duct
forms an epididymis, though this organ is reduced in birds.
Seminal vesicles are present in reptiles and birds. Mammals
have glandular outgrowths from the deferent ducts which augment seminal fluids but do not store sperm. They are commonly called seminal vesicles, but are better termed **vesicular
glands.** They are variable in size and form, and sometimes are
lacking.

Female Ducts.   Variation is less extreme in the patterns of
ducts of females, and urinary and genital systems are more independent of one another.

Kidneys of anamniotes are drained by opisthonephric ducts
and those of amniotes by ureters. These ducts convey only
urine in the female, but drainage of opisthonephroi is curiously
supplemented by accessory ducts in some sharks and urodeles. The mesonephric ducts leave vestiges in adults of some
amniotes.

It was noted above that the developing ovaries of TELEOSTEI fold to enclose pockets of coelom into which eggs rupture. Similar folding of the parts of the genital ridges posterior to the ovaries extends the same spaces as short oviducts which are apparently without counterpart in other vertebrates. This arrangement correlates with the prodigious numbers of small eggs released by most of these fishes.

Eggs of other gnathostomes rupture into the coelom before entering a duct system derived from the paramesonephric ducts. The anterior ends of the passages enlarge as **ovarian funnels** which have openings called **ostia**. The funnels of at least some species are thought to represent an anterior pair of pronephric tubules and the ostia the nephrostomes of those tubules. Much of the paramesonephric ducts, including all the anterior portions, become **oviducts.** These are lined by ciliated columnar epithelium with interspersed goblet cells. Eggs are

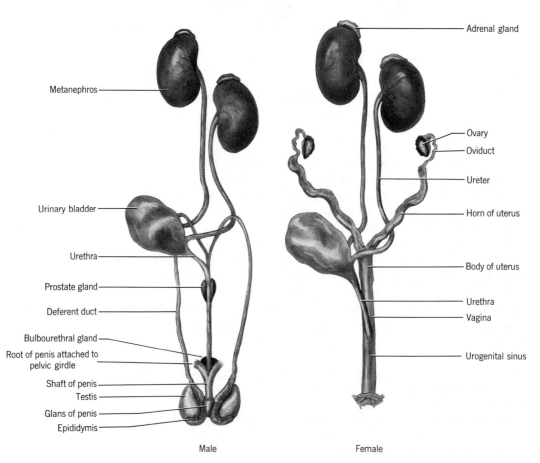

Male   Female

FIGURE 13-15
UROGENITAL SYSTEM OF THE CAT. Ventral views.

moved along by peristaltic contractions of smooth muscle in the walls of the ducts. Posterior portions of the paramesonephric ducts usually become more muscular to expel eggs and more glandular to provide nutritive or protective coats to eggs or to nourish unborn young. The physical and physiological readiness of the female tract to perform its functions vary markedly with the breeding cycle.

The ovarian funnels of CHONDRICHYTHYES lie relatively far forward in the body and often fuse in the midline. Oviducts lead into **shell glands** which envelop eggs in albumen and, in some species, a horny shell. Some sharks are ovoviviparous. They retain their developing eggs in expanded **ovisacs.** Nourishment of the embryo by yolk may be supplemented by a simple yolk-sac placenta.

DIPNOI and some PRIMITIVE ACTINOPTERYGII have paired oviducts apparently derived from paramesonephric ducts, though some doubt exists about the homologies of the ducts of most bony fishes.

AMPHIBIA have glandular and somewhat coiled oviducts. The lining of the coelom is provided with cilia which beat toward the ostia. Posterior portions of the tract apply coats of jelly to the eggs which are usually held temporarily in ovisacs, though many caecilians bear living young which develop in the oviduct.

The genital tracts of female REPTILES, BIRDS, and MONOTREMATA vary widely in detailed structure, but since all have large eggs they agree in general form. The right side of the tract tends to be larger in reptiles, whereas the same side is vestigial in birds. Ovarian funnels are large and pleated. The glandular upper part of the tract applies albumen to the eggs as they spiral along their course. Near the cloaca the tract enlarges as a combined shell gland and ovisac.

In female THERIAN MAMMALIA the genital tract is divided into three regions. First are the oviducts. Their ciliated epithelium is thrown into folds, but externally the ducts are relatively straight and slender in correlation with the small eggs, without albumen or shells, which they convey. The second region is the **uterus** which houses the fetus during pregnancy and provides the maternal contribution to the placenta. It is lined by a mucous membrane called the **endometrium.** When functionally active the endometrium is thick, soft, glandular, and highly vascular. Following each pregnancy or breeding cycle it regresses by sloughing or resorption. The uterus has a thick wall, the **myometrium,** composed of smooth muscle having circular, longitudinal, and oblique fibers. Posteriorly the uterus is closed by a muscular neck called the **cervix.** The third region of the tract is the **vagina** (= sheath) which receives the penis during copulation and serves as the birth canal. The vagina is soft and distensible. Its stratified epithelial lining may be glandular or cornified according to breeding condition and taxon.

The entire mammalian tract is paired in monotremes. In marsupials the terminal portion of the tract is fused to a single vagina, but anteriorly there are two vaginae (and sometimes a peculiar median vaginal sinus). In other mammals fusion of the embryonic primordia includes all of the vagina and usually extends to the uterus. If the uterus is completely double, it is said to be **duplex**, the condition found in monotremes, marsupials, elephants, many rodents, and some other groups. If the uterus is Y-shaped externally but nearly divided internally, it is said to be **bicornuate** (most ungulates, most carnivores, and others). If fusion is more nearly complete, yet does not include the anterior end of the organ, the uterus is **bipartite** (some members of several orders). Finally, if there is a single uterine chamber, the organ is **simplex** (most primates, some edentates). Degree of fusion does not follow clear evolutionary lines, and departures from the norm are not unusual within species.

Cloaca and Derivatives

The embryonic hindgut forms before urogenital ducts have developed. When the pronephroi are established, the nephric ducts extend posteriorly to enter the hindgut, and later the paramesonephric ducts do likewise. The common passageway for products of the digestive, urinary, and reproductive systems is then called the **cloaca** (= sewer). This primitive arrangement is retained by adults of hagfishes, elasmobranchs, dipnoans, amphibians, reptiles, and birds. The posterior part of the gut of adult lampreys, chimaeras, and bony fishes, by contrast, is no longer joined by the urogenital ducts and is called a **rectum**. Their nephric and genital ducts either exit from the body independently or join each other to exit at a common papilla. In either instance, urine and gametes are discharged posterior to the anus.

A different evolutionary path has been followed by mammals. In monotremes the embryonic cloaca is partly divided by a septum which wedges between the gut and allantoic stalk. The result is a dorsal **coprodaeum** (= dung + divide), a ventral **urodaeum** ("ur" = urine) which is joined by the ureters and paramesonephric ducts, and a common posterior **proctodaeum** ("proct" = anus). In therian mammals the embryonic septum continues to push back until a dorsal rectum is completely separated from the urogenital structures which then exit anterior to the anus. In males, urine and sperm are discharged by a common **urethra**. In females of most mammals, urinary and genital tracts exit by a common **urogenital sinus**. In primates and some rodents, however, fetal eversion of the common pas-

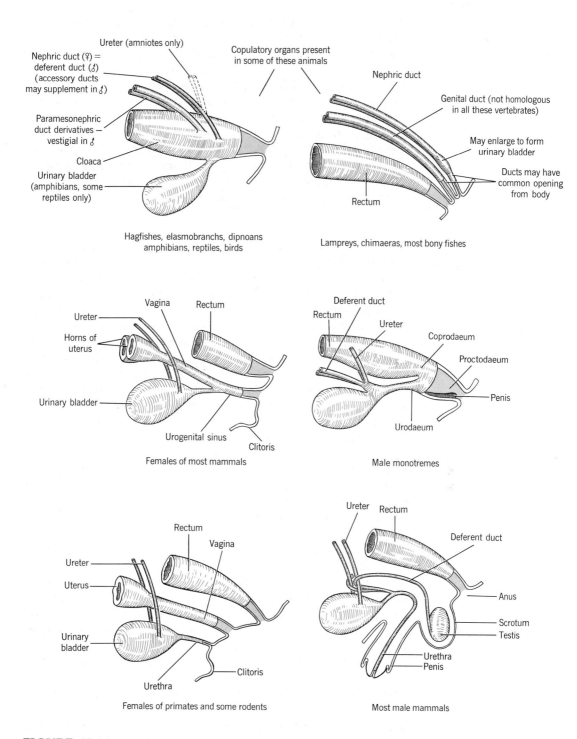

FIGURE 13-16

REPRESENTATIVE DIVISIONS OF THE CLOACA AND THEIR RELATIONS WITH UROGENITAL DUCTS AND THE URINARY BLADDER. Left lateral views.

sageway eliminates the urogenital sinus by separating an anterior urethral opening from a posterior vaginal opening.

Seminal vesicles and vesicular glands have been mentioned as common male derivatives of the nephric ducts, and similarly, albumen and shell glands have been mentioned as adjuncts of the oviducts. Additional accessory glands of the reproductive system (which unfortunately have in some instances been given the same names) may be derived from the urethra, urogenital sinus, and adjacent tissues. Products of these glands augment the seminal fluids, soften and lubricate the genital organs to facilitate copulation, and provide scents which act as sexual attractants.

Urinary Bladders

Most FISHES have urinary bladders, but even if much urine is passed, their bladders tend to be of small or moderate size because the aquatic environment makes long retention of urine unnecessary to accomplish sanitation. Enlargement of posterior segments of the nephric ducts provides temporary storage for dipnoans, primitive ray-finned fishes, and female elasmobranchs. The two ducts may fuse to form a single median bladder. The accessory ducts instead enlarge to form bladders in male elasmobranchs. Similar expansions of the posterior ends of the urinary ducts serve as bladders in lampreys and teleosts, but in this instance a pocket of the embryonic cloaca is considered to contribute to the adult bladder which is, therefore, largely a median structure.

AMPHIBIANS evolved a different kind of bladder of evolutionary importance. It is a large ventral outpocketing of the cloaca. Nephric ducts enter the cloaca dorsally so urine must cross the cloaca to enter the bladder. Almost certainly it was the enlargement and more precocious development of this bladder that produced the allantois of amniotes.

Some REPTILES (turtles, some lizards, *Sphenodon*) have an identical cloacal bladder, but this time it is formed by the retention in the adult of the base of the fetal allantoic stalk. Other reptiles and BIRDS understandably have no urinary bladder because they excrete a semisolid urine containing uric acid. Ureters enter the sides of the cloaca and urine mingles with fecal material before being discharged.

The ureters of MAMMALS join the ventrolateral surfaces of the embryonic cloaca and hence are associated with the urodaeum after the cloaca is divided. Parts of the urodaeum and allantoic stalk together form the urinary bladder. The bladder is lined by a distinctive kind of epithelium which thins as it stretches. Smooth muscle in the walls of the bladder contracts to empty the organ.

CYCLOSTOMES and most BONY FISHES lay their eggs in water where the attendant male promptly discharges his sperm over them. Successful fertilization is dependent on behavioral rather than on structural adaptations. Copulatory organs are lacking, though the two sexes of some species entwine their bodies during egg laying and rarely some kind of holding organ is present. Ovoviviparous and viviparous teleosts, however, require internal fertilization. Males of most such species have a margin of the anal fin modified as a copulatory organ called a **gonopodium.** It is enlarged, rigid, and movable. When inserted into the female tract it transmits sperm through a duct or groove. Several cottid fishes have instead an enlarged genital papilla which serves as a penis.

CARTILAGINOUS FISHES also have internal fertilization. This time it is the pelvic fins of males which develop copulatory organs called **claspers** (Figures 3-6 and 8-14). They are supported by the fin skeleton and may be of elaborate configuration. In some rays they include erectile tissue. Again, grooves convey sperm into the female cloaca.

Many AMPHIBIANS return to water to breed. Fertilization is usually external in anurans so copulatory organs are not needed. Fertilization is usually internal in urodeles but external genitalia are still lacking. The two sexes either press their cloacas together to transfer sperm or the male deposits packets of sperm which are later picked up by the cloaca of the female. Male caecilians use an evertable extension of the cloaca for internal fertilization by means of copulation.

Internal fertilization is necessary if copulation takes place out of the water, if there is internal development of the young, or if eggs are provided with shells before deposition. With few exceptions amniotes must have internal fertilization for the first and either the second or third of these reasons. REPTILES evolved two kinds of copulatory organs. Male lepidosaurs (except *Sphenodon*) have paired structures called **hemipenes** which lie concealed in long sacs opening to the outside of the body on each side of the cloacal aperture (Figure 9-7). During copulation the hemipenes are everted and introduced into the female cloaca. Most other reptiles have evolved the **penis,** which in this class is a grooved organ located internally on the floor of the cloaca. It is largely composed of two long, spongy, vascular bodies called **corpora cavernosa.** A **glans penis** caps the end of the organ. During sexual stimulation sphincters reduce the outflow of blood from these structures thus causing engorgement and enlargement. The groove then closes over and the penis protrudes from the cloaca to serve as an intromittent organ. Female turtles and crocodilians have a small **clitoris** which is the sexual homolog of the penis.

Sphenodon is the only reptile without a copulatory organ. This is perhaps a holdover of the ancestral condition. The penis of MONOTREMES is like that of reptiles except that the sperm channel is permanently separated from the cloaca.

Most BIRDS copulate by pressing the cloacas together for the transfer of sperm. Avian ancestors doubtless had a penis, however, because the ostrich and its relatives and ducks and geese—all relatively primitive birds—have a small penis.

At the indifferent stage of development THERIAN MAMMALS have a genital tubercle anterior to the cloacal opening. When the cloaca becomes divided, its ventral portion contributes to the urinary bladder anteriorly and forms the female urogenital sinus or male pelvic urethra posteriorly. The genital tubercle of females becomes the clitoris and is not penetrated by the urethra. In males the pelvic part of the urethra is extended by a phallic part which grows into the tubercle. The tubercle enlarges to become the penis. Corpora cavernosa and glans are present. The organ may be hidden under the skin but is usually at least partly external even when not erect. The glans is variously shaped and is forked in monotremes and marsupials to correspond with the divided vagina of the female. The penis may be stiffened by a bone, the baculum (Figure 8-27).

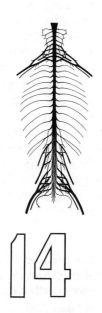

14

Nervous System: General, Spinal Cord, and Peripheral Nerves

Nerve cells are uniquely adapted for conducting stimuli from one place to another. They may be activated by sensory cells or by other nerve cells. A stimulus is usually carried to numerous other nerve cells before reaching the muscle or gland that will respond. Messages may be diminished, reinforced, selected, stored, blocked, or integrated within the system. Thus, the nervous system (with help from the endocrine glands) determines responses of the body to changes in the internal and external environments. It is the body's messenger and coordination system for most activities—for all activities that are rapid or complex.

The nervous system is of particular interest to morphologists because it is the most complex of the body, yet it is conservative in terms of change. Although it is too soft to be directly preserved in the fossil record, the size and shape of the brain, and the distribution of cranial nerves are remarkably well-recorded by fossil skulls. This record, comparative study of surviving vertebrates, and, to a lesser extent, developmental studies, are enabling morphologists to construct the outlines of the phylogeny of the system. The general habits of an animal can be accurately determined (where information is adequate) from the nervous system. It is by adding to knowledge of the fine structure and ultrastructure of this system that morphology can contribute the most to physiology, psychology, and behavioral sciences. The nervous system is an active field of research.

Elements of the Nervous System

Neurons and Neuroglia. Nerve cells, or **neurons,** range from short to uniquely long, yet are always small in bulk. Each has a **nerve cell body** which may be oval in outline or irregularly star-shaped. Within the nerve cell body is the nucleus and many distinctive granules which contribute to the very high rate of protein synthesis characteristic of nervous tissue. Most nerve cell bodies, particularly of the brain and spinal cord, support many filamentous processes and are termed multipolar. Neurons related to the nose, eye, ear, and lateral line usually have only two processes (i.e., are bipolar), and most related to spinal nerves have two processes which, however, exit from the nerve cell body at the same place and course together for a short distance before separating (pseudounipolar).

Each neuron has one to many processes called **dendrites** which receive impulses and transmit them toward the nerve cell body. Dendrites are up to 1 mm long. They branch in the direction away from the nerve cell body, sometimes achieving an exceedingly diffuse and complex pattern. There is a single process, the **axon,** which transmits impulses away from the nerve cell body. Axons may be short or, in large animals, 1m or more in length. They tend to branch less than dendrites, yet in some regions they give off collaterals and usually have twigs at their far ends where they communicate with other cells.

Between their short initial segment and their terminal twigs, most axons (and only axons) are sheathed. **Schwann cells** arrange themselves along developing axons located outside the brain and spinal cord. A fold of each cell then wraps around the axon much as a window shade wraps around its wooden core. Thus, the axon is ensheathed in lamellae derived from the cellular membranes of Schwann cells. These enclose lipids and proteins. The entire coating is called **myelin.** Myelin is essential

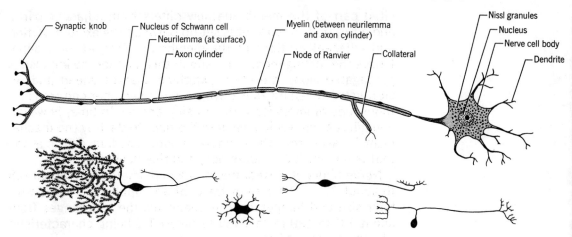

FIGURE 14-1
STRUCTURE AND SHAPES OF REPRESENTATIVE NEURONS.

for rapid conduction. It is interrupted at intervals along the enclosed **axon cylinder** to form **nodes of Ranvier.** Outside the brain and spinal cord, axons are also covered by a second sheath which is external to the myelin. This delicate transparent membrane is also formed from Schwann cells and is the **neurilemma.** (There are some long unmyelinated fibers in the peripheral nervous system—particularly in cutaneous nerves.)

Within the brain and spinal cord, both myelinated and unmyelinated parts of neurons may be in contact with cells called **neuroglia** (= nerve + glue). Three kinds of neuroglia (astrocytes, oligodendroglia, microglia) are distinguished histologically. They fill interstices between neurons and bind fiber to fiber. They also contribute to the energetics of neurons in various ways, including potassium transport, and possibly nutrition, excretion, regeneration, and repair. About half of the bulk of the brain is neuroglia.

Nerve Impulse and Synapse. A nerve impulse is an electrical phenomenon that passes as a wave along the surface membrane of a nerve fiber. The fluids bathing the inside and outside of the membrane differ chemically. The crucial difference is that there is about thirty times more potassium on the inside of the resting membrane and about ten times more sodium on the outside. The potassium leaks out through the membrane, but in some unknown way the membrane resists the entrance of sodium. The consequence is a difference in electric potential of about 60 μV across the membrane, with the inside negative. When the fiber is locally excited, the membrane suddenly, but briefly, allows sodium to rush in, and the inside of the membrane changes to a positive potential of about 50 μV. The excited part of the membrane now differs from adjacent parts, and a tiny eddy of current is set up between the excited portion and adjacent portions of the membrane. This, in turn, depolarizes the resting membrane and in this manner the impulse is propagated along the fiber, switching charges ahead and restoring them behind as it travels. An excised fiber can transmit thousands of impulses without using energy. In time, however, enough sodium would pass into the fiber to destroy the mechanism. To avoid this, the membrane pumps sodium out, in a way that is not yet fully understood, and this does use energy.

Nerve fibers of vertebrates range in diameter from about 0.5 to about 22 μ. The rate of travel of an impulse increases with fiber size and temperature. In mammals the rate ranges from about 1.0 to 120 m/sec, values below 1.5 being characteristic of unmyelinated fibers.

An impulse is propagated without decrement or is not propagated at all, and all conducted impulses are alike. It is by the frequency of impulses in each fiber, number of active fibers, and connections made by the neurons that the system decodes the messages. Other electrical charges are not conducted, especially in dendrites in the brain, but instead affect excitation.

There is no cytoplasmic continuity between neurons. The functional union of an axon of one neuron with a dendrite or nerve cell body of another neuron is called a **synapse.** Some neurons have few synapses, but others have thousands. The synapse transfers impulses only in the direction away from the axon. At its terminus, the axon enlarges to form a **synaptic knob.** The knob is separated from the adjacent dendrite or nerve cell body by a very narrow cleft (100–200 Å wide). In the knob are mitochondria and **presynaptic vesicles** which contain a

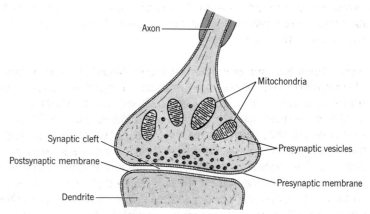

Axon

Mitochondria

Synaptic cleft

Postsynaptic membrane

Presynaptic vesicles

Presynaptic membrane

Dendrite

FIGURE 14-2
SYNAPTIC KNOB AND SYNAPSE.

chemical known as a transmitter. When the latter is released into the cleft by the arrival of a nerve impulse at the synaptic knob, it alters the permeability of the postsynaptic membrane to the ions bathing it, and thus causes propagation of the nerve impulse in the connecting neuron. At present, several transmitters are known including acetylcholine, noradrenalin, serotonin, and dopamine. These are found at different synapses, and their exact distributions and functions are still relatively unclear.

Junctions between axons and muscles or glands tend to be broader but apparently function in a similar manner.

Tracts, Nerves, and Ganglia. Nerve cell bodies of functionally related neurons tend to mass together, and their fibers, if long, tend to run parallel to one another in bundles. Within the cord such bundles are called **tracts.** Myelinated tracts are whitish and together make up the **white matter,** whereas nerve cell bodies and associated unmyelinated fibers, being of darker color, together make up the **grey matter.**

Outside the brain and cord, bundles of fibers are termed **nerves.** Aggregates of nerve cell bodies cause marked swellings on nerves and are termed **ganglia.** Opposite the appendages, adjacent nerves usually exchange bundles of fibers, thus weaving to form a **plexus.**

Within a nerve, the neurilemma of each fiber is surrounded by a delicate mesh of connective tissue, the **endoneurium,** and the entire nerve is wrapped in a stronger **epineurium.** In large nerves, groups of fibers are bundled together by the **perineurium.** The dissectionist is grateful that nerves and ganglia are tough—for their size. Although most single nerves are small, one may be surprised to see the combined bulk of the nervous system when it is revealed intact by an acid maceration technique which destroys all other tissues.

Some Divisions of the System. The parts of the nervous system are functionally interrelated to a remarkable degree. Nevertheless, it is convenient to recognize structural and functional divisions. Brain and spinal cord comprise the **central nervous system,** leaving nerves and ganglia to the **peripheral nervous system. Afferent,** or **sensory, fibers** of the peripheral nervous system carry impulses from receptor organs to the central nervous system; **efferent,** or **motor fibers** carry impulses from the central nervous system to effector organs. Nerves may be entirely sensory or motor, or may be mixed, having each kind of fiber. Within the central nervous system there are also **association neurons** (also called **interneurons**) which make up local circuits and are not themselves afferent or efferent.

Somatic fibers (sensory and motor) relate to the skin and its derivatives and to voluntary muscles. **Visceral fibers** (again sensory and motor) relate to involuntary muscles and glands of the various organ systems. It is useful to designate fibers as somatic sensory, visceral sensory, somatic motor, or visceral motor, because these types of fibers tend to be structurally independent.

Most organs innervated by visceral nerves receive two complete sets of fibers which elicit opposed responses. The sets have a measure of structural and functional independence from the remainder of the peripheral nervous system and are together called the **autonomic system.**

Development of Spinal Cord and Peripheral Nerves

Following chemical interaction with the underlying presumptive notochord, the dorsal ectoderm of embryos in the early gastrula stage of development thickens to become the **neural plate.** As the embryo begins to lengthen, the plate forms a long groove flanked by **neural folds.** Gradually the folds close over the groove and fuse together, thus completing the **neural tube,** or future central nervous system. The cavity within the tube is the **neurocoel;** in the adult it is relatively smaller and more cylindrical and is called the **central canal.** As the neural tube sinks below the surface, the ectoderm reestablishes a continuous layer above. In the angles between the neural tube and the skin ectoderm, **neural crest cells** are squeezed off. (These structures are all shown in Figure 5-4)

Each step in the development of the central nervous system occurs first at the anterior end of the embryo and progressively later and later in moving toward the posterior end. This is one of many developmental gradients of the body.

In the head of the embryo, the neural tube becomes relatively large and soon is compartmentalized into vesicles which foreshadow the regions of the brain. Further discussion of the development of the brain will be deferred to Chapter 15.

As seen in cross-section, the embryonic neural tube forms three layers having conspicuously different staining properties. In order, from neurocoel outward, these are the **ependymal, mantle,** and **marginal layers.** Some cells of the ependymal layer remain in place to become the thin, ciliated lining of the adult central canal. Most ependymal cells migrate outward to join mantle cells in forming both neurons and (a little later) neuroglia. These will be the gray matter of the adult cord. The marginal layer forms neuroglia and is penetrated by nerve fibers growing out of the deeper layers. It becomes the white matter of the cord.

As the cord grows in diameter, it enlarges every place except at the thin roof and floor of the neurocoel. Growth, therefore,

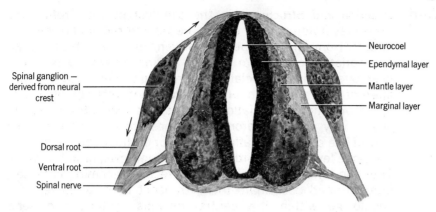

Neurocoel
Ependymal layer
Mantle layer
Marginal layer

Spinal ganglion — derived from neural crest

Dorsal root
Ventral root
Spinal nerve

FIGURE 14-3

STAGE IN THE DEVELOPMENT OF THE MAMMALIAN SPINAL CORD, SPINAL NERVES, AND GANGLIA. Cross-section. Arrows show direction of fiber growth.

establishes longitudinal surface grooves at these places known in the adult as the **dorsal median sulcus** (= furrow) and **ventral median fissure** (= a split).

Nerve cell bodies of the sensory fibers of spinal nerves are located in spinal ganglia located near the cord. These are derived from neural crest cells. Fibers from the developing ganglia grow outward to receptor organs and inward to penetrate the cord. Sensory fibers of cranial nerves and their ganglia develop in the same way except that there are contributions from the ectodermal placodes that will form certain sense organs of the head (see p. 395). The autonomic system also has ganglia. These are derived in part from neural crests and probably in part from cells that migrate out of the cord. Fibers of motor nerves grow out of the mantle layer of brain and cord.

Cells from the versatile neural crests also migrate to contribute to the formation of Schwann cells, myelin, and the neurilemma.

Throughout life, the nerve cell body actively produces cytoplasm which slowly flows outward along axon and dendrites. Neurons do not divide further after infancy; if one is lost, it is not replaced. If a nerve is cut, fibers severed from their nerve cell bodies degenerate whereas fibers that are still nourished slowly regenerate. They tend to feel their way along the empty sheath tubes and may reach their accustomed destinations. However, if the tubes are also destroyed, the regenerating fibers do not find their original end organs. In some instances (but seemingly not in others) anamniotes can then retrain the central nervous system to respond correctly to the new stimuli, but amniotes cannot.

Function and Structure. In the simplest possible reflex arc, messages from receptor organs are transferred within the cord directly from afferent fibers to efferent fibers which then send appropriate messages to effector organs. Nearly always, however, one or more association neurons are interposed between afferent and efferent neurons. Since each afferent fiber synapses with many association neurons, possible pathways run to any of very many effector fibers. These may be on the same side of the cord as the afferent fiber or on the opposite side, at the same level of the cord or at a different level, or in the brain. Thus the function of the cord is to receive incoming impulses, integrate and coordinate them, transmit them wherever they should go within the central nervous system, and send responses to the peripheral nervous system as appropriate.

The general structure of the spinal cord is best exemplified by a cross-section of the cord of an amniote. The gray matter is internal and has an irregular shape resembling the letter H. The upper arms of the H are the **dorsal gray columns** (or horns) and the shorter, broader, lower arms, the **ventral gray columns** (or horns). Nerve cell bodies of association neurons that synapse with somatic sensory fibers are on the medial side of a dorsal column. Cell bodies of association neurons synapsing with visceral sensory fibers are in a smaller, lateral, and slightly more ventral part of the dorsal column. Nerve cell bodies of somatic motor neurons fill the ventral columns. Nerve cell bodies of visceral motor neurons are in a small, intermediate, and lateral position. The **gray commissure,** just above and below the central canal, makes up the cross arm of the H and transmits fibers from one side of the cord to the other.

The external white matter is divided into right and left sides by the dorsal median sulcus and ventral fissure of the cord. Each half is further divided by the gray columns into three funiculi. The **dorsal funiculus** is between the dorsal column and the dorsal median sulcus. It transmits axons toward the brain. The **ventral funiculus** is between the ventral fissure and the ventral column. It transmits axons away from the brain. The **lateral funiculus** is between the dorsal and ventral columns and transmits fibers in both directions; those going toward the brain tend to be more superficial. The positions of the specific pathways vary among the vertebrates. Many of the fibers to and from the brain cross from one side to the other; the change is sometimes in the cord and sometimes in the brainstem. The reason for this crossing is a mystery.

The spinal cord is supported and protected by one or more layers of tissue called **meninges.** These are similar to, and continuous with, meninges of the brain and will be described in Chapter 15.

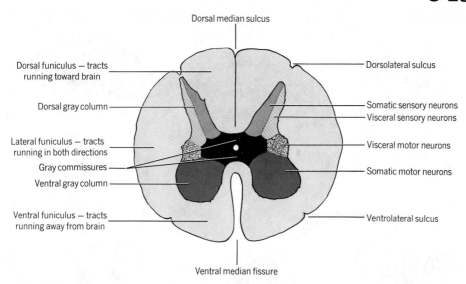

Dorsal median sulcus

Dorsal funiculus — tracts
running toward brain

Dorsal gray column

Lateral funiculus — tracts
running in both directions

Gray commissures

Ventral gray column

Ventral funiculus — tracts
running away from brain

Dorsolateral sulcus

Somatic sensory neurons
Visceral sensory neurons

Visceral motor neurons

Somatic motor neurons

Ventrolateral sulcus

Ventral median fissure

FIGURE 14-4
STRUCTURE OF THE MAMMALIAN SPINAL CORD.

Comparative Anatomy of the Spinal Cord. The spinal cord of
AMPHIOXUS is seemingly primitive in that the embryonic
neural folds do not completely fuse. Accordingly, the adult
neural canal communicates with the space around the cord by
a slit-like groove. Gray and white matter cannot be distinguished because nerve fibers of amphioxus are not myelinated.
There being little cephalic neural specialization, there are no
long spinal tracts. The cord is somewhat triangular in cross-
section.

CYCLOSTOMES, like other vertebrates, complete the neurulation process to enclose a central canal. The boundary
between gray and white matter is never sharp, and the lamprey,
at least, has no myelinated nerves. The cord is wide and its
ventral surface is concave where it fits against the notochord.

The cord of FISHES and AMPHIBIANS is subcircular in
cross-section. Gray and white matter are distinct. The configuration of the gray matter is various; ventral gray columns are
usually evident. Dorsal median sulcus and ventral median fissure are making their appearance among these animals as
increasing complexity of nuclei and tracts cause the cord to
enlarge. The diameter of the cord enlarges moderately opposite
the appendages where there is greater need for nervous integration.

AMNIOTES have a deep sulcus and fissure. The cord again
enlarges opposite the appendages, the **cervical enlargement**
being the more pronounced if the pectoral appendages are

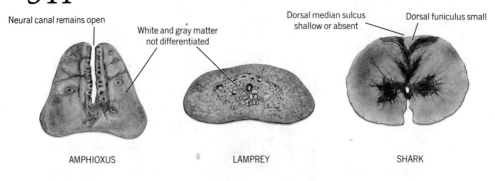

Neural canal remains open

White and gray matter not differentiated

Dorsal median sulcus shallow or absent

Dorsal funiculus small

AMPHIOXUS   LAMPREY   SHARK

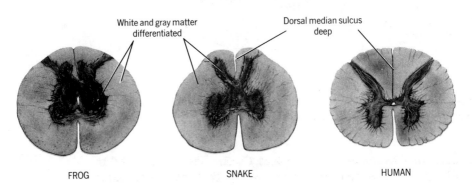

White and gray matter differentiated

Dorsal median sulcus deep

FROG   SNAKE   HUMAN

FIGURE 14-5

COMPARATIVE ANATOMY OF THE SPINAL CORD.

emphasized (bats, apes) and the **lumbar enlargement** more pronounced if the pelvic appendages are emphasized (ostrich, bipedal dinosaurs). In cross-section, the gray matter now is shaped like an H with thick arms. Adjacent to the gray matter in the cervical region of the cord there may be a narrow area where gray and white matter intermingle and fibers are short. This area is continuous with the **reticular formation** of like character in the brainstem.

A distinctive feature of BIRDS is the **glycogen body.** This is a conspicuous gland-like mass of tissue wedged into the cord of the lumbar area where the dorsal median sulcus opens to receive it. Its function has been debated but is not surely known.

Spinal cords of MAMMALS frequently have **dorsolateral** and **ventrolateral sulci.** The dorsal and ventral spinal nerve roots, respectively, join the cord along these grooves. The cords of anurans and some fishes are shorter than the canal within the vertebral column. This condition is not seen in reptiles and birds, but appears again in mammals (except monotremes). The embryonic cord fills its bony housing, but grows more slowly than the spine so that the adult cord ends in the lumbar region (except for a terminal filament), and the more posterior nerves must angle back to reach their destinations.

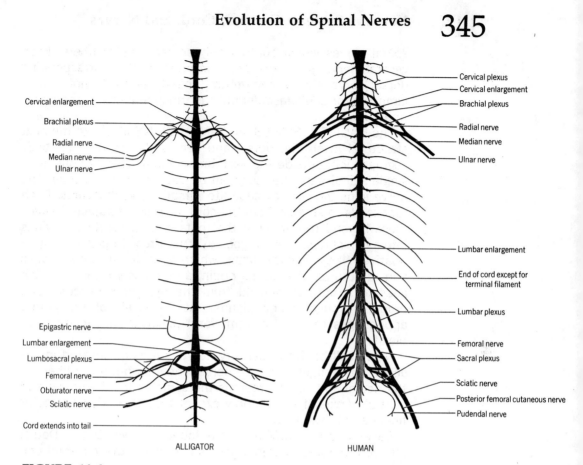

Cervical plexus
Cervical enlargement
Brachial plexus
Radial nerve
Median nerve
Ulnar nerve

Cervical enlargement
Brachial plexus
Radial nerve
Median nerve
Ulnar nerve

Lumbar enlargement

End of cord except for terminal filament

Lumbar plexus

Epigastric nerve
Lumbar enlargement
Lumbosacral plexus
Femoral nerve
Obturator nerve
Sciatic nerve
Cord extends into tail

Femoral nerve
Sacral plexus

Sciatic nerve
Posterior femoral cutaneous nerve
Pudendal nerve

ALLIGATOR   HUMAN

FIGURE 14-6

SPINAL CORD, PRINCIPAL NERVES, AND PLEXUSES OF A REPTILE AND A MAMMAL. Dorsal views.

AMPHIOXUS has a series of paired "dorsal" spinal nerves which contain three kinds of fibers: somatic sensory from skin and muscle, visceral sensory from internal organs, and visceral motor. Since most of the related receptor and effector organs are segmental, it is efficient for these nerves to be intersegmental and to run with the myosepta between muscle segments. All nerve cell bodies of sensory neurons are located within the cord. Consequently, there are no ganglia on the spinal nerves. The structures formerly called ventral spinal nerves are really segmentally arranged specialized muscle fibers which run to the surface of the spinal cord where their motor end plates are located. In this amphioxus is specialized.

The further evolution of spinal nerves follows a relatively simple and straightforward progression. LAMPREYS have intersegmental dorsal spinal nerves which are like the spinal nerves of amphioxus except that some of the sensory neurons have cell bodies outside the cord. They also have segmental ventral

Evolution of Spinal Nerves

spinal nerves which contain only somatic motor fibers. Each ventral nerve joins the cord anterior to the corresponding dorsal nerve. Dorsal and ventral spinal nerves do not join in lampreys (or cephalaspids) but run independently to their destinations.

The HAGFISH, FISHES, and AMPHIBIANS illustrate the next structural level. The dorsal and ventral nerves of each body segment join outside the vertebral column. This establishes a single spinal nerve per segment on each side of the body. The nerve joins the cord by separate **dorsal** and **ventral roots.** Each root tends to emerge from the cord as a row of adjacent twigs. Close beyond the union of the roots, the spinal nerve divides into a **dorsal ramus** which goes to structures of epaxial origin, a relatively large **ventral ramus** which goes to the appendages and structures of hypaxial origin, and a **visceral ramus** which goes to structures derived from the hypomere (visceral and branchial muscles and glands). At the levels of the paired appendages the ventral rami form plexuses of varying complexity.

Nerve cell bodies of sensory neurons are now located in the **dorsal root ganglion** of each nerve. The distribution of fibers in the dorsal and ventral roots is the same as for the dorsal and ventral nerves of the lamprey except that some visceral motor fibers exit in each root.

Several further advances are found in AMNIOTES. Dorsal and ventral roots of spinal nerves join inside the vertebral col-

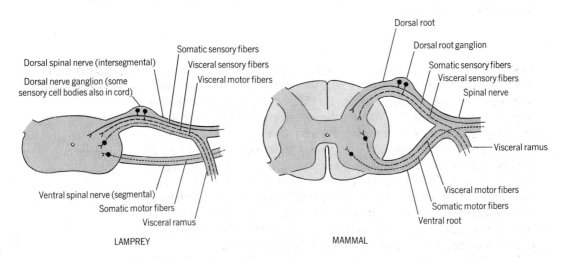

LAMPREY   MAMMAL

FIGURE 14-7

DISTRIBUTION OF NERVE COMPONENTS IN SPINAL CORD AND NERVES AT TWO EVOLUTIONARY LEVELS. More detail for the branching of spinal nerves and the distribution of visceral motor fibers is shown in Figure 14-10. Size of the nerves is exaggerated.

umn. Each dorsal root joins the cord at the same level as the corresponding ventral root rather than posterior to it. Usually all visceral motor fibers exit from the cord in the ventral root. This completes the gradual shift of these fibers from the dorsal spinal nerve and leaves the dorsal root with only sensory neurons. **Brachial** and **lumbosacral plexuses** tend to be more complex than for anamniotes, but the patterns of interweaving are various.

Origin and Nature of the Nerves. Spinal nerves are conveniently uniform in regard to occurrence, configuration of roots and branches, nerve fiber components, and relation to the central nervous system. This is not true of cranial nerves: A cranial nerve may be present in some vertebrates and missing in others (e.g., terminal nerve). A nerve may split in the course of evolution to become two (spinal accessory nerve from vagus nerve). Conversely, two nerves may fuse to become one (evolution of amniote trigeminal nerve). The same nerve that is a cervical spinal nerve of one vertebrate may be a cranial nerve of another (hypoglossal nerve). Also, fiber components believed to have been present in an ancestral nerve may become lost.

A further difference between spinal and cranial nerves stems from the nature of the segmentation of the central nervous system. The somites are already segmented when they first appear in the embryo, but the central nervous system is not. Motor nerves grow out of the cord at intervals to penetrate segmented somite derivatives. Sensory nerves and ganglia are derived from neural crests which become intersegmental as they are squeezed between the bulging somites. Thus, the segmentation of spinal nerves is regular but seems to be secondary. The head, by contrast, may in part never have been segmented, was in part segmented like the body but only in remote ancestors (derivatives of head myotomes and sclerotomes), and is in part segmented in an independent series (derivatives of visceral arches). In consequence, serial homology is less evident in cranial than in spinal nerves. Cranial nerves are numbered in spatial sequence, but it should be understood that in terms of fundamental nature the assignment of numbers is rather arbitrary.

These difficulties were a welcome challenge to morphologists at the turn of the century, and much attention was given to the analysis of cranial nerves. Various kinds of clues were used. Developmental evidence shows that sensory fibers have been lost from the accessory nerve: ganglia form in the embryo and then regress. Comparative anatomical evidence shows (for instance) that the trigeminal nerve of amniotes is a composite nerve: One of its principal branches joins the brain independently in some fishes. Physiological evidence shows the unique

Evolution of Cranial Nerves

nature of some nerves (only aquatic vertebrates have a lateral line system and it is served only by cranial nerves).

Putting together many clues of these kinds it is evident that cranial nerves are of three general categories. First, there are seven nerves (numbers 0, V in two parts, VII, IX, X, XI) that are in series with dorsal roots of spinal nerves, or better, with the dorsal spinal nerves of lampreys which, as described in the previous section, do not join their respective ventral spinal nerves. These nerves all join the brainstem at a lateral (not ventral) level. It is postulated that in the ancestral condition, each nerve carried the same components as dorsal spinal nerves: somatic sensory, visceral sensory, and visceral motor. A sensory ganglion is present close to the entrance of each nerve into the brain. Further, it is postulated that in the remote ancestral vertebrate, these nerves served branchial and pharyngeal areas: Each nerve had a major branch to each hemibranch bordering a gill slit, and minor branches to the adjacent pharyngeal wall and skin.

The second category includes four cranial nerves (III, IV, VI, XII) that are in series with ventral spinal nerves. All but one join the brainstem at the expected ventral level. (The exception is only partial; the nucleus of the trochlear nerve is ventral in the brainstem, but the fibers arch over the brain to emerge high on the opposite side.) These nerves carry somatic motor fibers. Ganglia are, of course, absent. The nerves innervate derivatives of head somites.

The third category has no counterpart in the spinal series because its nerves serve structures that are peculiar to, or centered on, the head: the nose, eye, ear, and lateral line system.

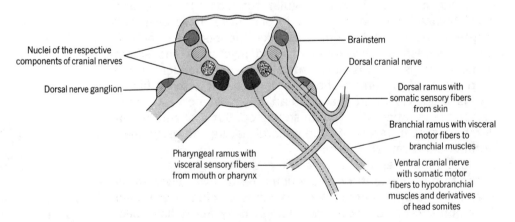

FIGURE 14-8

POSTULATED ANCESTRAL DISTRIBUTION OF NERVE COMPONENTS IN BRAINSTEM AND IN CRANIAL NERVES OF THE DORSAL AND VENTRAL SERIES.

(It is customary to place nerves I, II, VII, VIII, IX, and X in this category. However, VII, IX, and X are included because of their assumed relationship with the posterior part of the lateral line system, and there is now experimental evidence that the nerve serving this system is a separate unnumbered nerve with its own ganglion and separate projection into the brainstem. Complications!) These nerves are sensory. Their ganglia, unlike those of other sensory nerves are derived, at least in part, from ectodermal placodes. These nerves are usually considered to be somatic, though the designation is somewhat arbitrary and perhaps superfluous. Common practice is to call them "special" sensory nerves in recognition of their distinctive nature. To a degree, this is a category of leftovers because the nerves of the nose and eye are not serially homologous either with one another or with the other nerves of the category.

There is one further complication. The visceral motor nerves of the body are so distinctive in various ways that together they make up the autonomic system, which is described at the end of this chapter. Even at levels of the cord, it was the visceral motor fibers which were least stable, emerging sometimes in the dorsal root and sometimes in the ventral root. In the head, visceral motor fibers of the autonomic system join four cranial nerves (III, VII, IX, X). These include nerves of both the dorsal and ventral categories.

Counting seven nerves in the dorsal series, four in the ventral series, six in the special series (as qualified above), and four with autonomic fibers, and knowing that there are thirteen cranial nerves in all, it is evident that there is some doubling up. Thus, the oculomotor nerve is in the ventral series but has autonomic fibers tagging along; the facial nerve is in the dorsal series but may be augmented by both special and autonomic components.

This general information provides the basis for a more specific account of the evolution of each cranial nerve. The following account is clarified by Figure 14-9 and also Figures 7-23 and 15-3.

Structure and Evolution of Cranial Nerves. The TERMINAL nerve was not discovered until the other cranial nerves had been given numbers from I through XII. Hence, it has the number 0. It is regarded (on meager evidence) as being in the dorsal series. It may once have innervated tissues surrounding a gill slit anterior to the level of the mouth. If so, that slit was lost at the dawn of vertebrate history. The nerve runs to the nasal epithelium and sometimes to the vomeronasal organ, yet seems not to be olfactory in function. Having a ganglion, it is classed as a somatic sensory nerve, any ancestral visceral fibers having been lost. Clarification of its function awaits study by a competent neurophysiologist. The nerve is present in all vertebrates except cyclostomes, birds, and some mammals (including man). It is largest in elasmobranchs.

The OLFACTORY nerve (number I) is in the special series. It runs from the olfactory epithelium and vomeronasal organ (if present) to the olfactory bulb of the brain. It is unique in that its fibers are extensions of the receptor cells. Consequently it has no ganglion even though it is sensory. The nerve is present in all vertebrates, its size relating to the excellence of the olfactory sense. It is long if the rostrum is long and the olfactory tracts of the brain are short. Otherwise the nerve is short. Often it is divided into many twigs. It is paired in cyclostomes in spite of the unpaired nature of that animal's nasal pouch.

The OPTIC nerve (II) is again in the special series. It runs from the eye to the brain and is unique in that it develops as a tract of the embryonic brain (see p. 410). The associated ganglion cells are in the retina. The nerve is constant in all vertebrates. The two optic nerves may completely cross under the brain (teleosts, birds, and some other vertebrates), but often some fibers cross and some do not. In mammals, half the fibers of each nerve cross to the other side, an arrangement which in this class probably contributes to the coordination of eye movements.

The OCULOMOTOR (III), TROCHLEAR (IV), and ABDUCENS (VI) nerves are in the ventral series. They innervate the extrinsic muscles of the eye. All are somatic motor nerves, though the oculomotor is joined by autonomic fibers passing to muscles of the iris and ciliary apparatus of the eye. These nerves are constant in virtually all vertebrates.

The DEEP OPHTHALMIC (or profundus)nerve (V_1) is the second nerve of the dorsal series, though like the terminal nerve it does not relate to a gill slit in any surviving vertebrate. It runs to the skin of the rostrum. It is exclusively somatic sensory; any visceral fibers it may once have had are lost with the associated gill slit. The nerve is represented in all vertebrates, but is an independent nerve with its own ganglion only in ostracoderms, placoderms, and some primitive bony fishes. In other vertebrates it becomes a branch of the next nerve to be described.

The TRIGEMINAL nerve (V) has two branches in those vertebrates having an independent deep ophthalmic nerve. In other vertebrates the deep ophthalmic (in mammals simply called ophthalmic) (V_1), maxillary (V_2), and mandibular (V_3) nerves are the three branches which give the trigeminal nerve its name. (A superficial ophthalmic branch of the trigeminal may also be present, and unfortunately the facial nerve has a branch by the same name.) The maxillary and mandibular branches apparently represent a branchial nerve that once served the pre-

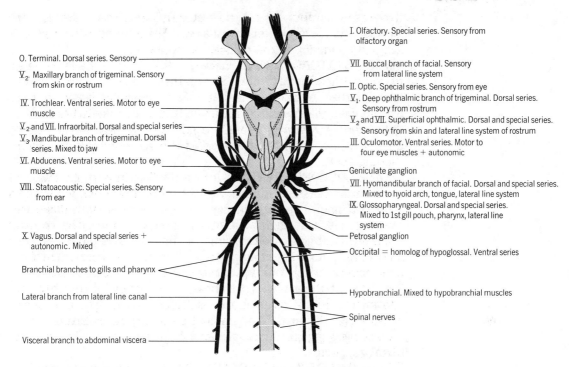

FIGURE 14-9

CRANIAL NERVES OF THE SHARK *Squalus*. **Ventral view. (Trigeminal and facial nerves are more independent in most other vertebrates.)**

mandibular gill slit next posterior to the one served by the ancient deep ophthalmic nerve. The trigeminal is, therefore, a composite nerve. The maxillary branch retains only somatic sensory fibers from the teeth, gums, and skin of the upper jaw. The mandibular branch similarly serves the lower jaw. It also has motor fibers to the various jaw muscles derived from the mandibular arch. These are striated, voluntary muscles, but since they are derived phylogenetically from the gut tube (see p. 198), the associated nerves are designated as visceral. The large semilunar (or Gasserian) ganglion is located where these branches all merge before entering the brain.

The FACIAL nerve (VII) is in part the nerve in the dorsal series that is associated with the spiracular cleft and derivatives of the hyoidean arch. It is the first nerve of this series to retain all the components of the ancestral dorsal nerves: somatic sensory to related areas of the skin, visceral sensory to much of the mouth and taste buds, and visceral motor to all muscles derived from the hyoidean arch. Included among the latter are the muscles of facial expression of man—hence the name of the nerve. The facial nerve of fishes also has a large component from the special series. This consists of the sensory fibers from the cranial part of the lateral line system. Finally, there is a component of the autonomic system serving tear glands and several salivary glands. The composite nerve is large. Its ganglion is the **geniculate ganglion.**

The STATOACOUSTIC nerve (VIII) (also called vestibulocochlear and auditory) serves the inner ear and is, therefore, in the special series. It has a **vestibular branch** to the organ of equilibrium and an **acoustic** (or cochlear) **branch** to the organ of hearing. There are ganglia by the same names. Since the ear is within the skull, the auditory nerve is short, though stout.

The GLOSSOPHARYNGEAL nerve (IX) is in part the nerve of the dorsal series associated with the first branchial gill slit and arch of fishes. Somatic sensory fibers are present in some vertebrates. Visceral sensory fibers run to part of the pharynx and some taste buds. Visceral motor fibers serve some small muscles. As for the previous nerve, there is a component from the special series of nerves innervating the lateral line system, this time at the back of the head, and another from the autonomic system, this time to a salivary gland. The composite nerve is usually small. Any somatic sensory fibers present have a **superior ganglion;** visceral sensory fibers have a larger **petrosal ganglion.**

The VAGUS (X) and ACCESSORY (XI) nerves can best be listed together because the latter is present only in amniotes and is derived by splitting away from the original vagus. The vagus spans the levels of several ancestral head somites and joins the brainstem in a linear series of twigs. These nerves are in part the last of the dorsal series, and as such are associated with all remaining branchial structures. A small somatic sensory branch of the vagus serves skin in the gill and ear region. Visceral sensory fibers come from posterior taste buds and the pharynx. Sensory fibers of the accessory nerve do not mature. Visceral motor branches serve muscles of the branchial arches and their derivatives (including some muscles of the shoulder). A large component of the vagus nerve of fishes innervates the part of the lateral line system that is on the body, and hence is from the special series of nerves (but see the parenthetical qualification on p. 349). Finally, the large, long, and important **visceral branch** of the vagus is the autonomic component of that nerve. Its fibers serve the heart, lungs (if present), and gut. Somatic sensory fibers have a **jugular ganglion;** visceral sensory fibers have a large **nodose ganglion.**

The HYPOGLOSSAL nerve (XII) is in the ventral series of nerves. It is an exclusively somatic motor nerve and innervates the hypobranchial muscles of throat and tongue. It is a cranial nerve in amniotes and some labyrinthodonts and a cervical nerve (called the hypobranchial nerve) in cyclostomes and fishes. It is derived from several postotic somites (the same for which the vagus was the dorsal nerve) and joins the central nervous system by a linear series of twigs.

The autonomic system is not isolated, structurally or functionally, from either the central or peripheral nervous systems. Hence, it is difficult to set its limits. It relates exclusively to involuntary functions of the body. Accordingly, the system includes only visceral fibers, and the visceral fibers serving the striated branchial muscles are excluded. Structures innervated by autonomic nerves are, therefore, the heart and vessels, some respiratory organs, glands, gut tube, urogenital organs, pigment cells, and intrinsic muscles of eye and skin. These structures are, of course, served by both sensory and motor fibers. There is nothing very remarkable, however, about the sensory neurons: like somatic sensory neurons they have their nerve cell bodies in dorsal root ganglia and certain cranial ganglia. Sensory fibers are, therefore, omitted from descriptions of the autonomic system, if not from its definition.

What, then, is distinctive about the visceral motor neurons that innervate involuntary organs? First, nearly all such organs are served by two sets of autonomic nerves—and the sets elicit antagonistic responses. These sets comprise the **sympathetic**

Autonomic System

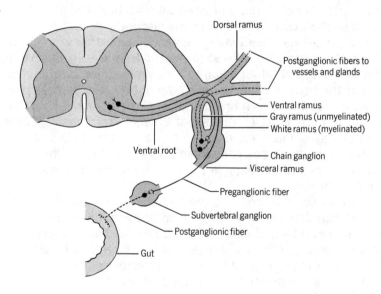

FIGURE 14-10

TYPICAL PATHWAYS OF THE SYMPATHETIC DIVISION OF THE AUTONOMIC SYSTEM OF AMNIOTES. Cross-section at posterior thoracic level. Other fibers run lengthwise in the body from the chain ganglia. Size of the nerves is exaggerated.

and **parasympathetic divisions** of the autonomic system. Second, every pathway includes a neuron having its cell body inside the central nervous system, and in addition (except in the adrenal medulla) a neuron (or in some instances several neurons) having its cell body outside the central nervous system. Cell bodies of the latter are in motor ganglia. Fibers between ganglia and the central nervous system are designated **preganglionic;** those between ganglia and end organs are designated **postganglionic.** Preganglionic fibers are myelinated; postganglionic fibers have little or no myelin.

The following descriptions of the two divisions of the autonomic system are based on amniotes (particularly mammals). Comments on the system in lower vertebrates will conclude the chapter.

The SYMPATHETIC division of the system is said to have a **thoracolumbar outflow** because that term indicates the levels at which its fibers emerge from the central nervous system in spinal nerves. Preganglionic fibers are relatively short. They reach the sympathetic ganglia by way of the visceral rami of spinal nerves (or the white part of the connection — see Figure 14-10). Most of the small but tough ganglia are **chain ganglia** arranged like two parallel strings of beads on the **sympathetic trunks** which are just ventral to the vertebral column. There are also three pairs of **cervical ganglia** against the carotid arteries in the neck, and about three unpaired **subvertebral ganglia** at the bases of the major arteries into the viscera.

A preganglionic fiber may synapse with a postganglionic fiber in the chain ganglion closest to its exit from a spinal nerve, or may go through that ganglion to a chain ganglion on the other side of the body, a chain ganglion at another level of the spine, or a cervical or subvertebral ganglion. Postganglionic fibers are relatively long. Most of them that emerge from chain ganglia reenter a spinal nerve and run with it and its branches to their destinations. Most postganglionic fibers that emerge from cervical or subvertebral ganglia run parallel with blood vessels to their destinations.

Postganglionic fibers, like the medulla of the adrenal gland, release **noradrenalin.** There are fewer terminal fibers than there are muscle and gland cells to be influenced. These nerve junctions, therefore, release relatively large quantities of noradrenalin which then spreads to adjacent effector cells.

In general, the sympathetic division of the system elicits responses of alertness, excitement, alarm, and the expenditure of energy as necessary to meet emergencies. Vegetative functions tend to be inhibited.

The PARASYMPATHETIC division of the autonomic system is said to have a **craniosacral outflow** because it emerges from the central nervous system with cranial nerves III, VII, IX, and X and

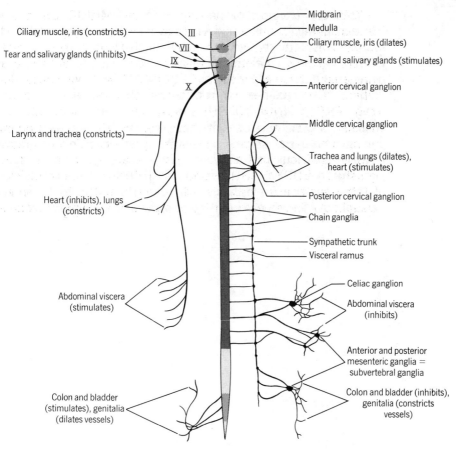

Ciliary muscle, iris (constricts)
III
VII
Tear and salivary glands (inhibits)
IX

X

Midbrain
Medulla
Ciliary muscle, iris (dilates)
Tear and salivary glands (stimulates)

Anterior cervical ganglion

Larynx and trachea (constricts)

Middle cervical ganglion

Trachea and lungs (dilates),
heart (stimulates)

Heart (inhibits), lungs
(constricts)

Posterior cervical ganglion

Chain ganglia

Sympathetic trunk
Visceral ramus

Celiac ganglion

Abdominal viscera
(stimulates)

Abdominal viscera
(inhibits)

Anterior and posterior
mesenteric ganglia =
subvertebral ganglia

Colon and bladder
(stimulates), genitalia
(dilates vessels)

Colon and bladder (inhibits),
genitalia (constricts
vessels)

PARASYMPATHETIC DIVISION   SYMPATHETIC DIVISION

FIGURE 14-11

DIAGRAM OF THE MAMMALIAN AUTONOMIC NERVOUS SYSTEM. General
distribution of fibers to vessels and skin are not shown.

with about three sacral spinal nerves. Preganglionic fibers are
relatively long. There are about four pairs of small parasympathetic ganglia in the head. These are located near the organs
served (eye, salivary glands, tear glands). Elsewhere, ganglion
cells are dispersed in the tissues of the viscera and are not
found by dissection. Postganglionic fibers are quite short.

Postganglionic fibers of parasympathetic neurons are like
most somatic fibers and preganglionic fibers of both divisions
of the autonomic system in releasing acetylcholine, though
again, the quantity is large.

The parasympathetic division of the system elicits responses
appropriate to quiet, vegetative activities such as digestion and
maintenance of resting levels of blood sugar.

The above account probably fits most AMNIOTES, though the autonomic system is well-known only in mammals. The system was late to evolve. AMPHIOXUS has visceral motor fibers running from each spinal nerve to the gut, but there are no ganglia outside the viscera. CYCLOSTOMES have autonomic fibers in the vagus nerve, but otherwise the system is rudimentary. CARTILAGINOUS FISHES send visceral motor fibers to vessels as well as gut and have some delicate autonomic ganglia under the spine. However, sympathetic and parasympathetic divisions of the system are scarcely, if at all, distinguished. In BONY FISHES and AMPHIBIANS the two divisions of the system are established in much of the body, though usually not in the more posterior organs of fishes and urodeles.

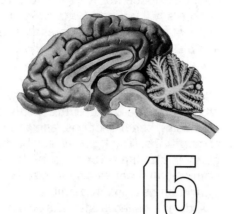

15

Nervous System:

Brain

T he brain is the most complicated, most challenging, and for many persons the most amazing organ of the body. It is under study by many researchers in various fields. Since understanding of its function must ultimately be based largely on its wonderfully intricate structure, the morphologist has a major role to play in the study of this master organ.

In order to gain an adequate background in brain structure for either teaching or advanced study it is desirable that "the pieces begin to fit together." A hasty or superficial review does not accomplish this purpose. Our objective, therefore, is to study the brain in enough depth to acquire basic vocabulary, learn general spatial relationships, and identify the major functional components. A little more repetition than for previous chapters will be intentional here in order to provide reinforcement of previous sections as further material is added to the story.

How the Brain Is Studied

The brain is studied in many ways. Descriptions are made of gross structure using entire or dissected brains, of fine structure using serial sections, and of ultrastructure using electron micrographs. Thick slices of the brain are stained to distinguish in striking color contrast the myelinated tracts from the unmyelinated cortex and nuclei. Histological techniques reveal individual neurons in detail.

A device called a stereotaxic instrument is used to position the head and brain in a precise standard position. The location of every minute part of the brain can then be expressed in terms of the three coordinates of space. Using the stereotaxic atlases which are now available for many experimental animals, the investigator can introduce microinstruments into any desired part of the living brain. Having done this, various techniques are used: Restricted areas are destroyed with electricity, freezing, or chemical agent, or very small bits of tissue are removed by suction. Impairment of function is then noted, and after a period of time, the animal is killed, and suitable staining techniques are used to reveal on sections of the brain the paths of degenerating nerve fibers. Alternatively, normal stimuli, microelectrodes, or chemicals are used to activate specific parts of the brain, and resultant behavior is observed. Animals may be taught to stimulate their own brains for the reward of pleasant sensations, and conscious human subjects have reported sensations and thoughts accompanying local stimulation of the brain during surgery. Following stimulation of one part of the brain, electric phenomena elicited in other parts are recorded using an oscillograph, multigraph pen recorder, or counters that translate data to tape for analysis by computer.

The impairment of function that follows accident or pathology is studied. Parts of the brains of experimental animals are severed and changes in behavior analyzed. Differences in the brains of different vertebrates are related to their widely different adaptations. Progressive maturation of the fetal brain is correlated with the onset of function. Natural electrical discharges of the organ (brain waves) are recorded during various activities. Culture techniques are used to study the growth and physiology of isolated neurons. The response of the brain to both the general and local administration of drugs is revealing. Finally, the science of cybernetics adapts mathematical models to brain function in an effort to elucidate mechanisms.

From all this come several overlapping levels of advancement in the study of this challenging organ. The first level is descriptive. Though not finished for any species and not even begun for many, this necessary level of investigation has produced hundreds of technical terms and shelves of much-labeled drawings. Study of the brain by students can be tedious if the emphasis is on description. Nevertheless, description serves as a basis for phylogenetic comparison. A second level of study relates structure and function. A third level identifies electrical circuits and produces complicated wiring diagrams. These last two levels have dominated brain research in recent years. Now it is gradually becoming apparent that the wiring diagram, however important, is conceptually inadequate. For some of the best-known circuits (e.g., that of vision), it seems that no part of the system is the exclusive site of any one function. Concepts of scanning mechanisms, holograms, and interference effects may prove useful for understanding memory. Such "mass action" of the brain points to further levels of analysis. The brain does not merely transmit, reject, or store the information in the 3 billion impulses that reach its 10^{10} cells in every waking second. It transforms the information, adapts it, and chooses among alternative responses in ways that surpass present comprehension.

Development of the Brain

By the time the neural folds close over the neurocoel in the later stages of neurulation, the future brain is already of greater diameter than the spinal cord. As soon as the tube is formed, the developing brain expands at three levels to form vesicles separated from each other by constrictions. These **primary vesicles** are the **forebrain** or **prosencephalon, midbrain** or **mesencephalon,** and **hindbrain** or **rhombencephalon.** The prosencephalon lies anterior to the notochord; the other vesicles are dorsal to the notochord.

There is reason to believe that the unknown ancestors of vertebrates had, as adults, brains similar to this present-day embryonic stage. At that time in evolution the brain may have been similar to the cord except for its response to sense organs peculiar to the head. It has commonly been believed that the prosencephalon was related to olfaction, the mesencephalon to vision, and the rhombencephalon to taste, equilibrium, and the lateral line system, but recent research indicates that this view may be too simplistic.

At the next stage of development, additional constrictions divide the brain further into five **secondary vesicles.** The anterior part of the prosencephalon becomes the **telencephalon,** largely through expansion of its lateral walls. These expansions will form the **cerebral hemispheres** of the adult. The posterior part of the prosencephalon becomes the **diencephalon.** The mesencephalon remains undivided. The rhombencephalon forms an anterior **metencephalon,** which forms the adult **cerebellum,** and a posterior **myelencephalon.** ("Encephal" = brain; most of the prefixes indicate position.)

Divisions between certain of these secondary vesicles are slight, and there is little functional and evolutionary basis for recognizing these vesicles. Nevertheless, it is a convenience to divide the brain into these parts for purposes of instruction.

The embryonic neurocoel is larger in the brain than in the cord. Within the vesicles it forms expansions called **ventricles.** The **lateral ventricles** occupy the cerebral hemispheres. If, as in fishes, the hemispheres are partly joined, then they share a common ventricle. The **third ventricle** is in the diencephalon. The neural canal expands within the mesencephalon of most vertebrates, but it is relatively restricted and tube-like in mammals and is then called the **cerebral aqueduct.** The **fourth ventricle** is located in both the metencephalon and the myelencephalon.

Most brains have nearly straight axes. Brains of birds and mammals, however, acquire three **flexures** in the embryo. The sharpness of each flexure depends on posture and cranial architecture, and does not relate directly to function. The **cephalic flexure** is in the mesencephalon and is concave ventrally. The **pontine flexure** is in the part of the metencephalon called the pons and is concave dorsally. The **cervical flexure** is within the posterior part of the myelencephalon and again opens

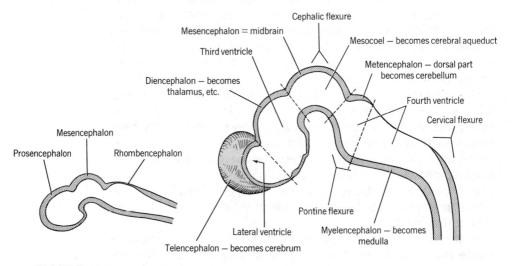

FIGURE 15-1

DEVELOPMENT OF THE MAMMALIAN BRAIN: stage of primary vesicles on left; stage of secondary vesicles on right. Brains cut in the sagittal plane.

ventrally. This flexure is most evident in bipeds that hold the head at an angle to the neck.

It is useful to divide the tube-like embryonic brain (and cord) into quadrants. The dorsolateral quadrants are called **alar plates** and the ventrolateral quadrants **basal plates.** In the cord, the dorsal gray columns, with their association neurons and axons of sensory neurons, develop from the alar plates, whereas the ventral gray columns, with their motor neurons, develop from the basal plates. In the brain, alar and basal plates become discontinuous, yet (as explained further below) they form neurons corresponding to their counterparts in the cord. The basal plate terminates at the diencephalon.

The three tissue layers of the embryonic spinal cord (ependymal, mantle, and marginal) are also present in the brain. The mantle layer is thick, and in the cerebral hemispheres and cerebellum of amniotes most of its inner cells migrate peripherally into the marginal layer. The vacated region will thus become white matter, and the invaded surface of the brain will become gray matter. This is the reverse of the spatial relationship elsewhere in the central nervous system where gray matter is internal.

Myelination is a slow process. It does not start until the fiber tracts are well-established and is not finished until well after birth.

More about the Organization of the Brain

In the previous section the brain was divided into five regions on the basis of development. Another division into three regions—**brainstem, cerebellum,** and **cerebrum**—is also useful. The central axis of the brain is the brainstem. It is the first region to form in ontogeny, the least variable, and the most like the spinal cord in structure. It receives all cranial nerves except the atypical terminal, olfactory, and optic nerves; relays impulses to the other two regions; and independently controls various vegetative functions of the body. Part of the adult metencephalon and all of the diencephalon, mesencephalon, and myelencephalon are included in the brainstem. The adult mesencephalon is usually called the **midbrain** and the adult myelencephalon the **medulla.**

The cerebellum and related pons (if present) are the principal adult derivatives of the metencephalon. They contribute to the coordination of motor functions. The cerebellum of amniotes and of some fishes is a conspicuous appendage of the brainstem covering much of the posterior part of its dorsal surface.

The cerebrum is the adult derivative of the telencephalon. Gradually this region of the brain enlarged and added new parts and functions until, in mammals, it dominates the brain both in size and control.

Within the spinal cord, nerve cell bodies mass in the dorsal and ventral gray columns, and these are continuous throughout the length of the cord. In the brain, by contrast, functionally related nerve cell bodies either mass at the surface of the cerebrum or cerebellum where they make up the **cortex** of those organs, or cluster in discontinuous masses within the brain. Such a cluster is usually called a **nucleus**, but may also be termed a center, ganglion, or body.

Similarly, within the cord, bundles of functionally related nerve fibers are usually called tracts. In the brain, such a bundle is also commonly called a tract, but unfortunately may have another name (fascicle, capsule, brachium, peduncle, lemniscus) depending on size, shape, and relationships.

Sensory fibers enter the brain from the cord and cranial nerves, and terminate in a nucleus of the brainstem or in the cortex of the cerebellum. Incoming impulses are usually passed from a first nucleus in the brainstem to one or more other nuclei or to the cortical areas (including the cerebral cortex). Each incoming impulse "comes to the attention of" many (usually thousands) of neurons in the gray matter of the brain. In this process of "considering" messages received, the nuclei are not mere relay stations for transmitting impulses; each is also an association center which processes, alters, and redistributes the messages in some way. The more complicated and the less automatic the ultimate response, the more likely that relevant impulses will reach the anterior part of the brainstem and cerebrum. Messages passing posteriorly may also be processed in "lower" nuclei before exiting from the brain in the spinal cord or cranial nerves. Further, tracts called **commissures** pass from one side of the brain to corresponding parts on the other side. These permit the integration of sensations and learning experiences from the two sides of the body. Crossings of sensory or motor fibers from one side of the brain or cord to the other are termed **decussations.**

Posterior Brainstem: Medulla through Midbrain

Nuclei of Cranial Nerves. It is again useful to refer to structure of the spinal cord as a frame of reference. There, somatic motor neurons are derived from the embryonic basal plates, whereas other neurons are from the alar plates. The same is true in the brainstem. In the cord, there is a more or less dorsal to ventral linear arrangement in the gray columns of somatic sensory, visceral sensory, visceral motor, and somatic motor neurons. This linear arrangement is roughly preserved in the brainstem. However, spreading of the dorsal parts of the alar plates of the medulla to accommodate the fourth ventricle causes the orientation of the linear series to change from dorsal–ventral to somewhat lateral–medial (Figure 14-8). Also, these neurons are pushed up close to the floor of the fourth ventricle as other structures come to occupy the ventral part of the medulla. Further, these components are continuous in the gray columns of the spinal cord, but are broken up into discontinuous nuclei in the brain. Finally, nuclei without counterpart in the cord are present in the brainstem to relate to the "special" senses of the head. These shoulder their way into position near their nonspecial equivalents.

It follows from this general plan that each cranial nerve tends to have a nucleus in the brainstem for each kind of component fiber it carries and that although corresponding nuclei of different nerves are at different anterior–posterior positions in the brainstem, they tend to be at roughly equivalent lateral-medial and dorsal–ventral positions.

These general relationships are useful for interpreting the anatomy of the nuclei of cranial nerves in specific animals, but nature contrives to complicate matters somewhat: One nucleus may serve two or more adjacent nerves and one ancestral nucleus may split into several nuclei. Some nuclei are long and some short; some large and some small. Jostled by neighboring nuclei and tracts, a nucleus may stray somewhat from its postulated ancestral position.

The oculomotor and trochlear nerves join the midbrain, and their nuclei are confined there. Nerves V–XII join the medulla, and most of them have nuclei there. Nuclei of the more anterior of these nerves commonly also (or instead) are located in the metencephalon, and a nucleus of the trigeminal nerve reaches forward to the midbrain.

To name the various nuclei of cranial nerves and to compare their numbers and positions in different vertebrates would not serve the objectives of this book. However, several representative nerve–nucleus relationships will be noted to illustrate the structural plan already presented: The oculomotor nerve has a medially situated nucleus for its somatic motor neurons and another smaller nucleus for its autonomic (visceral motor) fibers. The single nucleus of the trochlear nerve is found, as expected, just posterior to the somatic motor nucleus of the oculomotor nerve, even though the nerve itself exists from the brain at an unaccountably dorsolateral position. The large trigeminal nerve has a long somatic sensory nucleus in the expected dorsolateral part of the medulla. The visceral motor fibers of its mandibular branch (which innervate jaw muscles of branchial origin) have a midbrain nucleus which is relatively large in animals that chew their food. The facial nerve, having

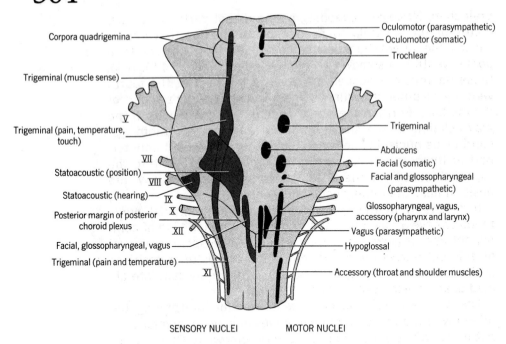

FIGURE 15-2

NUCLEI OF CRANIAL NERVES IN THE BRAINSTEM OF MAN. Dorsal view. Labels show nerve affiliations of nuclei, not their technical names.

several kinds of component fibers, has several nuclei in the brainstem. The statoacoustic nerve has separate "special" nuclei (or clusters of nuclei) for its vestibular and acoustic branches. The spinal accessory nerve shares one of its nuclei with the vagus from which it is derived.

Reticular Formation. The reticular formation occurs in all vertebrates. It is derived from the basal plates and is located in the central part of the brainstem from midbrain to medulla and, in reduced form, also in the anterior part of the cord. It consists of a diffuse, interlocking mass of nerve cell bodies and fibers; hence it is a mixture of gray and white matter. Its boundaries are indistinct (Figure 15-4). More or less sharply defined clusters of nuclei tend to form in its substance at several levels of the brainstem.

The reticular formation has sensory input from virtually all parts of the body and all senses, except olfactory. It in turn projects to the cerebrum, cerebellum, various cranial nuclei, and the cord. Its function can be expressed only in general terms: It is essential for consciousness. Stimulation awakens a sleeping animal and makes an awakened animal more alert. It contributes to the activities of both voluntary and involuntary muscles by facilitating, inhibiting, screening out noise, and coordinating stimuli.

Other Nuclei of the Posterior Brainstem.   There are numerous paired nuclei in the posterior brainstem other than those relating directly to cranial nerves. Several of the more prominent ones will be mentioned. The **olivary nuclei** arise from the alar plates, yet migrate down during development to the ventrolateral wall of the medulla. This complex of nuclei is present in all vertebrates. It is best developed in the most active representatives—mammals, birds, and some fishes. The largest component, the **posterior** (or inferior) **olive,** and the **anterior** (superior) **olive,** may each form a low but evident bulge on the sidewall of the medulla. In man the posterior olive, which is a motor coordination center, has the size and crumpled shape of a raisin, receives association fibers from some other nuclei of the brainstem and proprioceptive (muscle sense) impulses from the spinal cord (and some other input), and projects dorsally to the cerebellum. The median accessory olive is relatively well-developed in aquatic vertebrates with strong trunk and tail muscles.

The **ruber** (or red) **nucleus** and **substantia nigra** are located deep in the mesencephalon (and may extend into the diencephalon). The ruber nucleus is present in all vertebrates and is most developed in mammals, whereas the substantia nigra first appears in reptiles and is best developed in primates. Each nucleus may have evolved from the reticular formation. In man the ruber nucleus resembles a pea and the substantia nigra a lima bean. These are relay stations between the forebrain, on one hand, and posterior brainstem and cord, on the other. The ruber nucleus of carnivores (at least) plays a role in coordination of motor functions, particularly of flexor muscles.

The roof of the midbrain is called the **tectum** (= covering). In all vertebrates except mammals it consists primarily of a bilateral pair of conspicuous hemispherical eminences termed **optic lobes.** With mammals again excepted, the optic lobes are the primary center for the perception of vision (though even in fishes, the retina, diencephalon, and elsewhere in the mesencephalon are also involved in this complicated sense). The optic lobes are distinctly striated internally, and the visual image is projected onto the lobes point for point. The lobes are the most prominent feature of the brains of fishes that locate food by sight. They remain conspicuous in reptiles and birds.

The perception of vision is largely transferred by mammals to the cerebrum, though the midbrain tectum still has an important function: It tells the mammal where in space a visual object is, whereas the cerebrum tells what the object is. The

much smaller optic lobes are here called **anterior colliculi.**
Behind these are the **posterior colliculi** which are involved possibly in coordination of auditory reflexes. Taken together, these
four bumps are the **corpora quadrigemina.** There are commissures between the lobes of a pair.

Some of these brainstem nuclei, and others not named, are
sufficient to support various vegetative functions in normal circumstances. Even with other parts of the brain removed, experimental animals maintain heartbeat, respiration, swallowing,
and digestion.

Additional Features of the Posterior Brainstem.   The brainstem
transmits all impulses moving to or from other parts of the
brain. Obviously, much of its substance must consist of tracts
of nerve fibers. The pathways are complex, yet are well-known
for many vertebrates. Several tracts that are large and can
usually be seen on the surface of the brain will be identified
here.

Particularly large paired tracts of motor fibers run directly
from the cerebral cortex of mammals to the spinal cord without
interruption. Similar, but less prominent tracts are found in
other tetrapods and sharks. In mammals they emerge at the
surface of the brain on the ventrolateral wall of the midbrain
where they converge on each side to form the **cerebral peduncles** (= small feet, hence supports). The fibers are lost from
surface view in the metencephalon but continue in most
mammals as the prominent **pyramidal tracts** on the ventral surface of the medulla as they move to the lateral and ventral
funiculi of the cord.

The **trapezoid body,** which may be evident externally in
mammals (Figure 15-3), transmits fibers from the acoustic
nerve to the anterior olive and elsewhere. Large paired columns
of sensory fibers enter the medulla from the dorsal funiculi of
the spinal cord. These **gracile** and (more lateral) **cuneate fas-
cicles** flank the dorsal median sulcus of cord and posterior
medulla until they terminate in the **gracile** and **cuneate nuclei.**
On leaving the nuclei, the pathways in mammals move down to
the center of the medulla (thus contributing to the opening up
of the fourth ventricle dorsally) and continue to other nuclei in
the brainstem. In other vertebrates the projections of these
nuclei are less well-known.

The cerebellum, which arches over the fourth ventricle, joins
the brainstem by three pairs of peduncles (or brachia) which
flank the ventricle. If the cerebellum is large, the peduncles
are then prominent surface features. The **posterior peduncle**
carries fibers between the cerebellum and spinal cord, inferior
olive, and vestibular nuclei. The **middle peduncle** of mammals is
the largest and most lateral peduncle, partly hiding the others.

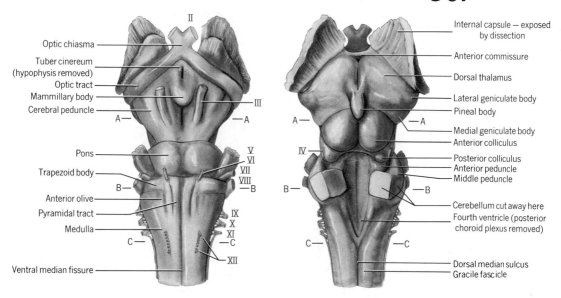

FIGURE 15-3

BRAINSTEM OF THE COW in ventral (left) and dorsal (right) views. Cerebrum and cerebellum removed by dissection. Letters show the levels of the cross-sections in Figure 15-4. (The brain of the sheep, commonly studied in the laboratory, is similar to the brain of the cow.)

It transmits fibers between the two sides of the cerebellum and from cerebrum to cerebellum. The **anterior peduncle** has fibers joining the cerebellum to the ruber nucleus and diencephalon.

The roof of the fourth ventricle is covered by the **posterior choroid plexus.** This is a delicate, much pleated, highly vascular membrane to which the brain contributes only its thin ependymal layer. The function of choroid plexuses is described at the end of this chapter.

Anterior Brainstem: Diencephalon

The anterior part of the brainstem differs from the posterior part in being derived entirely from the embryonic alar plates, in having no nuclei for cranial nerves and no reticular formation, and in relating to functions that are more highly evolved. The embryonic telencephalon forms the cerebrum which, in most vertebrates, is bilaterally paired and has no structures in the midline except for several commissures and the thin anterior wall of the ventricle. Consequently, the diencephalon forms the most anterior part of the brainstem. It is convenient to divide the diencephalon into three parts.

The narrow dorsal part of the diencephalon is the **epithalamus,** much of which is nonnervous in function. Most anterior is the **anterior choroid plexus** of the third ventricle. Posterior to the plexus is an evagination of similar structure (the

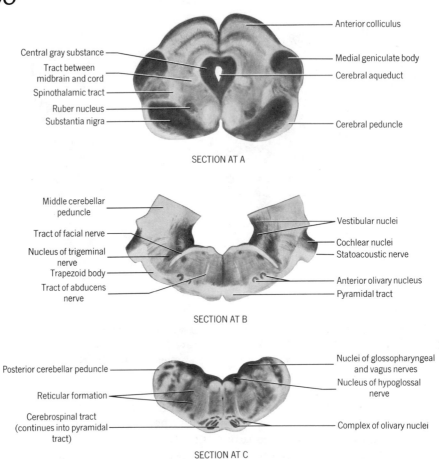

Anterior colliculus

Central gray substance

Tract between midbrain and cord

Spinothalamic tract

Ruber nucleus

Substantia nigra

Medial geniculate body

Cerebral aqueduct

Cerebral peduncle

SECTION AT A

Middle cerebellar peduncle

Tract of facial nerve

Nucleus of trigeminal nerve

Trapezoid body

Tract of abducens nerve

Vestibular nuclei

Cochlear nuclei

Statoacoustic nerve

Anterior olivary nucleus

Pyramidal tract

SECTION AT B

Posterior cerebellar peduncle

Reticular formation

Cerebrospinal tract (continues into pyramidal tract)

Nuclei of glossopharyngeal and vagus nerves

Nucleus of hypoglossal nerve

Complex of olivary nuclei

SECTION AT C

FIGURE 15-4

CROSS-SECTIONS OF THE BRAINSTEM OF THE COW at the levels shown by letters in Figure 15-3. The contrast between nuclei and tracts has been enhanced by staining.

paraphysis) which, however, does not mature in man. Next in line is the **habenular nucleus** or habenula. It is present in all vertebrates and contributes to the coordination of olfactory reflexes. Behind the habenula are two evaginations, the **parietal organ** and **pineal body.** These structures do not lend themselves to the organization of chapters by organ system. They are mentioned here because of their origin and relationships, in Chapter 16 because each may function as a sense organ (see Figure 16-17), and in Chapter 17 because one may function as an endocrine gland. The habenula has a small commissure, and a larger **posterior commissure** marks the posterior boundary of the epithalamus. Its fibers join some nuclei of the two sides of the diencephalon and perhaps of the mesencephalon as well.

Each thick lateral wall of the diencephalon is called a **thalamus.** This is the largest part of the anterior brainstem, being 4 cm long in man. The two thalami are bordered in part by the lateral ventricles and are separated from one another by the third ventricle, except where the large **middle commissure** or intermediate mass spans that cavity. Each thalamus is a compact oblong mass of many nuclei—30 or more are recognized. A ventral cluster of nuclei (**ventral thalamus**) is represented in all vertebrates and projects motor fibers posteriorly in the brain. A dorsal cluster (**dorsal thalamus**) projects sensory pathways (except olfactory pathways) to the cerebrum (to both its striatal and pallial parts, as defined below). The dorsal nuclei are most developed in tetrapods—particularly in mammals.

Two of the dorsal nuclei are of particular importance. These are located at the posterior end of the thalamus close to the tectum of the midbrain. One is the **medial geniculate body** which forms part of the auditory pathway and is joined by a pathway to the posterior colliculi of the tectum. The other is the **lateral geniculate body** which is a relay station in the primary visual pathway. Seen in cross-section, the lateral geniculate bodies of some mammals (like the optic lobes of many vertebrates) are striated into about six conspicuous layers. The visual field is projected onto these layers.

The thalamus is a relay center to the cerebrum and also is the last center short of the cerebrum where bodily functions are modulated. There is evidence that in several classes of vertebrates a level of awareness, including perception of both pain and pleasure, is located in the thalamus.

The ventral part of the diencephalon is the **hypothalamus.** It contains about a dozen pairs of nuclei which together integrate and largely control the autonomic functions of the body including water balance, temperature regulation, appetite and digestion, blood pressure, sleep and waking, sexual behavior, and emotions. Consistent with the dual nature of the autonomic system, each function has two centers in the hypothalamus—one for facilitation and one for inhibition.

On the ventral surface of the hypothalamus is the **optic chiasma** where the optic nerves converge and cross (usually with partial decussation of their fibers) before continuing up the sides of the brain as the **optic tracts.** The optic tracts terminate in the lateral geniculate bodies. Just posterior to the optic chiasma is an area called the **tuber cinereum** which encloses several nuclei of the parasympathetic system and subtends the **hypophysis.** The important hypophysis is in part of nervous origin but functions (partly under the influence of nuclei in the hypothalamus) as an endocrine gland. Accordingly, it is described in Chapter 17. Posterior to the hypophysis is a pair of small but evident lumps called the **mammillary bodies.** Within are nuclei of the same name which function in olfaction. These are present in all tetrapods and may be represented in fishes.

Cerebellum and Pons

The cerebellum is an ancient part of the brain, yet has enlarged and changed in the course of evolution much more than has the brainstem. It follows that it is a variable part of the brain and that portions of the cerebellum of mammals are relatively "new." The cerebellum develops from the dorsal part of the metencephalon and lags a little behind the development of the brainstem.

The cerebellum of cyclostomes and amphibians is small and smooth. Because it is merely a thickening of the wall of the brain tube, it has no cavity. The cerebellum is usually still smooth in fishes and reptiles, yet in them is prominent and encloses part of the fourth ventricle. In birds and mammals the organ is very large, lobed, and convoluted into tight **gyri** (convex folds) and **sulci** (concave grooves). Its solid walls then nearly exclude the fourth ventricle.

Unlike the spinal cord and brainstem, the gray matter of the cerebellum is in a thin superficial cortex. The central white matter branches to each lobe and gyrus. Because of its appearance in longitudinal section, this branching white matter is called the **arbor vitae** (= tree of life).

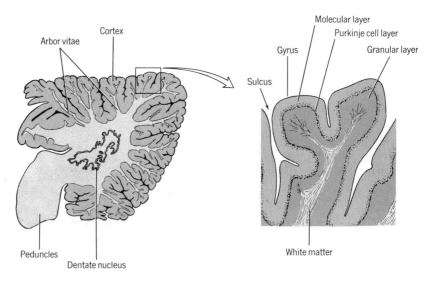

FIGURE 15-5
PARASAGITTAL SECTION OF THE HUMAN CEREBELLUM.

The cerebellar cortex of all vertebrates is divided histologically into three regions: a deep **granular layer** which has several types of cells, a middle **Purkinje cell layer,** and a superficial **molecular layer** which has scattered cells but consists mostly of synapses. Afferent impulses are relayed twice within the cortex before reaching the large and exceedingly branched Purkinje cells (see the most branched neuron of Figure 14-1). The arrangement of cells in the cortex assures that impulses will spread widely.

The cerebellum functions to control motor coordination and to maintain equilibration. It does not initiate motor activities, but processes those initiated elsewhere. It provides unconscious timing and integration of muscles that must contract together and of antagonistic muscles that must simultaneously relax. If the cerebellum is damaged, muscle activity loses control and precision.

In order to do this job the cerebellum needs extensive input from the sensory system. In primitive vertebrates the cerebellum has two parts. One is the **archicerebellum.** Its principal input comes from the labyrinth of the ear and the lateral line system by way of associated nuclei in the brainstem, though pathways from the cord may also exist. The archicerebellum persists in amniotes as the auricular, or flocculonodular lobes of the more evolved organ (Figure 15-12). The other part is the **paleocerebellum.** Its input is largely from proprioceptor (muscle sense) organs of trunk and limbs and is relayed by the spinal cord and posterior olivary nucleus. This part of the cerebellum is important for the maintenance of posture. Finally, when appendicular muscles became more prominent with the evolution of amniotes, and the cerebral cortex became larger and more dominating, the cerebellum responded by evolving an associated **neocerebellum.** Though not sharply set off, the neocerebellum is in the midline and pushes the paleocerebellum to the sides where, in mammals, it forms the hemispheres of the organ. The neocerebellum is important to voluntary locomotor activity and is particularly marked in birds. There is also input to the cerebellum from the senses of touch, sight, and hearing, and from the reticular formation.

The cerebellum has fewer efferent than afferent fibers. Outgoing impulses originate in the Purkinje cells and are relayed by **cerebellar nuclei.** The location of these nuclei is varied and complex in anamniotes. They are located in the base of the larger cerebellum of amniotes and in mammals split to become three or (in primates) four separate pairs of nuclei. The largest and most distinctive is the **dentate nucleus** which (like the inferior olive of the medulla) has a crumpled contour. There is a topographic relationship between the cerebellar nuclei and the regions of the cerebellar cortex. Efferent pathways run from these nuclei to the vestibular nuclei, reticular formation, ruber nucleus, and dorsal thalamus.

Afferent and efferent pathways between the cerebellum and cerebral cortex evolve as the newer, more dominant parts of the cerebral cortex evolve. Associated with this change is the appearance in some birds, and the prominence in mammals, of the **pons.** This addition to the ventral part of the brainstem at the level of the cerebellum receives in its **pontine nuclei** fibers from the cerebral cortex and relays impulses to the cerebellum through the middle peduncles. (The peduncles supporting the cerebellum were described above with the posterior brainstem.) The pons also transmits impulses from one side of the cerebellum to the other.

Cerebrum **General Structure.** The cerebrum is the adult telencephalon. The most anterior part of the cerebrum is always bilaterally divided. Nearly all of the organ is divided in amniotes and each half is then called a cerebral hemisphere.

The cerebrum is like the cerebellum in that it is of ancient origin yet enlarges and changes in the course of evolution. However, the changes are here more pronounced and do not form a single progression but instead follow separate trends leading to different end points in teleosts, birds, and mammals. The phylogeny of the cerebrum is one of the most striking in vertebrate evolution.

At the anterior end of each hemisphere is the **olfactory bulb** which receives the olfactory nerves. The remainder of each hemisphere is divided into two principal parts: The **corpus striatum** is in a ventral position. It is composed of various prominent nuclei and often swells to become large and globular. The **pallium** (= cover) forms the roof and side walls of the cerebrum. Its association centers tend to be spread out into a sheet. In vertebrates other than actinopterygians, the corpus striatum and pallium are in part separated by the lateral ventricles. Where pallium and corpus striatum merge, the boundary is indistinct in some vertebrates. Homologies of the various nuclei are sought using structural, comparative, functional, and histochemical evidence. The cerebrum is completed by various tracts and commissures. These parts will be discussed in order.

Olfactory Bulb and Tract. Axons of the conductive receptor cells of the nasal epithelium enter the olfactory bulbs. There they converge to synapse with two kinds of neurons within ball-like tangles of nerve endings called **glomeruli.** Axons of these second-level neurons enter the olfactory tracts which conduct impulses out of the bulbs. Another kind of neuron seems to provide local feedback circuits within the bulbs. There are fewer fibers in the outgoing tracts than in the incoming nerves, so the bulbs accomplish a summation of impulses. Third-and fourth-level neurons elsewhere in the brain are also involved in the complicated processing of olfactory input.

The size of these structures relative to the remainder of the brain varies over a wide range according to the importance of olfaction in the life of the animal (Figure 15-12).

Corpus Striatum and Basal Nuclei. The corpus striatum (= body + striped) is named for the appearance of part of its substance in some vertebrates as seen in section. For mammals, **basal nuclei** is an approximate synonym. The term **basal ganglia** is also used. The corpus striatum has three principal parts which are named according to function. It should be recognized, however, that historical misconceptions about sequence in their evolution have given emphasis to their distinctions. Also, the homologies assigned to the parts among the different classes of vertebrates are in some instances tentative.

One part of the striatum is called the **archistriatum.** It integrates olfactory and general somatic senses. The archistriatum of fishes consists of several indistinctly segregated nuclei called the **amygdaloid** (= almond-shaped) **complex.** Tetrapods retain the structure, and in mammals the corresponding amygdaloid nucleus is a globular mass which tends to be ventral to the other basal nuclei. Even in mammals it remains in part an association center for olfactory input.

The **paleostriatum** is represented in all vertebrates, with various names assigned to its parts. The homologous basal nucleus of primates is the **globus pallidus** (= ball + pale), the function of which is noted below.

The name **neostriatum** reflects the long-held belief that this part of the striatum evolved only in amniotes—a view that is no longer held. Some structures to which this name has been assigned may prove not to be homologs (particularly in birds), but the derivative mammalian basal nuclei are the long **caudate** (= tailed) **nucleus,** which arches over the others, and the **putamen** (= shell).

The corpus striatum is the highest integrative center of fishes (though, as we shall see, the pallium should not be discounted). The striatum appears to be enormous in reptiles and is usually considered to be their highest integrative center, though it now seems probable that part of the pallium is in fact incorporated into the "striatum." Paleostriatum and neostriatum are also large in birds and over these is a unique, thick, four-layered **hyperstriatum.** Recent evidence indicates that this is again derived from the pallium.

Considering their prominence, the function of the basal nuclei is poorly known for mammals. The caudate nucleus, putamen, and globus pallidus are here dominated by the cerebral cortex. They receive fibers from the cortex and thalamus. They project fibers to each other and to the thalamus and substantia nigra. Together they are thought to contribute to the function of muscle masses, as opposed to delicate control of individual muscles. Lesions of one or more of the basal ganglia result in specific motor disturbances, recognizable as different from those resulting from lesions of cerebellum or cerebrum.

(Two additional nuclei are commonly named in articles about the mammalian corpus striatum. Although phylogenetically and functionally more closely related to the caudate nucleus, the putamen may merge structurally with the globus pallidus to form the lentiform (= lentil-shaped) nucleus. Also classed as a basal nucleus is the flat, laterally placed claustrum. It may in fact be derived from the pallium rather than the corpus striatum. Although present in reptiles as well as mammals, its function is not clear.)

Pallium.   Three principal parts of the pallium are recognized: **paleopallium, archipallium,** and **neopallium.** Except in actinopterygians and mammals, the paleopallium is lateral to the ventricle and the archipallium is dorsal or median to the ventricle. The neopallium may be between the other parts or, apparently, in reptiles and birds, ventral or lateral to the ventricle in association with the striatum.

It was long believed that the telencephalon was initially associated only with olfaction. The paleopallium and archipallium were thought to relate only to olfaction and to be the only parts of the pallium of fishes. It is now known that much of the pallium of many fishes is *not* olfactory in function. Experiments have shown that parts of the pallium contribute to schooling, aggressive, and reproductive behavior and make learned responses possible. Accordingly, the neopallium is probably also ancient, though its limits are not clear at the fish level of evolution.

The architecture of the cerebrum of actinopterygians is unique. Pallium and corpus striatum are thick, merge together, and both lie lateral and ventral to a common ventricle. Paleopallium and archipallium are certainly present, and probably also the neopallium, but their spatial relationships are atypical, and thus it is difficult to homologize their nuclei with those of other vertebrates.

In reptiles and birds a thin pallium is stretched wide to arch over the now enlarged corpus striatum. As noted above, however, it now seems likely that part of the large striatum (the hyperstriatum portion in birds) is really neopallium. Birds do

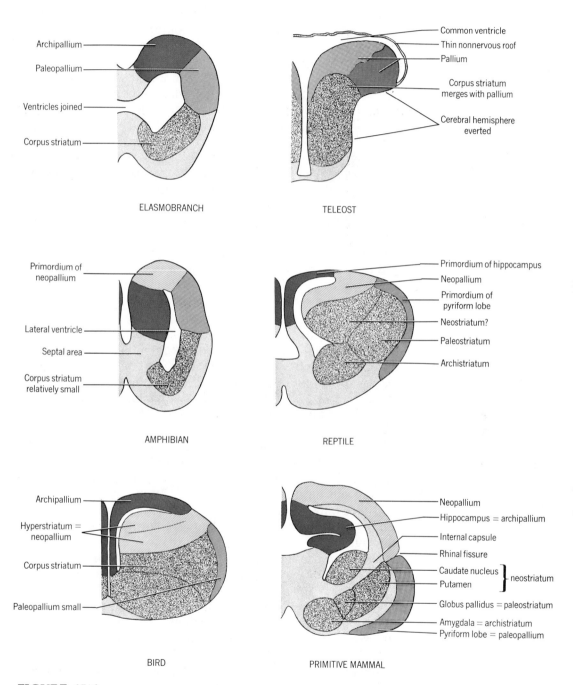

FIGURE 15-6
DIAGRAMMATIC CROSS-SECTIONS OF CEREBRAL HEMISPHERES SHOWING THE COMPARATIVE STRUCTURE OF THE CORPUS STRIATUM AND PALLIUM.

not seem to be seriously inconvenienced by the surgical removal of their thin superficial pallium, though there are indications that some capacity for memory is sacrificed. Contrary to common belief, some birds are more intelligent than many nonprimate mammals, and it has been proven that their ability to solve problems, and to remember how to solve new problems of a related kind, resides in the hyperstriatum.

In mammals the paleopallium and archipallium are wedged apart as the evolving neopallium enlarges between them. Anteriorly, the paleopallium includes the olfactory tracts. Posteriorly it is pushed down around the sidewall of the hemisphere to a position flanking the anterior brainstem. The resulting **pyriform** (= pear-shaped) **lobe** is prominent in mammals having a keen sense of smell. It is separated from the neopallium above by the **rhinal fissure.** The pyriform lobes are the olfactory cortex.

The archipallium is pushed by the neopallium in the other direction to the crown of the brain near the midline (reptiles, monotremes, marsupials) or over the edge onto the medial wall of the hemisphere (other mammals). As it moves, it rolls on itself lengthwise and sinks largely below the surface, thus forming a long arching band which impinges on the lateral ventricle. Its name, **hippocampus** (= sea horse), is suggested by its rolled appearance as seen in cross-section. The hippocampus of mammals does not function in olfaction. Although it is large and complex, its several functions are poorly understood. Together with the mammillary bodies and parts of the amygdaloid complex and thalamus it contributes to the **limbic system,** which functions in aspects of sexual and emotional behavior and memory.

All large mammals and some smallish ones have a convoluted neopallium (or cerebral cortex, or simply cortex); the gyri and deep sulci greatly increase total surface area. Intelligence relates to absolute brain size, relative brain size, and convolutions, but only in a general way; man's brain is not the largest, nor the largest in relation to body weight, nor even the most convoluted.

Gray matter is internal in the pallium of anamniotes but external in the cerebral cortex of amniotes. The various parts of the cortex can be distinguished histologically. Characteristic cell layers differ in the density, size, configuration, connections, and staining properties of their neurons. Cells in the six successive layers of the cortex of primates tend to be arranged in vertical columns, each having as many as 100 cells. Small mammals have more cells per unit volume of brain than large mammals.

The primate cortex is divided into regions for ease of reference (i.e., frontal, temporal, parietal, and occipital), and spe-

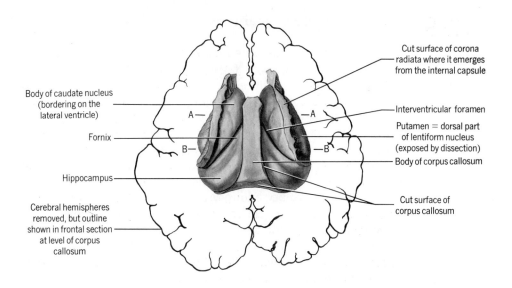

Cut surface of corona radiata where it emerges from the internal capsule

Body of caudate nucleus (bordering on the lateral ventricle)

Interventricular foramen

Putamen = dorsal part of lentiform nucleus (exposed by dissection)

Fornix

Body of corpus callosum

Hippocampus

Cerebral hemispheres removed, but outline shown in frontal section at level of corpus callosum

Cut surface of corpus callosum

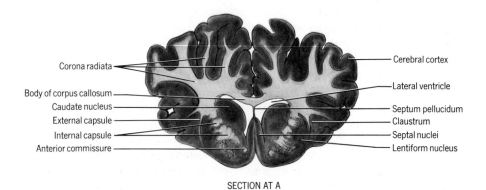

Corona radiata

Cerebral cortex

Body of corpus callosum

Lateral ventricle

Caudate nucleus

Septum pellucidum

External capsule

Claustrum

Internal capsule

Septal nuclei

Anterior commissure

Lentiform nucleus

SECTION AT A

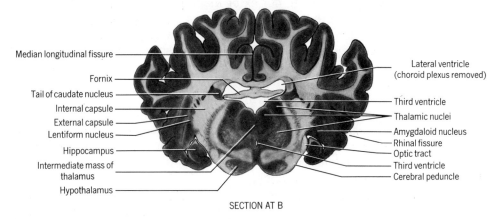

Median longitudinal fissure

Lateral ventricle (choroid plexus removed)

Fornix

Tail of caudate nucleus

Third ventricle

Internal capsule

Thalamic nuclei

External capsule

Amygdaloid nucleus

Lentiform nucleus

Rhinal fissure

Hippocampus

Optic tract

Intermediate mass of thalamus

Third ventricle

Cerebral peduncle

Hypothalamus

SECTION AT B

FIGURE 15-7

FOREBRAIN AND THALAMUS OF THE COW shown by a dissection seen in dorsal view (above) and by cross-sections (below) made at the levels indicated by letters. On the sections, contrast between nuclei and tracts has been enhanced by staining.

cialists have a standardized way of indicating subregions. In a band across the crown of the human brain at the posterior margin of the frontal region is the primary somatic motor cortex. Ultimate control of voluntary motor activity rests here, though this part of the cortex is "coached" and "advised" by other parts of the cortex and by the cerebellum, reticular formation, basal nuclei, ruber nucleus, and other centers.

Just behind the somatic motor cortex, at the anterior margin of the parietal region, is the primary somatic sensory cortex. Similarly, the primary auditory cortex is at the top of the temporal lobe, and the primary visual cortex (which develops in mammals as the optic lobes of the midbrain relinquish most of their function) is in the occipital region. These areas have a point-for-point relationship with function (though in some mammals there is overlap among adjacent points): Motor control of the thumb is at a specific place just below the forefinger point and above the neck point; tones are perceived at given points in treble to bass sequence; sight points are arranged in relationship to the visual field. If the face is relatively sensitive (as in man), the corresponding areas of the cortex are relatively large; if the hands are relatively sensitive (raccoon), the hand areas are large.

Although the above is true, it does not go far enough to indicate the complexity of cortical function. For unknown reasons, the two hemispheres of man are not symmetrical in their control of several functions; for example, speech, reading, and writing are vested in the left hemisphere whereas the analysis of music is in the right (of right-handed persons). The primary cortical areas noted above account for most of the cortex of lower mammals (marsupials, insectivores) but for only one-quarter of the cortex of man. In the parietal and occipital regions there are secondary and tertiary projection areas. These are not related to function on a point-to-point basis and have fuzzy boundaries. They code messages, store information, combine input, and provide spatial orientation. If they are impaired, a man might see well, yet confuse right and left; might walk well, yet get lost in a familiar place. Finally, the frontal region of the cortex relates to programs, intentions, orientation to goals, and sequence in the performance of activities. The entire cortex of experimental animals becomes relatively heavy if the animals live in a relatively diverse, stimulating, and "enriched" environment.

There is constant interaction among these parts of the cortex and constant interaction (sometimes both ways) between sensory input and motor output. Short-term memory, whatever it is, requires the frontal cortex and probably the thalamus. Long-term memory is different and seems to be stored by virtually all of the brain as a diffuse code. No one structure of the brain is

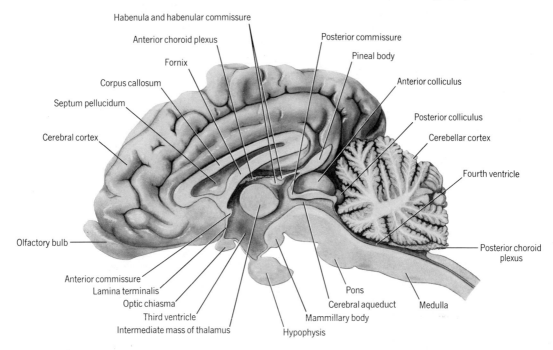

FIGURE 15-8
SAGITTAL SECTION OF THE BRAIN OF THE COW.

the exclusive site of any one function, and alternative circuits seem to be available for controlling any function.

All in all, several things remain to be learned about this organ!

Some Other Features of the Cerebrum. The tracts of the cerebrum are at least as constant as the nuclei. Association fibers, short and long, loop between the different parts of the cortex. The olfactory tracts divide into several paths as they merge into the brain: One crosses to the other olfactory bulb. Another stops in **olfactory nuclei** which send third-level neurons to the habenula and mammillary bodies of the diencephalon. Still another runs to the amygdaloid nucleus and pyriform lobe. All these structures function in the perception and integration of olfactory stimuli.

The conspicuous motor tract which exits from the primary motor cortex runs right through the corpus striatum as a broad white band called the **internal capsule.** It separates the caudate nucleus from the functionally related putamen. It is this band which continues, as the cerebral peduncles and subsequently as the pyramidal tracts, directly into the spinal cord. Collectively this is the **pyramidal system,** as contrasted to the **extrapyramidal system,** which includes the same cortical areas

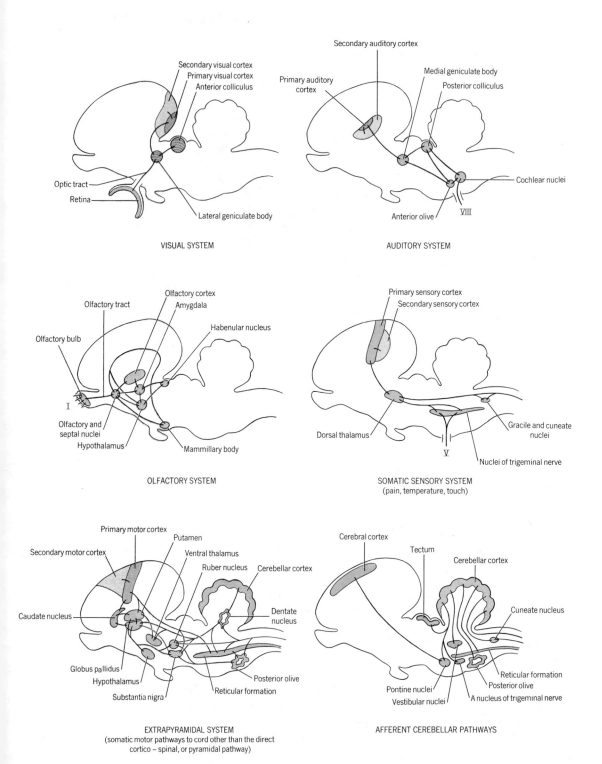

FIGURE 15-9

DIAGRAMS OF SOME PRINCIPAL COMPONENTS OF SOME FUNCTIONAL SYSTEMS OF THE MAMMALIAN BRAIN.

but also other parts of the cortex and efferent relay centers in the corpus striatum, thalamus, and elsewhere.

The cerebrum has several commissures. The small **habenular** and larger **anterior commissures** are derived from the archipallium and appear in the lower classes. Therian mammals have evolved a commissure, the **corpus callosum,** which in all but marsupials is the largest of all. It joins one neopallium to the other and is seen in the dorsal midline as a horizontal shelf when the hemispheres are pulled apart. If the corpus callosum is cut, something learned or experienced on one side of the body cannot be adequately used or acted upon by the other. The **fornix** is a conspicuous arching pathway positioned ventral to the corpus callosum anteriorly and lateral to the hippocampus posteriorly (Figure 15-7). It transmits fibers between the hippocampus and the hypothalamus.

Each lateral ventricle encloses a large choroid plexus which is rooted at the roof of the diencephalon with the smaller plexus of the third ventricle.

The brain of CYCLOSTOMES is primitive and also in some respects specialized and degenerate. The anterior part of the brain is foreshortened in response to crowding by the terminal mouth and dorsal nasal chamber. The large olfactory bulbs are separated by only a shallow constriction from thick cerebral hemispheres. Corpus striatum, paleopallium, and archipallium are somewhat vaguely delimited. Parietal organ, pineal body, and large habenula are all visible on the roof of the diencephalon. Optic lobes are evident in lampreys and small in blind hagfishes. The medulla is relatively large and supports a large everted posterior choroid plexus. The cerebellum is rudimentary, as would be expected for these sluggish, parasitic creatures.

Comparative Anatomy of the Brain

The brain of ELASMOBRANCHS is much advanced over that of cyclostomes and is also somewhat specialized according to species. Olfactory bulbs are large, widely separated, and enclose extensions of the lateral ventricles. Olfactory tracts are in evidence. Cerebral hemispheres are broadly joined in the midline and share a common ventricle. Corpus striatum and pallial areas are well-established. Present in the diencephalon are pineal body (but not parietal organ), choroid plexus, habenula, lateral geniculate body, thalamus, and large hypothalamus. The hypothalamus has **inferior lobes** and a thin **vascular sac** of questionable function flanking the hypophysis. These two structures are characteristic of fishes in general and are absent in tetrapods. Optic lobes are usually prominent and surround an expansion of the midbrain ventricle. The ruber nucleus has evolved. The cerebellum is large in active species, median in position, somewhat fissured in large sharks, and contains an expansion of the fourth ventricle. Archicerebellum and paleocerebellum are distinguished, and the cortex has differentiated into the three layers characteristic of higher vertebrates. Olivary nuclei and reticular formation are present in the medulla.

Brains of BONY FISHES are diverse. Those of Dipnoi resemble those of Elasmobranchii, whereas those of Actinopterygii have forebrain architecture shared by no other vertebrates. The brain of the surviving representative of the Crossoptergii is intermediate.

If the nasal epithelium is located far from the brain proper, then impulses span the interval in either of two ways: The olfactory nerves may be long, the olfactory bulbs close to the remainder of the brain, and the olfactory tracts quite short (the usual circumstance). Alternatively, the olfactory nerves may be short (as in carp and cod), the bulbs adjacent to the nasal chambers, and the olfactory tracts long.

The cerebral hemispheres of dipnoans evaginate in the usual manner, each becoming convex on its outer surface. Those of actinopterygians evert instead: The dorsal lip of the hemisphere curls outward so the hemisphere becomes concave on its outer surface. Pallium and corpus striatum are contiguous and thick. The homologies of the parts of the pallium are uncertain.

Conspicuous inferior lobes and vascular sac are present under the diencephalon as for cartilaginous fishes. Optic nerves pass one another at the optic chiasma or weave through one another in various patterns. Bulging optic lobes are usually the most prominent part of the brain of bony fishes. The cerebellum is smooth but usually large. It is more solid than in cartilaginous fishes. The cerebellum of actinopterygians has a distinctive anterior projection called the **valvula** which pushes into the ventricle of the midbrain, thus contributing to the separation of the optic lobes.

Large pathways relating to taste are present in the medulla. If food is tasted primarily on barbels and lips, a median **facial lobe** forms behind the cerebellum to house nuclei of the facial nerves. If food is tasted primarily within the mouth and pharynx, lateral **vagal lobes** form to house nuclei of the glossopharyngeal and vagus nerves. In the medulla of actinopterygians (and urodeles) there is a single pair of giant neurons called **Mauthner cells** which relate to the ear and lateral line system. They mediate escape reflexes to muscles used in swimming. The choroid plexuses of fishes are everted. Nuclei of the brainstem are complicated and specialized.

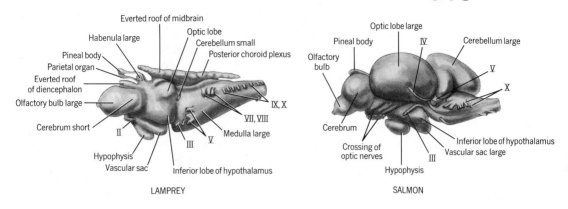

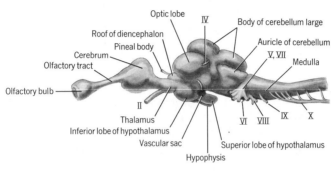

FIGURE 15-10

BRAINS OF REPRESENTATIVES OF THREE CLASSES OF FISHES.

The brain of AMPHIBIANS is remarkably unspecialized (particularly in urodeles) and is scarcely more advanced than that of cartilaginous fishes and dipnoans. The cerebral hemispheres are more separate from one another than in fishes, so they share little common ventricle. Primitive hippocampal and pyriform areas have formed, respectively, from the archipallium and paleopallium. The corpus striatum is small. The epithalamus is fish-like except that the pineal body is well-developed in anurans. The dorsal thalamus is beginning to enlarge. Incipient mammillary bodies are present in the hypothalamus. Optic lobes are of moderate (anurans) or small (urodeles) size. The cerebellum is rudimentary.

The brain of REPTILES is narrow, elongate, and nearly straight. Olfactory bulbs tend to be smaller than for fishes. Olfactory tracts are long. The cerebrum is large because of the expansion of the corpus striatum and associated neopallium.

The superficial parts of the pallium are thin. Gray matter has become external in the cortex. Relative sizes and positions of the divisions of pallium and corpus striatum indicate that there have been two trends in the evolution of the reptilian forebrain: one line (represented by turtles) in the direction of mammals and another (represented by crocodilians) in the direction of birds.

The parietal organ is functional in lizards (see Chapter 16). The dorsal thalamus is larger and more complex than in the lower classes, and the ventral thalamus has all the nuclei regularly recognized in mammals. Since most reptiles have excellent vision, the optic lobes are conspicuous. The midbrain still encloses an expanded ventricle as in anamniotes.

The reptilian cerebellum is smooth. It is largest in swimmers and rudimentary in snakes. Cerebellar nuclei are now within, rather than below the organ. Choroid plexuses are inverted.

The brains of BIRDS are relatively large, uniform, and distinctive. The organ is short and broad. There are marked cranial, pontine, and cervical flexures. Olfactory bulbs and tracts are evident in the scavengers, but in general are smaller than in other vertebrates. The avian cerebral hemisphere is surpassed in size only by that of some mammals. This is because of the enormous development of the corpus striatum with its associated "hyperstriatum" or neopallium. The parts of the cerebrum have a different configuration in birds that emphasize hearing (owl) than in birds that emphasize touch and manipulation with the beak (duck, snipe, parrot). The superficial parts of the pallium are exceptionally thin and have little function.

The dorsal thalamus is even more developed than in reptiles. Optic nerves, chiasma, and tracts are large. Optic lobes are particularly large and are layered within. They have connections from all sense organs and with the cerebrum. Squeezed between the cerebrum and cerebellum, the optic lobes have a uniquely lateral position.

The cerebellum is larger than in other vertebrates except some mammals. It is tightly convoluted. The medially placed neocerebellum being particularly prominent (because of its relation to flight), the organ is high and narrow. Related to the marked development of the cerebellum are the appearance of the pons under the brainstem and enlargement of the olivary nuclei within the broad medulla.

The olfactory bulbs and tracts of MAMMALS range from huge (aardvark, armadillo, anteater) to very small (primates). Although relatively smaller than in reptiles and birds, the

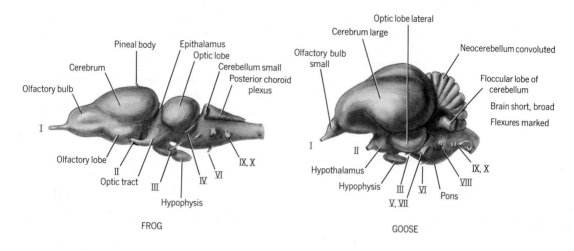

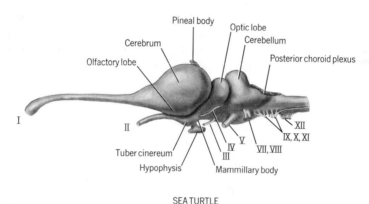

FIGURE 15-11

BRAINS OF REPRESENTATIVES OF THREE CLASSES OF TETRAPODS.

corpus striatum is prominent. It is represented by the basal nuclei of which the caudate nucleus and putamen, derived from the neostriatum, are relatively large. The extensive neopallium is the hallmark of the class. It dominates the entire brain both structurally and functionally. The hemispheres are smooth in most small mammals and convoluted in most large mammals. A new commissure, the corpus callosum, joins the hemispheres of therian mammals. The archipallium is represented by the large but somewhat enigmatic hippocampus. Pyriform lobes are lateral or ventral in position. They are extensive if olfaction is acute; otherwise they are restricted.

Thalamus and hypothalamus are highly differentiated. The midbrain is exposed in only a few mammals. Optic lobes, now called anterior colliculi, are small because the cerebral cortex has taken over much of their function. Posterior colliculi are present to complete the corpora quadrigemina of the midbrain tectum. The ventricle of the midbrain is restricted to a narrow cerebral aqueduct.

The mammalian cerebellum is large, much convoluted, and relatively broad. The advanced neocerebellum pushes the paleocerebellum to the sides where it forms lateral hemispheres. The cerebellar nuclear complex has differentiated into three or four distinct pairs of nuclei. A pons is prominent.

Support and Nourishment of the Central Nervous System

The central nervous system of cyclostomes and fishes is loosely surrounded by a protective fibrous envelope called the **primitive meninx.** Amphibians and reptiles have two envelopes (or meninges), an outer tougher **dura** and an inner **pia-arachnoid.** Birds are structurally intermediate between reptiles and mammals in having the beginnings of a third layer.

Mammals have three meninges. The strong **dura mater** (= hard + mother) is outermost. It is of mesodermal origin. In the vertebral column it is separated from the vertebrae by fat; in the cranium it adheres directly to the bones. The innermost

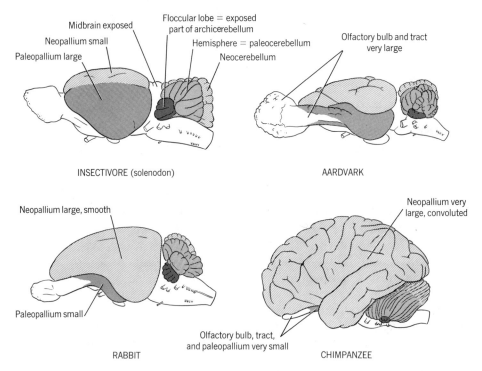

FIGURE 15-12

VARIATION IN THE STRUCTURE OF THE MAMMALIAN BRAIN. (Derivatives of the archipallium are hidden from lateral view.)

meninx is the thin vascular **pia mater** (= tender + mother) which adheres to the nervous tissue, following contours into every fissure and sulcus. Between the dura and pia is the non-vascular **arachnoid** (= spider-like), which is delicately fibrous and sends strands to the pia mater. Pia and arachnoid are both derived in part from neural crests and in part from mesenchyme.

Between the dura and arachnoid is a shallow, fluid-filled, **subdural space.** Between arachnoid and pia is a deeper **subarachnoid space.** The central canal of the spinal cord, the ventricles of the brain, and (in mammals) the subarachnoid space are filled with a considerable quantity of **cerebrospinal fluid.** This fluid is produced primarily by the vascular choroid plexuses of the brain, but also by other tissues—probably including the ependymal cells of the spinal cord. Each choroid plexus consists of ependymal layer joined to pia mater. Cerebrospinal fluid is clear, colorless, and similar to blood plasma except that it contains more chloride and virtually no protein. It flows slowly by secretion pressure and probably also by action of ciliated ependymal cells. Motion is toward the fourth ventricle from both the anterior ventricles and the central canal of the cord. The fluid escapes through holes in the roof of the medulla to the subarachnoid space. From there it reenters the bloodstream in various places, particularly through venous sinuses near the brain. Some cerebrospinal fluid also drains into lymphatics.

The pattern and extent of this circulation must be somewhat different in nonmammals (there being no subarachnoid space), but such vertebrates also have some cerebrospinal fluid outside the central nervous system: The choroid plexuses are usually everted into thin sacs of the fluid instead of inverted into their respective ventricles.

Such cerebrospinal fluid as bathes the outside of the central nervous system protects the system from damage when there

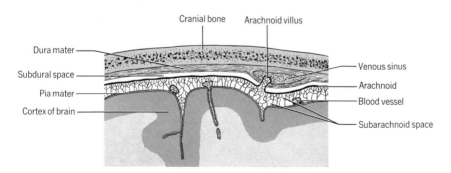

FIGURE 15-13
MENINGES OF THE MAMMALIAN BRAIN.

is a blow to the head or back. The fluid is also said to nourish the system, though the details of this function are not clear.

The structure and function of the circulatory system assure a rich blood supply to the central nervous system. Associated vessels are large and numerous, collateral systems are present, and in time of stress this system is given priority over others. Brain and cord need much blood for two reasons. First their metabolic rate is high and their need for oxygen is constant and great. Second, the exchange between blood and tissues may be less efficient than elsewhere. Nervous tissue is dense, interstitial spaces are minute, and there are no lymphatic vessels. The blood–brain barrier is distinctive in that respiratory gases and water pass quickly, but larger molecules leave the capillaries less readily than elsewhere.

16

Sense Organs

All cells of the body are responsive to their environments. Certain cells and organs, however, are specialized for monitoring the environment, and these comprise the sensory system. They stimulate the central nervous system, and it is there that sensation is perceived and integrated and where action, if any, is initiated.

Receptors range from mere nerve endings, through simple microscopic capsules, to the large and complex eye. Various classifications of these diverse organs are used: (*1*) general (widely distributed like pressure receptors) versus special (localized like the ear); (*2*) somatic (conscious reception of relatively superficial stimuli) versus visceral (unconscious recep-

389

tion of deep stimuli); (3) stimulated from internal sources (muscle tone, balance) versus external sources (cold, light); and (4) responsive to mechanical stimuli (touch, sound) versus electromagnetic stimuli (heat, light) versus chemical stimuli (taste, smell). None of these schemes is entirely satisfactory for our purpose of analyzing the major advances of vertebrate structure. The more complicated sense organs contribute interesting histories to the story of evolution and will be presented in approximate order of increasing complexity after giving brief attention to the relatively simple receptors.

Some Relatively Simple Receptors

The structural bases for the perception of such human sensations as hunger, fatigue, sex drive, and anxiety are unknown, and many vertebrates have sense organs, not shared by humans, for which functions can only be guessed. Surely, numerous kinds of receptors remain to be described. It is clear, therefore, that we can never know what it feels like to be a fish, and that much remains to be learned about sense reception — particularly in lower vertebrates.

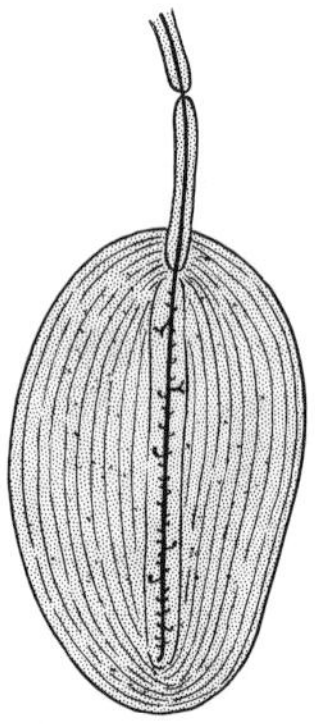

Various sensations, including touch, pressure, stretch, heat, cold, and general chemical sense, are at least sometimes received by naked nerve endings. There can be little to say about the phylogeny of these. Numerous kinds of **sense capsules** are found in the epithelial and connective tissues of tetrapods, though they are scarce in amphibians. These consist of variously modified nerve endings surrounded by small simple capsules having many configurations. They appear to monitor heat, cold, touch, pressure, and pain. The best-known is the **pacinian corpuscle** which is a marvelously effective tiny pressure transducer. Slight distortion of the 30–50 onion-like layers of the capsule sets up a **generator potential** in the nerve ending within the organ. If the potential reaches threshold intensity, the capsule greatly steps up the impulse at the first node of Ranvier. The resulting **action potential** is then transmitted along an axon.

FIGURE 16-1
PACINIAN CORPUSCLE.

It is difficult to identify the function of each kind of sense capsule, and it is possible that some kinds respond to more than one stimulus. Virtually nothing is known of their phylogeny, and homologies can rarely be made.

The **proprioceptive senses** apprise an animal of the relative positions of the parts of its body. Half awaking from deep sleep one may not know for an instant how the legs are flexed or the arms placed, but the slightest tensing of the muscles makes one's position known. Three kinds of proprioceptors are involved, all of which respond to tension and contribute to postural reflexes. They appear to be present in all tetrapods. The **muscle spindle** is a special muscle fiber wound up by a net of nerve endings and enclosed in connective tissue. It discharges when it is stretched. These organs are most abundant where muscle is merging into tendon. **Joint receptors** have complex nerve endings within the connective tissue of joint capsules. Similar **tendon organs** are located where tendons join bone. They discharge when muscles contract.

Several dissimilar kinds of sense organs are macroscopic yet relatively simple in structure. The **carotid body** is a small mass of cells of disputed origin which lie in the fork of the internal and external carotid arteries of tetrapods. Richly supplied with both nerves and blood, it detects fluctuations in the concentrations of carbon dioxide and oxygen in the bloodstream and sends impulses via the ninth cranial nerve to the parts of the brain controlling circulation and respiration. **Otoliths** are hard objects within the internal ear of some vertebrates. In addition to their function in equilibrium (see below) they probably act as accelerometers. Also, the large otoliths of several teleost fishes have been shown to have peizoelectric properties (i.e., converting force to electric potential); theoretically they could function as depth registers. **Pit organs** give their name to the American snakes called pit vipers. The conspicuous pits are located between the nostril and the eye. The floor of the blind pit is vascular and rich in nerve endings. The organ is a remarkably sensitive thermoreceptor. The background discharge is modified if the temperature increases or decreases by as little as 0.002°C. This enables the snake to detect within half a second the presence and position of warm-blooded prey that comes within striking distance. Pythons and boas either have series of similar, but smaller and less sensitive, pits in the scales surrounding the mouth, or they have thermal-sensitive skin areas.

FIGURE 16-2
MUSCLE SPINDLE.

Olfactory Organs. Seemingly unspecialized nerve endings at various places on the surface of the body of many vertebrates are sensitive to the chemical environment. Response of the human eye to smog and onions is an example. The principal chemical senses are olfaction and taste. Olfaction is a primitive sense. The earliest vertebrates appear to have had a keen sense of smell, and most of the forebrain was then devoted to olfactory signals.

Organs of Olfaction and Taste

Olfactory epithelium is localized in the nasal pits of fishes and in protected outpocketings of the respiratory passages in air breathers. Only part, and usually a small internal part, of the epithelium lining the nasal pit or passage contains olfactory cells. These are columnar cells each of which has about eight hair-like filaments on its free surface. These cells are unique in that they continue as axons into the olfactory bulb of the central nervous system. Supporting cells are dispersed among the thousands or millions of olfactory cells. A yellowish pigment is present in higher vertebrates which may enhance olfaction in some unknown way.

The combined area of all the filaments of the olfactory epithelium may exceed the surface area of the body, and it is on these filaments that dissolved chemicals are detected. The sense of smell is so wonderfully acute that the theoretical limit of sensitivity is reached by many animals—each filament can respond to a single molecule of certain odoriferous materials. Each kind of animal has its own spectrum of sensitivity, being very sensitive to some odors and little sensitive to others. The exact mechanism of response is not known, but it appears that the filaments can distinguish molecules having about seven different shapes, possibly by a submicroscopic lock-and-key mechanism. The almost infinite number of odors that can be distinguished would then be combinations of these in various relative concentrations.

Olfaction has not been detected with certainty in amphioxus. CYCLOSTOMES have a median olfactory sac which is ventilated by water passing in and out of the single nasal pouch. However, it is probable that the olfactory structures of the ancestors of these animals, and doubtless also of anaspids and cephalaspids, were paired as for other vertebrates. This is indicated by the bilobed nature of the sac, its innervation, and its ontogeny.

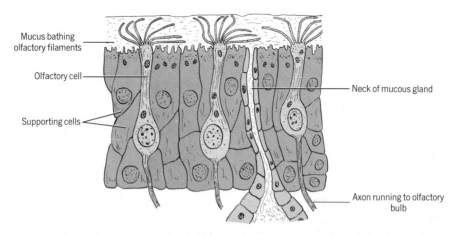

Mucus bathing olfactory filaments

Olfactory cell

Supporting cells

Neck of mucous gland

Axon running to olfactory bulb

FIGURE 16-3
SECTION OF THE OLFACTORY EPITHELIUM OF A TETRAPOD.

FISHES admit water into each nasal pit by one compressed opening (Selachii) or by two adjacent openings (Teleostei) constructed so that a partition of skin between them causes a continuous stream to pass over the olfactory epithelium. The nasal pit of cartilaginous fishes lies in the respiratory current, thus assuring ventilation even when the fish is stationary. The internal nostrils of Sarcopterygii may have evolved partly to benefit smelling in water (through increased ventilation) rather than breathing in air. The lining of each pit is pleated. The number and shape of the pleats relates to both species and age. These folds are assumed to control water flow in a beneficial way. The olfactory epithelium is largely confined to the clefts between the pleats.

AIR-BREATHING VERTEBRATES add mucous cells to the olfactory epithelium. This is necessary both to dissolve the particles to be smelled and to wash away material that has already been detected so that fresh samples of air can be examined.

FIGURE 16-4
SECTION THROUGH THE NASAL PIT OF A TELEOST.

The size of the nasal chamber increases, and its structure becomes more complex, as the secondary palate evolves. Most amniotes (turtles are excepted) increase the surface area of the nasal epithelium by folding it over several pairs of turbinates (= conchae), or scrolls of bone, which curl into the nasal chamber from its lateral walls (see Figure 10-4). Turbinates are relatively simple in most vertebrates but exceedingly complex in some mammals. The area of the olfactory epithelium may be increased in this way, but the primary function of the folded nasal epithelium is to clean, moisten, and sometimes to alter the temperature of respired air before it reaches the lungs. The olfactory sense is developed in only few birds (pelagic scavengers, vultures, and several other groups) and is weak in aquatic mammals which, unlike fishes, cannot ventilate the olfactory epithelium while submerged.

Tetrapods also depart from their ancestors among fishes in having a **vomeronasal organ** (or Jacobson's organ) which is a pair of pockets variously placed on the floor of the nasal chamber. There is some histological evidence that the fish "nose" is at least in part homologous with the vomeronasal organ and that the air "nose" is the innovation. However, the olfactory epithelium of all vertebrates relates to the olfactory bulb of the brain, whereas the vomeronasal organ relates to an accessory bulb dorsal and posterior to the olfactory bulb. The vomeronasal organ seems to be another organ for olfaction, though its function in mammals is obscure; having none ourselves we cannot tell what kind of sensation the organ elicits.

The vomeronasal organ opens into the nasal chamber of amphibians. Some salamanders have paired external nasolabial grooves which conduct fluids from the rostrum through the nostrils and into the vomeronasal sacs. In some reptiles, the vomeronasal organ connects with the mouth instead of the nasal chamber. Squamate lepidosaurs protrude the forked tongue and then insert the tips of the tongue into the paired vomeronasal organ to sense, in some way, the quality of adherent particles. In some other reptiles, and most mammals, the organ communicates instead with the small incisive foramen in the front of the palate. The organ is rudimentary or absent in turtles, crocodilians, birds, volant and aquatic mammals, and primates.

Taste Organs.   Taste organs are similar to olfactory organs in being chemoreceptors having epithelial hair cells. They differ from olfactory organs in being less sensitive by about four orders of magnitude, having more restricted response (there are fewer different tastes than odors), relating to different parts of the brain, synapsing with sensory nerve fibers rather than running directly to the brain, being of entodermal instead of ectodermal origin, and having the receptor cells aggregated into groups, or **taste buds.**

Each taste bud consists of supportive cells and 30–40 columnar taste cells all arranged in a barrel-shaped cluster. The buds may be dispersed or grouped on little hummocks of the epithelium called papillae. Papillae are of various sizes and shapes and may be arranged in various patterns—all of unknown functional significance. Olfactory epithelium is well-protected by its location in a sac or alcove deep in the nasal chamber. Taste buds, by contrast, are exposed and subject to wear. Accordingly, their cells are short-lived and are constantly replaced. Each bud is most sensitive to one of the four basic

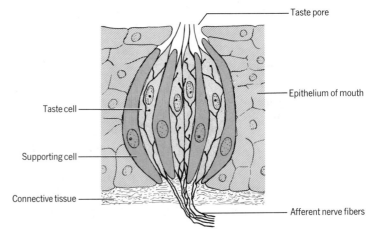

Taste pore

Epithelium of mouth

Taste cell

Supporting cell

Connective tissue

Afferent nerve fibers

FIGURE 16-5
SECTION OF A TASTE BUD.

tastes: salt, sour, sweet, or bitter. Nevertheless, all taste buds look alike, and the different vertebrates are not equally sensitive to these tastes. Being derived from epithelium at the levels of several visceral arches, taste buds are innervated by several cranial nerves: the seventh, ninth, and tenth.

Taste buds are abundant in the mouth and pharynx of cyclostomes and fishes, but they also may be located on the surface of the body, particularly on the heads and oral feelers of fishes that find food in sand, mud, or murky water. Amphibians have taste buds on the tongue, pharynx, and skin. Frogs have taste buds positioned to taste bits of tissue abraded by the palatal teeth and dissolved by enzymatic action of an oral secretion. Reptiles and birds, having dry skins and usually somewhat keratinous tongues, distribute most of their taste buds in the pharynx. These animals, especially birds, have a relatively poor sense of taste. Mammals also have taste buds in the mouth generally and on the pharynx, but concentrate them on the fleshy tongue.

Lateral Line System

The lateral line system and internal ear are so related by structure, function, and ontogeny that together they are called the acousticolateralis system. The lateral line system will be considered first because it is less complex than the ear. The system is present in fishes and in both larval and aquatic adult amphibians. It consists of thousands of microscopic organs called **neuromasts.** These are always freely distributed at the surface of the skin and usually are also located in shallow surface pits, along canals in the skin, or along horizontal tubes tunneling under the skin and even through dermal bones. Such tunnels always communicate with the surface by pores spaced along their length. The longest and most constant tube or canal is the **lateral line canal** — hence the name of the system. These ways of distributing and protecting neuromasts intergrade. Somewhat set apart are neuromasts located at the bottoms of **ampullary organs** (or ampullae of Lorenzini), which are deep, tube-like pits located in the skin and subcutaneous tissue of the head.

Early in development, several ectodermal placodes form on each side of the head and neck area of the embryo. It is probable that cells from neural crests play a part in their induction. One placode on each side forms the internal ear. Cells wander away from the other placodes to become the lateral line system and the ganglion cells of the nerves that will innervate that system. These are branches of the seventh cranial nerve which serves most neuromasts on the head, the ninth nerve serving a small area at the back of the head, and the tenth nerve serving the remainder of the system.

A neuromast looks somewhat like a taste bud. It is a clump of sensory hair cells and supportive cells. Each sensory cell has on its exposed surface a bundle of 20–50 sensory hairs called **stereocilia** and one larger **kinocilium** which stands at one edge of the stereocilia. Each neuromast that is free at the surface of the skin is capped by a tall, fragile, and transparent dome of extracellular material which protrudes into the water. This structure is the **cupula.** All sensory hairs of the neuromast are imbedded in the base of the cupula. Neuromasts opening into subsurface canals usually have cupulas also, but some of these, and all neuromasts within ampullary organs, are instead covered by a jelly which fills the surrounding space.

The function and comparative anatomy of free neuromasts and of those in canals and pits differ somewhat from the function and comparative anatomy of those within ampullary organs. We shall consider the former first. Any motion of the surrounding water causes a cupula to pivot, thus tilting the imbedded sensory hairs in the direction of motion. Each neuromast discharges electrical signals at a basic frequency which increases when displacement of the hairs is in one direction and decreases when displacement is in the opposite direction. There are two sets of neuromasts in each canal, one positively activated by motion of fluid along the canal in one direction and the other by motion in the opposite direction.

Collectively, free neuromasts and those of canals and pits form a system of distant touch that is so well-coordinated that

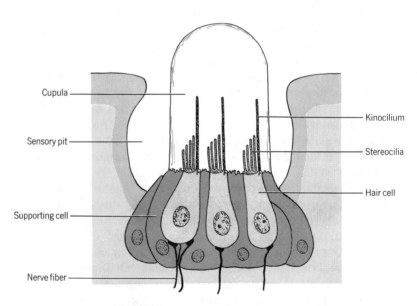

FIGURE 16-6
SECTION OF A NEUROMAST IN A SHALLOW PIT.

it gives both spatial and temporal information. Even when blinded, fishes can determine the position and motions of nearby prey and of other fishes. They can also sense the relative motion of their own bodies and the water. Sound waves from a distant source are not sensed. However, low-frequency sound waves from a near source cause displacement of the water and hence are heard (whatever they may "sound" like to a fish). The lateral line system probably contributes to obstacle avoidance and orientation of the body relative to currents, though these functions may be of lesser importance. Thermoreception has also been attributed to neuromasts but is considered by most researchers to be incidental or absent.

Cyclostomes and amphibians have no sensory canals on the head or body. Their neuromasts are free or in pits which tend to be linearly arranged. The lateral line system is well-known from the armor of ostracoderms and placoderms. Fishes have lateral line canals (sometimes doubled or branched) and a complicated but rather stable system of head canals. The canals are best developed in active fishes and least protected from the surface in fishes inhabiting quiet water. Most of the canals of chimaeras are open to the surface (see Figure 3-7).

Ampullary organs are found in elasmobranchs and several groups of bony fishes. These organs are primarily detectors of weak electric stimuli (see pp. 213–215). They sense the muscle potentials of nearby fishes (and of their owners). They also respond to thermal and mechanical stimuli. Thus, they are also organs of distant touch, though the principal stimulus is different from that of the remainder of the lateralis system.

Structure and Function of the Organ of Equilibrium. Details of **Ear** the origin of the ear are lost in antiquity, but the story can be surmised with reasonable confidence. It is probable that the lateral line system evolved before the ear and that of the two basic functions of the ear, equilibration and hearing, the former was perfected before the latter was well-initiated. It is desirable for an organ of equilibration to be on the head where it is close to the feeding apparatus and other special senses, the positions of which are of relative importance, and away from the oscillating locomotor mechanism. Further, it is desirable for the organ to be away from the surface of the body where signals of the type monitored by the lateral line system would be distracting. Evidently a portion of the lateral line system at the side of the head sank below the skin and was modified to become the new organ.

This probable phylogeny appears to be recapitulated by embryos today in that the internal ear forms from an ectodermal placode which, in fishes, is in the middle of the series of placodes that form the lateral line system. This otic (or auditory) placode develops under the inductive influence of the hindbrain, thus assuring its position at the back of the head. The placode sinks in, forming a pit, which then pinches away from the skin ectoderm to become the hollow **otic vesicle.** Gradually, the vesicle assumes the complicated shape which makes appropriate its adult name of **labyrinth.** Ganglion cells of the associated eighth cranial nerve are also derived from the vesicle.

The body of the labyrinth becomes more or less divided into a dorsal chamber, or **utriculus,** and a ventral chamber, or **sacculus.** A diverticulum from the utriculus is called the **endolymphatic sac,** and another from the sacculus is called the **lagena.** The labyrinth is completed by three looping **semicircular canals,** each of which joins the utriculus at both ends. The labyrinth and its canals are filled with the fluid **endolymph.**

In both utriculus and sacculus there is at least one patch of sensory and supportive cells. Each patch, called a **macula,** resembles a neuromast of the lateral line system, but is somewhat larger. Each sensory cell has the familiar clump of stereocilia and single, asymmetrically placed kinocilium. The hairs of each macula are imbedded in a modified cupula which is made heavy by deposits of calcium. These are in the form of crystals in some groups and in the form of solid masses called **otoliths** in others. The size and shape of otoliths is distinctive for many vertebrates.

When the head of the animal moves to a new position, the cupulas or otoliths tend to slide over the maculae in response to gravity, thus establishing the shearing force to which hair cells are sensitive (the kinocilium may plunge in and out as well as pivot back and forth). This makes the labyrinth an excellent position register.

The labyrinth quickly evolved a related function, which is the detection of angular acceleration, or rotation of the head. To understand the mechanism involved, imagine a straight garden hose filled with water. If one jerked the hose in the direction of its length, the inertia of the long column of contained water would cause it to tend to remain stationary while the hose itself moved. As a result, pressure would increase at the end of the hose away from the motion and water would tend to escape there. (If the hose were jerked sideways, there would be little tendency for relative motion of hose and water because there would be no long column of water in that plane.) Similarly, when a semicircular canal is rotated in its own plane, the canal moves slightly past the contained endolymph. At one end of each canal is a swelling, the **ampulla,** within which is a patch of sensory cells, here called a **crista,** which responds to the motion. Cristae have high, domed cupulas which extend across

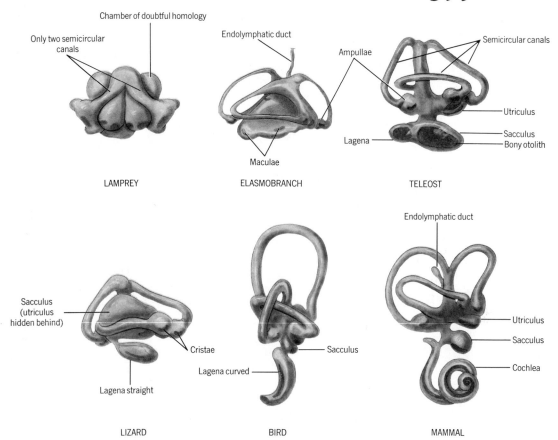

FIGURE 16-7
COMPARATIVE ANATOMY OF THE LABYRINTH. Lateral view of the right organ.

the canals, completely occluding the lumens. The hair cells are all oriented so as to respond to motion in the appropriate direction. (Hair cells of maculae are oriented in various directions.) Since one of the three semicircular canals lies in each plane of space, every rotational motion is registered.

In order to protect the delicate labyrinth, and to mask out distortions from irrelevant sources, the entire structure is surrounded by cartilage or bone. However, the labyrinth does not fit quite snugly inside its bony housing. The thin space between labyrinth and skeleton is filled (at least in places) with the fluid **perilymph,** which cushions the organ against harm and plays a role in the hearing function of the ear of terrestrial vertebrates.

Structure and Function of the Organ of Hearing. It was noted above that the lateral line system is responsive to low-frequency sound waves of close, or "near-field" origin which displace water sufficiently to be sensed by hair cells. The system is not sensitive to the pressure pulses of "far-field" sound. In order for far-field sound to be heard underwater, it is necessary that (*1*) pressure waves be received and amplified by a confined compressible medium (such as a gas), and (*2*) pressure changes in the gas be translated to motion changes in the endolymph bathing hair cells. The gas bladder of bony fishes was a ready-made resonator, and two groups of fishes independently evolved mechanisms for translating pressure waves from bladder to labyrinth, as described further in the next section. These fishes are known to hear well at frequencies below 1000 Hz (Herz is equivalent to cycles per second).

Most airborne sound waves do not strike the body with sufficient force to carry through soft tissues to a deep resonator such as the gas bladder (much of the force is simply reflected away). Consequently, in order for an animal to hear when out of water, it is desirable (though not always essential) that it have a thin surface membrane, the **tympanum,** or eardrum, that is free to vibrate against the low resistance of an air chamber on its inner side. It has been postulated that the rhipidistians among crossopterygian fishes learned to hold a bubble of air in the upper wing of the first, or spiracular pharyngeal pouch. If so, this may have improved hearing right away, but gradually the pouch became adapted as a true middle ear. The tympanum probably evolved from part of the overlying operculum, but this portion of the story is obscure and it is not certain that the tympanum is homologous in all vertebrates.

This evolutionary sequence is recapitulated by embryos of tetrapods: The extremity of the first pharyngeal pouch becomes the middle ear, its neck becomes the eustachian tube joining middle ear to pharynx, and the closing plate of the second pouch becomes the tympanum.

These advances accomplish the first requirement for hearing in air; sound waves are received by a tympanum vibrating freely against a middle ear cavity. There remains the second requirement, which is to translate this motion to the labyrinth. Merely to have the cavity of the middle ear impinge on the labyrinth would not be adequate, because sound waves (*1*) would be damped by the soft walls of the eustachian tube, and (*2*) would lose greatly in force in being translated from the gas of the middle ear to the fluid within the labyrinth. The first of these problems is solved by mechanically transmitting sound waves across the cavity. One or more bony ear ossicles accomplish this function. (The interesting origin of these ossicles from leftovers when the tetrapod jaw evolved was told in Chapter 10.) The second problem is solved in one, or both, of two ways. The footplate of the ear ossicle (or innermost ossicle, if there are three) vibrates against the **oval window,** which is an opening in the bony housing of the labyrinth. The area of the oval window is always much smaller than the area of the tympanum, thus increasing the force of vibration per unit area. Further, the outermost ossicles of mammals (incus and malleus) are constructed so as to mechanically decrease amplitude but increase force as they pivot. In all, force is increased about 14–20 times.

What remains is to translate motion of the innermost ossicle to shearing force over hair cells. When the footplate vibrates against the oval window it creates compressional waves in the perilymph beyond. This perilymph is in a tube which, in the first tetrapods, probably merely crossed the lagena of the sacculus wherein was a special auditory macula called the **basilar papilla.** In most amniotes this tube of perilymph, the **scala vestibuli,** extends away from the oval window and then turns sharply on itself to return as a parallel tube, the **scala tympani.** Between these two tubes is an extension of the lagena called the **scala media.** The three adherent channels lengthen and coil—slightly in birds and markedly in mammals—and are then called the **cochlea** (= snail shell) (Figures 16-7 and 16-10). Within the scala media is the basilar papilla or, if much lengthened in a cochlea, its derivative, the **organ of Corti.**

The basilar papilla or organ of Corti rests on the **basilar membrane** separating scala media from scala tympani. This organ has several to many rows of hair cells which, however, lack kinocilia in the adult. The stereocilia are embedded in a

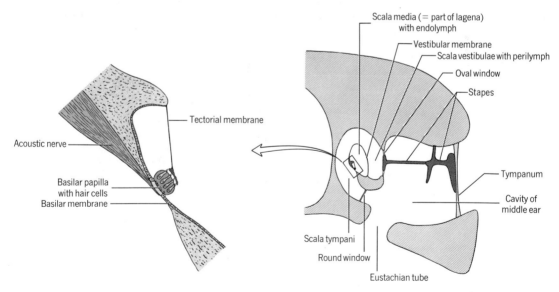

FIGURE 16-8

DIAGRAMMATIC SECTION THROUGH THE AUDITORY APPARATUS OF A LIZARD.

derivative of the ancestral cupula now called the **tectorial membrane.** Vibrations at the oval window cause traveling waves in the scala vestibuli which are translated across the thin vestibular membrane to the endolymph of the scala media, and thence across the basilar membrane to the scala tympani. These incompressible fluids can move within their bony walls because the scala tympani meets the cavity of the middle ear at a thin membrane covering the **round window.** Thus, when the tympanum and oval window move inward, the round window moves outward.

The basilar membrane and tectorial membrane pivot at different points, so their motions create shearing force between them which activates the hair cells. There is a systematic change along the length of the organ of Corti in pattern of hair cells, arrangement of stereocilia, and breadth of the organ. This is the structural basis for differential response at different levels along the organ; high frequencies are sensed at the base of the cochlea and low frequencies toward the tip. All in all, this is an amazingly complex but effective structure.

Comparative Anatomy of the Ear. The labyrinth, or inner ear, evolved very early in vertebrate history and, with many variations in configuration but none of basic design and function, has been retained by all vertebrates. The middle ear evolved as tetrapods evolved, and the external ear is scarcely found except in mammals.

CYCLOSTOMES, and at least some ostracoderms, have fewer than three semicircular canals (Figure 16-7). It is not known if

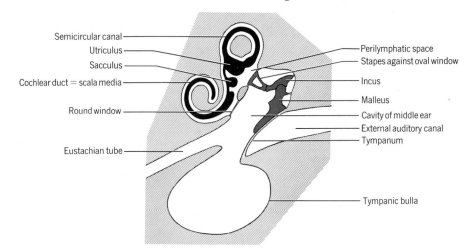

FIGURE 16-9

DIAGRAMMATIC SECTION THROUGH THE AUDITORY APPARATUS OF A MAMMAL. Somewhat distorted for two-dimensional representation. Bulla may be relatively much larger.

this is a primitive or degenerate condition. Utriculus and sacculus are not set apart, and the organ is compact. Some ELASMOBRANCHS retain a slender canal, the endolymphatic duct, where the embryonic otic vesicle pulled away from the skin ectoderm. Fine sand grains enter the sacculus through this canal and clump to form an otolith. Utriculus and sacculus are broadly joined.

The labyrinth of BONY FISHES is only partly surrounded by bone. Hard calcareous otoliths rest on the maculae, of which there may be three or four. The otolith of the sacculus is so large that it nearly fills the chamber. Otoliths of bony fishes are useful to systematists because they have distinctive shapes and annular growth rings. Two groups of fishes have achieved far-field hearing of low-frequency sounds by adapting the gas bladder as a resonator. One group (cods, herrings) have a pair of long extensions of the gas bladder which penetrate the brain cavity to impinge on extensions of the endolymphatic sacs of the labyrinths. The other group (minnows, catfishes, goldfishes) has modified processes of adjacent vertebrae into paired chains of three or four ossicles which pivot on the spine to convey vibrations from gas bladder to endolymphatic sac. There being only one resonator, these mechanisms probably do not permit directional hearing, though this conclusion has been questioned.

Ears of ancestral LABYRINTHODONTS probably were similar to those of such REPTILES as have not developed degenerate or highly modified ears. A large tympanum is either flush with the surface of the head or protected by a flap of skin (Figure 16-8). The eustachian tube may be broadly open to the pharynx. The single ossicle, or stapes, is commonly a slender but complicated structure having cartilaginous arms at its outer end. The lagena is somewhat lengthened but coils only in mammal-like reptiles. The basilar papilla is elongate and, like the tectorial membrane, is constructed in a variety of ways often departing considerably from the mammalian condition. The number of hair cells along the papilla gives some indication of the range of frequencies that can be detected and of the general sensitivity of the ear.

The ears of LIVING AMPHIBIANS and SOME REPTILIA are off the main line of descent in having specialized ears. Most anurans have short heavy stapes. These animals are most responsive to low-frequency sounds and do not hear sounds of more than 3000–5000 Hz. Caecilians, urodeles, some anurans, snakes, and some other reptiles lack a middle ear cavity and tympanum, but usually retain the stapes which is attached (usually by means of muscle, ligament, or cartilage) to the skin, squamosal, quadrate, hyoid, or even scapula. These ears may be sensitive to airborne vibrations and also are sensitive to

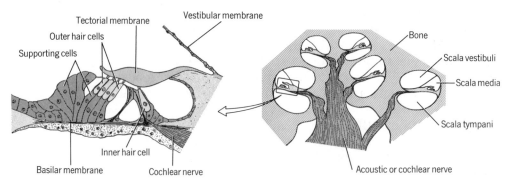

Tectorial membrane

Vestibular membrane

Outer hair cells

Supporting cells

Bone

Scala vestibuli

Scala media

Scala tympani

Inner hair cell

Basilar membrane

Cochlear nerve

Acoustic or cochlear nerve

ORGAN OF CORTI

SECTION OF COCHLEA

FIGURE 16-10

STRUCTURE OF THE MAMMALIAN INNER EAR.

vibrations conducted from the ground to the labyrinth by way of the bones of limbs or jaws. Some of these animals have even lost the true stapes.

The vestibular membrane is delicate in MAMMALS, but heavy in BIRDS. The basilar membrane of birds is much shorter and broader than that of mammals. The organ of Corti has one inner, and three or four outer rows of hair cells in mammals, but about ten times as many in birds. A coiled cochlea is present in each class (except in monotremes), making as many as five turns in mammals. Ears of mammals are distinctive in several ways: An external ear, or pinna, is usually present. This structure helps to funnel sound waves toward the tympanum, but may have other functions (cooling, recognition) and can be an inefficient funnel. The middle ear is housed in a bony tympanic bulla, and extensions of the middle ear cavity may penetrate adjacent bones as mastoid sinuses. There are, of course, three ossicles. Mammals commonly hear sound over the wide range of 20–20,000 Hz. Some volant and aquatic species emit and receive sound at very high frequencies as a means of echo-location. Loudness of sound is detected over about ten orders of magnitude. Very strong vibrations are damped by tiny muscles and by buckling of the chain of ossicles in order to prevent injury.

Eyes   All vertebrates have a pair of lateral eyes, and some also have an unpaired dorsal eye. The lateral eye is a precision organ which is second only to the brain in complexity. It is unmatched for range of adaptation, being modified to function in water or air, by day or night, at short range or far, and in habitats ranging from the sky to the depths of the ocean. From the eye alone, morphologists can determine much about the habits of any vertebrate. Evolutionary lineages are less easily established from eye structure, though some trends are evident. In several instances, prior knowledge of phylogeny helps to explain otherwise puzzling characteristics of the eye.

Structure and Function. Structure and function can profitably be studied in terms of the requirements of the eye as an optical instrument. A first requirement is for a firm housing—sharp images would not be possible on the filmplate of a camera having soft walls. The capsule of the eye is kept rigid by its outermost envelope, the **sclera,** which is stiffened by cartilage, or bone, or a tough mesh of collagenous fibers, or by a combination of these. Pressure of the fluid within the eye also contributes to keeping the envelope distended and firm.

Another requirement is for control of the amount of light entering the eye. This is accomplished by a curtain, called the **iris,** which surrounds the light window, or **pupil.** Delicate muscles within the iris cause it to pull back and enlarge the pupil when light is dim, and extend to narrow the pupil to a small circle or slit when light is bright.

There must be no random reflections of light within the eye, so the capsule is usually lined by a nonreflective black pigment. This **pigment layer** is just behind the light-sensitive cells where it can, in some vertebrates, temporarily surround the ends of those cells to screen them from excessively bright light. (There is directed reflection of light by some eyes—see below.)

An important requirement is for the formation of an image. When light enters a box through a small hole, a dim inverted image is formed, as shown by Figure 16-11. This is the principle of the pinhole camera, and it is actually used by several vertebrates (camel, some geckos) in bright light when the pupil is made very small. Usually, however, another principle must be followed. When a ray of light passes from one material into another having a different refractive index, it changes direction at the interface. The direction and amount of the change depend on the materials and on the angle the ray of light makes with the interface. A **lens** is a transparent object shaped so as to bend light into an undistorted image. The image is "upside down" both on the filmplate and the retina. The brain somehow turns the image over, or so it seems.

Over the front of the eye, the sclera merges into the transparent **cornea.** The cornea has about the same refractive index as water. Accordingly, corneas of eyes that function underwater do not bend light and can be of any shape. They are usually flat or streamlined. Rays of light entering the cornea from air are bent sharply. Hence, the cornea of terrestrial

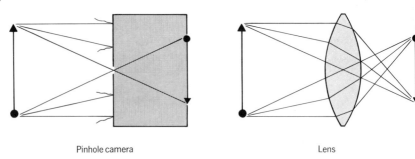

Pinhole camera   Lens

FIGURE 16-11
TWO METHODS OF FORMING AN IMAGE.

animals must be accurately curved to avoid the distortion of the image called astigmatism. The lens then completes the image formation initiated by the cornea.

When passing through an ordinary lens, central and marginal rays have slightly different focal points, thus causing spherical abberration. The lensmaker overcomes this problem by building fine lenses from several apposed elements. The eye overcomes it by having the cornea curve slightly less at its margins, and by having the lens more dense at its core. Further, light rays of different wavelengths bend slightly differently when passing through an ordinary lens, thus causing chromatic abberration. The maker of fine lenses combines crown and flint glass, which have compensating properties. The eye that requires a sharp color image screens out light having the shortest wavelengths (violet and blue) by using a color filter. Yellow pigment is added to lens or retina, or oil droplets of yellow or red color are distributed in the photoreceptor cells.

Pinhole cameras select light rays rather than bend them; every object is in focus regardless of its distance from the camera. When images are made by lenses, however, provision must be made for focusing the image where it will be recorded. In the eye, this is called **accommodation.** Some vertebrates (like the photographer) move the lens outward from a farsighted resting position to focus on near objects. Others move it inward from a nearsighted resting position to focus on far objects. Amniotes focus by changing the shape of a stationary lens, causing it to bulge for near vision and flatten for far vision. This is done by changing the tension in the **ciliary apparatus** which suspends the lens. Different amniotes use different mechanisms as told in a later section. The very small lenses of small vertebrates naturally have relatively great depth of focus and close near points, so little accommodation is needed.

The next requirement is for perception of the image. Photoreceptor cells are of two kinds, **rods** and **cones.** Each has a

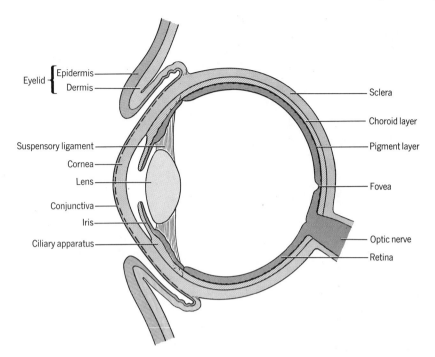

Eyelid { Epidermis — \
Dermis —

Suspensory ligament —

Cornea —

Lens —

Conjunctiva —

Iris —

Ciliary apparatus —

Sclera

Choroid layer

Pigment layer

Fovea

Optic nerve

Retina

FIGURE 16-12

GENERAL STRUCTURE OF THE EYE OF A DIURNAL MAMMAL.

nerve-like base, an inner segment with nucleus and mitochondria, and an outer segment which is long and cylindrical
for rods but shorter and conical for cones. The outer segment,
which is narrowly connected to the nourishing inner segment,
is intricately folded into a stack of hundreds of lamellae.

On these lamellae is **visual pigment,** which is instantly altered
chemically in the presence of light. This change is translated to
nervous impulses in a manner as yet not understood. Rods are
all alike. Their pigment is **rhodopsin** (or, in freshwater fishes
and tadpoles, a related pigment), each molecule of which
responds to a single quantum of light energy — the theoretical
limit. Cones all look alike, yet are of three classes having different pigments and responding to light of different wavelengths. Thus, cones are the basis for color vision. Rods are
more sensitive than cones by about two orders of magnitude
but do not record color. Nocturnal animals have only rods.

Beyond the rods and cones (in the direction of the inside of
the eyeball) are **bipolar cells,** and beyond these, **ganglion cells,**
whose dendrites extend through the optic nerve. There are also
two plexiform layers of nerve cells, one meshed into the system
at each end of the bipolar cells. This complex of nerve cells ac-

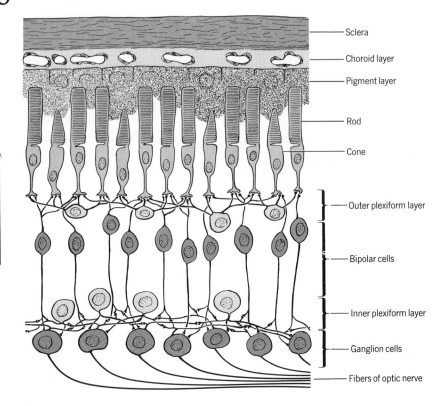

Direction of light

- Sclera
- Choroid layer
- Pigment layer
- Rod
- Cone
- Outer plexiform layer
- Bipolar cells
- Inner plexiform layer
- Ganglion cells
- Fibers of optic nerve

FIGURE 16-13

HISTOLOGY OF A MAMMALIAN RETINA seen in cross-section.

complishes summation of stimuli from different photoreceptors (particularly rods) and in other ways performs the first steps of forming the image perceived by the brain.

In order for the eye to have acuity of vision, as is usual for diurnal animals that rely on vision in feeding, moving about, or avoiding danger, the retina must record the image with a fine grain; photoreceptors must be tightly packed so different cells can respond differently to adjacent parts of the small image. In most diurnal vertebrates having color vision (most teleosts, frogs, most reptiles and birds, some mammals), the sharpest image is perceived at a place on the retina called the **area centralis** where the cells are most slender and closely packed (to as many as $10^6/mm^2$), where there are only cone cells, and where many animals have a pit called the **fovea** which bends light rays to enlarge the image by as much as 30%. Man has a "good" area centralis (we move our eyes when we read to keep the print focused there), but not the best: He needs a low-power binocular to gain the acuity of vision of a hawk!

If the eyes move independently in their orbits, their fields of vision may be independent or may at least partly overlap. If they have overlapping fields of vision, depth perception, or stereoscopic vision, is achieved. In mammals, nervous coordination prevents independent movement of the eyes.

Finally, the eye requires some structures that are accessory to its optical functions. A vascular **choroid layer** nourishes the eye. Extrinsic muscles turn the eyeball in its socket (Chapter 9). Glands and lids (sometimes including a third eyelid, or **nictitating membrane**) moisten and protect the eye in air. The eye appears to look through a hole in the skin, but the edges of the lids are in fact folds in the skin, not breaks, and the skin continues over the cornea as the thin, transparent **conjunctiva.**

This book cannot cover many of the specializations of the vertebrate eye, but two principal adaptations will be noted: The general structure of eyes adapted for vision underwater is presented in Chapter 23, and adaptations for vision when there is little light are presented here.

The many amphibians and mammals, and few reptiles and birds that are active at night, some fishes that frequent murky water, and many of the fishes that live deep in the ocean (where sunlight does not penetrate but where animals are luminescent) must be able to see in dim light. Their eyes are very large in front to gather in as much light as possible. The pupil opens very wide (though it may also close to a narrow slit by day). The lens is large, spherical, and placed well back in the eye to be close to the retina. The retina is relatively small, so the entire eye is deep in the optical axis, or even tubular. This arrangement mimics the slide projectionist who gains a bright image by placing his projector close to a small screen.

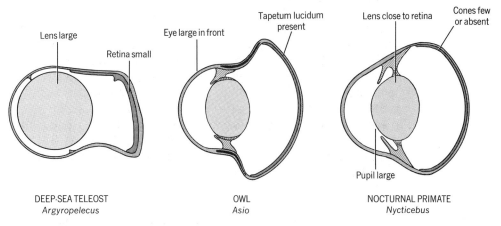

| DEEP-SEA TELEOST | OWL | NOCTURNAL PRIMATE |
| *Argyropelecus* | *Asio* | *Nycticebus* |

FIGURE 16-14
ADAPTATIONS OF THE EYE FOR VISION IN DIM LIGHT.

Cone cells are few or absent. The more sensitive rod cells are slender and closely packed. There is so much summation by the ganglion cells that light striking a thousand or more rods may trigger an impulse in a single nerve fiber. This enormously increases sensitivity to dim light, but slightly blurs the image. An area centralis is often lacking so the eye, which may be too large to turn much in its socket, need not be oriented exactly toward the objects perceived. The entire head is turned if need be — some owls can rotate the head 270°!

Finally, most nocturnal vertebrates have a mirror, or **tapetum lucidum,** behind the rods which reflects light back out of the eye, so light passes through the rods twice instead of once. This reflection is responsible for the nighttime eye shine of cats and owls when they are caught in the beam of a flash light. Analogous mirrors have evolved several times; reflection may be from crystals of guanine, lipid particles, or other pigmented layer, or supportive tissues.

Development and Origin.   Developmentally, the eye has three principal parts: retina and pigment layer, lens, and supportive tissues. First to appear is the primordium of the retina, which is an evagination of the sidewall of the diencephalon. This expansion, or **optic vesicle,** pushes toward the skin ectoderm, trailing an **optic stalk** behind. Next, the lateral surface of the vesicle sinks into the cavity of the vesicle, thus forming a double-walled cup resembling the bulb of a rubber syringe when indented on one side by the thumb. The outer wall of the cup becomes the pigment layer and the inner wall the retina. The edge of the cup forms the iris. The optic stalk becomes the optic nerve as axons from ganglion cells in the retina extend down its length. In a sense, the retina is a nucleus of the brain and the optic nerve a tract of the brain.

The lens forms from a placode of the skin ectoderm which develops only under the inductive influence of the underlying optic vesicle. It has been speculated that this placode is in series with those that form the lateral line system and inner ear, but evidence is meager. The **lens placode** becomes a hollow **lens vesicle,** but the cavity is soon diminished and then obliterated by the thickening of the inner wall of the vesicle.

Mesenchyme surrounding the optic cup differentiates into choroid, sclera, cornea, and ciliary apparatus. Developmentally, these are continuous with, and apparently also homologous with, the meninges of the brain.

The phylogenetic origin of the vertebrate eye is unknown; the first known vertebrates had already perfected the organ. This has not discouraged morphologists from speculating, however,

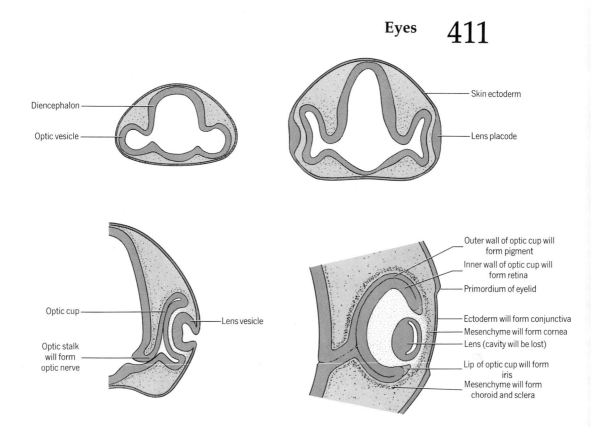

Diencephalon

Optic vesicle

Skin ectoderm

Lens placode

Optic cup

Lens vesicle

Optic stalk
will form
optic nerve

Outer wall of optic cup will
form pigment
Inner wall of optic cup will
form retina
Primordium of eyelid

Ectoderm will form conjunctiva
Mesenchyme will form cornea
Lens (cavity will be lost)

Lip of optic cup will form
iris
Mesenchyme will form
choroid and sclera

FIGURE 16-15
DEVELOPMENT OF THE EYE.

and nearly a dozen theories have been proposed since the 1870s. The most plausible theory is strongly supported by developmental, and weakly supported by histological and comparative anatomical clues. The neurocoel and ventricles of the brain are lined with ependymal cells which are usually ciliated. In the head region of many chordates from amphioxus to birds, these cells are light-sensitive. If phylogeny is repeated by ontogeny, then rods and cones evolved from ependymal cells of a part of the forebrain, and the eye could have been functional throughout its evolution. The outer segments of rods and cones are considered to be modified cilia.

(Light-sensitive cells are present in the skin of some fishes and amphibians. This is probably an ancient condition. Since future ependymal and retinal cells are at the surface of the body prior to neurulation, it is plausible that photoreceptors of the eye are related to those of the skin. A difficulty with this idea, however, is that neurulation is a developmental process that precedes the formation of sensory cells in the skin.)

Comparative Anatomy of Lateral Eyes. The hagfish, a scavenger in deep water, has degenerate eyes and is blind. The well-developed eyes of the LAMPREY are primitive in two respects: The conjuctiva is not fused to the cornea, and the ependymal layer is retained in the core of the optic nerve. Other distinctive features may or may not be primitive: The size of the pupil is fixed, the sclera is not stiffened by cartilage or bone, accommodation results when an extrinsic muscle pulls against the cornea, thus pressing the lens inward, and the lens is held in position by pressure only, not by a suspensory apparatus. As for primary swimmers in general, the eye is large and shallow along its optical axis, the lens is large and spherical, eyelids and glands are absent, and the extrinsic oblique muscles rotate the eyeball around its optical axis.

There is a wide range of eye structure within the large assemblage of fishes, but some generalizations can be made. ELASMOBRANCHS stiffen the sclera with cartilage. They accommodate somewhat by pulling the lens forward from its resting position with a small intrinsic muscle of ectodermal origin. A unique cartilaginous pedicel props the eyeball away from the back of the orbit; its function is not clear. Cones are few or absent. An area centralis is present. Crystals of guanine in the choroid cause it to reflect light as a tapetum lucidum. The chamber between lens and cornea is small.

BONY FISHES stiffen the sclera with cartilage and, in most teleosts, by several bony plates. The cornea is flat or streamlined. Cones are usually present, so color vision is typical. Usually, an intrinsic mesodermal muscle pulls the lens inward to accommodate for far vision (chondrosteans, and some other fishes, have no accommodation). Teleosts have a nutritive structure derived from the choroid, called the **falciform process,** which projects into the cavity of the eyeball. Many teleosts have an area centralis, and some have a fovea. A tapetum lucidum is occasionally present.

As larvae, AMPHIBIANS have fish-like eyes. At metamorphosis they develop eyelids and glands and an evenly curved cornea. A small mesodermal muscle within the eyeball moves the lens forward for near vision. The eyes of anurans are better developed than those of urodeles. Some have color vision as, in all probability, did many labyrinthodonts. (Caecilians and some salamanders are blind.)

REPTILES and BIRDS, most of which are diurnal, have the finest visual acuity and accommodation of all vertebrates. The eye is large, particularly in birds; eyes of hawks are larger than those of humans, and eyes of the ostrich are larger than those of elephants. The sclera is stiffened by a cartilaginous cup behind and by a ring of about 15 small, overlapping bones on the forward wall, where the eyeball might otherwise be distorted by the ciliary muscles. These muscles are striated (though of ectodermal origin) and, hence, faster in action than

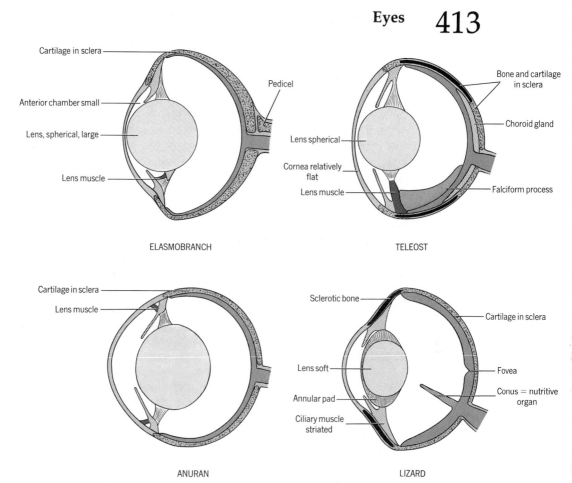

FIGURE 16-16

COMPARATIVE ANATOMY OF THE EYE.

those of other vertebrates. The small, soft, somewhat flattened lens has a peripheral **annular pad** which makes firm
contact with the ciliary apparatus. Accommodation is active
and instant; muscular contraction causes the lens to bulge
in front. Cones are usually numerous and color vision excellent (except in snakes and crocodilians). An area centralis is
present and there is one or in many birds and some lizards,
two foveas. There is a vascular projection into the cavity of
the eyeball. This is particularly large and complexly folded in
birds, where it is called the **pecten** (= comb.). Of the many
functions postulated for this structure, that of nutrition is
the most probable. Lacrimal gland and nictitating membrane are present. Iris musculature is striated. A tapetum
lucidum is rare. Some binocular vision is frequent, particularly
in predatory birds. (The eyes of snakes are in some ways
atypical because of derivation from burrowing ancestors for
which vision was unimportant. Amphisbaenians, some lizards,
and some snakes are blind.)

MAMMALS first evolved as small nocturnal creatures, and at that stage lost the perfection of eye structure of their reptilian ancestors. Some of the loss was later regained, but not all. There is no cartilage or bone in the sclera. Muscles of the ciliary apparatus and iris are smooth and, therefore, relatively slow. The shape of the lens is adjusted to focus the image, but the mechanism is inferior to that used by reptiles and birds; contraction of the ciliary muscles relieves tension on the suspensory apparatus which then allows the lens to bulge of its inherent elasticity. This process is relatively slow, particularly in old age, when it may also become incomplete, making near vision impossible (without eyeglasses). There is no pecten or corresponding nutritive organ. A tapetum lucidum is confined to several orders; a nictitating membrane is rare. Color vision appears to be partially regained in several orders but is complete only in primates and some rodents. An area centralis is present in some orders; a fovea is present only in higher primates. Binocular vision is frequent and coordination of the eyes is superior (The eyes of monotremes are in some respects atypical of the class.)

Dorsal Eyes. On top of the diencephalon there are, in the midline, two small evaginations: an anterior **parietal organ** (or parapineal) and a posterior **pineal body** (or epiphysis). One or both of these structures may be photoreceptive and is then termed the third eye.

Asymmetry of these structures in adult lampreys and embryo lizards, and the conformation of ostracoderm head armor, indicate that parietal and pineal organs may be derived phylogenetically from a bilateral pair of organs, the left member of which shifted forward to the midline while the right member slipped behind.

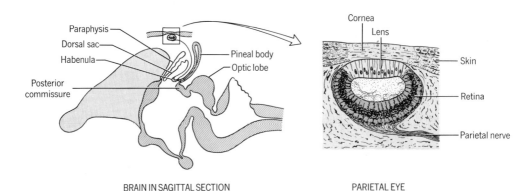

Paraphysis
Dorsal sac
Habenula
Posterior commissure
Pineal body
Optic lobe

BRAIN IN SAGITTAL SECTION

Cornea
Lens
Skin
Retina
Parietal nerve

PARIETAL EYE

FIGURE 16-17
DORSAL EYE OF A LIZARD.

The pineal organ is present in nearly all vertebrates. It has photoreceptive cells in lampreys, and endocrine properties are known for various vertebrates. Initially, at least, the endocrine functions seem to have been light-related (see p. 425). The parietal organ also has photoreceptor cells in lampreys, but in these animals is subordinate to the pineal eye. The parietal eye is functional in tadpoles and salamander larvae and in many lizards, where it may have a lens, retina, and tiny nerve. Histologically, its photoreceptive cells are closely similar to the cones of lateral eyes. Such light-associated behavior as sunning and activity rhythms is influenced by this organ in some lizards.

It is evident from the widespread presence of a parietal (or pineal) foramen in skulls of ostracoderms, placoderms, and early representatives of bony fishes, labyrinthodonts, and reptiles, that a third eye has been persistent in the evolution of vertebrates.

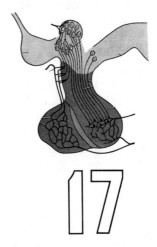

Endocrine Glands

Since endocrine glands utilize the circulatory system to transmit signals, their shape is of little importance, and a gland may usually be single, multiple, or diffuse without relation to function. Only the aggregate volume of cells is critical. Even the secretions have been remarkably constant over the ages, though there are some exceptions, and responses to the secretions are various. Endocrine glands have diverse developmental origins, are usually unrelated to one another in space, and are incompletely related in function. They do not comprise an organ system in the usual sense. Not much is yet known of endocrine function in most lower vertebrates.

417

General Nature of Endocrine Glands

The endocrine glands are ductless glands having secretions called **hormones** which are discharged into the circulatory system for distribution. Only certain tissues of the body are responsive to each hormone. Response may be morphologic, as for sex hormones that influence the development of secondary sexual characteristics, or physiologic, as for an adrenal hormone that influences kidney function. Many hormones act directly on tissues that are receptive to their "messages." Generally, however, the action is indirect. Thus, a hormone secreted by the hypothalamus of the brain and released by the hypophysis causes the ovary to secret a hormone to which the lining of the uterus responds.

Taken together, the endocrine glands function like the nervous system in several respects: Each controls and integrates bodily functions, each mediates control through the release of chemicals (the same chemicals in several cases), and each may accomplish interaction within its own system to coordinate its activities. Endocrine control differs from nervous control in tending to be slower and more sustained. The functions of the two systems merge in several instances.

All endocrine glands are small and highly vascular. They are often diffuse in lower vertebrates, but tend to be discrete in tetrapods. Most are constructed of cords of more or less cuboidal cells arranged among sinusoids and supported by a matrix of connective tissue. Several endocrine glands (neurohypophysis, urohypophysis) are instead constructed of thin attenuated cells, and one (thyroid) is constructed of follicles. The hormones of glands of mesodermal origin (gonads, adrenal cortex) are steroids, whereas hormones of glands of ectodermal or entodermal origin are proteins, peptides, or other derivatives of amino acids.

Structure, Function, and Comparative Anatomy of the Glands

Hypophysis. The **hypophysis,** or **pituitary gland,** is located under the brain (Figures 15-8, 15-10, 15-11). In mammals it is housed in a bony pocket of the basisphenoid bone (Figure 10-4). Although it is small, this gland is in both structure and function one of the most complicated organs of the body. Developmentally it has a surprising dual origin: The adult portion called the **neurohypophysis** forms from the part of the floor of the embryonic diencephalon termed the **infundibulum.** This structure may evaginate (most tetrapods) or remain nearly unfolded (amphibians, some fishes), in which case its name is not apt ("infundibulum" = funnel). The remainder of the gland, or **adenohypophysis,** forms instead from an invagination of the ectodermal part of the embryonic mouth cavity (the stomodaeum) called the **hypophyseal pouch,** or Rathke's pouch. This pouch and its derivatives are variously lobed in the different

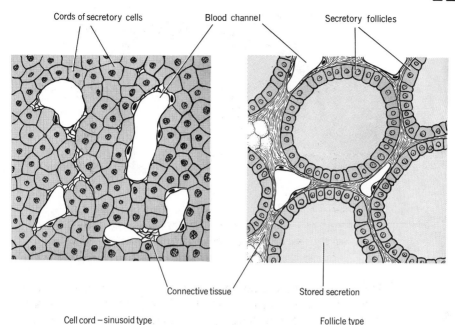

Cords of secretory cells Blood channel Secretory follicles

Connective tissue Stored secretion

Cell cord – sinusoid type Follicle type

FIGURE 17-1

TWO HISTOLOGICAL TYPES OF ENDOCRINE GLAND.

vertebrates. Its connection to the mouth is usually lost during maturation.

In general, the neurohypophysis has an anterior subdivision, the **median eminence** (not identified in forms below lungfishes), and a posterior or ventral subdivision, the **pars nervosa** (or posterior lobe of the gland). The adenohypophysis has several parts: The largest, most constant, and most active is the **pars distalis** (or anterior lobe). A **pars intermedia** is usually present but is lacking in birds. Other parts are of variable occurrence and doubtful function. They differ widely among the vertebrates, and homologies can be made, if at all, only by combining clues from embryology and histochemistry. The median eminence and pars distalis have a common blood supply; the pars nervosa has an independent blood supply.

The neurohypophysis is atypical of endocrine glands in that it is constructed largely of long parallel nerve fibers originating in the hypothalamus of the brain (Figure 17-4). Indeed, this part of the hypophysis functions by storing and releasing into the bloodstream hormones elaborated in the hypothalamus and transferred to the neurohypophysis by neurosecretion (of which, more below). Very different is the adenohypophysis which has cords of secretory cells of two kinds which branch without pattern among sinusoids. It is remarkable that only two kinds of cells can produce six or seven hormones.

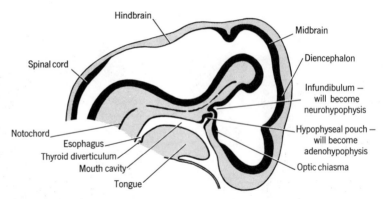

FIGURE 17-2
**EMBRYONIC ORIGIN OF THE HYPOPHYSIS shown by
a sagittal section of the head of a mammalian embryo.**

The pars intermedia secretes one hormone, **intermedin,** which influences the pigment of the skin. The pars nervosa releases two hormones, **antidiuretic** (which causes the kidney to hold back fluid) and **oxytocic** (which in mammals causes the letdown of milk and contraction of the uterus). The pars distalis usually secretes six hormones: **growth, thyrotropic, adrenotropic, follicle-stimulating** (which is active during the ripening of eggs), **luteinizing** (active during the formation of the corpus leuteum in the ovary), and **lactogenic** (which is poorly named because though it is needed for lactation it has other functions in non-mammals). The hypophysis is, in turn, influenced by the hormones of other glands, notably the ovary, thyroid, and adrenal.

It has been postulated that any of several structures of amphioxus may be homologous with the vertebrate hypophysis, but no conclusions can be drawn. In cyclostomes and fishes other than Sarcopterygii, the neurohypophysis is the more or less flat floor of the brain above other parts of the gland; a pars nervosa cannot be clearly distinguished. The atypical adenohypophysis of hagfishes is scattered as islets among other tissues. The adenohypophysis of Selachii is unique in having a ventral lobe of unknown homology. Fishes, and particularly ray-finned fishes, are distinctive for the way that the adenohypophysis and neurophypophysis interdigitate over a broad area.

In dipnoans and tetrapods, interdigitation between the neurohypophysis and adenohypophysis is much reduced, or, more often entirely lost. A pars nervosa forms from all or part of the infundibulum which is now usually evaginated from the floor of the brain. This causes the gland to subtend more from the brain. The pars intermedia is large in reptiles, small or absent in mammals, and absent in birds. When absent, its hormone, intermedin, may be produced by the pars distalis.

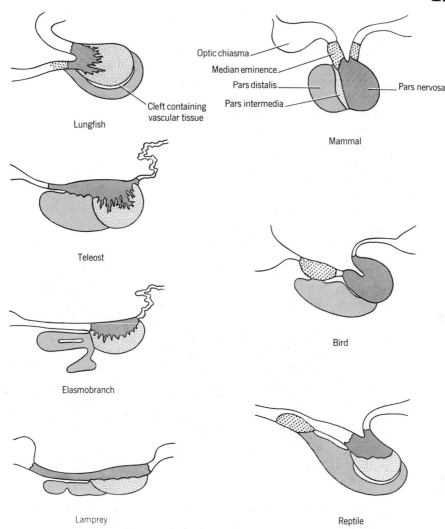

Lungfish

Optic chiasma

Median eminence

Pars distalis

Pars intermedia

Cleft containing
vascular tissue

Pars nervosa

Mammal

Teleost

Bird

Elasmobranch

Lamprey

Reptile

FIGURE 17-3
**COMPARATIVE ANATOMY OF THE HYPOPHYSIS as seen in sagittal section.
Anterior is to the left.**

Thyroid. The secretory cells of the thyroid are derived from a
midventral evagination of the entoderm of the embryonic
pharynx at about the level of the second pharyngeal pouch
(Figure 17-2). Surrounding mesenchyme contributes supportive tissues. The gland always consists of a cluster of rounded
follicles, and each follicle is lined by a single layer of cells
which are usually cuboidal (sometimes columnar when highly
active) and have microvilli on their free surfaces (Figure 17-1).
The viscous secretion product, called **colloid,** which fills the
follicles, stores **thyroglobulin.** This iodine-rich protein is converted by hydrolysis to either of two hormones, **thyroxine** and

(in much lesser quantity) **triiodothyronine.** These are amino acids which are linked to blood proteins for transport. The thyroid has an exceedingly rich blood supply for its size.

The gland begins to function early in ontogeny, contributing to the control of differentiation, growth, metamorphosis, the distribution of pigment, and sexual development. It has a profound affect on metabolic rate and may influence molt (amphibians and reptiles), feather shape, body temperature, and functions of the nervous, digestive, and excretory systems. The thyroid interacts with the hypophysis and in at least some instances (anurans) with the hypothalamus. The gland enlarges when diseased.

All vertebrates have a thyroid, and its origin traces back to cephalochordates. The endostyle on the floor of the pharynx of amphioxus, and also of the larval lamprey, ammocoetes, functions in the production and movement of mucus for filter-feeding. Nevertheless, there is evidence that part of the endostyle is the phylogenetic precursor of the thyroid: (1) The endostyle, like the thyroid, forms from a midventral evagination of the pharynx. (2) Part of it, like the thyroid, concentrates iodine from the blood. (3) At metamorphosis the endostyle of ammocoetes is partly converted to adult thyroid.

In cyclostomes and many teleosts the thyroid is relatively diffuse and is variously distributed near the ventral aorta, branchial afferent arteries, heart, gills, head kidney, spleen, brain, or eye. It is more discrete in other vertebrates but may be paired (amphibians, lizards, birds), bilobed (dipnoans, many mammals), or single (cartilaginous fishes, most reptiles). The thyroid of tetrapods is usually near the larynx, trachea, or bronchi.

Parathyroid. The secretory portion of the parathyroid glands differentiates from the epithelium of the third and fourth (and in reptiles also the second) pharyngeal pouches. Curiously it is the dorsal wings of the pouches that contribute in mammals, but the ventral wings in other vertebrates. It is clear from this and other evidence that the pharyngeal pouches share the potential for forming glandular tissue, yet the kind of gland formed by a specific region is not constant.

The gland consists of densely packed chief cells arranged in cords and clumps. (In mammals there may also be a lesser number of oxyphil cells of unknown function.) The parathyroid hormone is a polypeptide called **parathormone.** It affects the level of calcium, and less directly the level of phosphorus, in the blood. In the absence of the hormone, calcium disappears from the blood in a matter of hours, tetanus occurs in muscles, and death follows. Deficiency of the hormone leads to abnormalities of bones and teeth. A second hormone, calcitonin, has been attributed to the parathyroid of mammals, but its function is not yet clear.

Glandular tissue of unknown function has been identified in cyclostomes and fishes which may be homologous with the parathyroid. However, the gland is known with certainty only in tetrapods. It is usually divided into one or two pairs of small glands. These may be linear (birds), but are usually more or less globular. They are located in the throat area, commonly near, or even imbedded in, the thymus or thyroid (hence the name, parathyroid).

Interrenal Organ and Adrenal Cortex. There are two kinds of adrenal tissues, though only in mammals are they separated into distinct regions, the **cortex** and the **medulla.** The two tissues are different in function and embryonic origin. The cortical type of tissue is discussed here, and the medullary tissue is discussed in the next section. In bony fishes the tissue that is equivalent to the adrenal cortex is called interrenal tissue.

These structures are similar to the gonads in the steroid nature of their hormones and also in their embryonic origin. Cortical tissue is derived from mesoderm lining the coelomic cavity close to the place of origin of the genital ridges.

The secretory cells of cortical and interrenal tissues form cords which are arranged in three layers in mammals but have little or no organization in other vertebrates. About 50 cortical hormones are known in mammals, but many of these are also produced elsewhere in the body, some are readily converted to others, and only about a dozen are known to be physiologically active. Together these hormones are called **adrenocorticosteroids.** They are classed according to chemical structure and general function into four groups. One group (including cortisone, corticosterone, and cortisol) functions in carbohydrate and protein metabolism. A second group (including deoxycorticosterone) affects salt and water metabolism. The third group (including aldosterone) is typical only of mammals and relates to sodium and potassium metabolism. The last group (including adrenosterone) resembles the sex hormones. Most of these hormones interact with the pituitary, and many are involved in responses to stress.

Interrenal tissue of cyclostomes is scattered along the posterior cardinal veins and other vessels. In teleost fishes, interrenal tissue may be diffuse or discrete, but usually forms numerous small flecks located near or within the head kidneys. The gland is characteristically elongate and between the kidneys in cartilaginous fishes, elongate and adherent to the kidneys in anurans, and diffuse and adherent to the kidneys in urodeles. The cortical tissue of amniotes forms a pair of compact bodies located on or near the anterior ends of the kidneys (Figures 13-14 and 13-15).

Chromaffin Bodies and Adrenal Medulla. Tissue corresponding to the adrenal medulla tends to be much scattered in some vertebrates and is then termed **chromaffin tissue** because of its staining properties. Chromaffin and medullary tissue is innervated by preganglionic fibers of the autonomic nervous system. These nerves and glands are all derived from the ectodermal neural crests of the embryo, and all secrete **adrenalin** and **noradrenalin,** though the glands produce much more (particularly of adrenalin) than does the nervous system. The body responds to these hormones in many ways that better enable it to meet sudden emergencies (e.g., increased blood sugar and blood pressure, inhibition of smooth muscles).

The distribution of chromaffin tissue in the body corresponds to that of interrenal tissue but tends to be even more diffuse, particularly in fishes where it may occur along the postcardinal veins as well as near, on, or in the kidneys. Chromaffin tissue may lie near, but separate from, interrenal tissue (some fishes and some lepidosaurs), may be intermingled with interrenal or cortical tissue (some fishes, most amphibians and reptiles, birds), or may lie as a medulla within a covering of cortical tissue (most mammals). Even mammals, however, have chromaffin bodies or **paraganglia** associated with some sympathetic ganglia.

Gonads and Placenta. The development and structure of the gonads were presented in Chapter 13, but these organs should be mentioned again as endocrine glands. The ovary produces several **estrogens** (the principal ones are estradiol and estrone), **progesterones,** and in mammals, **relaxin.** Estrogens control the growth and development of the female genital duct system and are essential for reproduction. They also initiate and maintain secondary sexual characteristics, which are marked in some vertebrates and inconspicuous in others. The site of estrogen secretion appears to be the theca in mammals, but is in doubt for vertebrates having virtually no theca.

Following ovulation, the ruptured mammalian follicle is transformed into a temporary but pronounced gland termed the **corpus luteum.** This structure then secretes progesterone, a hormone which is essential for the final differentiation of the female reproductive tract in preparation for fertilization and pregnancy, and also for maintaining pregnancy. Structures similar to the mammalian corpus luteum form in the ovaries of various other vertebrates (sharks, teleosts, urodeles, birds, some reptiles) but seemingly not in others. It is therefore puzzling that all vertebrates have one or more progesterones.

Relaxin is the only gonadal hormone that is not a steroid. In some mammals, at least, it acts on the pelvic symphysis, mammary glands, and genitalia, readying them for their functions at delivery and thereafter.

The interstitial cells of the testis produce male hormones collectively called **androgens.** The principal androgens are **testosterone** and **androstenedione.** Androgens are required for the growth, differentiation, and function of the male genital ducts and copulatory organ (if present) and for control of secondary sexual characteristics and sexual behavior. All vertebrates have androgens, but interstitial cells have not been identified in all. It is possible that supportive cells in the seminal vesicles, or even the germ cells, produce androgens in some instances. To complicate the situation, males produce estrogens and females produce androgens, all from inadequately known sources, though the adrenals may be involved.

The mammalian **placenta** is a rich source not only of estrogen and progesterone, but also of a gonadotrophin and of prolactin, which is otherwise produced by the adenohypophysis. Prolactin is needed for milk production in mammals but is also present in lower vertebrates where it has varied functions.

All the gonadal hormones have complex interactions with the hypophysis, some relate to interrenal or cortical function or to the activities of the thyroid or pineal. Gonadal function is usually seasonal, the ultimate control often being photoperiod as mediated by the hypothalamus. The pathways are poorly known for fishes.

Miscellaneous, Possible, and Near Endocrine Glands. The origin and exocrine nature of the pancreas were noted in Chapter 10. Its islets secrete **insulin** which, in mammals, controls the deposit of glycogen in the tissues, and **glucagon** which controls its release. Among other vertebrates, however, there is great variation in response to these hormones—protein metabolism and solutes of the blood may be involved.

It is clear that one or more hormones are secreted by the **small intestine** (e.g., secretin, pancreozymin, cholecystokinin). These act on other parts of the digestive system to coordinate the digestive process. Although apparently present in all vertebrates, these hormones are virtually unknown except in mammals, and even in that class the cellular source of the secretion is unknown.

The tiny **pineal** organ atop the brain is a photoreceptor in the lower classes. In tetrapods it assumes endocrine functions which seem usually to involve the relations between homeostasis or reproduction and illumination. The pineal may influence activities of the hypothalamus and hypophysis. In tetrapods the gland secretes **melatonin,** a hormone which affects melanophores of frogs and the gonads of mammals.

Fishes, particularly teleosts, have in the tail a ventral swelling on the spinal cord. Histologically it resembles the neurohypophysis and hence is called the **urohypophysis.** It is suspected of endocrine function, possibly relating to osmotic regulation. The same is true of the **ultimobranchial bodies,** which are derived from the most posterior pair of pharyngeal pouches, and the **corpuscles of Stannius** located on the posterior parts of the kidneys of teleosts.

Not only are there these apparent endocrine glands of unknown function, but there are apparent "hormones" known not to be produced by glands. Thus, urea and carbon dioxide, derived from nonendocrine tissues, are distributed by the blood and convey "messages" to organs distant from their place of release. In order not to confuse the definition of hormone, these and similar substances are called **parahormones.**

Neurosecretion   Emphasis should be given to the close relationship noted above between certain parts of the nervous system and certain endocrine organs. Neurons in two or more nuclei of the hypothalamus actually extend into and constitute the neurohypophysis. Those reaching the pars nervosa elaborate the hormones of that part of the gland. These hormones are released into the general circulation. Those neurons reaching the median eminence release hormones which enter a miniature portal system running from the median eminence to the adenohypophysis. This portal system is lacking in fishes, but the interdigitation of neurohypophysis and adenohypophysis in many fishes may accomplish the same interaction.

The structure of the urohypophysis suggests that its nerve cells may also be secretory and that the swelling of the cord is in part the result of temporary storage of the product. The adrenal medulla and other chromaffin tissues do not resemble nervous tissue histologically, yet they behave as though they had evolved from postganglionic fibers of the autonomic nervous system. Nervous stimulation causes the medulla to secrete. The pineal gland of at least some mammals secretes its hormone on stimulation of the autonomic nervous system. Neurosecretion is also known in crustacea, insects, and other invertebrates.

These observations make it clear that the distinction between the nervous system and endocrine glands is not sharp—and that much remains to be learned about the relationship.

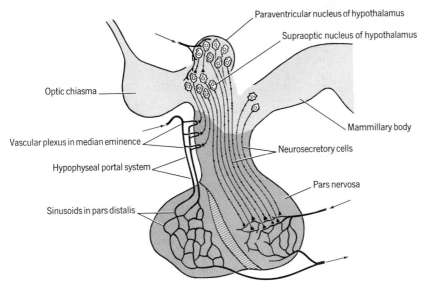

Paraventricular nucleus of hypothalamus

Supraoptic nucleus of hypothalamus

Optic chiasma

Mammillary body

Vascular plexus in median eminence

Neurosecretory cells

Hypophyseal portal system

Pars nervosa

Sinusoids in pars distalis

FIGURE 17-4

RELATIONSHIPS OF THE NEUROSECRETORY CELLS OF THE HY-
POPHYSIS. Anterior is to the left.

PART
III

STRUCTURAL
ADAPTATION:
Evolution in Relation
to Habit and Habitat

18

Structural Elements
of the Body

Part II of this book includes functional interpretation of structure, particularly for the respiratory, circulatory, and urinary systems and for the eye and brain. Emphasis, however, is on analysis of structure in relation to the long sweep of phylogeny; on conservative evolutionary changes common to all animals in such large taxa as classes and subclasses. Primitive and unspecialized characters are featured. Structures that are useful to animals of varied habits are stressed: Jaws are generally advantageous, two pairs of appendages proved to be a good general plan, a circulation divided into pulmonary and systemic circuits is superior for all tetrapods. The accumulation of such advances marks progress in evolution.

Animals as Specialists

431

Part III deals with the parallel influence of functional adaptation on different vertebrates; particular attention is paid to locomotor mechanisms in which parallels are seen most clearly. Animals with similar specialties are found scattered among the systematic categories, but are here brought together, their common problems are identified, and their various adaptations are interpreted from morphological clues.

Analysis is complicated by several factors. Different animals may do similar things in different ways: Squirrels and pottos both climb, but pottos slowly grasp the branches whereas squirrels run on the limbs or cling to them with sharp claws. No one kind of animal has *all* the structural modifications that are associated with its general habit. Further, one kind of animal may have several specialties: Frogs jump and swim, flying squirrels climb and glide, cormorants fly, swim, and dive. Also, the activities of animals are determined not only by structural adaptations but also by behavioral factors. Thus, gray foxes climb trees whereas red foxes do not. One cannot learn by dissection which climbs, or, indeed, that either climbs. Even without special structural adaptation, many animals can run, swim, climb, and dig somewhat. Conversely, animals may fail to move in ways for which they seem, on the basis of morphology, to be adapted. Thus, the adult gorilla has the structure that correlates with climbing by armswinging under the branches, but because it is very large, it seldom does so.

In spite of these complications, however, it is rarely difficult to determine the principal habits of a vertebrate animal from its structure. Some clues are subtle and some obvious, but all make sense. Their identification and interpretation can be very engaging. Animals are so good at their specialties!

This part of the book begins with two chapters presenting mechanical principles that relate to feeding, posture, and locomotion in general. Subsequent chapters analyze specific adaptations.

Properties of Supportive Materials

The materials of the body that accomplish support and movement are bone, cartilage, muscle, tendon, and ligament. (Soft organs are further supported by meshworks of collaginous fibers.) The suitability of these materials for the various requirements of the body depends on their properties.

Three important properties of living supportive tissues are not shared by any material available to architect or engineer: First, all display **growth** without interruption of function. They have remarkable capacity for repair of both major breaks and minor damage. This property protects these tissues from fatigue, or loss of strength with repeated loading, which is characteristic of nonliving supportive materials. The rate of repair is faster for muscle and bone than for cartilage, and in all the capacity diminishes with age. Second, all have amazing **capacity to adjust to circumstance,** slowly altering their substance and configuration in response to demand. It is common knowledge that muscular strength increases with exercise; the adaptability of other supportive tissues is mentioned later in this chapter. Finally, these properties taken together assure adequate **durability** for a lifetime of constant use. No manmade apparatus having even remotely comparable complexity of moving parts approaches the body in this regard.

A property of bone, cartilage, and muscle that conditions their other physical properties is **heterogeneity.** Disregarding minor impurities and imperfections, cast iron, ceramics, and glass are homogeneous, or uniformly the same everyplace and in all directions. Wood, by contrast, has grain which makes resistance to bending and splitting different in different planes. Similarly, bones have lamellae and osteons with specific orientations and may be compact or spongy (see p. 96). The internal cells and fibers within cartilage are unevenly arranged. There are several types of muscle fibers which occur in different proportions in different muscles (of which more in Chapter 19), and the relative proportions of muscle fibers to fat and connective tissue varies. Consequently, one cannot speak of *the* strength of, for instance, bone, but can report only the approximate strength of a given type of bone when loaded in a given way relative to the orientation of its components.

Cartilage and, particularly, bone are further heterogeneous in that each is a composite material consisting of two dissimilar components. In each instance, one component is an intricate meshwork of oriented collagenous fibers and sometimes of elastic fibers. The other component is a glycoprotein (for cartilage), or hydroxyapatite (for bone). The physical properties of the composite are unlike those of either component taken alone and are not the sum or average of the two taken together. Thus, collagenous fibers are soft, flexible, and very resistant to elongation, whereas hydroxyapatite is extremely hard, brittle, and resistant to compression. The composite of the two, bone, is more rigid than collagenous fibers, more flexible and resistant to fracture than the mineral, and more versatile in the kinds of loads it can withstand than either component alone. Engineers make composite materials from fibers of glass in resin, boron in aluminum, tungsten in copper, and others. Compound materials are often much stronger for their weight than are noncompound materials.

Before considering the strength of the structural elements of the body it is necessary to introduce several concepts and terms. **Force** is a push or pull which causes motion or must be resisted to prevent motion. Thus, the weight of an animal pressing on the ground is a force, the pull of a muscle on its insertion is a force, and the push of a fish tail against the water is a force. Force may be expressed as kilograms or pounds. (As noted on p. 482, it is more exact to express force in terms of acceleration imparted to a unit of mass. The newton, or equivalent unit, is then used. The more common usage adopted here serves our purpose and does not differ in concept.) Since forces of the body are concentrated at such places as insertions of tendons and contacts between bones, it is useful to consider **pressure,** or force per unit area. Pressure may be expressed as kilograms per square centimeter or pounds per square inch (or more precisely as the less familiar newtons per square centimeter).

Load is a general term referring to any force that is applied to a solid object. Adjacent bones of the legs and spine load each other; active muscles load related bones. In order for loaded objects to remain in equilibrium, equal forces must operate in opposite directions (this is an application of Newton's Third Law). Thus, as a tetrapod stands at rest, the downward force of its weight is opposed by an equal upward force by the ground. The weight is transmitted through the bones of the legs to the ground. When objects transmit loads, there are internal forces of one part of the object acting on adjacent parts. Force that results from transmission of a load is called **stress.** External loads cause internal stresses.

When any load is applied to any object, **deformation,** or **strain,** results. Strain is expressed as the relative deformation or change of shape that occurs when an object is stressed. Strain may be expressed as change in length, in volume, or in angle. Deformation may be permanent or temporary. Even moderate loads cause permanent deformation of modeling clay. Such materials could hardly support the animal body. If deformation is temporary, recovery may be almost immediate, as for bone, or somewhat slower, as for cartilage, tendon, and ligament. The capacity of a material to return completely to its original shape after a load is removed is called **elasticity.** The structural materials of the body have virtually perfect elasticity within usual load limits. Some elastic materials, like rubber, are much deformed by moderate force, whereas others, like steel, are only slightly deformed by great force. This property is expressed by the **modulus of elasticity,** which is stress divided by strain. If the modulus is high, the material is rigid; if it is low, the material is easily deformed.

The important property of **strength,** as applied to materials, is the capacity to resist force without breakage or permanent deformation. Strength varies, of course, with material, and is proportional to the cross-sectional area of the object (i.e., the more bone the greater the strength). As noted above, strength of heterogeneous materials varies with relative orientation of force to grain. Strength also varies importantly according to the direction of the applied force in relation to a surface; i.e., the interaction between adjacent objects differs depending on the direction of the forces acting between them. This is true both for a load applied to an actual external surface of an object and for a stress applied to an imaginary internal surface. Forces are of only two kinds: perpendicular to a surface or parallel to a surface. They can, of course, be applied at intermediate angles, but analysis of a kind presented in Chapter 19 shows that such forces can always be broken down into a perpendicular component and a horizontal component.

Perpendicular forces, in turn, are of two kinds: **Compression** makes an object shorter in the direction of the applied force (strain is then said to be negative) and **tension** makes it longer (strain is positive). Columns and pillars withstand compressive force; guy wires and cords that suspend objects withstand tensile forces.

Force applied parallel to a surface is called **shear.** Shear slides one part of a material crosswise to adjacent parts. Scissors cut by shearing. If a closed book is held between the palms of the hands and one cover is pushed or twisted relative to the other, the book is distorted by shear as the pages slip over one another.

Equipped with these concepts, let us now consider the strength of the supportive materials of the body. Fresh compact bone (not dry or embalmed or cancellous bone) loaded parallel to its grain has a compressive strength of 1330–2100 kg/cm^2 (19,000–30,000 lb/in.^{2}). About 170 students would somehow have to stand on a single 1 in. cube of compact bone

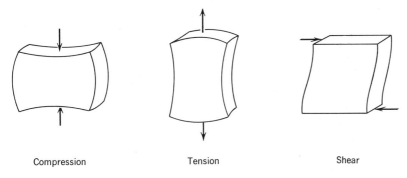

Compression   Tension   Shear

FIGURE 18-1
THE THREE PRINCIPAL KINDS OF FORCES and the distortions they tend to cause in solid objects.

in order to crush it! This is roughly four times the compressive strength of concrete. Values for cartilage vary, but are lower than those for bone. Tendons and ligaments, like string, merely crumple when compressed lengthwise.

Fresh compact bone loaded parallel to its grain has a tensile strength of 620–1050 kg/cm^2 (9000–15,000 lb/in.^{2}), or about half its compressive strength. Weight for weight, its tensile strength is nearly half that of steel. The tensile strength of cartilage is again less than that of bone. Tendon and ligament, however, although softer and lighter materials, have about the same tensile strength as bone.

The resistance of compact bone to shear may be as low as 500 kg/cm^2 (7150 lb in.^{2}) if stressed parallel to the grain, and as high as 1176 kg/cm^2 (16,800 lb/in.^{2}) if stressed crosswise to the grain. Cartilage, tendon, and ligament have less resistance to shear.

It might seem that resistance of tendon and ligament to tension, and of bone to all forces, are far in excess of demand. So they are for maintaining posture and engaging in moderate activity. In strenuous activity, however, forces exerted on the skeleton by individual tendons may reach several hundred kilograms, and excessive loads, as from a fall, occasionally cause tearing or breakage. It is clear from the above figures that shearing forces would be limiting in the body if they approximated usual compressive and tensile forces. Actually, pure shear is unusual, but bones may be sheared by twisting (i.e., rotating) at the same time they are compressed, and bending (i.e., bowing, or curving) forces, which are very common in the skeleton, combine shear, compression, and tension. The relative magnitudes of the kinds of stresses in the skeleton seem usually to be in proportion to the capacity of bone to withstand them: Compressive forces are largest and shearing forces are smallest. When bones do fail, any of the types of forces may have been responsible, though compressive fractures are least common.

Stress and Stress Lines

It will be easier to understand how the structural elements of the body are constructed for maximum effectiveness after considering the transmission of forces within homogeneous objects. When a solid cylinder resting on the ground is compressed by a uniform load, the downward force of the load is opposed by an equal and opposite upward force at the ground. These forces are represented by large arrows in Figure 18-2A. If the opposed forces are depicted instead by many arrows, each representing a unit of force, more information is included because even spacing of the arrows then shows that the pressure is uniform over the ends of the cylinder (part B). Within the cylinder, units of stress have the same magnitude and direction as the external pressure, so at any arbitrary plane they can also be represented by arrows, the number of arrows being proportional to the area taken. The lines we draw to represent the paths followed by units of force as they pass through an object are called **stress lines.** In this example the lines are straight and evenly spaced because loading is uniform (part C). They represent only compression and can also be called **compression lines.** The magnitude of compressive stress in the plane of the illustration is proportional to the height of the shaded rectangle (the units being arbitrary). At surfaces of the cylinder, and at horizontal planes within, there is no tension or shear. (In this, and the following examples, stress resulting from weight of the object itself is ignored for the sake of simplicity.)

Tensile force applied uniformly to one end of a cylinder or rod, and that which opposes the load at the other end, can also be depicted by arrows representing units of force. Again, straight lines show the paths of the forces within the object, but this time they are **tension lines** (represented by dashed lines in part D). The magnitude of tensile stress is again proportional to the height of the shaded rectangle. This model is closely approximated by stressed tendons and ligaments.

Bones are never cylinders evenly compressed over their ends in the direction of their long axes. If the end of a cylinder is compressed over a restricted area, then adjacent stress is great and is represented by compression lines that are close together (part E). As the lines pass away from the point of application of the load, however, they spread out until they are evenly distributed. The upper, outer parts of the cylinder are devoid of stress, but at the boundaries between stressed and unstressed areas, and also immediately under the load, stresses are complicated (and not figured). This model was approximated by the nearly solid, cylindrical, 2 m long femur of the great 54,500 kg (120,000 lb) dinosaur *Brontosaurus,* but for reasons given below, the long bones of tetrapods are rarely solid cylinders. We progress, therefore, to other applications of the concept of stress lines.

If a load is not applied perpendicularly to the end of a cylinder (as in the above examples) but is instead applied along an upper edge perpendicular to the axis of the cylinder, then the resultant stress must resist bending. Compression lines arch away from the load and come to run lengthwise in the opposite side of the cylinder at some distance away from the load (part F). Tension lines run lengthwise in the side of the cylinder near the load. The configurations of compression and tension lines near the load are complicated and somewhat dependent on the material, and hence are not illustrated. Shear is also present.

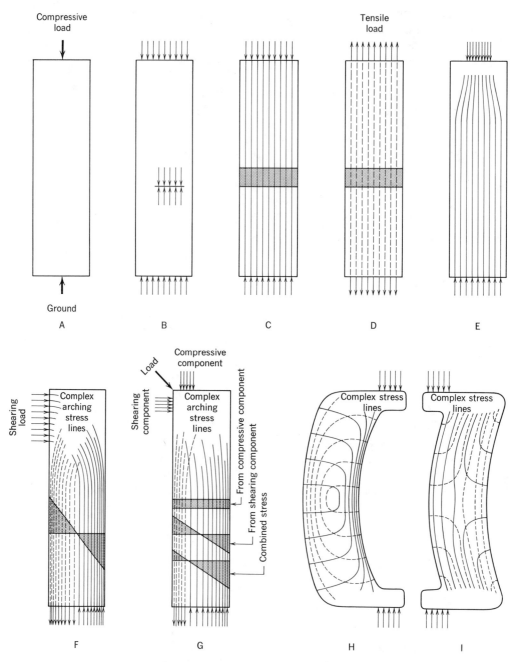

FIGURE 18-2

DIAGRAMS OF STRESS LINES WITHIN CYLINDRICAL OBJECTS.

Note that compression and tension are each greatest at their respective edges of the cylinder (stress lines are closest together there) and each diminishes to zero at the central axis of the cylinder. The stress at any intermediate point between edge and center is proportional to the height at that point of the relevant triangle, as shown in part F. If a tension load replaces the compression load, the pattern of stress remains the same, but the kinds of stresses are reversed. These models are approximated by the force of food against the jaw (compressive load) and by the forces of muscles on opposite edges of the summits of vertical neural spines (tensile load), though for reasons explained in the next section these bones are not cylindrical.

Most loads applied to long bones are neither parallel nor perpendicular to their long axes, but instead are at an intermediate angle, as when one bone loads another across a flexed joint, or a tendon inserts obliquely onto a bone. Such loads can be converted, however, to longitudinal (compressive or tensile) and transverse (shearing) stresses (see p. 463). Part G represents this more general situation when both compression and shear are present. The magnitudes of these stresses in the plane illustrated can be independently represented, respectively, by the heights of a rectangle and a pair of congruous triangles as before. Total stress can be represented by combining these figures as shown. It is seen that compression exceeds tension and that the axis of zero stress is no longer the central axis of the cylinder.

Finally, although the cylinder has provided a conveniently simple model thus far, few bones closely approach cylindrical shape with straight parallel sides. Most large bones have somewhat enlarged ends and curved shafts. Solid models of this general shape can be loaded either by compression or tension tending to bend them further (see part H) or straighten them (part I). The patterns of stress lines in the curved shafts are shown; those near the applications of the loads are intricate and are omitted. Note that stress is distributed throughout each shaft (except for one central focal point in each), but is greatest near the convex and concave edges of the shafts.

Small cavities, notches, and channels all weaken materials by causing local concentrations of stress. It has been shown by Currey that bones are constructed to minimize such loss of strength. It is the shafts, not the ends, that have the greatest strain, and these tend to be very smooth. Canals for blood vessels usually run at an angle to the long axis of a bone, and the lacunae housing osteocytes have their shortest axes at right angles to the long axis of the bone. These configurations reduce the concentration of stress. (The presence of lacunae in cellular bone is probably advantageous for stopping the spread of microfractures.) Further, bones tend to have at least small elevations or crests where tendons join them. This causes less concentration of stress within the body of the bone than would otherwise occur.

There are various ways of determining patterns of stress in simple objects such as the kinds described. When models (usually two-dimensional) are cut from photoelastic plastic, like Plexiglas, and then loaded as desired and photographed with polarized light, they show light and dark bands from which stress lines can be calculated. Models of bones have been tested in this way, but results are subject to the criticism that bones are neither two-dimensional nor homogeneous. Strain gauges affixed to objects show the magnitude of local stress. When a hard brittle lacquer such as Stresscoat is painted onto an object which is then loaded, the lacquer develops micro-cracks in a pattern that indicates distribution of stress at the surface.

Application of these methods is not easy. Engineers have difficulty determining the patterns of stress in solid, homogeneous objects of irregular but relatively simple shape when single loads are applied. The patterns in living bones are much more complicated. Thus, long bones are hollow, quite irregular in both internal and external contour, and are stressed simultaneously by many external forces which change constantly. It is impossible to accurately calculate detailed patterns of stress in living bones. Diagrams such as the often-published figures of stress lines in the proximal end of the human femur are misleading if not qualified as approximations. Nevertheless, as will be shown, the use and design of the structural elements of the body *do* correlate well with expectation according to their properties, including the general pattern of stress lines under usual loads.

Use and Design of Structural Elements

Tendons, Ligaments, and Cartilages. Tendons transmit the pull of muscles to bones, a function for which their flexibility and tremendous resistance to tension well suit them. They consist of tightly packed parallel bundles of collagenous fibers. Tendons that must move appreciably relative to adjacent tissues have sheaths and in some instances slip through lubricated channels resembling the spaces around movable joints of the skeleton.

The force of contraction of even a large muscle is usually concentrated by its tendon on a small area of the skeleton. This contributes to precision of movement and allows several muscles to act in different ways at about the same place. It is important that tendons enable muscles to move skeletal parts at a distance from their own positions. Weight distribution and body contours are thus controlled in ways that contribute to speed, endurance, and agility (of which more in subsequent chapters). Human fingers would be useless if encumbered by all of their own muscles.

Some tendons transmit tension around corners at movable joints in the manner of cords passing through pulleys. The living pulley may be a tunnel of bone (as where tendons of digital flexors pass around the proximal end of the tarsometatarsus of some birds), a bony projection forming a channel (as where tendons of abductors of the foot angle at the outside of the ankle of many mammals), or a ligamentous loop (as where tendons of digital extensors turn in front of the ankle). Loaded tendons that are straight sustain only tensile forces, but where a tendon bends, shearing forces also occur. Accordingly, tendons compensate by becoming thicker at such places. If forces other than tension are particularly severe (as where the tendon of the quadriceps muscle passes in front of the knee), small bones, which are better able to withstand such forces, interrupt the tendons. These are called sesamoid bones.

Some ligaments have virtually the same structure and properties as tendon, whereas others have less regularly oriented collagenous fibers and contain elastic fibers in various proportions. Because tendons always relate to muscles, they function only when muscles function. Ligaments, by contrast, function passively and are therefore superior where constant tension is

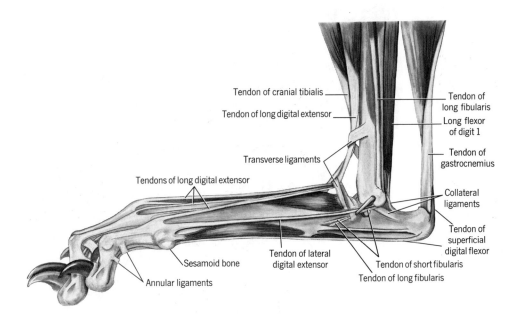

FIGURE 18-3

EXAMPLE OF RELATIONSHIPS AMONG TENDONS, LIGAMENTS, AND BONES shown by a lateral view of the left ankle and foot of the gray fox, *Urocyon*. (Drawn from a freeze-dried dissection and hence slightly shrunken.)

required. In the manner of lashings, ties, and elastics, they bind the skeleton together, limit the motion of some joints, and contribute to antigravity mechanisms.

Some ligaments are merely thickened portions of the capsules around movable joints. These merge with adjacent connective tissue and have indistinct margins. Other ligaments that also bind movable joints are tough and prominent. Ligamentous loops and sleeves guide tendons, particularly at joints. Some of these merge into the sheaths of the tendons.

The **nuchal ligament** is an example of an antigravity mechanism. This strong and extensible ligament (i.e., having a low modulus of elasticity) is prominent in large mammals with heavy heads and long necks. It extends from the summits of anterior thoracic neural spines to the back of the skull and neural spines of anterior cervical vertebrae. The head and neck are held in normal resting posture without muscular effort. A small muscular tug depresses the head to the ground, simultaneously stretching the ligament. When the muscles relax, the ligament shortens, thus elevating the head.

Ungulates have a suspensory mechanism which cushions their footfalls. Each foot is supported by an elastic ligamentous sling. This mechanism is described further on p. 507.

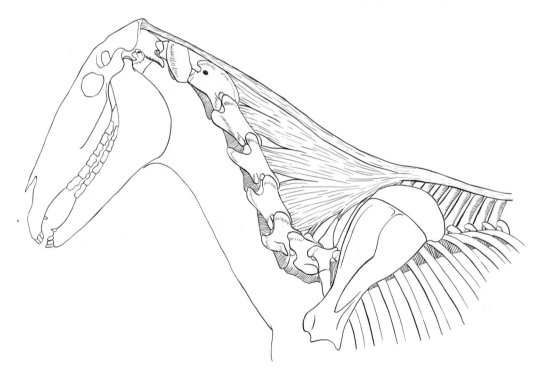

FIGURE 18-4
NUCHAL LIGAMENT OF THE HORSE, an antigravity mechanism.

Cartilage is found where moderate resistance to compression, tension, and shear (in different combinations according to circumstance) must be combined with firmness and some flexibility. It has the advantage over bone that it is lighter. The various types of cartilage (see p. 126) can be distorted in varying degrees, but all return to their original form when released. Elastic cartilage, the most flexible, supports the external ear, nose, and epiglottis of mammals. Fibrous cartilage provides a tough, but somewhat flexible cushion. It forms intervertebral disks and the pelvic symphysis of some tetrapods. It is also found at the insertions of some tendons. Hyaline cartilage forms the skeletons of embryos and elasmobranchs and parts of the skeletons of adults of many vertebrates. In tetrapods, it may substitute for bone where the greater strength of bone is not needed (e.g., in the carpus and tarsus of salamanders). It also stiffens the trachea and covers the articular surfaces of movable joints where its hardness, smoothness, and slight accommodation under pressure are advantageous. The principal kinds of cartilage may intergrade. Varying degrees of calcification may harden the hyaline cartilage of the sternal ribs of mammals, the epiphyses of amphibians, and the skeletons of elasmobranchs.

Bones that Resist Compression or Tension. In order to save on weight, bulk, and metabolic requirements, the supportive elements of the body are designed to provide adequate strength with minimum material. This principle is important for analysis of the skeleton.

Adequate strength to resist all stresses is provided to a mouse by even a slender skeleton. For reasons explained in Chapter 19, however, very large tetrapods would be unable to sustain the resultant loads if they were proportioned like enormous mice. Elephants, various extinct mammalian giants, and many dinosaurs are (or were) obliged to modify structure, posture, and behavior to minimize stresses on the skeleton and to resist the remaining stresses effectively. Figures given above show that bone can sustain more compressive force per unit of cross-section than any other kind of force. An animal could support itself with the least bone (and weight and bulk) if it could limit all loads to compression. This is not even possible for a static table, let alone a moving animal, yet the largest land animals do minimize stresses other than compression in ways that can be predicted from study of Figure 18-2. The column-like limbs of such animals have bones that are relatively cylindrical with heads nearly in line with their shafts, thus reducing bending forces (see Figure 19-14).

In Chapter 19 it is explained that the vertebral centra of most land tetrapods are subjected primarily to compressive

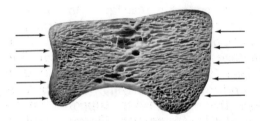

FIGURE 18-5

**LONGITUDINAL SECTION OF A CEN-
TRUM SHOWING ORIENTATION OF TRA-
BECULAE PARALLEL TO COMPRESSIVE
FORCES. The specimen is a lumbar vertebra
of a caribou,** *Rangifer.*

forces acting on their opposing ends. If large enough for the necessary muscle attachments and leverages and also solid, they would be much stronger (and heavier) than needed. They are not solid, but neither are they hollow. Spicules of bone called **trabeculae** brace the inside of each centrum, and these are largely oriented lengthwise in the direction of the predicted compression lines (see Figure 18-5).

The arm bones of gibbons are stressed primarily by tensile forces as the animals swing under tree branches. Few bones of few animals, however, are loaded primarily by tension; if constant or frequent tension must be withstood, a ligament, being as strong for its weight and less likely than a thin bone to fracture if bent sideways, is substituted. An example is the sacrotuberous ligament shown in Figure 18-6. Bones are heavily stressed locally by tensile forces where tendons insert on them. This most often occurs near the ends of bones where bending forces are also common.

Bones That Resist Bending in One Plane. When a solid cylinder resists a bending force applied in one plane (e.g., the plane of the paper in parts F and G of Figure 18-2), the resultant stresses are concentrated in that plane and are greatest at the surface of the cylinder. Accordingly, although a cylinder is useful for sustaining compression alone, or tension alone, it is not economical of material for resisting bending in one plane: Too much of the material is not stressed and hence is wasted. The engineer uses instead an "I-beam" as a girder. This beam has upper and lower bars of steel which are stressed when loaded, and a wall to hold the bars apart. (Which bar is compressed and which is tensed depends on the relation of the load to the support. See Figure 18-7). Bones are never designed as simple I-beams, yet the same principle of construction explains the dumbbell-like distribution of material sometimes seen in cross-sections of bones.

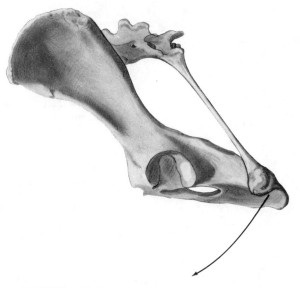

FIGURE 18-6

AN EXAMPLE OF A LIGAMENT THAT WITHSTANDS FREQUENT TENSION is the sacrotuberous ligament of the dog which resists the tendency of the innominate bone to rotate on the sacrum in the direction shown by the arrow when muscles that swing the leg to the rear pull on the ischium.

The carpenter's beam and joist provide another useful analogy. Lumber is used that has rectangular, but not square, cross-section, and is always oriented so that the longer dimension is parallel to the load (i.e., usually is vertical). The reason for this is that resistance to bending is equal to a constant times the width of the beam (dimension transverse to the load) times the square of its height (dimension in line with the load): $R = cwh^2$. If one dimension is twice the other, then the beam is about twice as strong on edge as it is flat; if one dimension is three times the other, the beam is three times stronger when on edge. Clearly, the animal body should "know about" this, and it does. The zygomatic arch is a bony beam turned on edge to the muscles acting on it. The same is true of most of the neural spines. The pygostyle of birds is a blade of bone oriented parallel to the air resistance transmitted to it by the tail feathers. The lower jaw at the level of the teeth is a modified beam of expected orientation in relation to loads imparted by teeth and muscles.

Resistance of a beam to bending also varies inversely as the square of its length. For this reason, bony beams are not long; other kinds of construction resist loads that must be held at

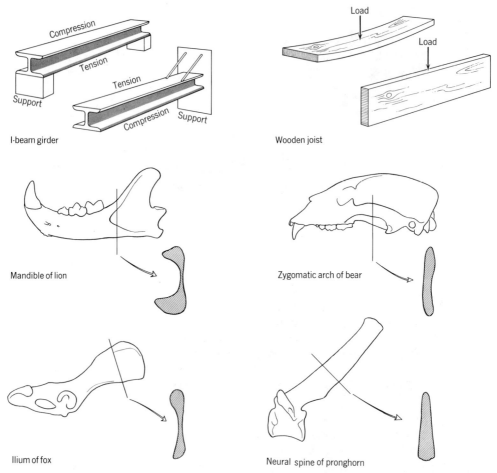

FIGURE 18-7
BONE STRUCTURE ANALOGOUS TO THE I-BEAM (left) AND JOIST (right).

some distance from the support of the mechanism. The nuchal ligament is one example; others will be described later.

Bones That Resist Bending in Several Planes. We have seen that a flat beam effectively resists bending when loaded edge on, but is weak when loaded flat-side on. The long bones of the appendages of tetrapods must resist bending in many directions; hence, they cannot be flat. The cylinder, discarded as wasteful of material when bending forces are in one plane, is here an effective model because compressive and tensile forces can concentrate at opposite edges of the cylinder no matter what the direction of loading. It is still true, however, that stresses are least toward the central axis of the cylinder (see again parts F–I of Figure 18-2). Therefore, the most strength for the least material is achieved by a hollow cylinder. This explains the hollow shafts of long bones.

Controlled bending of the vertebral column of a fish occurs at the intervertebral joints. The centra themselves must resist the bending forces produced by axial muscles. The spine as a whole functions as a somewhat flexible bony tube, the discontinuous cavity of which is the spaces between the markedly amphicoelous centra.

Returning to tetrapods, since resistance of a tube to bending varies inversely as the square of its length, and since the long bones are stabilized by muscles acting at their ends, they would be most subject to fracture at the centers of their shafts if there were no provision to the contrary. The shafts of such bones compensate by being a little thicker at midlength, by increasing their diameter, or both (Figure 8-16).

Although long bones are subject to bending forces in various directions, force may be greatest in one plane. The bone then compromises between the beam and tube by becoming slightly oval in cross-section, with the long axis of the oval in the direction of the dominant load and with thicker walls on the sides toward and away from the load. The phalanges of bats tend to have this configuration. It sometimes happens that stress is greatest along the concave side of a curved bone (Figure 18-2, part H), and the wall of the bone is then thickest there. However, with numerous forces acting on a bone, one cannot assume that the shaft is asymmetrically stressed just because it curves.

It would be impossible for a morphologist to accurately determine the directions and relative magnitudes of the forces acting on a bone during the daily activities of an animal. The amazingly adaptable living skeleton *does* determine the magnitudes and directions of its predominant loads. Bones seem to have built-in sensors which monitor strain and "report" to the mechanisms of bone destruction and growth. When dominant strain is moderate to severe, the bone is slowly remodeled in such a way as to reduce the strain. The nature of the sensing device is not yet clearly understood but apparently involves bioelectric phenomena (piezoelectric effect and semiconductor effect) within the bone.

It is evident from Figure 18-2, parts F–I, that since most muscles insert near the ends of the long bones, and since forces transmitted by one bone to another across a flexed joint are not parallel to the shaft of either, stress lines arc across the ends of those bones. It follows that the ends of long bones should not be tubes. Since solid ends would be stronger (and heavier) than needed, the most economical design is a network of interconnecting trabeculae and thin sheets of bone that follow the stress lines, and this is what we find. Stress lines change somewhat as loads change so the body adopts lines that are a compromise of the more usual loads. (The arching of the trabeculae is usually evident, as for the lower trabeculae in Figure 18-8, but rarely are as regular as in the head of the human femur which is so commonly figured. The same is true of the rather stylized Figure 18-10.) Adaptability is again evident. The trabeculae and sheets become oriented as the young animal begins to move about, and their orientation changes if an injury alters the usual loads. The spongy nature of the ends of long bones also provides that they can function as shock absorbers.

Where several bones function as a unit in sustaining usual loads, stress lines, and hence trabeculae, also traverse those

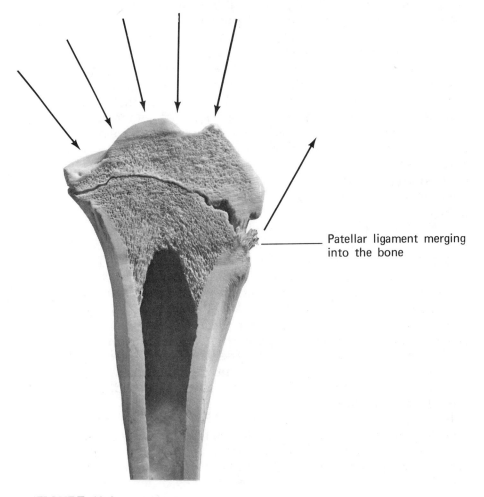

Patellar ligament merging into the bone

FIGURE 18-8
LONGITUDINAL SECTION OF A COW TIBIA SHOWING ORIENTATION OF TRABECULAE IN RELATION TO SOME OF THE FORCES ACTING ON THE BONE.

bones as a unit. The human tarsus is an example. It functions as a beam loaded in the middle by the tibia (though we shall see later that this is not all of the story).

Tendon to Muscle; Tendon and Ligament to Bone. A tendon is usually stronger than the contracting muscle with which it is associated. The union of tendon to muscle sometimes is a little less strong than the muscle, but only a little. The tendon may appear to end where the muscle begins, but it branches and pervades the muscle, its fibers merging with those of the perimycium and endomycium. In pulling on its own fibrous framework the muscle also pulls on its tendon.

Muscles that take origin from large areas of bone (e.g., supraspinatus) may gain sufficiently firm attachment by merging their connective tissue with the periosteum of the bone. Tendons and ligaments concentrate so much force on such small areas that a stronger attachment is needed. Imagine the difficulty of joining with glue the end of a flexible cord to a hard smooth material with enough strength to sustain loads of 900 kg/cm^2 even though the angle of attachment changes! Insertions of ligaments and tendons do tend to be weak points in the bone-muscle system, yet the body surpasses human technology in solving the problem.

The collagenous fibers of tendons are not attached to bone; they merge into it (Figure 18-8). Fibrous bone forms all of the skeleton of small animals. Large animals have osteons throughout most of the skeleton, where compressive forces dominate, but retain fibrous bone at the insertions of tendons. Fibers of tendons penetrate the bone and there become indistinguishable from its fibers. (Fibrous cartilage may intervene. Also, calcification may merge into a tendon at the tendon–bone junction.)

There remains an apparent source of weakness. Consider a large tendon of circular cross-section that inserts at right angles to the surface of a bone. If all bundles of the tendon are of equal length, they share the load equally when tension occurs. If the angle of insertion changes, as would be expected if the tension causes motion, the relative distance from muscle to bone decreases on the side of the tendon now forming an acute angle with the bone and increases on the side forming an obtuse angle. One might expect that fibers on the long side of the tendon would carry all the stress and would give way one by one. This does not happen, primarily because the collagenous fibers of a tendon, although parallel in the body of the tendon, weave at its insertion, thus distributing the load throughout the insertion. Further, the strength of a tendon usually has a large safety factor; not all of it need support the pull of its associated muscle. Fibers of a resting tendon are a little wavy; a pull that straightens them on the long side transmits some tension also to the short side if the difference in length is slight. Further, the angle of insertion rarely changes very much.

The elastic fibers of ligaments are largely replaced by collagenous fibers at their insertions. Thus, the insertions of ligaments are like those of tendons, though details for each differ according to size of animal, general angle of insertion, and specific location.

Kinds and Functions of Joints. Joints between bones are classified on the basis of both structure and function, though the two are, of course, related. A first structural category is the immovable joint or **synarthrosis** (= together + joint). The bones may be joined only by connective tissue, which is the rule for membrane bones (e.g., on the roof of the skull), or only by cartilage, which is usual for replacement bones (e.g., at the base of the skull and between shafts and epiphyses of long bones). The cracks between bones joined by synarthroses are called **sutures.** These joints are places of growth; sutures must remain open for growth to occur. When the growth period terminates, synarthroses of birds and mammals tend to ossify and thus become obliterated one by one on a schedule characteristic of each species. Most sutures of marsupials, however, and some sutures of many other mammals, remain open for life.

Synarthroses are further characterized by the configuration of the suture, and this relates to function. If the suture is approximately straight and the bones have nearly squared-off edges, then a **butt joint** is formed, as between the two nasal bones, and between bones of the basicranium of most mammals (Figure 18-9). Butt joints can withstand compression but little shear or bending.

If the same square-edged bones were instead joined by overlapping, the union, a **lap joint,** might be somewhat stronger. However, as human builders have learned, when a lap joint is compressed or tensed, the area of contact is not evenly stressed. When glue is used it tends to give way at the leading edges and to break toward the middle. If the overlapping edges taper instead so that the two members remain in line, the union is called a **scarf joint.** Here the entire area of contact is evenly stressed by most loads and strength is much improved. Scarf joints (also called squamous joints) often join thin flat bones. They occur between some bones of the mandible of reptiles, and in most vertebrates join various of the skull bones. Cetaceans, whose skulls must withstand the force of waves, have more cranial scarf joints than other mammals.

A synarthrosis that is very effective for withstanding compression and shear between hard structures that are not thin

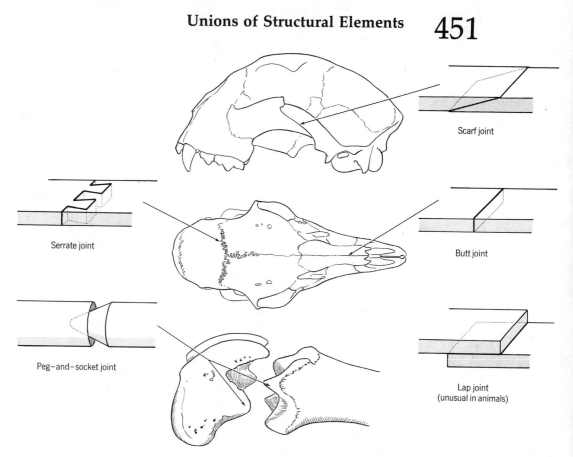

FIGURE 18-9

SOME KINDS OF SYNARTHROSES. Above, skull of a cheetah, *Acinonyx;* **middle, skull of a deer,** *Odocoileus;* **below, distal epiphysis and shaft of the femur of a young wolf,** *Canis.*

and flat is the **peg-and-socket** (also called gomphosis). Such joints join thecodont teeth to the jaw bones and often the jugal to the maxilla. Most epiphyses of long bones join their shafts by complex joints including several pegs and sockets, sometimes relatively deep (distal end of femur of mammals) and sometimes shallow (proximal end of tibia). Another synarthrosis is the **serrate joint** which has such an irregular suture that the adjoining bones interlock repeatedly throughout the union. This firm type of joint is found between roofing bones of the cranium of some tetrapods, particularly of amphisbaenians (which dig with the head) and artiodactyls (which support horns or antlers).

These various synarthroses may occur in combination and may intergrade.

A second, and intermediate, structural category is the **amphiarthrosis** (= both + joint) which allows some motion in response to compression, tension, or twisting, yet is tough. The surfaces of the adjoining bones may be covered by hyaline cartilages which, in turn, are joined by a pad of collagenous fibers or by fibrous cartilage. The union between the bones of such a joint is called a **symphysis** instead of a suture. Examples are the mandibular symphysis of many vertebrates, the pelvic symphysis, which allows more motion in females toward the end of pregnancy than in males, and the joints between most vertebral centra, which provide for motions of the spine. The function of joints between centra is conditioned by configuration (for terminology, see pp. 158 and 159). Joints between procoelous and between opisthocoelous centra allow adequate motion in any direction, withstand compression, and resist dislocation better than platyan centra. Procoelous and opisthocoelous vertebrae are, therefore, common in the necks of tetrapods and in their tails if the tail is strong. Platyan centra are usually restricted to the trunk, where shearing is minimal.

In another kind of amphiarthrosis called a **syndesmosis,** the bones are joined by moderately thick zones of collagenous fibers or by ligaments, and somewhat greater motion is allowed. Examples are the unions of radius to ulna and of fibula to tibia in certain mammals having slight play between these pairs of bones. However, syndesmoses are more characteristic of various other classes. Thus, such joints are common among bones of the protrusible upper jaws and movable opercula of bony fishes. The flat surfaces of cranial bones of some amphisbaenians are joined by two sets of fibers which criss-cross. When the bones tend to be pushed past one another, one set of fibers is under tension and the other is slack. When the bones are pulled the other way, there is a little slippage and then the other set of fibers comes into play.

The last general structural category of joints is the freely movable joint or **diarthrosis** (= two + joint). The articulating surfaces of the bones are covered by extremely smooth hyaline cartilage (Figure 18-10). In fetal life the **joint cavity** develops which is necessary for movement of one bone on the other. Where not bordered by the cartilage-covered bones the cavity is enclosed by a **joint capsule.** The capsule may be thin and membranous but is usually at least partly tough and fibrous, containing both collagenous and elastic fibers. The capsule is lined by a cellular **synovial membrane** which is more or less folded. It contains fat cells and in some joints fat pads which encroach on the cavity and help cushion its changing configuration as the joint moves. Ligaments binding a diarthrosis may be within or partly outside the capsule, or may be inside the joint cavity, as at the hip and knee of mammals. Tough pads of fibrous cartilage called **menisci** (singular, meniscus) occur inside the cavities of several joints (knee and jaw articulation of mammals) where they seem to firm and guide the turning bones, though their association with these joints and not with others remains a puzzle. Bathing joint cavities is a small quantity of **synovial fluid,** apparently produced by the synovial membrane. This clear or yellowish fluid is similar to tissue fluids and contains mucin. It is more or less viscid according to the joint. Its function is to nourish the hyaline cartilage (which is devoid of blood vessels) and, importantly, to lubricate the joint.

No matter how congruously two surfaces may be shaped and how carefully polished (e.g., two flat pieces of glass), when they rest together they touch only at microscopic elevations. When the dry surfaces slide over one another the microelevations grate, thus producing heat, friction, and wear. If a lubricant is introduced between the surfaces, fewer solid-to-solid contacts are made. Much of the shearing force takes place between molecules of the lubricant, and friction and wear are reduced. This kind of lubrication is used in man-made machinery and is called **boundary lubrication.** It occurs in diarthroses when there is no motion and just as motion starts or stops. Some wear takes place; white cells in the synovial fluid remove from the joint capsule the microscopic fragments of cartilage that result.

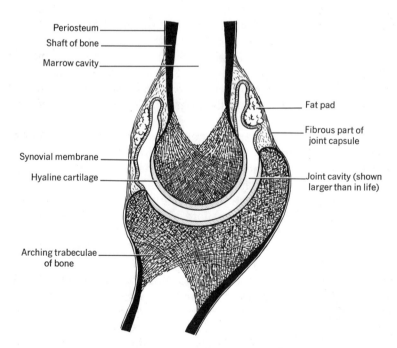

FIGURE 18-10
STRUCTURE OF A DIARTHROSIS.

When a lubricant can be kept thick enough (about 50 μ) to hold two polished surfaces completely apart, even though they are under considerable pressure, then as they slide over one another the only shearing is within the lubricant—and there is no wear. This is **fluid film lubrication.** To achieve it, the surfaces must not be quite congrous so the lubricant is shaped like a wedge and pressure must force the lubricant into the joint. Engineers have learned to lubricate critical bearings in this way, and diarthroses probably have fluid film lubrication during motion. The moving bones do not quite touch anyplace and come very close to it at only two or three places at a time. Consequently, healthy joints have negligible friction and remarkably little wear.

Diarthroses are classified according to both function and shapes of articulating surfaces. Function and shape are correlated, but several shapes may serve similar functions, and one general shape may serve several functions. Thus, terminologies tend to be either inadequate or inconsistent. Further, there are intergrades and combinations. Nomenclature does not substitute for interpretation.

A **hinge joint** has a more or less cylindrical head which rotates in a corresponding socket. Motion is primarily around one axis, as for a door hinge. The articulation of the mandible to the skull of carnivores is a simple hinge joint. The head of a hinge joint may have splines which mesh with grooves in the socket, or it may have bulges and waists which are reciprocated by the socket. These devices further limit motion to one plane and resist dislocation. Examples are the elbow joint, ankle joint of many mammals, and joints between phalanges, particularly of runners.

Some hinge joints can further be designated **snap joints.** These are stabilized by their ligaments in the open and closed positions and are unstable in intermediate positions. Any hinge joint revolves around an axis which lies within the convex member and is transverse to the plane of motion. Ligaments holding the joint together are lateral in position. If one end of such a ligament inserts exactly at the pivot, then its length remains constant as the joint moves and tension does not change. If, however, a ligament crosses over the pivot to insert slightly beyond, then its length and tension are reduced as the joint moves in either direction from an intermediate position. The joint then snaps into the open or closed positions. Snap joints are found at the elbow and ankle (hock) of various large mammals. The mechanism provides some passive support in the standing posture. In order to better secure the union, snap joints usually have at each side of the hinge either two liga-

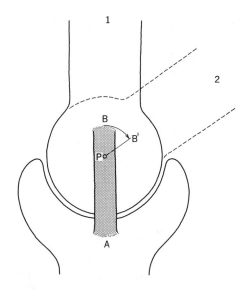

FIGURE 18-11

MECHANICS OF A SNAP JOINT. As the upper bone rotates hinge-like from position 1 to position 2 on the lower bone, which is fixed, insertion B of ligament AB moves to B′ in an arc around the pivot of motion, P. Since distance AB′ is shorter than AB, the joint is unstable in position 1.

ments that cross or a single broad ligament that twists. (Can you say how the ligaments of a hinge joint could be arranged to snap the joint into a single given position, open, closed, or intermediate?)

A **ball-and-socket joint** has a hemispherical head which turns in a nearly congruous socket. A wide range of motions, including rotation, is implied, and the shoulder and hip joints are the usual examples. The union of the occipital condyle to the atlas of archosaurs can also be cited. A shallow socket (shoulder) allows more excursion than a deep socket (hip). Joints between procoelous and between opisthocoelous centra, and unions of the heads of ribs with their facets, are ball-and-socket in structure, but they have less range of motion and are usually not so designated, even if the joint is a synarthrosis. A modification of the ball-and-socket joint is the peg-and-socket diarthrosis which allows only rotation around the long axis of the peg. This unusual kind of joint joins the avian quadratojugal to the quadrate and functions with the mechanism that moves the upper part of the bill (see p. 630).

Somewhat similar to the last named joint in function is the **pivot joint,** which allows rotation of one bone around its own long axis. During pronation and supination of the manus the radius pivots on the ulna; its disk-shaped head revolves in the radial notch. Similar in function, though entirely different in structure, is the pivot joint between the mammalian atlas and axis vertebrae. (Examples of ball-and-socket, hinge, and pivot joints are seen in Figure 8-23).

If the convex head of a bone is biaxial instead of hemispherical, and fits into a biconcave socket, an **elipsoid joint** results. Motion is in two planes (e.g., flection–extension and abduction–adduction) and rotation is prevented—very different from a pivot joint. The human radius-to-carpus joint is an example (see also the slow loris, Figure 20-16). The same in function, though different in structure, is the **saddle joint** found between the heterocoelous cervical vertebrae of birds (see the pelican, Figure 8-2). The articulatory surface of the anterior vertebra is convex horizontally and concave vertically, whereas that of the posterior vertebra is concave horizontally and convex vertically.

Another family of joints has the name **plane joint.** The articulating surfaces are more or less flat and permit various motions depending largely on the nature of associated ligaments. Contact between the bones may be maintained if they glide over one another as do the pre- and postzygapophyes of vertebrae. Usually the articulating surfaces are not quite flat, so that gliding motions force them apart. Some motions separate flat-ended bones to a surprising degree; the joints between the carpal bones of large mammals are a striking example (see the vicuna, Figure 20-16).

The patella has a curved surface which slides in the patellar groove of the femur. Lumbar vertebrae of artiodactyls have postzygapophyses which are rolled into scrolls and prezygapophyses which are trough-shaped. These joints limit some kinds of motions. They are no longer so "plain," yet have no special name.

Still other kinds of joints defy current terminology yet invite attention. Thus, nature has designed the mammalian knee joint without regard for orderly classification: The femoral condyles largely rotate on the platform of the head of the tibia, but they also roll over it. Motion is mostly around one axis, as for a hinge joint, but not entirely so, and there is also slight rotation of the tibia around its axis. Many birds can move the upper bill on the braincase (of which more in Chapter 25). The joint usually consists merely of a zone of thin flexible bone—a type not named in human anatomy texts. Anurans have either a syndesmosis or diarthrosis between the sacral vertebra and the arms of the pelvic girdle. The joint may function as a hinge in the vertical plane or may allow side to side bending. In some frogs these motions are possible to a degree, and the vertebral column can also telescope forward and backward on the pelvic girdle under the control of apposed sets of muscles.

Mechanics of Support and Movement

Force of muscular contraction is a property of great importance in the analysis of bone-muscle systems. It is commonly stated that the maximum force a muscle can exert is equal to the force of contraction of one of its fibers times the number of fibers, and therefore is proportional to the cross-sectional area of the muscle. This is only a rough approximation. It assumes that all muscle fibers are parallel to each other and to the axis of the tendon of insertion, that all fibers are of the same size, and that all have equal and constant force of contraction.

The first condition, all fibers parallel, is approximated by some strap-like muscles that do not taper at origin or insertion

Force and Work of Muscles

(coracomandibularis of shark and thyrohyoid of cat). The fibers of most muscles gather somewhat at the ends of the muscles where they angle in toward the emerging tendons. The strength of a muscle that is partly or entirely pinnate (see p. 192) is roughly proportional not to a level cross-section of the muscle as a whole but to a section that cuts all fibers at right angles, regardless of their orientation. Such a cut is easy to conceive, but difficult to accomplish.

The second condition, all fibers of the same size, is rarely met. Striated muscle fibers are classed as red, white, and intermediate. Red fibers owe their deeper color to a greater amount of the pigment myoglobin. Most muscles contain each kind of fiber, but the proportions vary among different muscles of the same animal and between the same muscles of different animals. White fibers have about twice the diameter of red fibers and hence are less closely packed. Each type of fiber tends to be larger in large animals.

The third condition, equal and constant force of contraction by all fibers, never pertains. Muscles contract with most force when stretched a little beyond resting length; as they shorten, force falls off (see the length–tension curve, Figure 19-1). Further, not all fibers of a muscle contract simultaneously, even in a maximum effort. The motor units take turns (a motor unit is one motor nerve fiber and the muscle fibers it activates). The proportion of all units that contract in a maximum effort is subject to alteration by conditioning and physiological factors.

As a consequence of these variables it is never possible to determine accurately from nonliving material the actual or rela-

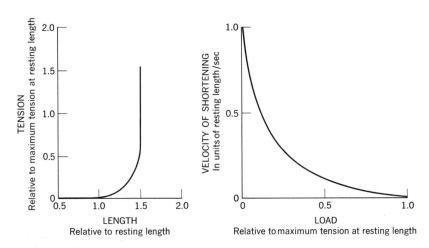

FIGURE 19-1

LENGTH–TENSION CURVE (left) AND LOAD–VELOCITY CURVE (right) FOR SKELETAL MUSCLE.

tive strengths of muscles. Moreover, cross-sectional areas of muscles are difficult to measure and are influenced by degree of stretching and method of preservation. Morphologists are turning more and more to experimentation with living material. However, it is also difficult to measure forces in live bone-muscle systems under conditions that closely resemble normal behavior. Often the investigator is obliged to settle for approximations of actual circumstances. This need not cause dismay. The living body is so complex, adaptable, and variable, that when the researcher can learn how the mechanism works, and the order of magnitude of the parameters involved, he has learned much.

Returning, then, to strength, a striated muscle can deliver to its tendon a force of about 3 kg/cm^2 (42 lb/in.^{2}) of its cross-sectional area taken at right angles to the fibers, but let us remember that this is no more than an approximation.

With this relationship in mind, consider again the contrast made in Chapter 18 between a mouse and a hypothetical animal shaped like a mouse but having the size of an elephant. In each instance, the load on the muscles is proportional to the mass of the body, which is proportional to the cube of any linear dimension. Thus, as body length increases x times (from ordinary mouse to giant mouse), body weight increases x^3 times. But the forces the muscles can exert are roughly proportional to areas and will thus increase only x^2 times. If body size were significantly increased without altering body proportions, the load on the locomotor system would increase faster than its capacity to provide support. This important principle is basic to some interpretations of the structure of vertebrate giants made later in this chapter.

The **contraction distance,** or amount of shortening possible for a striated muscle having approximately parallel fibers may be as much as 30% of its resting length, or nearly 60% of its stretched length. The longer a muscle, the greater the contraction distance. Some muscles are maximally stretched and contracted (e.g., muscles of distensible tongues), but since muscles are physiologically most effective at intermediate lengths, leverages and behavior are usually such as to restrict the shortening of muscles well within their theoretical maxima.

Work is force times the distance through which it acts. Since, in living systems, loads usually change as structures move, it is difficult to measure work accurately. To the extent that force is roughly proportional to the cross-sectional area of a muscle, work is roughly proportional to its volume (or mass). If a muscle contracts isometrically (i.e., without shortening), it does no work although it does expend energy. The concept of force times time is then more useful than that of work for estimating nutritional requirements and onset of fatigue.

Force Vectors and Their Resolution. Forces are **vector quantities.** That is, they have both magnitude and direction. Each property is important to the analysis of bone-muscle systems and each can be represented graphically by an arrow called a **vector.** The arrow is usually placed so that its tail is at the point of application of the force, e.g., the insertion of the tendon (alternatively, the head of the arrow could be placed at the insertion). The orientation of the arrow represents the direction of the force, and its length represents the magnitude of the force according to any arbitrary scale (e.g., 1 cm = 10 kg). In Figure 19-2, part A shows the long head of the triceps muscle inserting on the olecranon process of the mammalian ulna. If the force of contraction is approximated from the cross-sectional area of the muscle, or better, by direct measurement from the live muscle, then the force of contraction can be represented by the vector F_1 in part B.

The medial head of the triceps (part C) inserts at the same place by the same tendon and can similarly be represented by

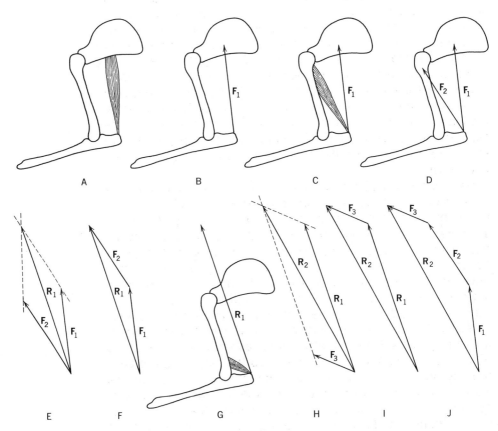

FIGURE 19-2
FORCE VECTORS AND THE RESOLUTION OF FORCES.

the vector F_2 of part D. (If the second muscle inserted near the first but not by a common tendon, the forces could still be considered to act at a common point if extensions of their lines of action intersected.)

Having two muscles pulling on the same point but in different directions and with different tensions, it becomes important to learn the magnitude and direction of their common, or net effect. What, we ask, is the vector of the single hypothetical muscle which, acting alone, would load its insertion in just the same way as the actual muscles acting together? The desired force is called the **resultant** of the given forces and its derivation is called a **resolution of forces.** The resultant is usually determined graphically by drawing a **parallelogram of forces** as shown in part E: F_1 and F_2 form two sides of a parallelogram which is completed by drawing dotted lines. The diagonal, R_1, is the desired vector of the resultant force.

Since opposite sides of a parallelogram are equal, an alternative graphical solution is to place F_2 so that its tail is at the head of F_1 (or vice versa), and then to draw the line that makes the third side of a triangle (i.e., the triangle that is half of the parallelogram constructed above). That third side is again the desired vector, R_1 (see part F).

A third muscle, the anconeus, also inserts at the same place (part G), its vector being F_3. To determine the resultant force, R_2, of all three muscles contracting simultaneously, one can either draw the diagonal of the parallelogram having vectors R_1 and F_3 as sides, or one can close the triangle having R_1 and F_3 as sides (parts H and I). Either method is a two-step solution, since R_1 had first to be determined from F_1 and F_2. The problem can also be done in one step by joining vectors F_1, F_2, and F_3 tail to head (in any sequence) and then closing the polygon (part J). The closing line is again R_2, or the vector of the resultant force of all three muscles acting together.

The magnitude and direction of a resultant force can be calculated somewhat more precisely, if need be, by trigonometric methods.

The determination of the magnitude and direction of the force of contraction of a pinnate muscle provides an application of the resolution of forces. Consider the stylized, flat pinnate muscle shown in Figure 19-3 A. It can be regarded as two muscles pulling in different directions on a common central tendon. The force of contraction of each side taken alone is in the direction of the fibers of that side and has a magnitude that is roughly proportional to the cross-sectional area of all the fibers of the side (or to the distances AB or AC if the muscle is of uniform thickness). Vectors F_R and F_L of part B represent the forces of the right and left sides of the muscle. Completing the parallelogram of forces, R is found to be the vector

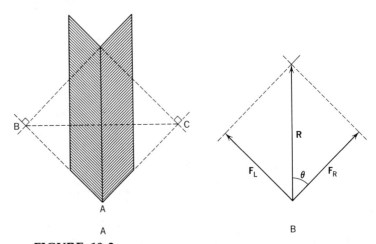

FIGURE 19-3
APPLICATION OF THE RESOLUTION OF FORCES TO
A STYLIZED, FLAT, PINNATE MUSCLE.

of the resultant force. (Alternatively, $\mathbf{R} = 2\mathbf{F} \cos\Theta$.) The value of $\mathbf{R}$ is greatest when the muscle fibers of each side insert on the central tendon at an angle of about 45°. Note, however, that the angle of insertion must change as the muscle shortens.

In order to have the same force of contraction as this pinnate muscle, a strap-like muscle with parallel fibers would need to have the much greater width BC (Figure 19-3A). Such a muscle having the same overall length as the pinnate muscle would have much longer fibers, however, and could move its insertion farther. Clearly, a pinnate muscle can develop great force for its overall width, but has a short contraction distance. Pinnate muscles, unlike muscles with fibers parallel to their tendons, do not become wider during contraction—an advantage if the muscle must function in a confined space. Further, pinnation allows a muscle to have an irregular shape.

The central tendons of some pinnate muscles are ossified; examples are found in the "drumstick" of the turkey. Similar splints of bone are found in the neck and jaw muscles of certain birds and were present in the backs of some large dinosaurs. Among mammals, pinnation is seen in most of the flexors of the limbs and in the mylohyoid. The human deltoid is a complexly pinnate muscle as is the subscapularis of some mammals (see Figure 21-8).

The reader may care to work out for himself the direction of the resultant force when a fan-shaped muscle, such as the pectoralis, trapezius, or latissimus dorsi, contracts as a unit (which it does not necessarily do). If a magnitude is assigned to the force of each of several parts of such a muscle, then the magnitude of the resultant force can also be determined.

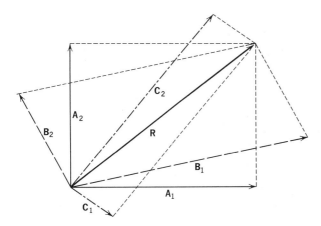

FIGURE 19-4
THREE PAIRS OF VECTORS, A_1 and A_2, B_1
and B_2, C_1 and C_2, ALL HAVING THE
COMMON RESULTANT, R.

Components of Forces. Just as two or more forces can be combined into one resultant force, so a given force can be broken down into two or more components. And just as an infinite series of pairs of forces can have the same resultant force (Figure 19-4), so conversely a given force can have an infinite number of pairs of components. However, it is usually desirable to specify the direction of each desired component, and there is then only one solution.

Consider the vector of the mammalian triceps muscle (one head or all heads in combination) as shown in Figure 19-5A. When the muscle contracts, the ulna turns counterclockwise on the humerus. The insertion of the muscle swings in an arc around the pivot of motion. The radius of the arc is the distance from the pivot to the insertion. At any instant in time the direction of motion of the insertion is in the direction of the tangent to the arc that passes through the point of insertion. (A tangent of a circle touches the circumference of the circle at one point and is perpendicular to a radius drawn to that point. At successive instants in time the tangent will change as the joint moves because the insertion will move along the arc.)

Since the insertion usually does not move in the direction of the pull of the muscle, it is important to learn the magnitude of the part, or component of the pull that *is* in the direction of motion. Two components must be selected such that their resultant is the given force of the triceps. One component will be selected to include all the force in the direction of motion, and, as just explained, this must be along the tangent passing through the insertion. The other component must be selected so as to cause no motion at all, either clockwise or counterclockwise. Only one direction fits this requirement and that

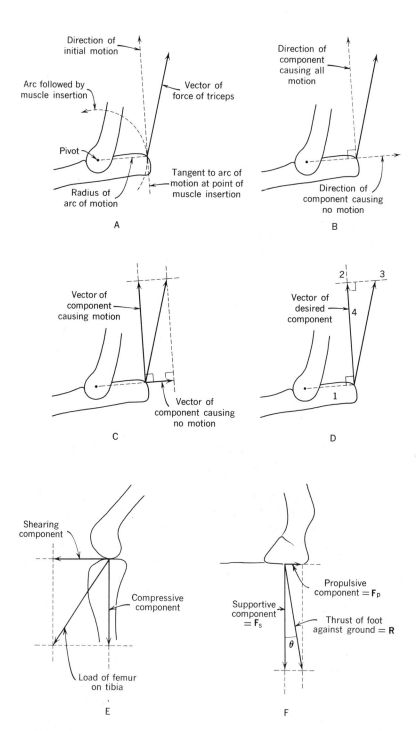

FIGURE 19-5

DETERMINATION OF A DESIRED COMPONENT OF A GIVEN FORCE.

is normal (or perpendicular) to the arc of motion at the point of insertion (see part B). (The vector of this component is an extension of the radius of the arc at the insertion.) Having the directions of the two components, it is simple to derive their magnitudes graphically by completing the parallelogram of forces that has the vector of the force of the triceps as a diagonal. Since the components have been selected to be perpendicular to one another, the parallelogram is a rectangle (part C).

Because the component in the direction of motion is the only one of concern in this instance, and because the desired parallelogram has only right angles, the problem can be solved more directly as follows (part D): (*1*) Draw a line from the estimated pivot of motion to the insertion of the muscle. This is a radius of the arc of motion. (*2*) Draw the line perpendicular to this radius that passes through the point of insertion. This is the tangent giving the direction of the desired component. (*3*) Draw the line perpendicular to the tangent that extends from the tangent to the tip of the vector of the given force. This line is the side of the rectangle of forces that is opposite to (and therefore equal to) the component of force selected to cause no rotation of the ulna on the humerus. (*4*) The vector of the desired component in the direction of motion is now that part of the tangent between the point of insertion of the muscle and the intersection with the perpendicular drawn in the previous step. It only sounds complicated; one can draw the diagram in less time than it takes to read about it.

In Chapter 18, it was noted that a load applied diagonally to the end of a long bone can be converted to a longitudinal compressive component and a transverse shearing component which tends to bend the bone. The directions of the desired components are determined by the nature of the problem and are at right angles to one another. Completion of a rectangle of forces establishes their magnitude (part E).

When the foot of a running animal thrusts against the ground, it both supports and propels the animal. The direction of the supportive component is vertical (in opposition to the pull of gravity) and the direction of the propulsive component is horizontal in the line of travel. If the thrust of the foot at a given instant can be determined, then it is easy to solve for the magnitudes of the components (part F). Similarly, one can calculate the components of an applied force that are necessary for analyzing centrifugal force (Figure 20-18), lateral undulation of snakes (Figure 21-11), friction (Figure 22-4), and the forward and lateral components of the diagonal thrust of the tail of a fish against the water (Figure 23-7).

Components of forces can be determined trigonometrically as well as graphically. In this instance the calculations are more simple and direct than for the resolution of forces because only right triangles need be used. Thus, in Figure 19-5F, $F_p = R \sin\Theta$ and $F_s = R \cos\Theta$.

Note that as the angle between the given force and a component increases from 0 to 90°, the magnitude of the component decreases from that of the given force to zero. In terms of mechanics, muscles are most effective for pivoting bones when they pull in the direction of motion. By inserting onto its central tendon at an angle, a pinnate muscle increases the number of its fibers and hence its force of contraction. However, as the angle of insertion increases, the effective component of the force decreases. The optimum angle (usually a little less than 45°) represents a compromise between these opposing factors.

Bone-Muscle Systems as Machines

A **machine** is a mechanism that transmits force from one place to another, usually also changing its magnitude. Thus, when a screwdriver is used to pry the lid off a can, moderate downward force applied to the handle produces great upward force at the tip of the tool against the lid. Similarly, when the triceps muscle pulls up on the olecranon process, a downward force is produced at the forefoot (Figure 19-6, parts A and B). All bone-muscle systems are machines. It is useful to designate any input force applied to a machine as an **in-force** (F_i) and any output force derived from a machine as an **out-force** (F_o). In the body, in-forces are applied by the pull of tendons or tensed ligaments, by gravity, and by external loads; useful out-forces are ultimately derived at the teeth, feet, digits, and elsewhere. For now, we will consider only simple machines having one in-force and one out-force.

Lever Arms and Torques. An in-force may be transmitted to an out-force by a crankshaft, hydraulic device, pulley, lever, or other mechanism. Most feeding and locomotor systems of the body transmit forces by levers, and only these will be considered here. A **lever** is a rigid structure, such as a crowbar or bone, that transmits forces by turning (or tending to turn) at a pivot. Each force is spaced from the pivot by a segment of the lever called a **lever arm**; the **in-lever arm** (l_i) extends from the in-force to the pivot, and the **out-lever arm** (l_o) extends from the pivot to the out-force. In the example of the screwdriver used to pry open a lid, l_i extends from the hand on the handle to the lip of the can, and l_o extends from the lip to the tip of the tool pressing on the lid. In the other example, l_i is the olecranon process and l_o the forearm from elbow joint to forefoot.

The product of a force times its lever arm is called a turning force, or moment, or **torque** (τ). Every functioning lever includes at least two torques, one for the in-system and one for the out-

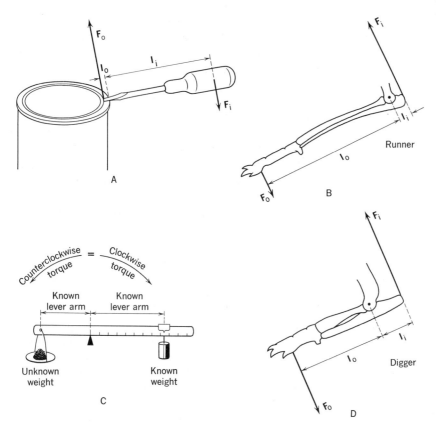

FIGURE 19-6

PRINCIPLES OF IN-FORCES AND OUT-FORCES, LEVER ARMS, AND TORQUE. (In- and out-torques are not adjusted to be in equilibrium.)

system. Thus (as qualified below), $\tau_i = F_i l_i$ and $\tau_o = F_o l_o$. Torques are expressed in gram-centimeters, or equivalent units. When $F_i l_i > F_o l_o$, the lever rotates in the direction of F_i; when $F_i l_i < F_o l_o$, the lever rotates in the direction of F_o; when $F_i l_i = F_o l_o$, the system is in equilibrium and there is no motion.

When a lever system is in equilibrium, any of the variables can, of course, be easily obtained if the other variables are known. This is the principle of a beam balance: The product of a known weight times its lever arm (the calibrated scale) is adjusted until it equals the product of the unknown weight times a fixed lever arm (Figure 19-6C). Likewise, when tetrapods stand, the forces of all postural muscles are adjusted so that in-torques equal out-torques and support is maintained without motion. Other examples are given later in this chapter.

It is important that the student of body mechanics be able to solve the equation $F_i l_i = F_o l_o$ for any of the variables and to understand the relation of each to the others. Thus, if it is desirable for a mammal that digs to produce a large out-force at the forefoot when the triceps contracts, then since $F_o = F_i l_i / l_o$, it is seen that the animal can increase the out-force by increasing F_i or l_i, or by decreasing l_o. The adaptations of many diggers include all of these (compare B with D in Figure 19-6).

Actual versus Effective Forces and Lever Arms. There is one further qualification. The product of force times lever arm is only equal to the torque of the system when the force and lever arm are at right angles to one another. There are two ways to ensure that this condition is met. Each is simple and either may be the easier to apply to a dissection or experimental animal, so each will be presented.

The actual length of the lever arm of an in-force is the straight-line distance from the insertion of the relevant muscle to the pivot of the motion caused by contraction of the muscle. This is called the **actual lever arm** (l_a). As noted in the section on components of forces, this line (a radius of the arc traveled by the insertion as it moves) is at right angles to the effective component of force that is in the direction of motion. If we call the force of the muscle the **actual force** (F_a) and the force of the effective component of the actual force the **effective force** (F_e), then the required condition is met when $\tau = F_e l_a$ (Figure 19-7A). (To indicate that these variables relate to the in-torque, we can write $\tau = F_{ie} l_{ia}$.) Using this method, one must always calculate the effective force before calculating the torque.

The alternative method uses instead the actual force, regardless of the relation of its line of action to that of its effective component. The appropriate, or **effective lever arm** (l_e), is then the perpendicular extending from the line of action of the muscle to the pivot. This perpendicular strikes F_a short of the insertion when the angle between F_a and l_a is acute; it strikes a projection of F_a beyond the insertion when the angle between F_a and l_a is obtuse (contrast B and C of Figure 19-7). Now $\tau = F_a l_e$, and also for any given position of the joint, $F_e l_a = F_a l_e$. The two methods are compared in part D. Can you position the forces so $F_a = F_e$ and $l_a = l_e$?

Relations of In-Force to Out-Force. Thus far my examples have shown the pivot to be located between the in-force and out-force. The torques are then in different directions, one clockwise and one counterclockwise around the pivot. This relationship is common in the body and is usual with extensor muscles (see Figure 19-8A). Two other arrangements are possible, each having the in- and out-force on the same side of the pivot and turning in the same direction: either the out-force can be closer to the pivot than the in-force (a less common arrangement, but two examples are figured), or it can be farther from the pivot than the in-force (a usual arrangement with

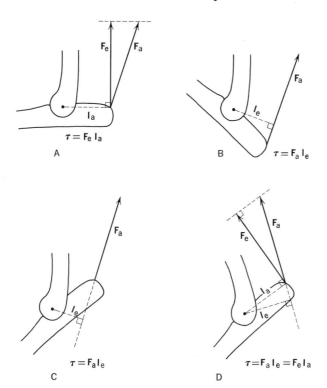

FIGURE 19-7

USE OF ACTUAL AND EFFECTIVE FORCES AND LEVER ARMS IN CALCULATING TORQUE.

flexor muscles as shown by part C). Note that in the first arrangement (or first-order lever), the lever arms are independent and that either can be the longer, though in the body, the out-lever arm is usually much longer. In the other arrangements the longer lever arm (which is the in-lever arm for second-order levers and the out-lever for third-order levers) includes all of the shorter lever arm—they "share" part of the lever.

Figure 19-8 shows that one joint and one muscle can function as different kinds of levers. When one walks in sand, the foot (to one's distress) even functions as two kinds of levers at the same time. The important thing is not to memorize diagrams but to learn to identify the pivot, in- and out-forces, and in- and out-levers in any specific bone-muscle system.

Summation of Torques; Two-Joint Systems. More than one in-force may tend to turn the same lever. In order to determine the net effect, one determines the torque of each independently, adds all that tend to turn the lever in one direction and subtracts the sum of all (if there are such) that tend to turn it

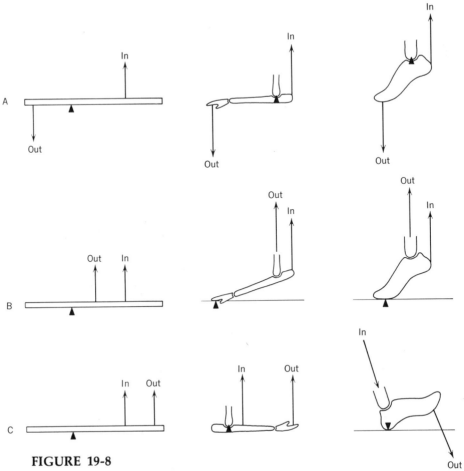

FIGURE 19-8

THREE WAYS THAT IN-FORCES AND OUT-FORCES CAN BE ARRANGED IN LEVER SYSTEMS.

in the other direction. Multiple out-forces, as when several teeth simultaneously crush food, are treated in the same way. In strychnine poisoning all muscles contract, but the limbs are held in rigid extension because in general, extension torques exceed flection torques.

If the weight of a part of a living machine is itself to be considered, as it should be in making an accurate calculation, then in addition to torques resulting from the action of muscles, one must allow for the torque from the pull of gravity on the lever (e.g., forearm or thigh). The lever responds to gravity as though all its weight were concentrated at its center of gravity. The weight of the lever times its effective lever arm, calculated using the center of gravity, is the desired torque. The center of gravity of an irregularly shaped object is difficult to locate by calculation but is easy to locate experimentally if the object can be isolated: The object is suspended (e.g., by a string) successively from two points on its surface. The point where extensions of the lines of support intersect is the center of gravity.

Two-joint muscles, which pass over two diarthroses, are common (e.g., gastrocnemius, gracilis, biceps, long head of triceps). Each can move either joint, both, or neither, depending on the actions of other muscles and loads. In any event, contraction moves, or tends to move, both joints, so the muscle simultaneously provides in-force to two lever arms. Tendons of some digital flexors pass over three or more joints. Such systems are virtually impossible to analyze in detail, either theoretically or experimentally, and even for one instant in time. However, practical approximations of the mechanics of a unit of activity (a digging stroke, a running stride) have been undertaken by monitoring the principal out-forces, learning the approximate strengths and activities of relevant muscles (by electromyography, by preparing length–tension curves, and in other ways), determining usual postures (by various methods including x-ray cinematography), and measuring lever arms. We should not expect anything so wonderfully complex as the moving body to lend itself to simple analysis.

Balance and Counterbalance. The mechanics of the support of the trunk of tetrapods has long been a subject for speculation and analysis. Various kinds of bridges have been proposed as analogs. The curved spines of many small mammals resemble the arch bridge (see Figure 19-9A). This analogy is near the truth (see below) but fails because the spines of many large animals do not arch. The truss bridge, which has a rigid superstructure (part B), fails as an analog if the spine is likened to the roadbed; the spine is in fact under longitudinal compression, whereas such a roadbed is under tension. If instead, the spine is likened to the top profile of a truss bridge and the abdominal musculature is likened to the roadbed, then the distribution of stress would be correct, but the rigid framework of braces would be missing. This suggests the analogy of a closed oblong bag (the trunk with its enclosed coelomic cavity) filled with fluid. Such a bag could be supported at its ends and would have peripheral stresses like those of the truss bridge. However, if pressures were sufficient to prevent sagging, the living bag would not be adequately flexible.

The analogy of the cantilever bridge (part C) has often been cited, since its relevance to body support was favored by D'Arcy Thompson in his widely known book, *On Growth and Form,* originally published in 1917 (see References at the end of this book). The roadbed, which is this time under compression, is equated to the spine, and the top profile of the bridge is equated to the long epaxial muscles. The criss crossing braces

Mechanics of Body Support

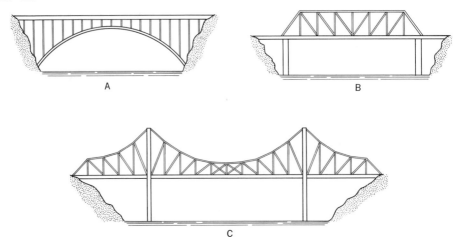

FIGURE 19-9

TYPES OF BRIDGES PROPOSED AS ANALOGS FOR THE SUPPORT OF THE BODY.

of the bridge are equated to the neural spines and short epaxial muscles. It is here, as the Dutch morphologist Slijper and the German morphologist Kummer have pointed out, that the analogy partly fails, because the appropriate structures are not constructed to provide enough rigidity for constant support. It does not follow that no support can be provided in this way when muscular activity contributes tension.

Another characteristic of the cantilever bridge does apply to the support of the body of many tetrapods. The functional unit of a cantilever bridge is from midspan to midspan. During construction, the overhanging cantilevers are built out equally on each side of each tower. (A cantilever is a projecting member that is supported at only one end). Thus, the load on one side is counterbalanced by the load on the other. In different terms, the mass of one cantilever times the distance of its center of mass from the tower is a torque that is equal in magnitude but opposite in direction to the similar torque of the cantilever on the other side of the tower, hence there is equilibrium and the tower does not fall. The principle of balance and counterbalance is widely used by tetrapods (Figure 19-10).

Bows and Bowstrings from Bones and Fibers. Another analogy for the support of the body, proposed in 1798 but neglected until presented by Slijper in 1946, is more nearly valid than any of the bridge analogies. It equates the trunk vertebrae to a bow, which may be either curved like the archer's bow (many small mammals) or nearly straight with the ends bent down like the violinist's bow (salamanders, crocodilians, lizards, many large mammals) (see Figure 19-11, parts A and B). The principal string of this living bow is the ventral abdominal muscula-

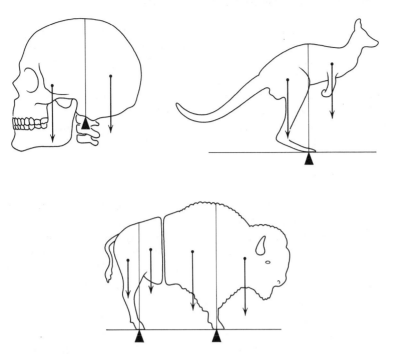

FIGURE 19-10

THE PRINCIPLE OF BALANCE AND COUNTER-BALANCE IN THE SUPPORT OF THE BODY.

Archer's bow

Violinist's bow

A

B

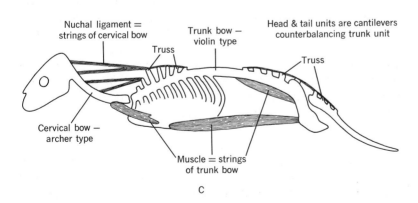

Nuchal ligament =
strings of cervical bow

Truss

Trunk bow —
violin type

Head & tail units are cantilevers
counterbalancing trunk unit

Truss

Cervical bow —
archer type

Muscle = strings
of trunk bow

C

FIGURE 19-11

BOW ANALOGS AND COMBINED FACTORS IN THE SUPPORT OF THE BODY.

ture, scalenus and related muscles, and intervening sternum. The psoas and quadratus muscles under the lumbar vertebrae form a shorter secondary bowstring. The spine cannot sag if tension is maintained in these bowstrings.

The cervical vertebrae of the larger mammals form another bow, this time inverted. The strands of the nuchal ligament form multiple bowstrings. The ligament can be stretched by muscles, thus decreasing the arch of the bow to depress the head.

All the analogs proposed have the shortcoming that they are passive, static systems, whereas the body is active and dynamic. The trunk cannot remain balanced on the limbs as a bridge is balanced on piers; the length and orientation of animal cantilevers are subject to change; the loads on animal bows and bowstrings shift in magnitude and direction.

Further, there has been too much tendency for morphologists to choose among the analogs instead of combining them. The trunk can be regarded as having a bow, the spine, which is prevented from collapsing by muscular bowstrings which, in turn, may be relieved in part by the counterbalance of a trussed cantilever (a heavy, extended tail) and in part by the counterbalance of an inverted-bow cantilever (the neck) loaded at its extremity by the head (Figure 19-11C). We may assume that when a tetrapod stands at rest, this complicated system remains in equilibrium with only slight sustained tension in a few muscles and intermittent corrective tension in some others.

On p. 449 the human foot was likened to a beam loaded at the center. So it is, but it is also a bow, touching the ground at the ball of the foot and the heel, that is prevented from collapsing by muscles and by a bowstring (the plantar aponeurosis) which can be tightened by lifting the toes. On p. 445 the zygomatic arch was likened to a beam, and so it is, but it is often also an arch, firmly anchored at each end, and strengthened by a bony string, the basicranial axis, and sometimes also by the orbital ligament which suspends the arch from the postorbital process above. These interpretations are subject to experimental investigation and are sure to receive further attention.

Stops, Slings, and Locks. The factors contributing to support that have thus far been presented relate primarily to the axial part of the body. Since the legs are not rigid vertical columns but instead are hinged, and usually bent struts, it is important to learn how they can support the body. During periods of activity, muscles provide support as well as motion. During periods of inactivity the animal may avoid a support role for the legs by resting sprawled on its ventral surface (amphibians, reptiles), by crouching (rodents, rabbits), by sitting (primates), or by lying down (carnivores, many artiodactyls). Nevertheless, some large tetrapods stand for long periods; their limbs must then support them without excessive muscular effort. Various factors contribute.

As described in Chapter 18, joints, together with their ligaments, are constructed to limit both the kind and extent of motion that can occur. In large mammals, no joint distal to shoulder and hip (and these to only a limited degree) can flex in the transverse plane; passive support is provided against lateral bending. Also, hyperextension in the sagittal plane is prevented by either ligamentous or bony stops: Elbows, knees, hocks, and digits cannot collapse by "bending backwards."

The tendency to collapse by flection may be reduced or avoided in any of several ways other than muscular contraction. The largest tetrapods (living and extinct) stand (or stood) with limb segments vertically aligned, or nearly so, thus balancing one bone on the next and reducing bending torques (of which more in the next section). Ungulates have more or less cuboidal podials and flat-ended metapodials which balance over one another in rows like a child's building blocks. During flection of the "knee" (corresponding to the human wrist) the squared-off carpals separate by pivoting, hinge-like, on their posterior margins (see vicuna, Figure 20-16). When weighted, this may lift the center of gravity very slightly. If so, a little work must be done to initiate flection, which is thus resisted when the animal stands at rest (this analysis needs verification).

The human knee (and probably some upright limb joints of other animals) can be extended just beyond vertical. When weighted, there is then a slight torque in a direction in which the joint cannot bend further. This prevents bending in the opposite direction and provides the standing body with a lock against flection. Some resistance to flection is also provided by snap joints (see p. 454 for a description). This resistance may help to prevent collapse of a joint that is otherwise nearly in equilibrium.

Ungulates also have a sling mechanism which prevents collapse of the fetlock joint between metapodial (or cannon bone) and proximal phalanx. This joint is always bent when weighted; the angulation depends on the load. It is thus an important shock absorber. Because it is bent, it must be supported. This is accomplished by a remarkable ligamentous sling which extends from the posterior surface of the proximal end of the metapodial, down under the fetlock joint (where it is anchored to sesamoid bones), and then around the proximal phalanx to insert on the anterior surfaces of the distal phalanges (Figure

Mechanics of Support and Movement

19-12). As the foot is heavily loaded by shifting weight when at ease, or by impact when moving, the ligament stretches, thus allowing the joint to flex sharply. In doing so, the ligament stores potential energy which is released by returning the joint to a neutral angulation when the load is diminished (see also Figure 20-17).

Ungulates, unlike elephants, usually have flexed stifle joints (corresponding to the human knee) and moderately flexed hock (or heel) joints. Passive support to relieve extensor muscles is provided to perissodactyls and some artiodactyls by a mechanism which is effective, simple, and subject to voluntary control. The tibia on one side and the almost completely tendinous superficial flexor muscle of the digit (or digits) on the other side form the long arms of a parallelogram which is completed above by the distal end of the femur and below by the cal-

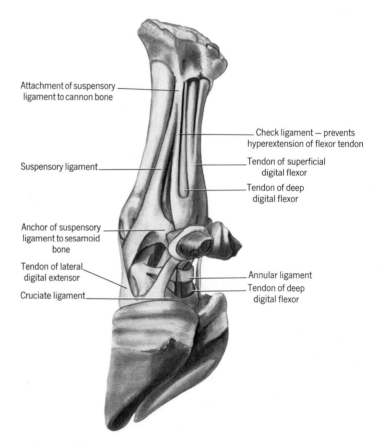

Attachment of suspensory ligament to cannon bone

Check ligament — prevents hyperextension of flexor tendon

Suspensory ligament

Tendon of superficial digital flexor

Tendon of deep digital flexor

Anchor of suspensory ligament to sesamoid bone

Tendon of lateral digital extensor

Annular ligament

Cruciate ligament

Tendon of deep digital flexor

FIGURE 19-12

STRUCTURE OF THE RIGHT FOREFOOT OF THE COW, INCLUDING THE SUSPENSORY MECHANISM. (Drawn from a freeze-dried dissection and hence slightly shrunken.)

caneum and other tarsal bones (Figure 19-13). All angles of this parallelogram must change simultaneously. The stifle and hock cannot flex independently, and if any of the angles is prevented from changing, the entire system remains rigid. This is accomplished by a locking device at the stifle. The ridge flanking the patellar groove on the medial side is enlarged and ends proximally in an eminence. The patella is anchored to the tibia by three strong ligaments which just permit the patella to be pulled (by the quadriceps muscle) behind this eminence when the joint is extended. When the patella is thus seated, the ligaments prevent flection of the joint; the patella must be pulled down into the patellar groove before the limb can be flexed.

These support mechanisms are representative. Numerous others are known (some that relate to specific adaptations will be described in subsequent chapters) and doubtless still more await identification.

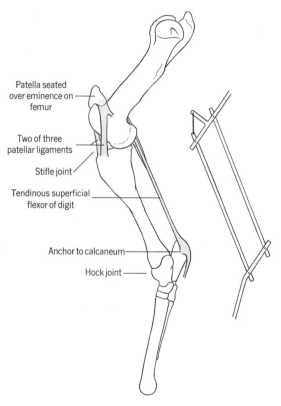

Patella seated over eminence on femur

Two of three patellar ligaments

Stifle joint

Tendinous superficial flexor of digit

Anchor to calcaneum

Hock joint

FIGURE 19-13

SUPPORT MECHANISM OF THE HIND LEG OF MANY UNGULATES shown by a medial view of the right leg of a horse.

Vertebrate Giants. An evolutionary trend toward large body size has been common within vertebrate lineages, and as the body weight of a terrestrial animal approaches 900 kg (2000 lb), marked adaptations for support become necessary. Beasts having such adaptations are said to be **graviportal** (= heavy + to carry).

The selective advantages of large size include (*1*) virtual freedom from predation (except, alas, by man); (*2*) the ability to roam over large areas in search of food, drink, shelter, or breeding grounds; (*3*) the capacity to use and produce energy more slowly than small animals, so that relatively little food is required per unit of body weight and so food of low nutritive quality may be sufficient; and (*4*) a low surface-to-volume ratio and a high capacity for heat production, enabling the larger animals to heat up and cool off slowly.

The largest arthrodire was 9 m (30 ft) long. The largest living fish is the tropical whale shark which reaches about 18 m (60 ft) in length. The basking shark attains 12 m, and one ray has a fin span of 7 m. A Siberian freshwater sturgeon is said to have reached 1360 kg (3000 lb), and the swift swordfish attains a length of 4.5 m and a weight of 454 kg (1000 lb).

Gigantism has not characterized the amphibians, but the largest labyrinthodont was thickset and about 4.5 m long, and a living salamander grows to 2 m. There have been giants in many reptilian lineages: An extinct sea turtle was nearly 4 m long, and a marine lizard (a mosasaur) was 10 m long. Plesiosaurs also reached 9 m in length and were quite bulky. The longest snakes probably reached 12 m. The largest reptiles were archosaurs. A crocodile attained about 13 m, and many dinosaurs of the order Saurischia were gigantic: Bipedal *Tyrannosaurus,* the largest of all carnivores, was 9 m long: *Diplodocus,* with its long neck and tail, was probably the longest land animal at 26.6 m (87.5 ft); *Brachiosaurus,* which was almost as long but stouter, may have been the heaviest at an estimated 63,000 kg (70 tons). The largest bird was the extinct elephant bird which is estimated to have weighed 436 kg (965 lb); the ostrich weighs up to 135 kg.

An extinct wombat was the size of a large rhinoceros. Bull elephant seals have reached 3500 kg, and bull walruses weigh as much as 1260 kg. Four extinct mammalian orders (Amblypoda, Embrithopoda, Notoungulata, Astrapotheria) included bulky land herbivores ranging to 3.3 m in length and having the proportions of rhinoceroses or elephants. The extinct ground sloth was the size of a small elephant, and an armadillo-like edentate (a glyptodont) was 2.8 m long and massive. Steller's sea cow, exterminated by man, is said to have reached 7.5 m in length and 4000 kg in weight.

The hippopotamus rarely attains 4.5 m and 4500 kg (5 tons). A bull giraffe may weigh 1900 kg. The various groups of perissodactyls have all tended to large size: The draft horse, at about 630 kg (1400 lb) was exceeded by the largest of the long-legged, clawed chalicotheres, the massive, horned titanotheres, and the rhinoceroses. The living Indian rhinoceros reaches 4000 kg, and the extinct, hornless *Baluchitherium* was the largest of all land mammals—it stood about 5.3 m (17–18 ft) high at the shoulder! The bull African elephant, largest of living land animals, stands 4 m high and weighs as much as 7000 kg (7.7 tons). Human greed has nearly exterminated the largest of all vertebrates, the blue whale, and few large specimens remain. This magnificent creature reached 23-30 m (75–100 ft) A 27 m specimen weighed 136,400 kg (150 tons)!

Aquatic giants support their bodies effortlessly by floatation. Terrestrial giants reduce their requirements for support by avoiding unnecessary oscillations (the spine is usually stiff), jolting (they rarely jump or gallop), and exertion (the limbs are little flexed in locomotion). Further, they reduce demands on their skeletons by modifying structure and posture so that compressive forces, which can be withstood with minimal tissue, are increased and bending forces are decreased. Limb bones are oriented vertically. Their columnar shafts are straight, and articulatory surfaces are in line with the shafts. The olecranon process is deflected backwards so that the elbow joint can open completely. The heads of the humerus and femur face more nearly upward than for smaller tetrapods, and the acetabulum faces more nearly downward. Even the scapula and ilium tend to be broad and vertically oriented. (Graviportal animals are also said to be rectigrade, which means straight-walking.) Proximal limb segments are long; distal segments are short. Radius and fibula are large and free. The feet are broad. Usually, all five toes are retained, are subequal in length, and radiate over a pad that cushions and evenly distributes the great weight.

These modifications shift part of the support role from muscle to bone. Ligamentous slings and guys, and modifications of muscle mechanics to favor force and reduce velocity, contribute further to economy of effort.

Velocities and Lever Arms. Movement is change of position. Speed is rate of change of position. **Velocity (v)** is speed in a given direction. An animal may be said to be able to swim or run at a stated speed because it can move in any direction;

Mechanics of Motion

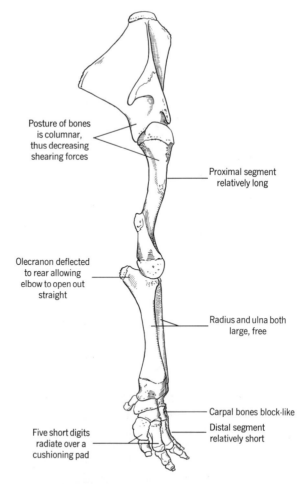

Posture of bones
is columnar,
thus decreasing
shearing forces

Proximal segment
relatively long

Olecranon deflected
to rear allowing
elbow to open out
straight

Radius and ulna both
large, free

Carpal bones block-like

Distal segment
relatively short

Five short digits
radiate over a
cushioning pad

FIGURE 19-14

**SOME GRAVIPORTAL ADAPTATIONS of
the right foreleg of a 12 year old Indian elephant,** *Elephas.*

muscles and bones, however, are said to attain certain velocities because their directions of motion relative to associated
structures are always restricted. Velocity is expressed as centimeters per second, or equivalent units, and direction is stated
or implied. Velocity, like force, is therefore a vector quantity
and can be represented by an arrow (see p. 460).

Just as parallelograms can be drawn to resolve several forces
acting together, so they can be drawn to resolve several constant or instantaneous velocities acting together. When a man
swims straight out from the bank of a river, his velocity relative
to the water and the velocity of the current relative to the bank
combine to move him diagonally downstream at a rate exceeding
either velocity taken alone. Similarly, if the distal end of the
humerus of a running animal is extended on the scapula at the

instantaneous velocity v_1 (Figure 19-15A) as the body is moving relative to the ground at velocity v_2, then **R**, the diagonal of the parallelogram, is the resultant velocity of the distal end of the humerus relative to the ground.

Likewise, a velocity can be broken down into components. As for forces, a useful application establishes components at right angles to one another by constructing a rectangle of velocities as in part B, which shows the glide of a pigmy scaly-tailed squirrel.

It is often desirable to combine forces (or torques) that act independently on the same lever; two muscles can pull harder than one. Velocities, by contrast, cannot be applied independently to different parts of the same lever; disregarding the inertia of the system (which can be an oversimplification—see the load–velocity curve, Figure 19-1), two muscles cannot pull faster than one. The velocity of any point on a lever is determined only by its distance from the pivot and the angular velocity or rate of turning of the lever as a unit. Thus, the relative velocities of different insertions acting together on one bony lever are determined by their positions and not by inherent differences in rate of shortening of associated muscles.

We have learned that the relation of in- and out-forces to their respective lever arms is expressed by the formula $F_i l_i = F_o l_o$. Therefore, $F_i = F_o l_o / l_i$ and $F_o = F_i l_i / l_o$. Out-force increases with the length of the in-lever and decreases with the length of the out-lever: The digger needs a long olecranon and short forearm. It follows from the previous paragraph that **in-** and **out-velocities** (v_i and v_o) are also related to their respective lever

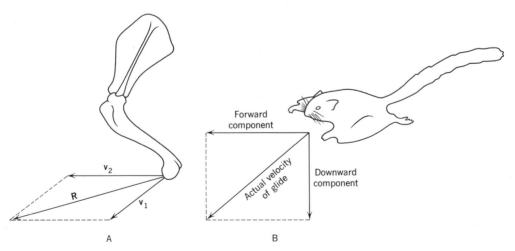

A   B

FIGURE 19-15
THE CALCULATION OF RESULTANTS AND COMPONENTS OF VELOCITIES.

arms, but in the reverse way: $v_i l_o = v_o l_i$, so $v_i = v_o l_i / l_o$ and $v_o = v_i l_o / l_i$. For high velocity at the foot, the olecranon should be short and the forearm long. The same lever cannot enhance both the force and the velocity of the same muscle. The problem of providing an animal with both low and high "gear systems" is discussed on p. 498. (Optimum structure is determined not only by these mechanical factors but also by related physiological factors.)

It is often desirable to know the speed of an animal as a whole. It may also be useful to know the maximum velocity of a part of the body (as of the center of gravity of a jumper at takeoff) or the velocity at a given instant (as of the foot of a runner when it strikes the ground), but the parts of moving animals seldom attain constant velocities and average velocities are not often useful. Therefore, other variables must also be considered.

Mass and Acceleration.   Objects remain at rest or in uniform motion unless acted upon by external forces. (This is Newton's First Law.) They have the inherent tendency to remain at rest or in motion at constant velocity. This tendency is called **inertia**. It is not expressed in numerical terms, but experience tells us that heavy objects have more inertia than light objects. Actually, the critical factor is not weight, which is determined by the pull of gravity, but **mass** (m), which represents quantity of matter and remains the same in space as on earth. The vertebrates are on the earth where mass is measured by weighing, so the difference is not of practical importance, but nevertheless, it is preferable to use the concept of mass.

The capacity of an object moving in a straight line to overcome resistance is called **linear momentum** (M) and is equal to $m v$. Momentum is conserved: When one system loses momentum, another system gains an equal amount. Thus, when a bird flies to a perch, the momentum lost by the bird is gained by the perch, which may cause it to sway.

The relation of mass (m) to force (**F**) and change in velocity per unit time, or acceleration (a), is stated by Newton's Second Law and is simply $F = ma$, when appropriate units are selected. Acceleration is expressed as centimeters per second per second, or equivalent units, and obviously relates to velocity (**v**), time (t), and distance (s). If an object is at rest when acted on, then the basic relationships are $v = at$, $s = vt$, and $Ft = mv$. Other equations can be derived by substitution. If an object is already in motion when acted on, then the equations must be modified somewhat. Derivations and interpretations are found in texts on elementary mechanics.

These relationships show that if an object is to be accelerated or decelerated rapidly (e.g., the body of a bird at takeoff or the tongue of a feeding chameleon), it should be light in order to avoid excessive forces; if an object is heavy (body of an elephant), relatively more time is needed to achieve maximum velocity; greater velocity is attained when a force acts on an object for a relatively long time (a reason for the very long hind legs of the jumping frog and tarsier).

These formulas and examples apply only to rectilinear motion. Since animals and their moving parts often do not travel in straight lines, other factors must now be considered.

Curvilinear and Rotational Motion. When a tetrapod jumps, or momentarily lifts all feet from the ground while running, or folds its wings for an instant in flight, then (disregarding wind resistance) its center of gravity tends to continue to move with the initial direction and velocity (according to Newton's First Law), but is simultaneously accelerated in a different direction by gravity. One component of its motion, the rate of fall, is not constant. Hence, the resultant motion is not rectilinear. While unsupported, the jumping body as a whole moves in a parabolic curve regardless of any motions of its appendages.

When a running animal turns sharply, its velocity also changes even though its speed may remain constant (because velocity has both magnitude and direction). There is then a sideways, or centrifugal, force out of the curve of the turn which must be balanced by an inward centripetal force in order to prevent skidding.

These are examples of curvilinear motion. Such motion relates particularly to jumping, dodging, turning in flight and throwing and will be noted further in subsequent chapters.

Man-made machines that move on land are supported and propelled by wheels. Vertebrates, instead, use moving parts (legs, fins, jaws, spines) that swing back and forth. Such parts may be thought of as segments of wheels which constantly reverse their direction of rotation. This is inefficient because the appendages must constantly be accelerated and decelerated. However, oscillating parts do have the advantage of versatility: the functional length and period of oscillation of a limb can be modified at will to adapt to rough ground or angle of terrain.

Turning wheels and the oscillating levers of moving animals rotate around axes; the mechanics of rotational motion apply. Consider an extended limb that is rapidly pivoting at the hip or shoulder during the supportive phase of a running stride. Its **angular velocity** (ω) is the rate of rotation, or degrees turned in a unit of time. Its **angular acceleration** (α) is the rate of change in angular velocity. Its **moment of inertia** (I), which corresponds to mass in rectilinear motion, is an index of the resistance of the limb to acceleration. This equals the mass (m) of the limb times the square of a constant (k) called the radius of gyration, the value of which depends on the distribution of the mass of the limb. The value is greater for long limbs than short and greater for limbs that are heavy distally than for limbs that are heavy proximally. The turning force, or torque (τ) of the limb is $I\alpha$. **Angular momentum** (L) equals $I\omega$, and like linear momentum it is conserved. By flexing a limb, or spine, or tail so as to concentrate its mass closer to its axis of rotation, k, and hence I, are decreased. Because L does not change, ω increases. Human divers and gymnasts learn to control the spin of their bodies in this way; cats, jerboas, and gibbons do likewise.

Complicating factors make it extremely difficult to apply these, and related formulas, to the quantitative analysis of actual animal movements: Articulated segments of the body commonly rotate around independent axes at the same time that they turn as a unit. Each segment then has its own values for each variable, and the values of the variables for the segments taken together change constantly. Further, one segment may rotate simultaneously around more than one axis (e.g., extension with supination, or flection with adduction).

In general, however, study of the formulas $I = mk^2$, $\tau = I\alpha$, and $L = I\omega$ makes it easy to interpret many morphological adaptations. It is seen that the mass and distribution of mass of an oscillating structure are critical to its function unless the part has little mass (legs and jaws of very small tetrapods), or moves slowly (legs of the sloth), or overcomes external resistance far in excess of resistance offered by its own mass (foreleg of a burrowing mammal). If an oscillating structure moves fast, then the muscles that move it, being heavy, should not be located in the fastest moving part of the structure (distal part of leg or wing) but instead should transmit their forces by levers or long tendons from regions of relatively slow motion (shoulder of antelope, breast of bird). Forces can be reduced and angular velocities increased if the distal parts of oscillating structures are light and slender. Very large animals eliminate such oscillations as can be avoided (e.g., flection and extension of the spine).

Effort, Energy, and Efficiency.   In order for an animal to produce relatively large forces, torques, velocities, momentums, or moments of inertia, it must pay the price of a relatively large expenditure of effort. The effort required of the body is proportional to the sum of the energies produced in its various mechanical systems. Some kinds of energy (electromagnetic, chemical) do not directly concern us here even if they are involved in muscle physiology. The production of heat energy is a

desirable consequence of the locomotion of small homeo-
thermous vertebrates in cold weather, but in general represents
wasted energy. Of direct relevance to locomotion are potential
energy (E_p) and kinetic energy (E_k).

Gravitational potential energy is energy of position. In
climbing a tree, a flying squirrel does work against gravity, thus
gaining potential energy for its glide. Birds gain potential energy as they fly high above the ground. The limbs of running
animals gain potential energy when lifted. Quantitatively,
$E_p = mgh$, where m is mass, g is the pull of gravity, and h is the
height.

Elastic potential energy is exemplified by a loaded spring, a
drawn archer's bow, a stretched ligament, or an inflated lung. It
equals $\frac{1}{2} ks^2$, where k is the spring constant, or restoring force
per unit displacement, of the particular material stretched, and
s is the displacement. Some of the supportive mechanisms
described earlier in this chapter rely on elastic potential energy
to restore equilibrium after a system has been disturbed.

The body expends no effort when potential energy is released. Sometimes the body expends no effort to store potential energy: The soaring bird is lifted by updrafts; the suspensory mechanism of an ungulate's foot may be stretched by an
external load. Often, however, the body must use muscular effort to derive needed potential energy.

Kinetic energy is energy of motion. In rectilinear motion (the
charging rhinoceros and striking falcon), $E_k = \frac{1}{2}mv^2$. In rotational motion (the beating wing, snapping jaw, and swinging
limb), $E_k = \frac{1}{2}I\omega^2$. Since in each instance velocity is squared, it
is clear that high speed in animals (as in automobiles) is costly.
Animal swimmers and runners do not sprint for long.

Energy is never destroyed: The E_p of the flying squirrel at the
top of a tree becomes E_k as it glides down; the E_k of a jumping
salmon becomes E_p as the fish comes to rest in a higher pool.
When a pendulum swings, it changes E_p to E_k on the downstroke and E_k to E_p on the upstroke. Disregarding friction and
air resistance, no new energy need be introduced to keep it
going. Oscillating appendages of moving tetrapods do the same
to a degree—the E_p of lifted limbs and elevated wings is recovered in each cycle of motion—yet much energy is lost to the
environment (largely as frictional heat), so the muscles of the
body must constantly work to maintain the motion. The rate at
which a pendulum naturally swings, and the rate at which it
can be forced, both relate to its length (though the formulas
are different): The greater its length the longer its period. This
is a reason that each man can walk most comfortably at a certain rate, and why long-legged animals tend to walk with long,
slow strides, whereas short-legged animals tend to walk with
short, fast strides. However, the parallel between the pendulum

and the animal limb can be overrated: Limbs are usually moved faster than their natural periods. Their functional lengths, and hence natural periods, are shorter on the recovery stroke than on the support stroke; the relative durations of these strokes must relate to rate of travel, not to the physics of pendulums.

The **efficiency** of a mechanical device is the ratio of the useful energy taken out (usually as force or work that produces potential or kinetic energy) to the energy put in (usually as fuel). The efficiency of an inorganic machine can be exactly measured: For a hydraulic turbine it might by 90%, for a steam engine 20%, and for many engines less. With practice, rhythmic human muscular effort, as measured by an ergometer such as a bicycle-peddling apparatus, has an efficiency of 25% or a little more; the most efficient speed is always less than the maximum speed. However, when considering normal behavior of animals, efficiency is a less exact, though equally important concept. Thus, the cheetah's way of running is more extravagant of energy than that of the blackbuck, yet the cheetah overtakes the blackbuck often enough to survive, and in that (somewhat different) sense is the better runner. The weight of the hydraulic turbine is unimportant, but for flying machines, animal or man-made, the ratio of useful energy output to weight is more important than its ratio to fuel input. The ratio of thrust to weight is greater for the flight muscles of birds than for aircraft engines, which in turn, rate far ahead of steam engines. The skill of the piano player is limited by neuromuscular control, not the conversion of food to mechanical energy. The efficient animal performs the needed activity, however demanding or exacting, with sufficient economy of effort to make the activity advantageous.

Running and Jumping

Animals that travel far, fast, or easily on the ground are said to be **cursorial.** Quadrupedal cursors evolved from walkers, and in general are either predators or medium-to-large-sized herbivores. Animals that jump or hop are said to be **saltatorial.** Saltators are often bipedal, and if the hind legs are used in unison for a succession of jumps, kangaroo-fashion, the gait is called a **ricochet.** Saltators have evolved many times among small vegetarian tetrapods living in relatively open habitats, and among small arboreal vertebrates. Most cursors are at least fair jumpers. Saltators either move easily on the ground or are excellent climbers.

Advantages of Speed and Endurance

Cursorial and saltatorial animals have a number of selective advantages:

(1) Cursors are able to forage over large areas. A pack of African hunting dogs may range over 3800 km^2 (1500 mile^2), and a mountain lion works a circuit some 160 km long.

(2) Cursors can seek new sources of food and water when familiar supplies fail. Africa's big game animals may travel great distances, and individual arctic foxes have wandered 1300 km.

(3) Cursors can take advantage of seasonal variation of climate and food sources. Some herds of caribou migrate 2500 km each year.

(4) Predators run to overtake prey, exploiting superior speed or relay tactics according to their habits.

(5) Prey species run to escape predators. They are commonly about as swift as their pursuers (the latter rely partly on surprise) and may have superior endurance. Small prey species may be master dodgers; when chased, a kangaroo rat bounces in a different direction on nearly every hop. When startled on land, a frog can reach the safety of the water with several jumps.

(6) Animals may leap to clear obstacles or to see over obstructions. The long springy legs of the Peruvian maned wolf are said to enable it to keep mice in view in tall pampas grass. Rabbits make "spy-hops" to check up on pursuers.

(7) Arboreal saltators jump to climb and are able to move from branch to branch with great agility.

Cursorial and Saltatorial Vertebrates

AMPHIBIANS. None is cursorial. Many Anura, particularly semi-aquatic and arboreal species, jump well, although they lack control and endurance. Many species can jump about 1 m. At the annual frog-jumping contest at Angels Camp, Cal., the record for the bullfrog is an average of 1.94 m (6 ft 4⅜ in.) for three consecutive jumps, and W. Rose reports that a South African frog can average 3.28 m (10 ft 9⅛ in.) for three jumps.

REPTILES. Many lizards can run at 8 or more km/hr for short distances. The collared lizard can run 24 km/hr (15 mile/hr). More than two dozen species of lizards are known to be bipedal when running fast. Many thecodont dinosaurs were excellent cursors.

FIGURE 20-1

EXAMPLES OF A SALTATOR (above), TWO BIPEDAL CURSORS (center), AND A QUADRUPEDAL CURSOR (below) showing long hind legs and, for the bipedal cursors, balance of the body over the legs.

Green frog
Rana

Ostrich
Struthio

Basilisk
Basiliscus

Sand lizard
Uma

BIRDS. The pheasant and road-runner are faster than most birds, yet they run only about 25 km/hr. The ostrich has been credited with 80 km/hr (50 mile/hr), which is probably as fast as any biped has ever run.

MARSUPIALS. The marsupial wolf and rabbit-like bandicoots are cursors, and the pouched mice, kangaroos, and wallabies are ricochetal. One kangaroo cleared a 2.7 m (9 ft) fence.

PRIMATES. Man is the most cursorial of primates. Athletes have run 200 m at 37 km/hr (23.2 mile/hr) and 30 km at 19.5 km/hr (12.2 mile/hr). Some lemurs and monkeys make long jumps while climbing. The tarsier, which weighs only about 120 g, is a prodigious leaper, and galagos can leap vertically more than 2 m. Man has high-jumped 2.28 m (7 ft 6½ in.) and long-jumped 8.36 m (29 ft 2½ in.).

RABBITS AND RODENTS. Jack rabbits are able to run 64–72 km/hr, and a leap of one rabbit measured 7.12 m (23 ft 4 in.). Some of the larger rodents (agouti, cavy) are good runners. Kangaroo rats and mice, jerboas, springhares, and some other

Kangaroo
Macropus

Springhare
Pedetes

Kangaroo rat
Dipodomys

FIGURE 20-2

EXAMPLES OF RICOCHETAL MAMMALS shown at takeoff and landing and midway in a very high jump. Note body and leg proportions and use of tail for balance.

rodents are ricochetal except when moving slowly; the habit evolved four or five times within the order.

CARNIVORES. The cats and dogs are cursorial and many other carnivores are moderately so. The whippet runs 55 km/hr, the coyote 69 km/hr, and the red fox 72 km/hr. One fox, running before hounds, covered 240 km in 1½ days. For rewards of food, a terrier climbed, on a treadmill, 1500 vertical m/hr for 17 hr. The cheetah seldom runs more than ½ km, but for such distances is the fastest of animals. This remarkable cat can probably run 110 km/hr (70 mile/hr).

UNGULATES. Some extinct rhinoceroses were slender, long-legged, and probably very fast. The horse has run 0.4 km at 69.7 km/hr (¼ mile at 43.3 mile/hr) and 80 km at 18.2 km/hr (50 mile at 11.3 mile/hr). One Mongolian wild ass is reported to have run 26 km at the surprising rate of 48 km/hr. Several antelopes run at 85–95 km/hr. The pronghorn has been paced with a car at 98 km/hr (61 mile/hr). The camel has traveled 186 km (115 mile) in 12 hr. The impala jumps 2.4 m (8 ft) high. Cursorial ungulates are also found among extinct mammals; the litopterns evolved horse-like form and speed before horses.

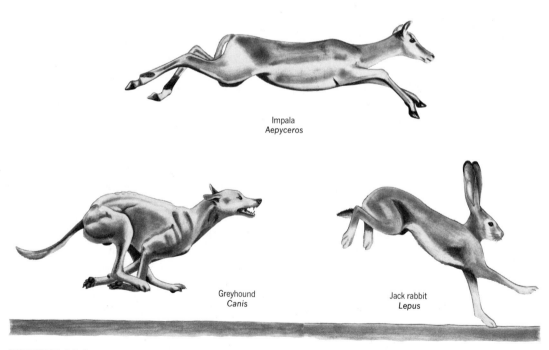

Impala
Aepyceros

Greyhound
Canis

Jack rabbit
Lepus

FIGURE 20-3

EXAMPLES OF MAMMALIAN CURSORS showing flection and extension of spine and legs, passage of hind feet outside forefeet, and varied positions of the shoulder.

General Requirements of Cursors

Saltators and bipedal cursors tend to be more specialized than quadrupedal cursors. Accordingly, we shall start with an analysis of adaptations for running on four legs. In order to run well an animal must (*1*) overcome the inertia of its body to attain speed; (*2*) overcome the movement and inertia of the legs and any other oscillating parts with every reversal in the direction of their motion; (*3*) support the body without benefit of wheels; (*4*) compensate for forces of deceleration, including the action of the ground against the feet as they come down; (*5*) control its course; and (*6*) maintain these functions as long as required.

A full cycle of motion of a running or walking animal is called a stride. Speed equals length of stride times rate of stride. The giraffe emphasizes length, and the wart hog emphasizes rate. Great speed requires both. Endurance is speed sustained through economy of effort. Economy of effort is dependent on the ways that parts of the body are moved and on the magnitude and distribution of masses. For convenience we shall consider these related factors one at a time.

Length of Stride

A galloping race horse covers about 7 m per stride. The faster, but smaller, cheetah covers at least as much distance — about ten times its shoulder height and nearly six times its chest–rump length. How do runners manage such long strides?

Length and Proportions of Legs. The longer the leg, the longer the stride. One might think, therefore, that speed could be increased merely by enlarging the body uniformly in all dimensions. Many cursorial animals are large; however, this must be for other reasons, because enlarging body size also reduces speed in several ways. To contribute to speed, it is necessary to make the legs *relatively* long in relation to the other parts of the body (Figure 20-4).

For reasons explained on p. 484 and noted later, the distal segments of the legs usually lengthen more than the proximal segments. One cannot offer an exact formula, but cursorial ungulates usually have a radius that is as long, or a little longer, than the humerus, and a tibia that is as long, or sometimes markedly longer, than the femur. The foot skeleton, which in walkers and climbers is shorter than the corresponding middle limb segment, is, in cursorial ungulates, about equal to, or even longer than, the middle limb segment. It is the metacarpals or metatarsals that lengthen most (Figure 20-5). The carpals never lengthen, and the tarsals lengthen only in several jumpers (Figure 22-6).

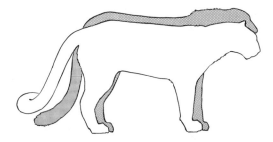

FIGURE 20-4
THE RELATIVELY LONGER LEGS OF THE
CURSOR shown by the very fast cheetah,
Acinonyx, **standing behind the moderately fast**
mountain lion, *Felis.*

(Among birds, cursorial ability is not easily determined from the length or proportions of the legs. All birds evolved from small reptiles that hopped in trees, and the distal segments of the legs are nearly always elongated. The legs are longest among wading birds, not runners. Further, although cursorial lizards have relatively long hind legs, their limb proportions do not follow the general rule stated above. Seemingly, this is because their legs move in eliptical arcs lateral to the body.)

Foot Posture. We have just seen that runners have relatively long leg bones. But it is the *effective* length of the leg — the part that contributes to stride length — that is important, and this can be further increased in other ways. Man's foot, however long, does not contribute to the length of his leg unless he rises on "tip-toes." Bears, opossums, raccoons, and most other vertebrates that walk well but seldom run, have similar feet. Such feet are called **plantigrade** (= sole + walking).

Running dinosaurs, birds, carnivores, and extinct ancestors of hoofed mammals increase effective leg length by standing on what corresponds to the ball of the human foot. These animals are **digitigrade** (= finger + walking).

Perissodactyls, artiodactyls, and the cursorial representatives of several extinct orders of mammals have, like the ballet dancer, further increased effective leg length by standing on the tips of the digits. This foot posture is **unguligrade** (= hoof + walking), which is why these animals are called ungulates.

Where foot posture and limb proportions are each modified for the cursorial habit, the enhanced length and slenderness of the leg skeleton is striking (Figure 20-5).

Role of the Shoulder. The effective length of the forelimb of many runners is increased further by altering the structure and

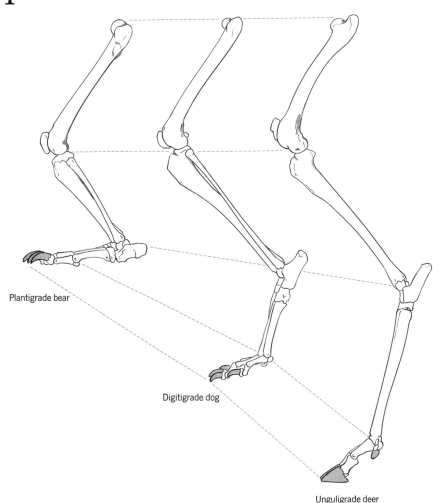

Plantigrade bear

Digitigrade dog

Unguligrade deer

FIGURE 20-5

CONTRAST IN PROPORTIONS AND FOOT POSTURE in the left hind leg skeleton of a noncursor (left), moderate cursor (center), and highly specialized cursor (right).

function of the shoulder. In amphibians, reptiles, birds, and some mammals, the position of the shoulder joint is virtually immobilized by the clavicle and coracoid (if present) which run like struts from the sternum to the scapula. In most mammals, however, the scapula is free to move somewhat, and cursors increase this freedom by (1) reducing the clavicle to a vestige (carnivores) or abandoning it altogether (ungulates), and (2) reorienting the scapula so that it lies not flat against the back of a broad chest (as in man) but flat against the side of a deep, narrow chest, where it is free to rotate in the same plane in which the leg swings. The shoulder joint then moves in the sagittal plane, which is the equivalent of lengthening the leg by moving its pivot from the shoulder joint to a point part way up the scapula (Figure 20-6). Inability of the cursorial dinosaurs to use the shoulder in this way may have contributed to their "preference" for the bipedal habit.

Role of the Spine. Cursorial lizards undulate the spine in the horizontal plane as they run; the pivoting of each girdle is timed to help advance the related unweighted leg and to help swing the weighted leg to the rear.

The smaller quadrupeds among mammals—particularly carnivores—have their legs positioned under the body instead of to the side, and consequently undulate the spine in the vertical plane; the back advances like the body of a measuring worm at the same time that the legs are swinging back and forth. The body of the animal is longer when the back is extended than when it is flexed. Were the animal to extend its back while its body is suspended in the air, the hindquarters would move backward as the forequarters move forward, and the center of gravity of the body would not be affected (an expression of Newton's Third Law). However, the galloping animal extends its back only when its hind feet are on the ground. Muscles of the legs pressing on the ground (and friction at the ground)

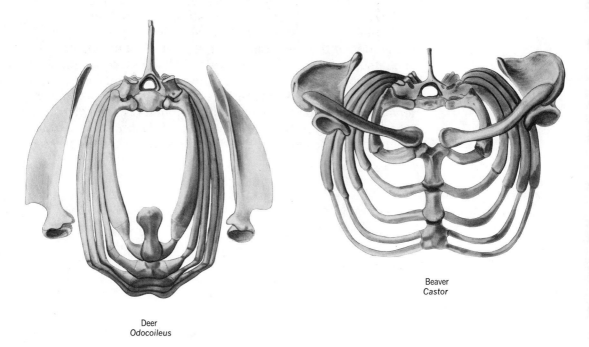

Beaver
Castor

Deer
Odocoileus

FIGURE 20-6

CONTRAST BETWEEN THE SHAPE OF THORAX, POSITION OF SCAPULAS, AND DEVELOPMENT OF CLAVICLES IN A CURSOR (left) AND A NONCURSOR (right). Only the first five thoracic segments are shown.

prevent the hindquarters from moving backward (decelerating), so all of the increase in body length is added to stride length. Similarly, the forelimbs prevent, or at least reduce, deceleration of the forequarters as the back is flexed; therefore, the shortening of the body is also added to stride length. The cheetah is so adept at this maneuver that it could theoretically run nearly 10 km/hr without any legs at all! (Flection and extension of the spine, and its timing relative to the thrust of the legs, also assures that the speed of the animal, i.e., of its center of gravity, is slightly greater than the speed of the associated girdle during the time that a pair of legs is propelling the body.)

The extra rotation of the hip and shoulder girdles that is added by flection and extension of the spine increases the swing of the legs proper so that the limbs reach out farther, front and back, and strike and leave the ground at more acute angles than they would if the spine were held rigid. Again, this increases stride length.

Unsupported Intervals. There is another important way to lengthen stride. Running gaits usually include periods of suspension when all feet are off the ground. The distance the body moves forward while it is unsupported is added to the length of the step to give the length of the stride. The bear (which runs fairly well, but is hardly cursorial) scarcely gets all its feet off the ground at the same time when it runs. Most ungulates have one unsupported period in each stride. The galloping canid has two unsupported periods, one when the body is flexed and another when it is extended. The proportion of the stride interval during which all feet are off the ground is nearly twice as great for the cheetah as for the horse. Bipedal runners are also unsupported for much of the duration of the stride. The hop-

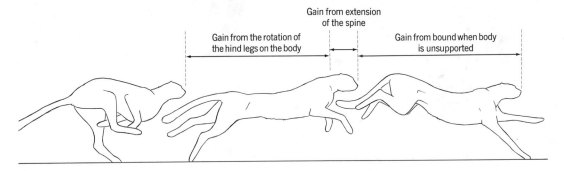

FIGURE 20-7
SOURCES OF THE LENGTH OF STRIDE OF A FAST-RUNNING CHEETAH shown for half of a cycle. Each factor is repeated in the other half cycle except that flection of the body substitutes for extension.

ping African springhare may have its feet off the ground 85% of the time!

Muscle Mechanics: The Most for the Least. Muscles can move the joints through wider angles when they insert close to the joints than when they insert farther away. Thus, in Figure 20-8, muscle A moves the foot only over distance a, whereas muscle B, contracting an equal distance, moves the foot over the greater distance b. We shall see that this arrangement also benefits the rate of the stride, so it is characteristic of cursorial animals to have limb muscles that insert close to joints.

The short-legged mouse can travel faster than the long-legged **Rate of Stride** frog. Speed is the product of length of stride times rate of

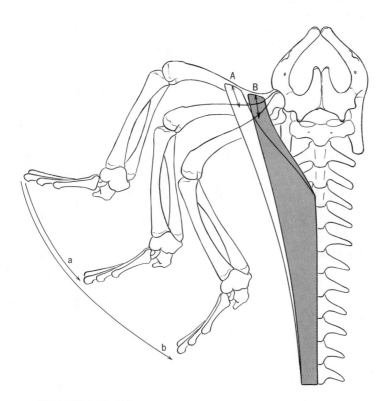

FIGURE 20-8
EFFECT OF DISTANCE OF MUSCLE INSERTION FROM THE JOINT TURNED ON THE RATE AND DEGREE OF ROTATION. The drawing shows in ventral view the caudofemoralis muscle, base of tail, pelvis, and right leg skeleton of the tegu lizard, *Tupinambis.* **(Position B approximates the actual condition. A small insertion of the same muscle onto the knee is omitted.)**

stride, and the mouse takes many strides for each plop of the frog. Fast runners must take long strides but take them rapidly. The galloping race horse completes about 2¼ strides/sec, and the fast-running cheetah completes about 3½ strides/sec. How are cursorial animals adapted to take rapid strides?

Rate of Muscle Contraction. As an animal gathers speed, it increases the rate of its stride by causing its muscles to contract faster. One might infer that cursorial animals as a group would have evolved the ability to contract their muscles faster than other animals. It is true that there is some variation in this ability among species and individuals, but for various reasons the muscles of large vertebrates tend to contract slower than equivalent muscles of small vertebrates. Earlier in this chapter it was noted that the larger the animal, the longer its legs and stride, and hence the faster it might be expected to run. Now we see that longer stride and slower stride tend to cancel one another as body size increases, so that speed remains about the same. We must look elsewhere for ways to increase rate of stride.

Muscle Mechanics: High "Gear Ratio." High gear is used when a car is driven fast: The car is made to move rapidly in relation to engine speed. Cursors have evolved higher "gears" than most other animals. It was noted above that muscles can move joints through wider angles when they insert close to joints than when they insert farther away. Following principles presented on p. 482, $v_o = v_i l_o / l_i$, and therefore out-velocities are also increased by moving insertions close to joints (i.e., by reducing the value of l_i). Thus, it was shown in Figure 20-8, that muscle B, contracting at the same rate as muscle A, and having equal loading, moves the foot through the distance b in the same time interval required for muscle A to move the foot through the shorter distance a. Because it inserts closer to the joint, muscle B can move the foot a greater distance in the same time or an equal distance in less time. The muscle also gains a physiological advantage by doing the job without shortening as much.

The ratio l_o / l_i for limb muscles of cursors and saltators tends to be larger than for corresponding muscles of unspecialized tetrapods, and much larger than for those of diggers and swimmers. "Larger" and "much larger" are useful terms only when speaking in generalities; with specific bones at hand for analysis, more specific values are required. Figure 20-9 shows

FIGURE 20-9

CONTRAST BETWEEN THREE BONE-MUSCLE SYSTEMS OF MAMMALIAN CURSORS (left) AND NONCURSORS OF THE SAME ORDERS (right). The ratio of the in-lever (l_i) to the out-lever (l_o) is greater for cursors.

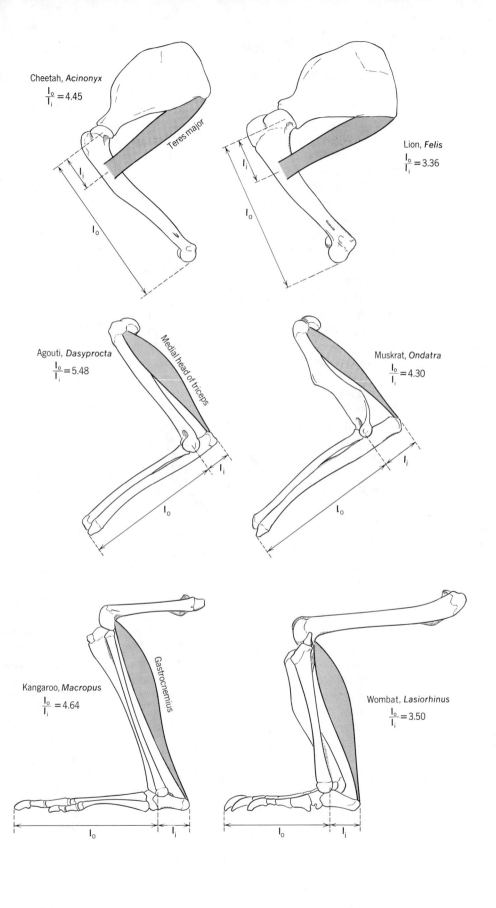

Cheetah, *Acinonyx*
$$\frac{l_o}{l_i} = 4.45$$

Teres major

Lion, *Felis*
$$\frac{l_o}{l_i} = 3.36$$

Agouti, *Dasyprocta*
$$\frac{l_o}{l_i} = 5.48$$

Medial head of triceps

Muskrat, *Ondatra*
$$\frac{l_o}{l_i} = 4.30$$

Kangaroo, *Macropus*
$$\frac{l_o}{l_i} = 4.64$$

Gastrocnemius

Wombat, *Lasiorhinus*
$$\frac{l_o}{l_i} = 3.50$$

characteristic differences between cursors and noncursors in regard to representative bone-muscle systems.

(Since different kinds of animals emphasize different mechanisms to attain similar ends, it is preferable, when assessing cursorial ability, to consider the mechanical advantages of several muscles, or to compare the same leverages in related animals, as in Figure 20-9: for the teres major muscle, the swift horse has a lower l_o/l_i value than the slow porcupine, but for the gastrocnemius muscle it has a higher value than even the kangaroo.)

As was noted on p. 482, there is a reciprocal relationship between the velocity and the force that a given muscle can produce through a series of skeletal levers. Like the car, which in high gear can move fast but is correspondingly handicapped in climbing grades, the cursorial animal can add further speed only by sacrificing power (work per unit of time). However, although runners do not need great power, like race cars they retain some relatively low gears. For example, Figure 20-10 shows that the semimembranosus, with its relatively long effective lever arm, is a low-gear system relative to the middle gluteus muscle system which has the same action but a much shorter effective lever arm. The animal may use the lower gear to initiate the swing of the femur, and then, like a car driver, shift to the higher gear to speed the action. Finally, cursors have reduced requirements for force by reducing the loads on their muscles. We will come to that shortly, but first, there is another way of increasing rate of stride.

Summation of Velocities.   Having achieved leverage that is optimum for speed, there is little further that the musculature, acting on a single joint, can do to increase the velocity of the particular action it controls. If these muscles become larger, or if additional muscles are added to help them, force is increased, but the velocity of the action remains about the same. (Several men can lift more weight together than one can lift alone, but several equally skilled sprinters cannot run faster together than one of them alone.) But if different limb muscles simultaneously move different joints in the same direction to achieve a greater total motion, the independent velocities they produce are added to derive the total velocity at the foot. (When a man walks down an escalator, his motion in relation to the steps and their motion in relation to the building are added to give his rate of advance in relation to the building.) The trick is to move as many joints as possible in the same direction at the same time without interfering with the support role of the limbs. We have already seen that cursorial vertebrates add an extra pivot to the limbs by abandoning the flat-footed, plantigrade foot posture in favor of digitigrade or unguligrade pos-

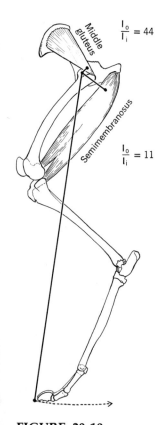

$\dfrac{l_o}{l_i} = 44$

$\dfrac{l_o}{l_i} = 11$

Middle gluteus

Semimembranosus

FIGURE 20-10

HIGH-GEAR AND LOW-GEAR MUSCLE SYSTEMS shown by a lateral view of the left hind leg of the vicuna, *Vicugna*. The muscles are diagrammatic.

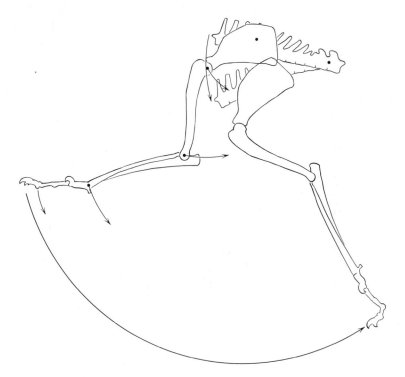

FIGURE 20-11

PRINCIPLE OF THE SUMMATION OF INDEPENDENT VELOCITIES shown by the travel of the leading forefoot of a running cheetah from the initiation of the backswing to the instant the foot leaves the ground. Arcs show approximate amount of rotation around the respective pivots (indicated by spots), not relative velocities.

ture. Further, their scapulas can rotate through 20–25° on the chest. Finally, cursorial carnivores time the flexing of the spine in such a way that the chest and pelvis are always rotating in the direction of the swinging limbs. The net benefit to speed is considerable.

The relationship between body size and the requirements for strength of body framework was presented on p. 459, where it was shown that if body size were increased without altering body proportions, then the load on the locomotor system would increase faster than its capacity to provide support and power. This principle is significant to the analysis of structure of large cursorial vertebrates. It explains why elephants (and probably dinosaurs — in spite of speculation to the contrary) cannot gallop or jump, why some small runners, such as foxes, can travel as fast as race horses

Mass, Endurance, and Design for Economy of Effort

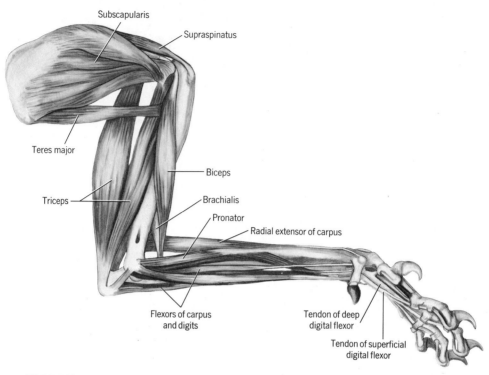

FIGURE 20-12

**MUSCLES OF THE MEDIAL SIDE OF THE FORELIMB OF THE FASTEST TE-
TRAPOD, THE CHEETAH,** *Acinonyx.* **(Drawn from an air-dried dissection and hence
somewhat shrunken.)**

without having marked structural adaptations for speed, why
no saltatorial animal is as large as typical ungulates, and why
the larger cursors must have great adaptation for speed and
endurance in order to run at all. The adaptations of the larger
runners must include not only many already discussed, but
others that reduce the loads on locomotor structures, providing
economy of effort. The evolutionary process has been so effective at fulfilling these requirements that some large animals
can both move swiftly, and run at somewhat less than maximum rates for long distances. What are the elements of design
for economy of effort?

First, many oscillating motions are reduced or eliminated.
The legs must swing back and forth—no help for that—but the
feet are not lifted so high, the back is relatively stiff, and the
shoulders and pelvis are not bounced up and down so much.
The gallop, which has one or two periods when the body is unsupported, is sometimes abandoned in favor of the smoother,
steadier trot or running walk. Unsupported periods can only
follow vertical acceleration, and this is costly in energy.

Next, the mass of the limbs is reduced in several ways. Adductors and abductors of the legs are reduced or are converted to move the limbs in the direction of travel. Muscles that manipulate the digits, rotate the forearm, or twist the feet inward or outward are also reduced or even eliminated. The ulna and fibula, which in other animals take part in foot-twisting and rotation mechanisms, are reduced as these functions are lost. The ulna must be retained where it completes the elbow joint, but distally may become a sliver which fuses to the radius. The fibula may become an attenuated splint (even in small mammalian cursors), and in artiodactyls is sometimes represented only by a nubbin of bone at the ankle.

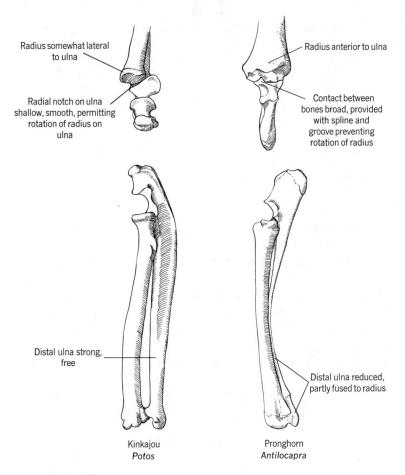

Radius somewhat lateral to ulna

Radial notch on ulna shallow, smooth, permitting rotation of radius on ulna

Radius anterior to ulna

Contact between bones broad, provided with spline and groove preventing rotation of radius

Distal ulna strong, free

Distal ulna reduced, partly fused to radius

Kinkajou
Potos

Pronghorn
Antilocapra

FIGURE 20-13

CONTRAST BETWEEN THE FOREARM SKELETONS OF A NONCURSOR (left) AND A CURSOR (right) shown by the left radius and ulna in lateral view (below) and the same bones at the elbow seen from above and behind (above).

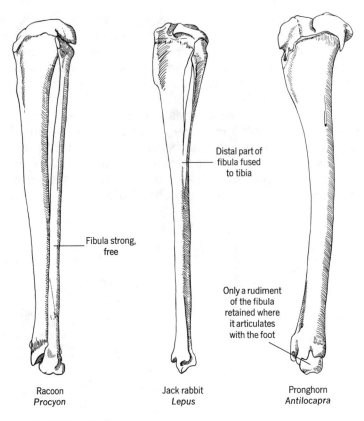

Distal part of
fibula fused
to tibia

Fibula strong,
free

Only a rudiment
of the fibula
retained where
it articulates
with the foot

Racoon
Procyon

Jack rabbit
Lepus

Pronghorn
Antilocapra

FIGURE 20-14
CONTRAST BETWEEN THE FIBULAS OF A NON-CURSOR (left) AND TWO CURSORS (center and right)
shown by the left leg in lateral view.

Further, since the forces that must be developed in os-cillating systems with every reversal in the direction of their ac-tion are also in proportion to the square of their angular veloc-ities (see again p. 485), the loads on muscles causing such motions can be reduced not only by reducing the masses of the systems but also by reducing their velocities. This is a principal reason why it is the lower limb segments that lengthen in the evolution of the long legs of the runner: When devoid of the muscles and bones related to twisting, rotating, and digit manipulation, those segments are relatively light; fleshy parts of the limbs are kept close to the body, where they do not move so far, and hence not so fast, as the more distal segments.

Other elements in design for economy of effort provide fur-ther saving of weight without loss of strength. The feet of un-specialized vertebrates tend to be broad and pliable. Their metapodials are rounded in cross-section and are well-sep-arated. Some runners (some dinosaurs, carnivores) provide more strength by crowding these bones together into a compact unit; each bone becomes somewhat square in crosssection. Still more strength can be provided with the same weight (or the same strength with less weight) if the skeletal material is distributed among fewer bones; hence, some of the best runners and jumpers (kangaroo, jerboa, perissodactyls, artiodactyls), and particularly the large ones, tend to lose the lateral toes and to fuse the basal elements of the remaining toes into a single bone of compound origin. This process has produced the cannon bone of ungulates and, in response to a hopping habit, the tarsometatarsus bone in the ancestors of all birds. The result is a slender, light, strong foot (Figures 1-4 and 20-15).

To compensate for the bracing lost as bones and muscles of the lower limbs are reduced or eliminated, and to guard against dislocations, cursorial vertebrates have evolved joints modified to function as hinges, allowing motion only in the line of travel. This has been done by introducing (or enlarging) interlocking

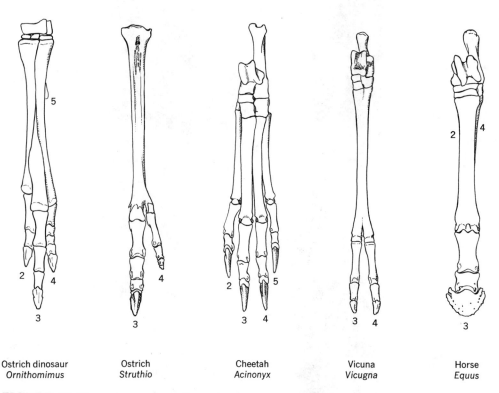

Ostrich dinosaur	Ostrich	Cheetah	Vicuna	Horse
Ornithomimus	*Struthio*	*Acinonyx*	*Vicugna*	*Equus*

FIGURE 20-15

LENGTHENING, COMPACTION, AND FUSION OF METATARSALS, LOSS OF LATERAL DIGITS, AND SQUARING OR FUSION OF TARSALS in the left hind foot of selected cursors of three classes. Digits are numbered.

Jerboa
Jaculus

Flying squirrel
Petaurista

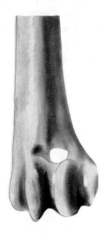

Jack rabbit
Lepus

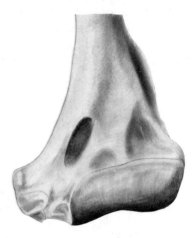

Kinkajou
Potos

Vicuna
Vicugna

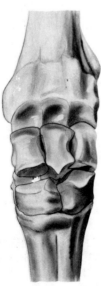

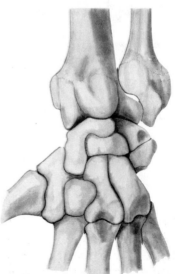

Slow loris
Nycticebus

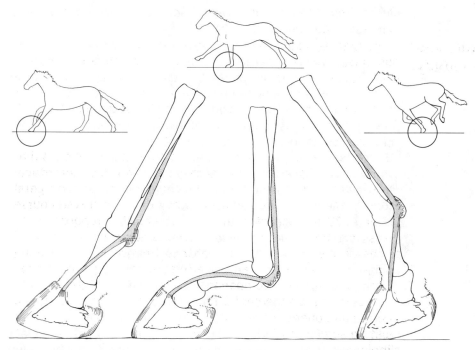

FIGURE 20-17

SPRINGING ACTION OF THE PRINCIPAL SUSPENSORY LIGAMENT IN THE FOOT OF THE HORSE.

splines and grooves in the joints (digits, ankle or hock, elbow) or by substituting flat or cylindrical shapes for spherical shapes at the articulations (wrist and, to lesser extent, shoulder).

Finally, hoofed mammals have evolved still another mechanism for relieving the muscles. When the foot of the running animal strikes the ground, the impact bends the joint between the phalanges and the metapodial bones (fetlock joint). This stretches the ligaments which are called suspensory, or springing, ligaments (Figure 19-12). Since ligaments are elastic, the energy of deformation is recovered as the system is unloaded, thereby relieving muscles in straightening the joint and giving an initial upward impetus to the entire body (Figure 20-17). Nearly 700 kg are required to break the major springing ligament in the front leg of the horse. The evolution of these ligaments has been traced in horses from the scars they left on fossil bones. They evolved from foot muscles

FIGURE 20-16

CONTRAST BETWEEN SELECTED JOINTS OF SALTATORS AND CURSORS (left) AND CLIMBERS (right) showing that the former have a deeper patellar groove at the distal end of the femur (above), more marked trochlea and grooves at the distal end of the humerus (center), and more block-like carpals at the wrist (below). Each drawing is an anterior view of the left side of the body.

Stability and Maneuverability

at the time the animals were becoming large and fast residents of the open plains.

Large grazers and browsers that are relatively free from predation need not be particularly agile. Water buffalos, camels, giraffes, and rhinoceroses cannot start, wheel, dodge, and stop quickly. Small antelopes and certain carnivores, on the other hand, can maneuver their bodies with almost unbelievable skill. A fairly light and supple body, alertness, and rapid neuromuscular coordination are primary requisites.

A consequence of the mechanics of curvilinear motion is that when a moving object turns, there is an outward, or centrifugal force which must be opposed by an equal inward, or centripetal force if the object is to avoid skidding out of its curved course (Figure 20-18). Centrifugal force is directly proportional to mass and the square of velocity, and is inversely proportional to the radius of the turn. In order to turn sharply, a running animal must lean into the turn, adjusting the angle of its body so that the outward component of the thrust of its feet onto the ground equals the centrifugal force. This is opposed by the inward component of the force of the ground on the feet. There must be sufficient friction between feet and ground to prevent slipping—a requirement easily met by hoof and claw on rough ground, though a turning mouse may slip badly on a polished floor. Some animals can turn in several body lengths, even when running fast. Their bodies may make an angle of only 25–30° with the ground as they spin around.

Any force, F, that acts against the side of an animal (e.g., centrifugal force, wind pressure, or the jostling of another animal) tends to upset the animal with a torque of Fh, where h is the height of its center of gravity above the ground (Figure 20-19). In order for stability to be maintained, Fh must be less than Mw, where M is the mass of the animal and w is half the width of the animal's stance. It follows that an animal that stands or moves straight ahead (keeping feet of both sides of the body on the ground at once) is most stable if it is broad and short-legged like the hippopotamus. Cursors must be long-legged, however, and often must lift both right or both left feet at once to move fast. Further, as the previous paragraph explained, they must lean into turns when running. This is difficult for a highly stable animal. It follows that stability must be sacrificed for maneuverability. Agile cursors control lateral forces actively by adjusting posture rather than passively by virtue of body proportions.

Further, in order to turn sharply, a galloping mammal must lead with its inside front leg. That is, the foot toward the inside of the turn must strike the ground after its opposite in each couplet of footfalls. This provides that successive footfalls are

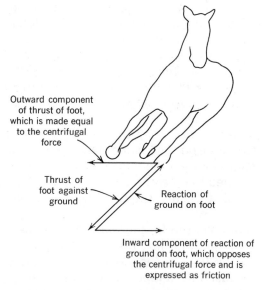

Outward component
of thrust of foot,
which is made equal
to the centrifugal
force

Thrust of
foot against
ground

Reaction of
ground on foot

Inward component of reaction of
ground on foot, which opposes
the centrifugal force and is
expressed as friction

FIGURE 20-18
FORCES RELATED TO STABILITY WHEN TURNING.

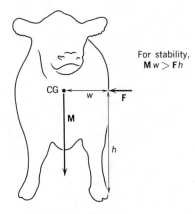

For stability,
Mw > **F**h

CG

w

F

M

h

FIGURE 20-19
SOME FACTORS IN-VOLVED IN STABILITY.
(The center of gravity is represented by CG.)

in the direction of the turn and the animal can balance over its support. The fast carnivores, unlike ungulates, can change lead immediately before the forefeet are brought down. The hapless antelope cannot evade the cheetah's dash by dodging, for the cheetah follows almost in its tracks.

Gaits   A regularly repeating sequence and manner of moving the legs in walking or running is called a gait. Gait selection relates to rate of travel and to the size and structure of the body. If the footfalls of a pair of legs, fore or hind, are evenly spaced in time, as in pacing, walking, and trotting, the gait is said to be symmetrical. In walking gaits, each foot is on the ground more than half the time; in running gaits, each foot is on the ground less than half the time.

In the pace, the two feet on the same side of the body swing more or less in unison. The camel family and some dogs pace naturally when moving moderately fast, and some harness horses are trained to race at this gait (Figure 20-20). All pacers are long-legged; the gait would be unstable for short-legged animals.

At the usual walk, the four footfalls are independent; a forefoot strikes the ground next after the hind foot on the same side of the body (lateral sequence) in most tetrapods except primates, because interference between forefeet and hind feet is then avoided. The same gait can be done at the run but is unusual except for certain show horses. Gaits having independent footfalls provide the most stability.

In the trot, which like the pace is usually performed at moderate speed by mammals, the two feet on opposite sides of the body swing more or less in unison. Since a line between the supporting feet passes close under the center of gravity, the gait is favorable for animals with broad bodies or, as for lizards, with legs splayed to the side of the body.

Most primates, and several other tetrapods do a diagonal sequence walk; a forefoot strikes the ground next after the hind foot on the opposite side of the body. The hind foot passes to the side of the forefoot to avoid interference. Some lemurs and small artiodactyls do the same gait at the run.

Galloping and bounding gaits are said to be asymmetrical because the footfalls of the two feet of a pair are unevenly spaced in time (Figure 20-21). The foot of a pair which strikes the ground first in each couplet of footfalls is called the trailing foot; the other is the leading foot. As noted earlier in this chapter, asymmetrical gaits increase length of stride by introducing periods of suspension when all feet are off the ground. The suspension may come when the legs are gathered under the animal (horses), stretched out fore and hind (deer), or both (cheetah, pronghorn). Rabbits, weasels, and many other mammals of comparable size place the forefeet on the ground alternately, but place the hind feet on the ground more of less in unison. This gait is the half bound. Many squirrels, mice, and other small mammals place each pair of feet in unison, thus doing the bound. Several artiodactyls may place all four feet in unison—a gait called the pronk. In the half bound and bound each hind foot must be swung forward lateral to its corre-

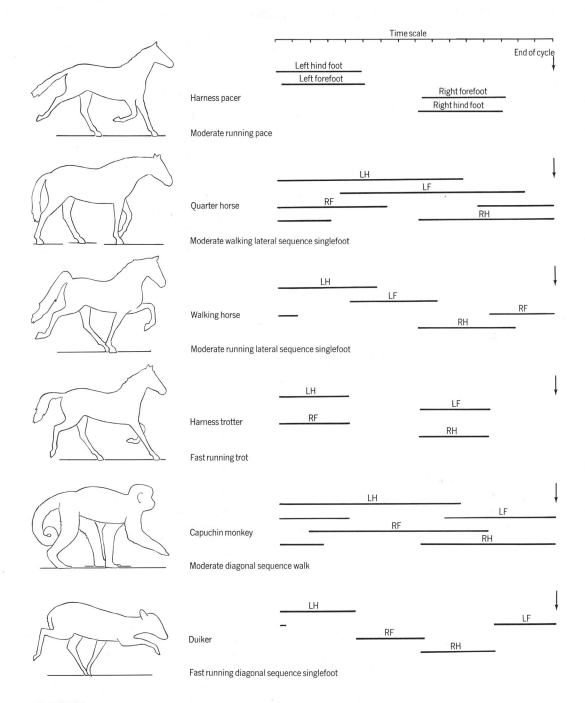

FIGURE 20-20

REPRESENTATIVE SYMMETRICAL GAITS OF TETRAPODS. All drawings show position of the body the instant the left hind foot strikes the ground. In moving from top to bottom of the page the two forefeet rotate counterclockwise relative to the hind feet. The gait diagrams show by the length of the lines the duration of contact of the respective feet with the ground. Each diagram shows one complete cycle starting with the instant the left hind foot strikes the ground.

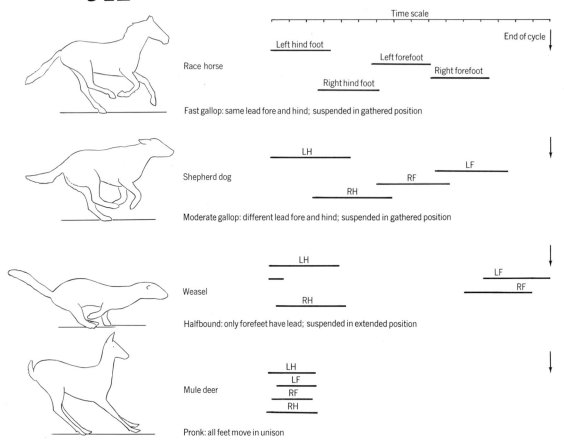

FIGURE 20-21
**REPRESENTATIVE ASYMMETRICAL GAITS OF MAMMALS. All drawings show
position of the body the instant the left (and trailing) hind foot strikes the ground. Notations for the gait diagrams are the same as for Figure 20-20.**

sponding forefoot to avoid interference; i.e., the hind feet have
a wider track than the forefeet.

**Saltators and
Bipedal Cursors**   Ricochetal mammals can accelerate faster from rest and can
alter both the speed and direction of their motion faster than
their quadrupedal relatives. These are important escape mechanisms which are purchased at the price of efficiency. Since
the body must repeatedly be lifted against gravity, much energy is required. Saltators do not maintain fast progression for
long.

The height (h) to which an animal can jump is limited only by
its takeoff velocity ($\mathbf{v}$): $h = \mathbf{v}^2/2g$, where g is the pull of gravity. A
250 g galago has been accurately observed to jump vertically
2.26 m (7 ft 4¾ in.) from a crouch. Since its center of gravity
was lifted more than 2 m, this performance is remarkable and

must be close to a record for any animal, although a rat kangaroo, in jumping 2.4 m high, did about as well. The human high-jumper runs up to the takeoff in order to translate horizontal velocity to vertical velocity and yet lifts his center of gravity only a little more than 1 m. The takeoff velocity needed to lift the center of gravity 2 m is 625 cm/sec; that to lift it 1 m is 442 cm/sec.

The range (R) of an animal's jump depends on takeoff velocity and the angle of takeoff (θ): $R = (\mathbf{v}^2\ 2 \sin \theta)/g$. Theoretically, maximum range is attained when θ is 45°, which is about the angle adopted by frogs and galagos for long jumps. The vertical jump of the galago as described above is the equivalent of a standing long-jump of about $4\frac{1}{4}$ m. The human long-jumper runs up to the takeoff, and because at maximum speed he cannot change the direction of his motion enough to take off at 45°, he takes off instead at 25–35°, achieving a record jump of 8.36 m (29 ft $2\frac{1}{2}$ in.). Kangaroo rats and springhares usually ricochet with a takeoff angle that is also below 35°.

The acceleration required to achieve takeoff velocity is $\mathbf{v}^2/2s$, where s is the distance through which the applied force acts. This, in turn, is the difference in the functional length of the springing hind leg between its initial flexed and final extended positions. As their proportions attest, long hind legs are a great asset to saltators.

The mass (m) of the body does not enter directly into the formulas for height and range of jump. However, the force that must be applied to the ground to accelerate the body to takeoff velocity equals $m\mathbf{v}^2/2s$. As mass increases, the power plant increases in rough proportion, so that within limits, animals of different size but equal proportions can perform about the same. Nevertheless, the body cannot endure excessive forces, and when animals widely disparate in size are compared, complex physiological considerations must be considered. Locusts can thrust against the ground with a force eight times their body weight, and galagos and tarsiers must do almost as well. Men and kangaroos manage only about two times their body weight, but it is not possible to decide if the locust or kangaroo is the better jumper.

Relative lengthening of the hind limbs and of their distal segments is more extreme for saltators than for cursors. The tibia may be one and one-half or even two times as long as the femur (see kangaroo, Figure 20-9). Several tarsal bones may lengthen (frogs, tarsier, galago — see Figure 22-6). Lengthening of the hind limbs of ricochetal mammals is the more striking because the forelimbs are not modified for speed. They are usually used for slow progression and for handling food, so they do not become vestigial but often they are reduced in size. Some bipeds that swing the legs alternately, increase the length

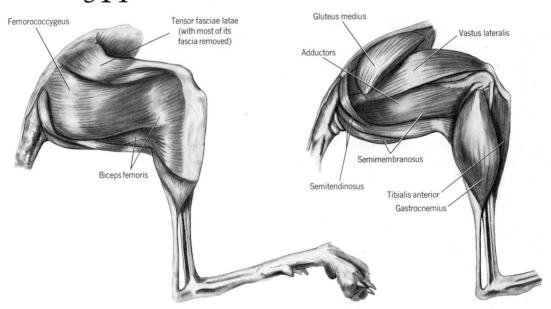

FIGURE 20-22

RIGHT HIND LEG OF A RICOCHETAL RODENT, the jerboa *Allactaga,* **showing proximal position of musculature, lengthened distal limb segments, reduced lateral digits, and some principal superficial muscles (left) and deeper muscles (right).**

of the stride by oscillating the pelvis around the long axis of the vertebral column—man is a notable example.

Bipedal reptiles share with ricochetal mammals the need for a long tail. It is elevated while running so that it will counterbalance the fore part of the body and thus bring the center of gravity over the hind feet. Bipedal lizards are unable to run if part of the tail is amputated; rabbits could not become ricochetal with so short a tail; absence of a tail correlates with man's upright posture. Most ricochetal mammals also use the tail as a prop, thus providing a third point of support when standing on the hind legs. Although once a jumper has left the ground its center of gravity must move in a predetermined parabolic path until it touches the ground again, a saltator can change the orientation of its body in midjump by lashing its tail. The kangaroo rat can even reverse its field in the air, ready to bounce back along its own track on the next hop. The tail is often tufted with hair; its greater distal weight and air resistance then improve its effectiveness for controlling its jump.

Concentration of weight over the supporting hind legs is also achieved by shortening somewhat the presacral part of the spine, particularly in frogs and mammals. The lumbar region of the spine of ricochetal mammals is robust and has prominent neural spines. The thoracic region is slight and has small neural spines. The cervical vertebrae of ricochetal rodents tend to fuse, apparently to reduce bobbing of the head.

Further characterization of the feet of springers that are also climbers is found in Chapter 22.

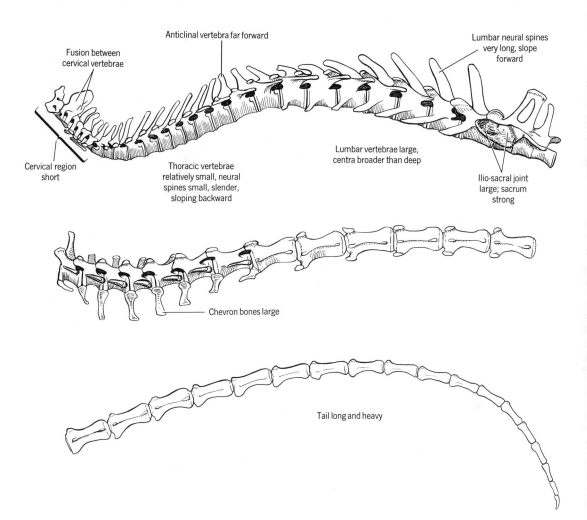

FIGURE 20-23

SOME CHARACTERISTICS OF THE VERTEBRAL COLUMN OF A RICOCHETAL MAMMAL shown by the springhare, *Pedetes.*

21

Digging; and Locomotion without Appendages

Vertebrates that are morphologically adapted to dig effectively are said to be **fossorial.** Those that spend most of their lives underground are **subterranean.** However, the distinction between fossorial and nonfossorial animals is not sharp. Some vertebrates live in burrows dug by other animals, and many tetrapods can dig somewhat even without marked structural adaptation. Thus, the alligator lizard pushes its way into litter to hide, the thrush scratches leaves away to uncover food, the caribou paws snow from the lichens it eats, and the elephant scrapes holes with its forefeet to reach ground water where surface sources have gone dry. This chapter will stress the more highly adapted diggers.

517

The fossorial habit may have evolved in every class of vertebrates, though little is known of digging by ostracoderms and placoderms. Fossorial adaptations evolved independently in many orders of fishes and mammals and more than once in several orders. Legless progression is not confined to diggers, but evolved as an adaptation for burrowing.

Advantages of Digging

Fossorial vertebrates have various advantages:

(1) Digging establishes microhabitats that are suitable for resting, aestivating, or hibernating. The burrow is cooler and more humid than the desert air (most desert rodents are active on the surface of the ground only at night and the remainder must retreat periodically to cool off), warmer than the winter storm (mountain chipmunks could not sleep above the snow without freezing), and relatively safe from lightning fires (many small forest tetrapods survive the flames).

(2) Many diggers, and most that are small, secure from the ground foods such as insects, insect larvae, earthworms, roots, and tubers. Several predators dig to secure smaller diggers as food.

(3) Diggers can store food underground where it is safe from other animals and the weather, and where it will be available during another season. Many kinds of rodents make large stores of dry grass or seeds; the arctic fox buries caches of birds and eggs.

(4) Nearly all fossorial vertebrates escape from predators by retreating underground. Many do not venture far from their holes and scamper back on the slightest sign of danger.

(5) Digging provides protected nests and dens in which to lay eggs or rear young. Many vertebrates, from the 10 g shrew mole to the 60 kg aardvark, raise their families underground.

Several Ways of Digging

Vertebrates dig in various dissimilar ways. A first method may be called **cover-up digging.** The animal merely covers itself with sand or soft mud. It does not make an open hole or progress through the substrate, and rarely digs deep. The animal may cover itself to escape from the environment at the surface but usually does so either to lie in wait for prey (with protruding eyes uncovered) or to escape or hide from predators. The digger may shuffle into sand or mud (some fishes and anurans), vibrate its body for the few seconds it takes to submerge in sand (some desert lizards), run or swim rapidly and then dive into sand (several lizards, various fishes), or sway its body from side to side thus creating a trench into which it sinks (several snakes). Most of these animals have flat bodies and somewhat modified sense organs. Nevertheless, their adaptations are largely behavioral, and cover-up diggers are usually not said to be fossorial.

Soil-crawling is the term I shall give to a second general technique which is used by synbranchiform eels, caecilians, some salamanders, amphisbaenians, legless and short-limbed lizards, and burrowing snakes (Figure 21-1). The animal moves through soil that is usually soft or sandy but may be quite firm. Often the substrate closes behind the animal so a permanent burrow is not established, but the soil may instead be compacted, thus opening a burrow. A high degree of specialization is required. The body is long and slender, which reduces the amount of soil that must be displaced. Legs are reduced or absent. The head is relatively firm and usually serves as the digging instrument.

Another method of digging is **scratch-digging.** (The term "rapid-scratch" has been used but is not always apt.) By alternately flexing and extending its limbs in the manner of a dog with a bone to bury, the animal cuts and loosens the soil with its claws and pushes or flings it to the rear with the pads of its feet. Some turtles, some birds, carnivores, ground squirrels, and a variety of other mammals dig in this way. Nests, dens, and burrows are excavated, often in hard soil.

A fourth method, **chisel-tooth digging,** is followed by gophers, mole-rats, and various other rodents. Huge gnawing incisors and powerful jaw and neck muscles are used to dislodge the soil, which is then moved with head or feet. Hard soil can be excavated, but somewhat damp or otherwise tractable soil is usually preferred.

A fifth method is **rotation-thrust digging.** (The term "lateral-thrust" has been used but is less apt.) The method is best exemplified by the true moles. A mole placed on the surface can dig out of sight in the damp soil it prefers in about 6 sec. These subterranean digging machines have broad shovel-like forefeet and short powerful forelimbs. There is no pronation and supination of the forearm, and (in sharp contrast to scratch-diggers) movement at the elbow merely positions the hand without providing a forceful stroke. The power for digging comes from the rotation of the uniquely short but broad humerus around its own long axis. This is accomplished by the relatively enormous teres major muscle. (The teres major is assisted by the latissimus dorsi, subscapularis, and posterior part of the large pectoralis. The recovery stroke results from contraction of the supraspinatus, spinodeltoid, and cleidohumeralis, and particularly of the anterior part of the pectoralis. The peculiar biceps also counter rotates the humerus as it flexes the forearm. The wrist moves only as a hinge. Rotation of the humerus pushes the radius forward which, in turn, straightens the hand.)

Several extinct genera of insectivores belong to a family having mole-like habits but a less modified shoulder. The echidna holds its wide humerus horizontal to the ground and rotates the bone around its long axis when it walks. It is probable that these strong animals are also rotation-thrust diggers. Tortoises have adapted the biceps to rotate the humerus in a comparable way when digging.

Some vertebrates dig by more than one method. Many rodents are both scratchers and tooth-chiselers; several reptiles either scratch with short forelimbs or fold them away and use soil-crawling. Also, various diggers do not fit into this classification (e.g., the dipnoan which digs into the mud using body and fins, the jaw fish which moves and arranges rocks with its powerful jaws, and the young crocodile which bites chunks of clay from a muddy bank). Some vertebrates make themselves uninvited guests in burrows (vacant or occupied) dug by their more fossorial cousins, doing only house cleaning and slight alterations on their own.

The digging habits of many fossorial vertebrates are as yet poorly known. Less has been written about the biomechanics of digging than of most other locomotor specializations.

Fossorial Vertebrates

AGNATHS and FISHES. The flattened bodies and dorsal eyes of cephalaspids and antiarchs indicate that they were bottom feeders. Perhaps some were cover-up diggers when it was best to get out of sight. Skates and rays are dorsoventrally flattened and may cover themselves up lightly, leaving eyes and spiracles free. Flounders are bilaterally flattened and lie on one side. The eye of the down side migrates around to the up side during ontogeny. Light cover-up digging may supplement protective coloration in making these fishes hard to see. Among truly fossorial bony fishes are some gobies and catfishes, the jaw fish, and various eels. Cichlid and centrarchid fishes dig breeding holes in the substrate. Synbranchiform eels dig deep and extensive burrows. Dipnoans dig vertical burrows into the mud to wait out times of draught.

AMPHIBIANS. Caecilians (Figure 4-2) are soil-crawlers. Some short-legged urodeles are also soil-crawlers which wriggle through litter and loose soil. Other urodeles, including the ambystomids, or mole-salamanders, have stout bodies and legs with which they dig. Frogs commonly dig into mud, and one desert toad digs a deep hole with its hind legs if it cannot find a suitable rodent burrow in which to wait for cool humid nights.

REPTILES. Snakes evolved their legless form as a result of burrowing. Most subsequently returned to the surface to crawl, climb, or swim, but some snakes remained in the ground or returned there. Amphisbaenians and legless lizards have adopted similar habits and convergent structure. Many skinks and other lizards also burrow, and the tuatara retreats underground. All turtles bury their eggs using the hind feet as scoops. Tortoises have modified the forelimbs as effective digging tools.

BIRDS. No bird has evident structural adaptations for digging, but shearwaters, puffins, an owl, and several other birds nest in holes or burrows which they either appropriate from mammals or laboriously construct themselves with beak and feet.

MONOTREMES and MARSUPIALS. The platypus and the echidnas are powerful diggers; the platypus exposes its claws by folding away the webs used when swimming. The marsupial mole is among the most highly adapted of subterranean mammals; it is a small scratcher. One bandicoot digs, and the wombat, which weighs about a third as much as a man, may excavate a burrow 30 m long.

INSECTIVORES. About a dozen genera of true moles (family Talpidae) are extremely effective rotation-thrust diggers. Moles can progress just under the surface at 2 body lengths/min and 200 body lengths/day. The unrelated golden moles (family Chrysochloridae) are among the most specialized of subter-

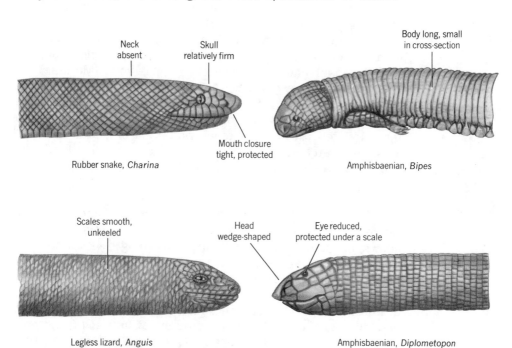

Neck absent   Skull relatively firm

Mouth closure tight, protected

Rubber snake, *Charina*

Body long, small in cross-section

Amphisbaenian, *Bipes*

Scales smooth, unkeeled

Head wedge-shaped

Eye reduced, protected under a scale

Legless lizard, *Anguis*

Amphisbaenian, *Diplometopon*

FIGURE 21-1

REPRESENTATIVE SOIL-CRAWLERS showing some adaptations for this mode of digging.

ranean scratch-diggers. Hedgehogs, mole shrews, and tenrecs also make burrows.

EDENTATES, PANGOLINS, and the AARDVARK. Armadillos (nine genera), pangolins, and the aardvark are the most powerful of scratch-diggers. Anteaters do not burrow but rip into termite nests and the soil to secure insects.

CARNIVORES. The six genera of badgers and the ratel are fossorial; some of them dig out burrowing rodents for food. Various canids excavate dens or dig to cache food, yet have scant structural adaptation for digging.

RABBITS and RODENTS. The plains pika and the less cursorial of the rabbits are moderately good diggers, though they are not highly adapted for the habit. The order Rodentia has more fossorial representatives than any other; only some of them are mentioned here. The mountain beaver digs in stream banks. Ground squirrels, marmots, and prairie dogs (all of the family Sciuridae) are avid scratch-diggers. Many of them make extensive burrow systems in hard soil. Gophers (eight genera in the family Geomyidae) are subterranean rodents of North America which make large burrow systems. Kangaroo rats, jerboas, and springhares are all ricochetal, yet manage to dig daytime retreats in sandy soil. Four genera of African mole-rats or blesmols (family Bathyergidae) are expert chisel-tooth burrowers, and the fifth genus of the family is a scratcher. The Mediterranean mole-rat (Spalacidae), bamboo and root rats (Rhizomyidae), and all but one of the mole voles (Muridae) are chisel-tooth burrowers. The Asian mole-rat, or zokor (Cricetidae), has enormous foreclaws and is a scratch-digger. The related shrew mouse, burrowing mouse, and mole mouse, and the unrelated tuco-tuco (Ctenomyidae), all of South America, are also scratchers. Their neighbor, the coruro (Octodontidae), is probably a chisel-tooth burrower.

In Chapter 20 it was easy to indicate the skills of vertebrate runners and jumpers. Relevant speeds and distances are known for many, and man, being something of a runner and jumper himself, appreciates the prowess of those animal champions. Comparable data are meager for diggers, however, and no digging is done in the Olympic Games. Can you imagine an event that would determine which human contestant could first dislodge 30 times his body weight of firm soil using only his finger nails, and then, using his hands and feet, transport

FIGURE 21-2

FURTHER EXAMPLES OF DIGGERS showing some of their adaptations.

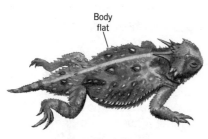

Body flat

Horned lizard, *Phrynosoma*

External ear rudimentary

Long-clawed ground squirrel, *Spermophilopsis*

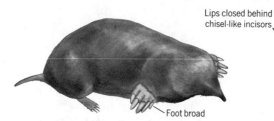

Foot broad

Eastern American mole, *Scalopus*

Lips closed behind chisel-like incisors

Blesmol, *Cryptomys*

Eye vestigial

Limbs short

Marsupial mole, *Notoryctes*

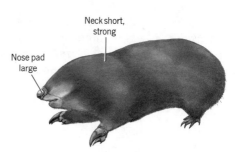

Neck short, strong

Nose pad large

Golden mole, *Amblysomus*

Limbs powerful

Giant armadillo, *Priodontes*

Claws long, strong

Pangolin, *Manis*

all the dirt 7 body lengths distant and pile it onto a platform that is as high as he can reach? Well, the little tuco-tuco does this daily, not in competition but in his morning rounds. And, who knows, in the animal olympics the tuco-tuco might not make it past the gophers, mole-rats, and blesmols to the finals.

General Requirements of Diggers

Fossorial vertebrates must be effective in meeting certain requirements: (1) All diggers must exclude sand, dust, and earth from the mouth, eyes, ears, respiratory passages, and cloaca or anus. (2) All except some cover-up diggers must maneuver within the soil or inside confined spaces. Many must find their way, detect and avoid predators, and for some, locate mates and protect eggs in complete darkness. (3) All that do not dig in sand, litter, or soft mud must break or loosen the soil. (4) All that make open burrows must either transport and dispose of loose soil or compact it. How do the different diggers meet these requirements?

Keeping Dirt Out of Mouth, Sense Organs, and Lungs

Large diggers (wombat, aardvark, badger) probably can keep their mouths out of the dirt, at least most of the time. Fossorial amphibians, reptiles, and insectivores have close registration of the closed jaws; the margin of one jaw often fits into a groove in the other to make a tight seal. The lower jaw of some burrowing reptiles is recessed behind the upper jaw. The furred lips of chisel-tooth diggers meet behind the protruding incisors, thus enabling the animal to exclude dirt from the mouth at the same time it is gnawing.

Diurnal vertebrates that burrow to nest, rear young, or hibernate, yet do their foraging above ground have eyes of normal size (tortoises, hedgehogs, ground squirrels, canids). Nocturnal rodents that burrow to escape daytime heat have large eyes (kangaroo rats, jerboas). Presumably these animals close their eyes as needed when digging. Possibly some have evolved improved mechanisms for cleansing the eyes. Snakes and certain lizards have fused but transparent eyelids. The real specialists have small to minute eyes (monotremes, armadillos, pangolins, gophers, African mole-rats, root rats, tuco-tuco) or vestigial eyes which may differentiate light from dark but form no image and often are hidden under the skin (caecilians, amphisbaenians, marsupial mole, true moles, golden moles, Mediterranean mole-rat).

Many burrowing amphibians and reptiles have no external auditory canal. If there is a tympanum, it is thick. Sound reception is often mediated by a special mechanism involving the skin, jaw, or other structures. The external auditory canal of burrowing mammals tends to be small. It is ventral in position in insectivores. The angle of the canal relative to that of non-fossorial vertebrates, and the possible presence of a valvular closure, should be examined.

The external nares of diggers tend to be small. It is probable that they can be closed by many burrowers. The outer part of the nasal passages is narrow and slopes upward in digging reptiles. The openings can be closed or at least constricted by muscular valves or erectile tissue and may be covered by a fold. The armadillo is able to suspend breathing for 3 or 4 min while digging vigorously in dust and sand. Some shrews have peculiar diverticula from the lungs which seem moderately effective at trapping and disposing of foreign material.

Living and Maneuvering Underground

Maneuvering in the confines of a narrow burrow is facilitated in several ways. Some fossorial amphibians and reptiles and all fossorial mammals are short-legged. (We shall see that short legs are also favored by considerations of muscle mechanics.) Although compact, the trunk tends to be flexible, and some of the animals can turn very sharply by doubling into a ball and then unrolling in the new direction. Some fossorial vertebrates (e.g., snakes, amphisbaenians, gophers) have such loose skins that the animal can to a degree turn within its skin and then let the skin follow. All burrowers are adept at backing up.

Some diggers have special reasons for having long tails: The ricochetal kangaroo rat uses his for balance, the pangolin uses his as a grasping organ when climbing, and the armadillo uses his as a prop when digging. Otherwise, the evolutionary process seems often to have found a tail to be in the way underground and it tends to be short, even in burrowing snakes and some other legless diggers (though some amphibians and lizards are exceptions). Some small subterranean vertebrates have little or no tail (caecilians, marsupial mole, golden mole, Mediterranean mole-rat, and various others).

Virtually nothing is known about how subterranean animals find their way in the absolute darkness of a deep burrow system. Fossorial reptiles seem to smell or hear their prey. Are the vibrissae of the mammals particularly effective? Do they use odors or a tactile memory? Is the tail, when retained, particularly sensitive, as has been claimed?

Many burrow systems are nearly saturated with water vapor. Since digging is hard work, the animals have a problem in dissipating excess heat. A study of six unrelated genera of rodent diggers showed them to have a low basal metabolic rate and a wide range of thermoneutrality (temperature at which oxygen consumption is minimal). The gopher studied radiated heat from its nearly hairless tail, and one genus of African mole-rat is entirely naked.

Digging; and Locomotion without Appendages

Selection and Maintenance of Tools. When a man shovels dry sand or forest litter, he must expend energy to transport the material but virtually none to break or loosen it. It is already loose because its particles scarcely adhere to one another. Similarly, animals that use the cover-up and soil-crawling methods of digging usually confine themselves to sand, litter, or soil that is loose enough to be compacted or pushed aside without first being broken up. Accordingly, they may have no tools for breaking soil.

When a man shovels damp earth he must break it free before he lifts it, but it is not difficult to do so. It is sufficient for him to push the blade of a shovel into the earth with his foot and then pry a little with the handle. A large shovel blade is satisfactory. Similarly, rotation-thrust diggers confine their activities to moderately soft soil. They compact, break, and cut the soil by pushing on it or scraping it with their stout claws. Since this is not particularly difficult, a large "shovel blade" is effective. Moles have broad claws and very wide forefeet. Similarly, amphisbaenians dig by ramming the wedge-shaped head into the soil and then (according to species) lifting the head to compact the soil into the wall of the burrow.

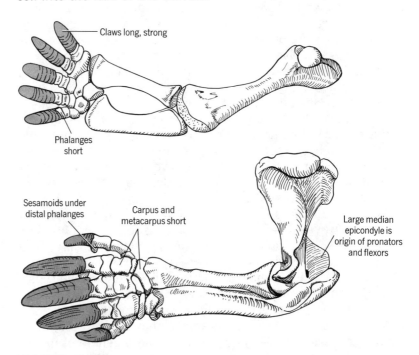

FIGURE 21-3

SOME ADAPTATIONS FOR DIGGING shown by dorsolateral views of the left forelimb skeletons of the tortoise, *Gopherus* **(above), and echindna,** *Tachyglossus* **(below).**

The man who digs in dry compacted soil must expend much energy on breaking the soil prior to moving it. The shovel cannot be forced into the undisturbed material, so a pick is first used to loosen it. The pick is effective because it delivers great force to a restricted area; i.e., with each blow it applies high pressure to a limited area. Chisel-tooth and scratch-diggers among vertebrates tend to avoid rocky and otherwise intractable soils, yet many of them burrow in remarkably hard earth. They gnaw with their incisors or scratch with their sharp claws, thus applying great pressure to a small area before going on to another spot. The blade must be long and strong to do its job. The badger has five subequal claws, the longest of which is half the length of its forearm between elbow and wrist joints. The mountain beaver, tuco-tuco, gophers, and ground squirrels emphasize three or four claws, the longest of which may (in some gophers) be three quarters of the length of the forearm. The anteaters, marsupial mole, and golden moles emphasize one or two claws, the longest of which may (in some golden moles) be longer than the forearm. The "shovels" of moles can exert against soft soil more force in relation to body weight than can the "blades" of ground squirrels, but the ground squirrel exerts against harder soils about twice as much pressure as does the mole.

The tools of diggers are subject to tremendous wear. Burrowing reptiles compensate by molting successive generations of the epidermis, thus exposing new surfaces to the substrate. The upper incisors of some gophers grow out at the rate of 248 mm/yr. The lower incisors, which are maneuvered more as the condyle of the mandible slips in its loose groove, may grow at the rate of 445 mm/yr. These rates are $2\frac{1}{2}$–3 times those recorded for some nonfossorial rodents of comparable size. Likewise, the center foreclaw of a gopher grew out at 90 mm/yr, and that of a tuco-tuco at 72 mm/yr.

The incisors of rodents and of the wombat have enamel only on their forward surfaces. The softer dentine wears away behind, thus providing a self-sharpening mechanism.

The spadefoot toad digs with a different kind of tool. There is a horny epidermal tubercle at the edge of the hind foot which is used as the animal progresses backward into loose soil.

Design for Large Out-Forces.   Fossorial vertebrates that dig in firm soil must be capable of applying great force against the substrate. Therefore, unlike cursors and climbers, they are constructed so that their relevant bone-muscle systems (particularly of the forelimb) produce large out-forces (F_o). In Chapter 19 it was shown that $F_o = F_i l_i / l_o$, where F_i is the in-force and l_i and l_o are, respectively, the in- and out-levers.

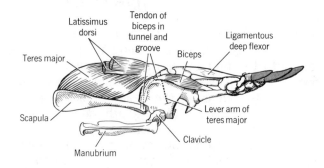

Latissimus dorsi

Tendon of biceps in tunnel and groove

Biceps

Ligamentous deep flexor

Teres major

Scapula

Lever arm of teres major

Manubrium

Clavicle

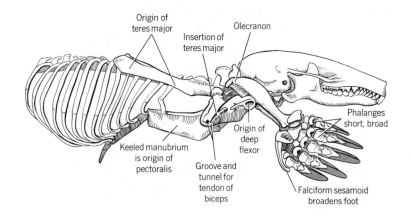

Origin of teres major

Insertion of teres major

Olecranon

Phalanges short, broad

Keeled manubrium is origin of pectoralis

Origin of deep flexor

Groove and tunnel for tendon of biceps

Falciform sesamoid broadens foot

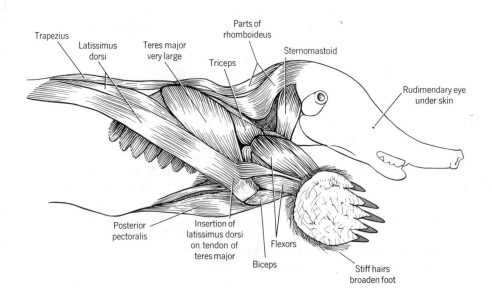

Trapezius

Latissimus dorsi

Teres major very large

Parts of rhomboideus

Triceps

Sternomastoid

Rudimendary eye under skin

Posterior pectoralis

Insertion of latissimus dorsi on tendon of teres major

Flexors

Biceps

Stiff hairs broaden foot

It is evident that one way to increase F_o is to reduce l_o. Consequently, the more expert diggers all have short legs and necks. In sharp contrast to the limbs of cursors, the limbs of diggers have relatively short distal segments. The radius is nearly always shorter than the humerus, and the manus, exclusive of the terminal phalanx with its claw, is markedly shorter than the radius. Although strong, the metacarpals may be very short (tortoises, echidna, moles) and the proximal phalanges may be even broader than long (echidna, pangolin, anteaters, moles).

A second way to increase an out-force is to increase the related in-lever. Accordingly, muscles used in digging tend to insert far from the joints they turn. The insertion of the mammalian deltoid muscles commonly extends more than halfway down the length of the humerus (away from the shoulder joint which pivots). Part of the latissimus dorsi of the golden mole increases its in-lever to the shoulder joint by shifting its insertion from the proximal part of the humerus (the usual position) nearly to the elbow joint, and similar leverage is gained by the unusual thoracic head of the triceps of the armadillo. The wide median epicondyle of the humerus, which is a feature of all scratch-diggers, increases the in-lever of the pronator of the forearm. A relatively proximal origin on the humerus of the long supinator muscle increases its in-lever and enables it to flex the manus as well as supinate the forearm. A relatively long pisiform bone at the carpus increases the in-lever of one of the flexors of the manus. Crucial to the special mechanism of the rotation-thrust digging of moles is a very wide flaring tubercle for the insertion on the humerus of the enormous teres major muscle. This carries the insertion away from the central long axis of the bone and thus increases the in-lever that makes possible powerful rotation of the humerus around its own long axis (Figure 21-4).

These kinds of adaptations of diggers are particularly striking when in-levers are expressed as fractions of their related out-levers. Thus, in measuring representative skeletons of 27 genera of expert diggers belonging to seven mammalian orders I find the olecranon process (the in-lever of the triceps) to be about $\frac{1}{5}$ (a ground squirrel), $\frac{1}{3}$ (gopher, African mole-rat), $\frac{1}{2}$ (aardvark, pangolin, mole), $\frac{2}{3}$ (Mediterranean mole-rat, armadillos), or even $\frac{3}{4}$ (marsupial mole, golden moles) the length of the ulna distal to the pivot at the elbow joint. A complete listing (if the data were available) would doubtless show that diggers differ from their nonfossorial relatives in having larger values of l_i/l_o for every bone-muscle system used in digging.

FIGURE 21-4

STRUCTURE ASSOCIATED WITH THE ROTATION-THRUST DIGGING OF MOLES. Above, ventral view; center and below, lateral views. Above and center, the Western American mole, *Scapanus;* below, the Old World mole, *Talpa.*

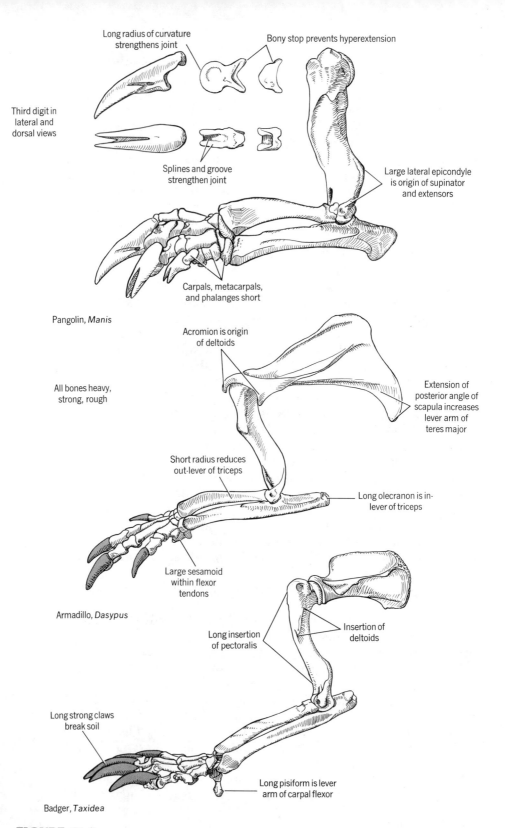

FIGURE 21-5

SOME ADAPTATIONS FOR SCRATCH-DIGGING shown by lateral views of left forelimb skeletons.

FIGURE 21-6
FOOT OF THE SPADEFOOT TOAD, *Scaphiopus,* **showing the blade-like tubercle used for digging.**

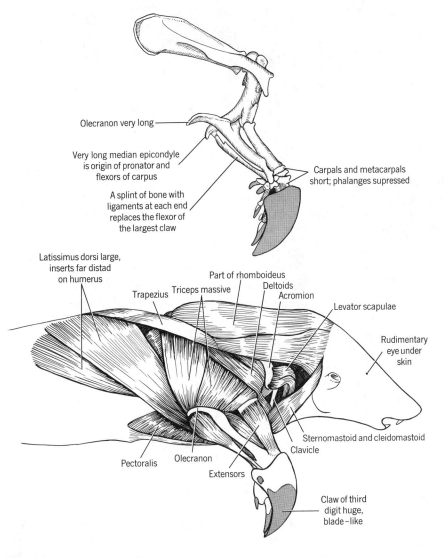

Olecranon very long

Very long median epicondyle is origin of pronator and flexors of carpus

A splint of bone with ligaments at each end replaces the flexor of the largest claw

Carpals and metacarpals short; phalanges supressed

Latissimus dorsi large, inserts far distad on humerus

Part of rhomboideus

Deltoids

Acromion

Trapezius Triceps massive

Levator scapulae

Rudimentary eye under skin

Sternomastoid and cleidomastoid

Clavicle

Pectoralis Olecranon

Extensors

Claw of third digit huge, blade-like

FIGURE 21-7
STRUCTURE ASSOCIATED WITH THE SCRATCH-DIGGING OF THE GOLDEN MOLE, *Amblysomus.*

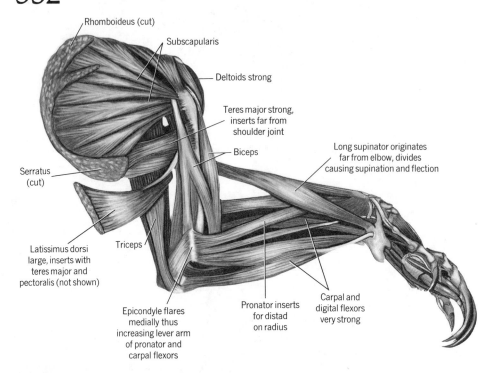

Rhomboideus (cut)

Subscapularis

Deltoids strong

Teres major strong, inserts far from shoulder joint

Long supinator originates far from elbow, divides causing supination and flection

Biceps

Serratus (cut)

Latissimus dorsi large, inserts with teres major and pectoralis (not shown)

Triceps

Epicondyle flares medially thus increasing lever arm of pronator and carpal flexors

Pronator inserts for distad on radius

Carpal and digital flexors very strong

FIGURE 21-8

STRUCTURE ASSOCIATED WITH THE SCRATCH-DIGGING OF THE GIANT AN-TEATER, *Myrmecophaga,* **shown by a medial view of the left forelimb. (Drawn from an air-dried dissection and hence somewhat shrunken.)**

A third way to increase out-force is to increase in-force. The relevant muscles of diggers are enormous. To accommodate such muscles, origins and insertions are large. This, together with their proportions, makes the forelimb bones of diggers rugged and rough. The median epicondyle of the humerus (origin of flexors of digits) and deltoid crest (insertion of deltoids) are particularly prominent. The posterior angle of the scapula may be enlarged to accommodate the origins of the teres major and long head of the triceps. The anterior segment of the sternum of true moles and golden moles is long and deep to receive their great pectoral muscles. Chisel-tooth diggers have large areas of origin and insertion for their powerful jaw muscles. Diggers that push dirt with their heads (many amphisbaenians, golden moles, marsupial mole, Mediterranean mole-rat) have a large flat occipital area for the insertion of strong neck muscles (Figure 21-9).

There is another factor in design for large out-forces. When starting a burrow, some soil-crawlers loop the body over the head thus weighting it so it can be thrust into the soil. Large scratch-diggers hunch the back over the forelimbs and may

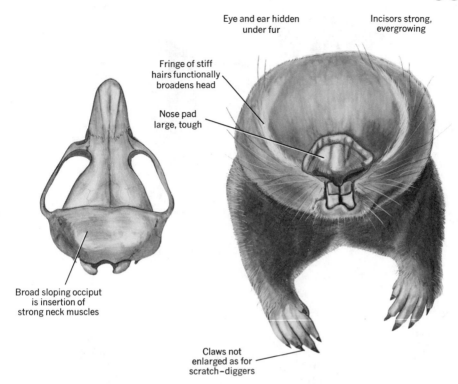

Eye and ear hidden under fur

Incisors strong, evergrowing

Fringe of stiff hairs functionally broadens head

Nose pad large, tough

Broad sloping occiput is insertion of strong neck muscles

Claws not enlarged as for scratch–diggers

FIGURE 21-9

SOME ADAPTATIONS OF A CHISEL-TOOTH DIGGER THAT MOVES DIRT WITH ITS HEAD, the Mediterranean mole-rat, *Spalax.*

prop the body with the tail, thereby applying the weight of the body to the digging tools. Subterranean diggers, however, are uniformly small and hence light. (They can then feed on the small food found in the earth, keep within one stratum of the soil, and avoid rocks and large roots.) In order to prevent motions of digging from merely pushing them away from the soil, they must force their bodies against their digging tools. Rotation-thrust diggers brace against one side of the burrow with one forepaw while digging with the other. To make this possible, the forelimbs are positioned laterally, opposite to one another. Human weight lifters lift overhead about twice their own body weight. It has been shown that by its thrust, a mole can move as much as 32 times its own body weight!

If an amphisbaenian presses up with its head it presses down with its "chest." As chisel-tooth diggers press up with their lower incisors they press down with their forefeet. Small burrowers also brace themselves with their hind legs. Xenopid frogs can at the same time lengthen the back at the sacroiliac joint, thus forcing the head forward in the mud. In response to the use of the hind legs for bracing, the innominate bones of

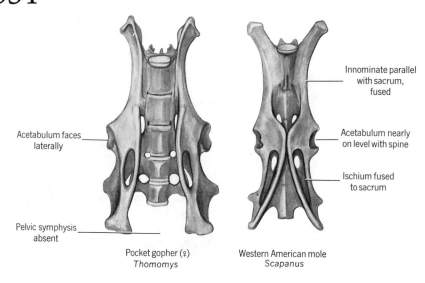

Acetabulum faces laterally

Pelvic symphysis absent

Pocket gopher (♀)
Thomomys

Innominate parallel with sacrum, fused

Acetabulum nearly on level with spine

Ischium fused to sacrum

Western American mole
Scapanus

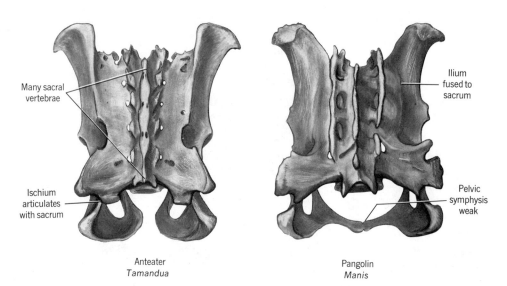

Many sacral vertebrae

Ischium articulates with sacrum

Anteater
Tamandua

Ilium fused to sacrum

Pelvic symphysis weak

Pangolin
Manis

FIGURE 21-10

SOME CHARACTERISTICS OF THE PELVISES OF FOSSORIAL MAMMALS. Ventral views above; dorsal views below.

the mammals tend to be nearly horizontal (in line with a forward thrust). The hip joint is relatively far dorsal to be on a level with the spine. This reduces compressive forces at the pelvic symphysis, which is nearly always weak and sometimes absent (moles, anteaters, pangolins, and some gophers). The innominate bones are firmly sutured with, or fused to, a relatively large number of vertebrae, and the sacrum is long.

Resistance to the Force of the Soil on the Body. Soil-crawlers reduce or resist the force of the soil on the body in various ways. The skins of burrowing fishes are commonly particularly rich in mucous glands. Scales are absent (various fishes, amphibians) or, if present, smooth and unkeeled (reptiles) to reduce the friction of soil against the body. The head is short and narrow. The skull is relatively solid; most sutures are obliterated and others are serrate. There is no neck or shoulders. The reduction and loss of legs is itself a major accommodation to streamlining. In order to displace as little soil as possible the trunk becomes long and slender—in other words, snake-like. Unspecialized lizards have about 23 presacral vertebrae whereas fossorial lizards may have 60, and amphisbaenians more than 100 presacral vertebrae. Caecilians and snakes may have 250 or more such vertebrae. Cervical and lumbar ribs are usual.

Vertebrates that move along narrow burrows must avoid snagging and abrading themselves against burrow walls. Subterranean diggers have small external ears or none at all. Fur is lax, often short, and sometimes nearly upright so it can brush in any direction. At least some of the mammals clean their fur by shaking the body, dog fashion. Others groom away mud and dirt with the paws.

There must be provision against dislocation and hyperextension of joints of the forelimb and manus when a digit snags on a rock or root. Hyperextension of the phalanges of echidnas, pangolins, and anteaters (at least) is prevented by squared articulatory surfaces or bony stops that limit rotation of the joints (see Figure 21-5). Dislocation of the digits is prevented in pangolins and anteaters (and others?) by large areas of bone-to-bone contact at the joints and by deep interlocking splines and grooves, recalling those of the phalanges of ungulates. The scapula of gopher tortoises is braced against the plastron more firmly than is that of turtles. Protection is also afforded by generally heavy and rugged construction, and in some instances by structural unity (i.e., common firmness) of the palm (tortoises, echidna, moles). Some ligamentous checks against dislocation have also been described.

Diggers that excavate dens or burrows must transport and dispose of earth after they loosen it. Many snakes use coils of the body to sweep sand and other debris out of burrows. Lizards and turtles sweep with their feet. Some reptiles create an air space under the body so they can breathe in spite of an overburden. As mammals break the soil with forefeet or teeth it is pushed underneath or beside the body. From time to time the animal kicks this loose soil back out of the way with the hind

Transporting and Disposing of Soil

feet. When the burrow behind becomes choked with soil it is time to take it away, and several methods are used. Some diggers back up in the burrow vigorously kicking the dirt back as they go. The hind feet are used simultaneously (tuco-tuco, African mole-rats, armadillo) or alternately (hedgehogs). Other burrowers turn around to push the dirt forward using the forefeet only (moles); forefeet, chest, and chin (gophers); or nose and the top of the head (gopher tortoise, Mediterranean mole-rat). The gopher tortoise may push dirt with the upturned front of its plastron. Leaf-nosed and hog-nosed snakes probably also use their heads as shovels, as do synbranchiform eels and amphisbaenians. Most dirt is moved to the surface, but some is used to plug abandoned side tunnels of the burrow system.

Whichever foot, fore or hind, is used for moving the soil, it is made broad in one or more ways: The toes may be webbed (toads, sea turtles and others, moles, golden moles), the pad of the foot may be widened by cartilages or bones placed lateral to the first digit (mountain beaver, moles, gophers, tuco-tuco, etc.), or the pad of the foot may be fringed with stiff hairs (nearly all subterranean mammals). The Mediterranean mole-rat similarly increases the effectiveness of its broad flat head as a dirt pusher by adding lateral fringes of stiff hairs (Figure 21-9).

Some fossorial vertebrates find it necessary to tamp the earth of a surface mound to prevent back-slip, or of the walls of a burrow to make them smooth. The Mediterranean mole-rat has a broad, tough, muscular nose pad which it uses assiduously for this purpose. Golden moles and the marsupial mole may use their nose pads in like manner. One of the African mole-rats has a peculiar way of extending its hind feet on a surface mound of earth and then vibrating them at about 22 cycle/sec to tamp the soil.

Locomotion without Appendages

Terrestrial locomotion without limbs has evolved independently several times among amphibians and reptiles, but is best exemplified by snakes and amphisbaenians. Several thousand species are involved, so the adaptation must be considered highly successful. Legless progression evolved in each instance as an adaptation for burrowing (or possibly in some instances for crawling among sticks and stems), but for snakes it has secondarily proved to be advantageous also for crawling on the surface of the earth, climbing, and swimming. The manner of progression of snakes has long attracted attention, yet it could not be analyzed until the development of high-speed motion picture photography. Gray and his associates at Oxford, and Gans and his associates in New York, have now shown that there are four principal ways in which legless vertebrates move, though more than one way may be used at a time.

The first, and most common, is **lateral undulation.** The body is thrown into serpentine loops, right and left. The animal locates with its coils several projections such as pebbles or plant stems. The body then presses sideways (not downward) against these objects in a direction that is obliquely backward in relation to the direction the snake is to move. The mechanical analysis for the forces at any one projecting object resembles that for the action of a fish tail: The thrust of the snake against the object (F_t in Figure 21-11) is opposed by an equal and opposite force (F_o) exerted by the object against the body. This force has a forward component (F_f) in the direction that the snake as a whole is moving, and a lateral component (F_l).

However, unlike the tail of the fish which continues to have water to thrust against as the tail changes position, the coil of the snake must maintain contact with the stationary object if it is to continue to thrust against it. Consequently, the undulations of a slender fish form traveling waves, whereas the undulations of a snake form standing waves. Each coil stays in the same place; the body follows exactly in the trail established by the head.

A snake cannot move forward by sliding a coil past a single object against which it pushes as it goes. This is because that particular coil does not move in the direction in which the

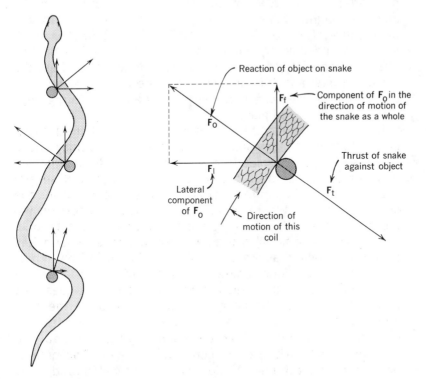

Reaction of object on snake

F_f ← Component of F_o in the direction of motion of the snake as a whole

F_o

Thrust of snake against object

F_l

Lateral component of F_o

F_t

Direction of motion of this coil

FIGURE 21-11
DIAGRAM OF LATERAL UNDULATION OF A SNAKE.

snake as a whole is moving, but instead moves along its own long axis as shown in Figure 21-11. That direction being at right angles to F_o, F_o can have no component in the coil's direction of motion. If F_o (or a component thereof) were in the direction that a second coil moves past a second object, then the snake could push or pull past the second object by pushing the first coil against the first object. For continuous motion, the snake requires three or more objects which cannot all be on the same side of the body. The action is more efficient if few objects are used (3–5 unless the snake is very long and slender). The lateral components of the various thrusts add up to zero, and (disregarding friction for the moment) the sum of the forward components is the propulsive force of the animal. It is seen that all the forces are interconnected. As the "neck" contacts new objects and the tail slides away from others, or as any object slips, the animal must instantly adjust both the magnitude and direction of the thrusts of all active coils. One must be impressed by the complexity of the feedback mechanism and nervous control required.

Snakes are unable to progress along narrow burrows by lateral undulation because then they cannot thrust obliquely backward with their coils. Further, they cannot progress by this method on smooth surfaces because they must thrust laterally, not vertically. There is, of course, a vertical force from the animal's weight. This causes friction, which in this instance is undesirable. The coefficient of friction (see p. 550) is minimized by the smooth nature of the snake's ventral scutes. There is also sliding friction against the projections where the snake thrusts, and this also counters the propulsive force of the animal. Snakes can move very freely among rotating pegs placed on lubricated glass.

The second method of legless progression is **rectilinear movement.** The skin of many snakes fits loosely over the body and is very distensible. Muscles that slant back and down from the ribs to the scutes cause the ventral skin to bunch at several regions so that the scutes overlap. Between these regions the skin is stretched. Where scutes are bunched they rest on the ground, where stretched they are lifted clear of the ground. One by one, additional scutes are drawn into each bunched region from behind as others are stretched away in front. Thus, the scutes move along somewhat in the manner of the feet (or prolegs) of caterpillars—each starts and stops as it goes. The bunched scutes thrust obliquely backward against the substrate, and friction is required to prevent slipping. Muscles that slant backward and up from the scutes to the ribs haul the body along within its skin in continuous motion. The body is held in a straight line. Motion is symmetrical and directly ahead. It is slow. This kind of motion may be used for stalking prey or for moving in a narrow tunnel, but otherwise it is infrequent.

The third way of moving without limbs is **concertina movement.** The snake draws itself into one or more S-shaped coils. The posterior coils then press downward and backward against

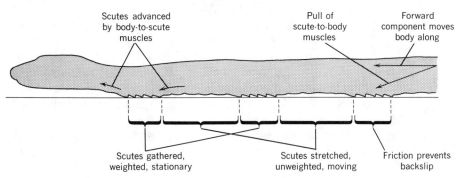

Scutes advanced by body-to-scute muscles

Pull of scute-to-body muscles

Forward component moves body along

Scutes gathered, weighted, stationary

Scutes stretched, unweighted, moving

Friction prevents backslip

FIGURE 21-12

DIAGRAM OF RECTILINEAR MOVEMENT OF A SNAKE.

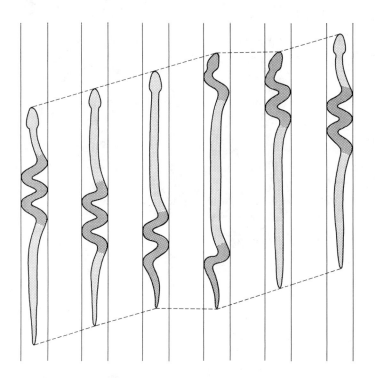

FIGURE 21-13

DIAGRAM SHOWING SUCCESSIVE POSITIONS IN THE CONCERTINA MOVEMENT OF A SNAKE PROGRESSING IN A TUNNEL. Dark shaded parts of the body are stationary; light shaded parts of the body are moving.

the substrate, relying on friction to prevent slipping. The forward component of this thrust is used to advance the head and anterior part of the body, which is held clear of the ground to avoid resistance from sliding friction and to increase the static friction of the stationary posterior coils by increasing the loading on them. Before stability is threatened, the anterior part of the body touches the substrate, builds coils, and ceases motion, so that it, in turn, can draw up the posterior part of the body before the cycle is repeated. The firmness of the base provided by the stationary coils is increased if coils are forced outward to wedge the animal within a tunnel, or between rocks or crevices in bark, or, if they are forced inward, to constrict a branch. Concertina movement is common, particularly among climbing and burrowing snakes, and often is combined with lateral undulation.

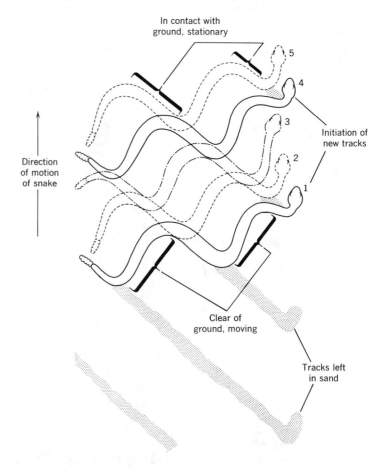

In contact with
ground, stationary

Direction
of motion
of snake

Initiation of
new tracks

Clear of
ground, moving

Tracks left
in sand

FIGURE 21-14

DIAGRAM SHOWING SUCCESSIVE POSITIONS IN
THE SIDEWINDING OF A SNAKE.

The last method is **sidewinding,** a kind of progression that probably evolved from the concertina method and is adapted for fast travel over loose or sandy soil. Figure 21-14 shows that the snake makes a series of tracks which are more or less straight lines, parallel to one another, angled to the direction of travel, and each is about as long as the animal. The snake contacts two or three tracks at a time. Parts of its body are within the tracks and parts are arching between tracks. As successive segments of the body are laid down to extend one track, successive segments are released from the previous track. The parts of the body that are along tracks are stationary, whereas those that span between tracks are moving and are held clear of the ground. Periodically, the head and neck reach forward to initiate new tracks. The entire action is very rapid. Most snakes can move in this way, but desert species are the most adept.

22

Climbing

Tetrapods that are adept at climbing may be called **scansorial,** though the term is less commonly used than the companion words "cursorial" and "fossorial." Climbers are often **arboreal,** but this term means living in trees and does not directly indicate manner of locomotion; most birds are arboreal, yet few climb.

Many tetrapods are expert climbers and many more climb moderately well on occasion. The climbing habit evolved independently more times than can be traced, and several times in each of several orders. Adaptations for climbing by primates and by some of the other more strikingly modified climbers have been analyzed. However, the adaptations of many small scansorial tetrapods have scarcely been studied.

543

Advantages of Climbing

The selective advantages of climbing include the following:

(*1*) Climbers can secure in shrubs and trees such foods as leaves, shoots, flowers, fruits, cambium, honey, spiders, insects, and birds' eggs.

(*2*) Many climbers avoid predation by remaining off the ground or by returning to the safety of rocks or vegetation when danger threatens. Also, climbing affords vantage points from which to look out for danger.

(*3*) Several predators follow their prey into the trees: The fisher captures tree squirrels; the arboreal viper lies in wait for scansorial rodents. The leopard hauls his kill into a tree partly to keep it safe from jackals and hyenas when he is not in attendance.

(*4*) By climbing, many animals find sheltered places to rest during the part of the day when they are inactive. Similarly, they find or make safe secluded nests in which to rear their young.

(*5*) Where ground vegetation is dense climbers may be able to travel more freely and rapidly in the open upper story of the trees than they could on the ground.

(*6*) Animals that glide must climb to reach takeoff points.

Vertebrates That Climb

FISHES and AMPHIBIANS. Several kinds of air-breathing fishes move about on land and may scramble into low vegetation using strong mobile fins and perhaps fin spines as well. None, however, is really scansorial. Among amphibians, the many species of tree frogs in at least seven families are expert climbers. There are many arboreal salamanders (family Plethodontidae), some of which are so skilled that they can walk along a string.

REPTILES. Many lizards and snakes are climbers; some rarely descend from shrubs or trees. Noteworthy are the chameleons (family Chamaeleontidae), geckos, (Gekkonidae), various iguanids (Iguanidae) including the anoline lizards, and the tropical tree snakes.

BIRDS. Excluding birds that merely perch in trees or forage by flitting from twig to twig, some remain that are truly scansorial. These are the woodpeckers, woodhewers, creepers (of two families), nuthatches (of three families), parrots, crossbills, some of the ovenbirds, and the hoatzin. In addition, rock nuthatches and wall creepers climb on rocks. The nuthatch climbs with feet only, the woodpecker with feet and tail, the parrot with feet and beak, and the hoatzin with feet and wings.

MARSUPIALS and INSECTIVORES. Eleven of the twelve genera of opossums are fine climbers and all seventeen genera of pha-

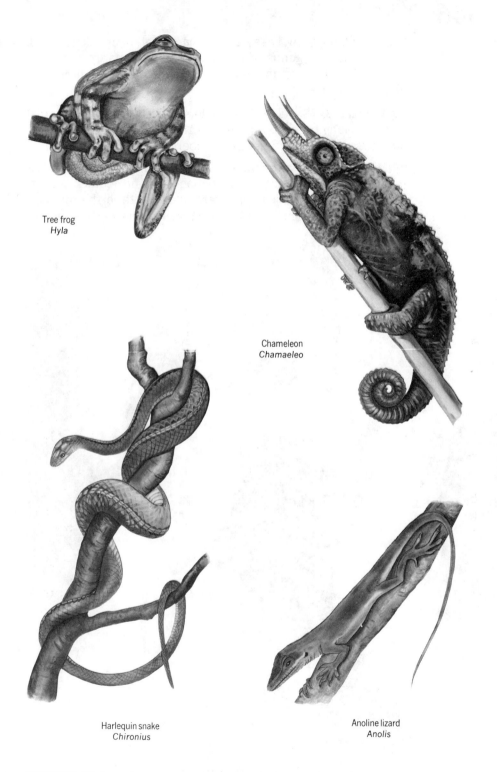

Tree frog
Hyla

Chameleon
Chamaeleo

Harlequin snake
Chironius

Anoline lizard
Anolis

FIGURE 22-1
REPRESENTATIVE SCANSORIAL AMPHIBIANS AND REPTILES.

langers climb with ease. Some marsupial mice climb, and, surprisingly, one genus of kangaroo has secondarily become a climber. Among insectivores, the five genera of tree shrews are highly scansorial.

COLUGO and BATS. Like other gliders, the colugo is an arboreal climber. Many bats roost in trees without doing much climbing, but several genera are nimble climbers.

PRIMATES. Nearly all primates are skilled climbers, and the few that climb little or not at all (baboons, some lemurs, gorilla, man) had arboreal ancestors. Particularly noteworthy for their structural adaptations or climbing behavior are certain lemurs

Nuthatch
Sitta

Parrot
Pyrrhura

Flicker
Colaptes

FIGURE 22-2
REPRESENTATIVE AVIAN CLIMBERS.

(family Lemuridae); indris (Indridae); lorises, pottos, and galagos (Lorisiidae); tarsiers (Tarsiidae); spider and woolly
monkeys (Cebidae); langurs (Cercopithecidae); and gibbons
and the orangutan (Pongidae).

EDENTATES and PANGOLINS. The two smaller genera of anteaters are climbers, as are the two genera of sloths. Some
pangolins are arboreal.

RODENTS. Scansorial rodents are in general less structurally
modified than are the better climbers of other orders. Nevertheless, there are more climbers in this order than in any other.
The common names of many are unfamiliar or poorly established. Climbing rodents include tree squirrels, "flying"
squirrels, chipmunks (Sciuridae); scaly-tailed squirrels (Anomaluridae); vesper rats, harvest mice, gerbil mice, red-nosed mice,
tree mice, pine mice, wood rats (Cricetidae); climbing rats and
mice, forest rats and mice, tree rats, cloud rats (Muridae); dor-
mice (Gliridae); New World porcupines (Erethizontidae); and
echimyid rats (Echimyidae).

CARNIVORES, HYRAXES, and UNGULATES. Climbing carnivores
are little modified structurally for the habit, yet they include all
members of the raccoon family, some members of the weasel,
bear, mongoose, and cat families, and two kinds of foxes. One
genus of hyrax is remarkably skilled at climbing smooth tree
trunks, and others scramble among rocks. Excepting the occasional acrobatics of some goats, no ungulate climbs trees,
yet mountain goats and sheep, chamois, tahrs, and the klipspringer are master rock-scramblers.

Climbers have two basic requirements: (1) They must propel
themselves on a uniquely discontinuous and three-dimensional
substrate, and (2) they must avoid falling, both when moving
and at rest, under particularly difficult circumstances.

Of course, runners and diggers must also propel themselves
on a somewhat uneven substrate and must also avoid falling,
so it is not surprising that the structural and behavioral adaptations of climbers differ in degree, but usually not in kind, from
the adaptations of many other animals. Indeed, many climbers
are not markedly modified for their locomotor habit. The tree
mouse looks about like the terrestrial mouse, the scansorial
and terrestrial species of murine opossums are similar, and the
climbing fennec and nonclimbing red fox are constructed in
nearly the same way. Further, of all the locomotor specializations, ability to climb combines with most other specializations:
The climbing leopard also sprints, the tree kangaroo also hops,
the anteater also digs, the tree frog also swims, the colugo also

Requirements and Basic Mechanisms of Climbers

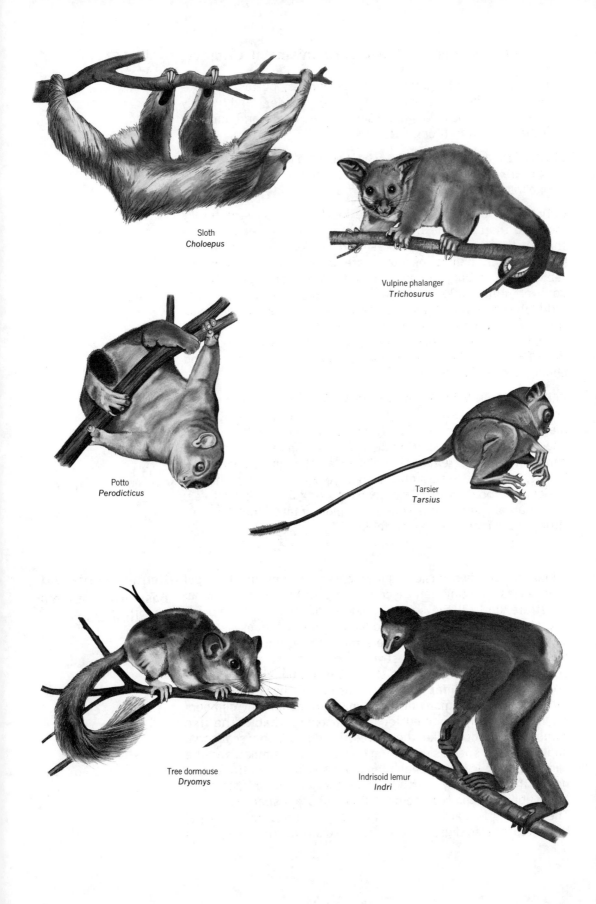

Sloth
Choloepus

Vulpine phalanger
Trichosurus

Potto
Perodicticus

Tarsier
Tarsius

Tree dormouse
Dryomys

Indrisoid lemur
Indri

glides, and the parrot also flies. Nevertheless, some climbers (tree frogs, various salamanders, geckos, anoles, several bats) do utilize unique mechanisms, and scores of others have modified more familiar mechanisms to a striking degree.

The requirement that climbers propel themselves over a discontinuous substrate leads to adaptations for leaping, springing, swinging, and reaching and pulling. The requirement that climbers avoid falling under difficult circumstances leads to adaptations for grasping, balancing, bracing, cushioning, applying suction, clinging, hooking, and adhering. These adaptations will be analyzed, but since they are numerous, and combine in many ways (springing with grasping, swinging with hooking, running with clinging, walking with adhering, running and leaping with grasping and cushioning, etc.), it will be useful to discuss first the few basic principles a climber can utilize to remain in contact with sloping rocks, branches, or twigs.

The Role of Friction. When a monkey stands on a horizontal log with each foot directly under its girdle, the downward force of the body is countered by an equal upward force of the support. There is compression at the interface between foot and log, but virtually no shear. Accordingly, there is no tendency for the foot to slip. The thrust of the foot against the log (**T**), and the force that is normal (i.e., perpendicular) to the surface of the log (**N**) are the same (Figure 22-4A). This would also be true of a standing horse, and the monkey is at the moment no more climbing than the horse.

If the monkey walks along the horizontal log, then the foot pushes against the surface at an angle to the vertical. This thrust can be divided into a component that is normal to the surface and one that is shearing, or parallel to the surface (**P** in Figure 22-4B). (See pp. 460, 461 for the method of dividing forces.) If the monkey is not to slip, **P** must be countered by an equal frictional force **(F)**, which is the mechanical resistance to motion of the foot along its support. Likewise, the horse is able to walk only because friction opposes the propulsive forces along the ground.

If the monkey stands on a branch that is inclined to the horizontal, then **T** can be divided into **N** and **P** even though there is no motion (part C), and friction is required to keep the foot from sliding down the branch. If the monkey walks up the inclined branch, **N** is decreased and **P** is increased (part D). Consequently, friction must also increase if the monkey is not to slip. The climbing monkey and terrestrial horse have now parted company. It is evident that adaptations for climbing include mechanisms for maximizing friction.

How can that be done? Friction is a complicated phenomenom which does not lend itself to exact analysis. The max-

FIGURE 22-3

REPRESENTATIVE MAMMALIAN CLIMBERS.

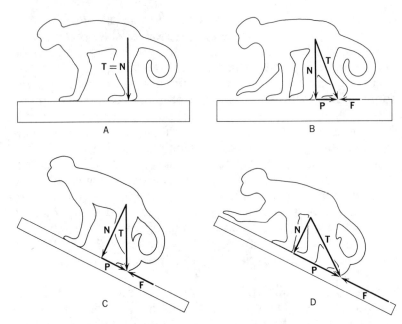

FIGURE 22-4
THE ROLE OF FRICTION IN PREVENTING SLIPPAGE.

imum friction that can be developed before sliding starts depends on the kinds and textures of the materials in contact, and on the force acting between them. The approximate relationship for dry surfaces is $F = \mu N$, where μ is the coefficient of friction. The value of μ might be roughly 0.6 for footpad on bark. (Since the footpad is somewhat elastic, F is in fact proportional to a fractional power of N, and μ decreases as N increases. Actual values are not available for the materials involved.) If sliding starts, μ decreases. As drivers may learn, it is easier to prevent a skid than to stop one.

It is obvious from the formula $F = \mu N$, that climbers could increase F by (1) selecting substrates that would give high values for μ, (2) evolving integumentary surfaces that would increase μ, and (3) developing mechanisms for increasing N. Methods 1 and 2 invite further study. Presumably, extra care is taken when the substrate is wet because lubrication greatly reduces friction (and alters the formula for its calculation). As we shall see, climbers are efficient at increasing N in various ways. These do not include, however, marked increase in body weight. Climbers are of medium to small size so they will not break the branches that must support them and so they can be agile.

As the formula shows, maximum friction tends to be independent of apparent area of contact. Even flat, polished surfaces actually touch at only a limited number of microscopic high points which constitute a small fraction of the visible area of apparent contact. If the visible area of contact is decreased, pressure on the remaining area is increased and more microscopic points are forced into contact, thus maintaining about the same area of actual interaction between the surfaces. This is why the klipspringer (an antelope) can perform feats of rock climbing on the tips of very tiny hoofs. Nevertheless, some climbers have large footpads. This reduces abrasion per unit area of the integument and, because large pads that are also flexible tend to touch the substrate in several planes, increases stability by preparing the foot to resist, with friction, disrupting forces coming from various directions in space. (Large pads may also increase interlocking—see below.)

The Role of Interlocking. If the flat surface between an object and its support is inclined to the horizontal, then, as noted above, there is a force, **P**, parallel to the surface which must be opposed by an equal frictional force, **F**, if slipping is to be avoided. Force **P**, however, has a horizontal component (**H** in Figure 22-5 A), and slipping could also be prevented by a counterforce (**H′**). If the support has a side wall, this counterforce is provided (part B). Interlocking has then substituted for friction in providing stability. When the claw of an iguana, parrot, or chipmunk lodges in a crevice of a rock or branch, this method of support is operative.

If the object instead contacts its support on a dozen or so small sloping surfaces and on as many small side walls, then the same interlocking principle applies (part C). The numerous stiff tail feathers of a woodpecker and the horny plates under the tail of a scaly-tailed squirrel exemplify this kind of support when they press into rough bark.

Intermeshing of fine points of contact is also one basis for frictional force, so as interlocking surfaces become quite small (e.g., scales on the foot of a small lizard, or "fingerprint" ridges

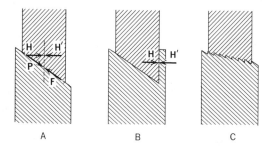

FIGURE 22-5
THE ROLE OF INTERLOCKING IN PRE-VENTING SLIPPAGE.

on the finger balls of a galago), the boundary between friction and interlocking as methods of support becomes fuzzy. A cushion-like footpad that contacts both microscopic and macroscopic projections on a branch utilizes both friction and interlocking to prevent slipping. The hook-like claw of a sloth might be prevented by friction from slipping along a smooth pole, but if its sharp edge were to cut slightly into the pole, interlocking would also contribute support.

The Role of Adhesion. When one smooth metal slides over another, great pressure and high temperature at the microscopic points of contact may cause molecular bonds to form. This is another basis for frictional force. Ordinarily, however, dry materials cannot be brought close enough together at enough points, even if polished and clean, for intermolecular forces to establish a significant amount of bonding between them. If a thin layer of a suitable adhesive is introduced between the materials, then they adhere because the adhesive wets (i.e., broadly contacts) each surface and is attracted to each. Thus, a glass coverslip adheres to a vertical windowpane when bonded by a thin film of water. Common experience tells us that sticky materials are better adhesives than water, and the fingerpads of some amphibians secrete a sticky material which bonds them to vertical leaves and stems. (Hard-setting adhesives such as epoxy and glue do not form instant bonds and cannot be quickly released. Climbers could hardly use such hard-setting adhesives for the bonds they require.)

The ability of the dry-footed geckos and anoline lizards to walk upside down on glass long defied explanation. Friction and interlocking are out (despite published claims to the contrary). There is now experimental evidence that the highly specialized structure of their toes (see below) enables them to establish intermolecular attractions (or Van der Waals forces) with the substrate; that is, adhesion without glue or sticky adhesive. The explanation makes the fact no less miraculous!

Adaptations for Propulsion

Walking, Running, Leaping, and Springing. Animals that commonly walk or run along more or less horizontal branches (iguanas, tree mice, anteaters) have no problems of propulsion not shared by their terrestrial relatives. The feet, and sometimes the tail, may be modified to grip the substrate, but the remainder of the body is not distinctive.

Some climbers commonly, though not exclusively, propel themselves by leaping from one support to another. The jump may be somewhat upward, but usually is outward or partly downward. The animal may be moving when it leaps, and the body is more or less horizontal at takeoff. Examples are certain arboreal snakes; numerous lizards; tree squirrels; capuchin, howler, vervet, and proboscis monkeys; langurs; and arboreal mangabeys. These climbers (except the snakes) tend to have long limbs, slender bones, and muscle mechanics similar to that of cursors, though without relative shortening of proximal limb segments. The back is relatively long, strong, and flexible.

Several primates propel themselves primarily by springing. When at rest they tend to hold the body in a vertical position, and they are often stationary before takeoff. The jump may be in any direction including steeply upward. These highly specialized animals are the tarsiers, galagos, indris, and two genera of lemurs. The smaller of these prodigious springers can jump 2 m straight upward, and the larger ones may jump 10 m out and down from one tree to another. All have the long hind legs, flexed knee posture, and general limb mechanics of saltators (see pp. 512, 513), although the femur is not shorter than the tibia. The foot (unlike that of jerboas and kangaroo rats) must be adapted for gripping the substrate, but in tarsiers and galagos it is also much elongated for springing by lengthening of the navicular and calcaneum bones.

Reaching and Pulling. Many of the more adept climbers propel themselves entirely or in part by reaching from one support to another and pulling themselves along. The orangutan, pottos, lorises, and (in their upside-down way) sloths are noteworthy examples, though some frogs, chameleons, opossums, phalangers, various monkeys, the colugo, echimyid rats, and many others also use this method of propulsion. There is some evidence that orangutan-like climbing was in the ancestry of man.

These animals must meet three principal requirements. The first is a long reach. Hence, (with birds excepted) the limbs are longer than for any other locomotor adaptation except gliding, springing (hind limbs), and flying (forelimbs). Proximal and

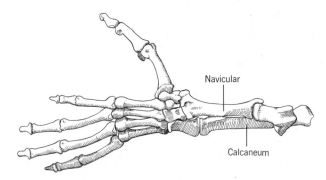

FIGURE 22-6
LENGTHENED TARSUS OF A SPRINGER shown by the left foot of the primate *Galago.*

middle limb segments are subequal in length. The feet are large, yet must respond more to the needs of gripping than propulsion, so they do not lengthen as they do in cursors and terrestrial saltators.

Various tree snakes provide for reaching in a very different way. The zygapophyses of the anterior vertebrae together with a complex musculature prevent dorsoflection of the spine so the body will not sag as it is extended from one support toward another.

The second requirement is flexibility and agility. To gain strength with a wide range of movement, the heads of humerus and femur are not only spherical in curvature but also represent larger portions of complete spheres than is usual. (This contrasts with the conditions in cursors, diggers, and flyers.) The girdles, even of some scansorial reptiles, are modified to allow freedom of movement. Toward this end, the scapula and clavicle of the mammals tend to become modified in ways which, being even more extreme for swingers, are described in the next section. To assure maximum pronation and supination of the forearm, the ulna and radius are free and about equally developed, the proximal head of the radius is round, the radial notch on which the radius rotates is evenly curved and is lateral in position (not anterior as for cursors), and a styloid process at the distal end of the ulna forms a pivot around which the carpus turns. Similarly, the fibula is free and relatively large. The wrist joint is ellipsoid, not hinge-like. Considerable rotation, adduction, and abduction may be possible within the tarsus. Splines and grooves are relatively little developed at limb and foot joints.

Third, these climbers require appropriate bone-muscle mechanics. Marked strength is not needed, so muscles and muscle attachments are not prominent, and the bones are light and slender. Many of these animals (and particularly the sloths) commonly assume postures in which extensor muscles do not oppose gravity in the usual manner. Hence, these muscles are less developed, and their in-levers are shorter than is characteristic in terrestrial mammals. Thus, the in-lever of the triceps (the olecranon process) ranges from about $1/8$ (an opossum) to $1/12$ or less (lorises, sloths) of the out-lever (the length of the remainder of the ulna). Flexors, pronators, supinators, and abductors are better developed.

Swinging. Some primates propel themselves by swinging under the branches. Unlike other methods of propulsion, only

FIGURE 22-7

FEATURES OF THE APPENDICULAR SKELETON OF CERTAIN CLIMBERS shown by the left leg of a sloth, *Choloepus* (left), and left arm of a spider monkey, *Ateles* (right).

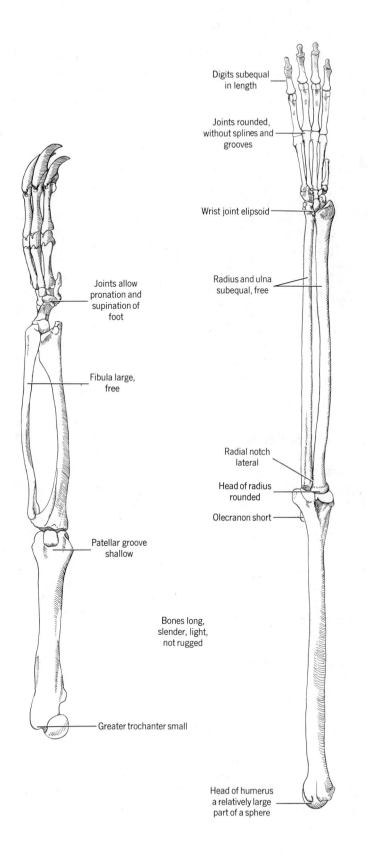

Digits subequal
in length

Joints rounded,
without splines and
grooves

Wrist joint elipsoid

Radius and ulna
subequal, free

Joints allow
pronation and
supination of
foot

Fibula large,
free

Radial notch
lateral

Head of radius
rounded

Olecranon short

Patellar groove
shallow

Bones long,
slender, light,
not rugged

Greater trochanter small

Head of humerus
a relatively large
part of a sphere

the forelimbs are used. Swinging is best exemplified by gibbons and the related siamang, though it is used on occasion by spider, woolly, howler, langur, colobus, and proboscis monkeys and by the chimpanzee, orangutan, and (more rarely) gorilla. (Climbers that swing have long been called brachiators, but unfortunately primatologists have been so inconsistent in the inclusion or exclusion of animals that swing only occasionally that the term is now ambiguous.)

The climber leans, or virtually falls forward, reaches for an overhead support with one hand, swings pendulum-fashion down under the support, flexes or rotates the suspending arm a little on the upswing to add impetus, and stretches the other arm to another support which may be reached only after a period of free-floating travel. The gibbons and siamang scarcely climb in any other way and are remarkable acrobats.

Swingers, like springers, are highly adapted for their specialty. They have the same adaptations for reaching, agility, and use of arms under tension as the reach-and-pull climbers, only the modifications are more extreme. The hind limbs are long relative to the trunk, particularly in gibbons and spider monkeys, but the forelimbs are disproportionately long, becoming even twice or more as long as the trunk in the orangutan, gibbons, and siamang. The fossa on the humerus which accommodates the very short olecranon process is deep so the elbow can be completely straightened. The architecture of the shoulder enables it to withstand tension and to move sideways and overhead to a unique degree: The clavicle, acromion process, and associated deltoid muscle are prominent. The glenoid cavity is oriented forward. The latissimus dorsi, pectoralis, biceps, and long head of triceps brace the shoulder against tension. Also strong are the trapezius, which pulls the acromion process in toward the neck, and the anterior serratus, which pulls the posterior angle of the scapula out toward the side of the chest. The insertion and orientation of these muscles are also modified so that together they rotate the scapula on the trunk to a unique degree. This raises the arm on the body, and, during the swing, probably rotates the body on the outstretched arm.

Supination of the supporting forearm coupled with twisting of the trunk advances the leading (nonsupporting) shoulder during a swing, adding reach and force to the action. To help accomplish this the supinator is strong, its in-lever is increased by bowing of the radius, the sternum is broad, and the chest is broad rather than deep.

In contrast to cursors and quadrupedal leapers, swingers have short compact backs so the trunk can swing as a unit. The lumbar area contributes the least to locomotion, so it tends to be inflexible, has relatively few vertebrae, and these vertebrae

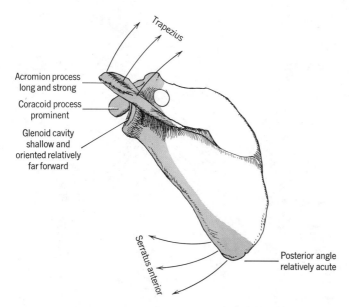

FIGURE 22-8

ADAPTATIONS OF THE SCAPULA FOR REACHING AND SWINGING shown by the spider monkey, *Ateles.* Shading indicates areas of relatively heavy bone which includes the lever of the mechanism that pivots the scapula.

have short centra. The zygapophyses of mammalian lumbar vertebrae are constructed to allow flection and extension but to limit rotation of the spine. The zygapophyses of anterior thoracic vertebrae do allow rotation. The transition occurs within one vertebra. This vertebra (the diaphragmatic vertebra) is farther posterior in swingers (which rotate the spine with each swing) than in cursors and leapers (which flex and extend but do not rotate the spine).

Some climbers are slow-moving (chameleons, lorises) whereas many runners, leapers, and swingers are very quick. The latter require remarkably rapid and precise neuromuscular responses. The morphological bases for such control may be assumed to include the relative prominence of cerebellum, olivary nuclei, ruber nucleus, primary motor and sensory cortex, and optic tracts and centers. The eyes are large and face forward to provide overlapping fields of vision, and hence depth perception.

Adaptations for Maintaining Contact with the Substrate

Grasping. When the fingers and palm of a chameleon, potto, or man encircle a twig or pole and grip firmly, muscular effort creates forces which are normal to the surface of the support. These forces increase frictional resistance to slipping: the

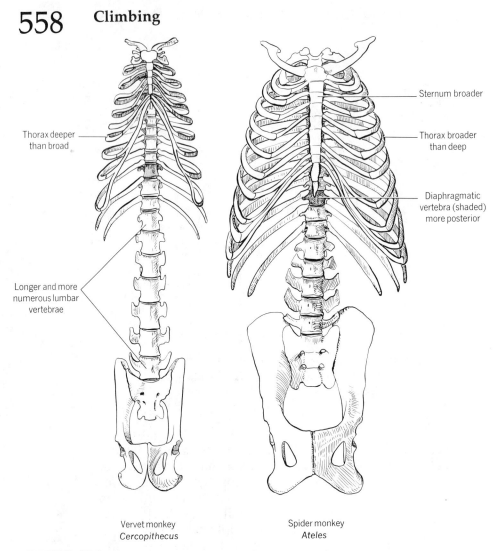

Thorax deeper than broad

Longer and more numerous lumbar vertebrae

Sternum broader

Thorax broader than deep

Diaphragmatic vertebra (shaded) more posterior

Vervet monkey
Cercopithecus

Spider monkey
Ateles

FIGURE 22-9

CONTRAST BETWEEN THE TRUNK SKELETONS OF A MONKEY THAT LEAPS (left) AND ONE THAT SWINGS (right).

tighter the grip the more resistance. An animal with strong digital flexors can thus supplement the normal forces that it can develop using only its body weight. Further, by grasping it can resist slipping in any direction.

Grasping is a particularly versatile and effective way of maintaining contact with the substrate, and the different climbers have independently evolved many grasping mechanisms. Various snakes grasp branches with coils of the body to provide support. The first digit is opposed to the others in one or both pairs of feet of some tree frogs, salamanders, birds, opossums, and many primates. The second digit of the hand tends not to be very effective in a power grip (even in man) and in the potto

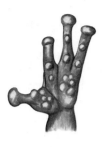

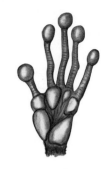

| Tree frog
Hyla | Geckonid lizard
Gecko | Potto
Perodicticus | Tarsier
Tarsius |

FIGURE 22-10

LEFT HANDS OF SOME CLIMBERS.

| Arboreal salamander
Bolitoglossa | Chameleon
Chamaeleo | Parrot
Pyrrhura | Indrisoid lemur
Indri |

FIGURE 22-11

LEFT FEET OF SOME CLIMBERS.

and lorises has become short and weak. The koala, and some other phalangers, grasp between the second and third digits, as do the chameleons with the hind foot. The forefoot of these lizards grasps between the third and fourth digits, and echimyid rats do the same. Parrots and some other avian climbers oppose digits two and three with digits one and four. The palms, soles, and digits of graspers are naked and sensitive.

Terrestrial mammals often irretrievably lose one or more lateral digits. If their descendants become secondarily arboreal, those digits cannot be used for grasping, and nature has produced interesting compensations: The two-toed anteater can depress its heel into strong opposition to its remaining digits, and one Central American porcupine can fold the pad of its foot lengthwise and forcefully grasp between the lateral edges of the footpad.

The tail often evolves into a grasping organ. Such a tail is said to be **prehensile** and is characteristically long, strong, sensitive, and curled at the end. Animals with prehensile tails include some salamanders; chameleons, and several other lizards; some arboreal snakes; opossums; some phalangers; capuchin, spider, and woolly monkeys; anteaters; pangolins; the kinkajou; various rats and mice; and one porcupine. Most of these animals curl the tail ventrally, but the porcupine curls its tail dorsally. Prehensile tails tend to be flexible at the base and to have short broad vertebrae near the end. Often they have naked pads where they grasp.

Balancing, Bracing, Cushioning, and Sucking. Climbers that move quickly must be effective balancers. It follows from principles presented on p. 508 that scansorial mammals that walk or run on top of branches can increase stability by lowering their center of gravity. Arboreal salamanders, snakes, and lizards keep their center of gravity very low. Tree squirrels, rats and mice, marmosets, and the kinkajou either have short legs or flex the legs to hold the body low. Birds that forage on tree trunks have short legs to keep their center of gravity close to their support. The girdles of the chameleon permit the legs to come vertically under the body (unusual in reptiles) so the animal can balance over narrow stems. Climbers that swing or hang under their support using hook-like appendages (see below) are in stable equilibrium—like rocking chairs they tend to maintain position. Most climbers have long tails which contribute to the maintenance of balance; climbers without tails swing, hang, or move slowly.

Numerous climbers use their tails as braces, struts, or props. Woodpeckers, woodhewers, and creepers prop themselves with their stiff tails; the terminal segment of the spine (pygostyle) and its musculature enlarge for the purpose. Some species of tarsier have a naked and roughened area at the base of the tail to make it a better prop. The scaly-tailed squirrel has horny ventral scales on its tail.

The appendages of most climbers have broad, soft, cushion-like pads. These are often roughened by small grooves or fingerprints which increase friction and interlocking. Footpads are particularly well-developed in primates, anteaters, and porcupines. Fingerpads are conspicuous in tree frogs, some arboreal salamanders, opossums, phalangers, indris, pottos, galagos, and tarsiers. Climbers with prehensile tails have tail pads. Feet having such cushions tend to be broad and loose. Metapodial bones are well-spaced, round in cross-section, rounded on their distal ends, and devoid of splines at the joints. Phalanges are similarly rounded except that the terminal phalanx may be somewhat spatulate. Claws or nails are positioned so as not to interfere with the action of the pads.

The suction cup of human technology is usually a shallow cup of rubber, with the rim pressed against a smooth surface. The elasticity of the rubber keeps the rim tightly pressed against the surface and thus reduces pressure inside the cup so that atmospheric pressure presses the cup against the surface. Shearing force is resisted by friction. Disk-winged bats of two genera have suction cups on knuckles and ankles which function in the same way. Elastic tissue, not muscular tension, maintains suction within the disk once it is seated. The tree hyrax clings firmly to smooth vertical tree trunks without grasping or using claws. It is claimed that muscles pull inward on a cleft at the center of the soft footpad to create suction. Glands keep the pad moist to establish a good seal.

Clinging and Hooking.   Most climbers that do not grasp use strong, much-curved claws to cling to the substrate. The tips of the claws interlock with cracks and crevices. (Because it lacks claws, a salamander has hook-like terminal phalanges.) Usually the weight of the animal is insufficient to maintain firm interlocking, so one set of claws must be pulled against another. The two feet of a pair may sprawl wide on opposite sides of the body (rock lizard), or the hind feet may be turned toes-backward so foreclaws can pull against hind claws (a tree squirrel). In addition, scansorial birds pull with one or two claws of the foot against the other claws of the same foot. The digital flexors are

FIGURE 22-12
SUCTION DISK ON THE WING OF THE BAT, *Thyroptera.*

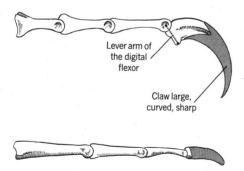

Lever arm of
the digital
flexor

Claw large,
curved, sharp

FIGURE 22-13
ADAPTATIONS OF THE AVIAN FOOT FOR CLIMBING shown by third toes of a flicker, *Colaptes* **(above), and for contrast, of the nonclimbing, nonperching merganser,** *Mergus* **(below).**

strong and the terminal phalanx is designed to provide a good in-lever (Figure 22-13).

Some climbers modify the appendages as hooks and swing or hang under their support. Sloths and pangolins use 1–3 very long, strong, curved claws as their hooks. Bats and the colugo use 5 subequal claws. Primates that swing use the four fingers of the hand together as a hook. The hand is very long, and the phalanges are curved to conform to the round cross-section of the branches. Prehensile tails can be used as hooks as well as grasping organs. In all these instances, the flexor tendons of the hook are short enough to passively prevent the hook from opening up: a dead gibbon can be hung by an upstretched arm; a dead spider monkey can be hung by the end of its tail.

Adhering. The expanded finger balls of tree frogs are not only friction pads. They enclose glands having a sticky secretion which these little frogs use to glue themselves to rocks, leaves, and stems. They usually press their moist tummies against the substrate so these will also adhere.

Many tropical salamanders have webs between their well-spread digits. The result is a large common adhesive pad. To break contact, the pad is curled up from its margin or lifted from the back. Frogs may have an "extra" segment of cartilage or bone just proximal to the terminal phalanx. This seems to assist the animal in feeling about with the tips of its toes for the best spot to make contact with the substrate.

The gecko has sharp claws with which to cling to rough substrates. If a smooth support is moderately inclined, the animal can maintain its position by friction. On steep smooth surfaces and overhangs, however, this creature brings into play one of nature's most remarkable adaptive mechanisms. Under each toe are 16–21 broad imbricated scales or lamellae. On the exposed surfaces of the lamellae of each toe are up to 150,000 hair-like setae ranging from 30 to 130 μ long. Each seta branches into about 2000 bristles, and each of these has a saucer-like endplate measuring about 0.2 μ in diameter. There are in all some 100 million of these endplates which touch the substrate at points on their rims. (All these numbers vary by species.) A slight shearing force (from body weight or muscular pull) is required to give the setae the S curve which positions the endplates on the substrate. Blood sinuses under the lamellae cushion the toes so a maximum number of endplates can reach into any irregularities of the surface. Collectively there are so many close contacts that the animal adheres by surface tension.

The contact is firm: When one investigator tried to pull a large gecko from a vertical pane of glass, the glass broke. Adhesion to vertical glass continues even when the animal is

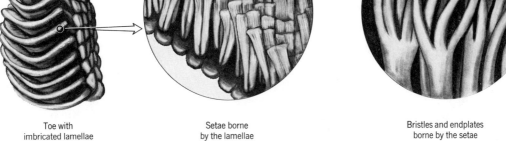

| Toe with imbricated lamellae | Setae borne by the lamellae | Bristles and endplates borne by the setae (Diameter of field is 2.5 μ) |

FIGURE 22-14

ADAPTATIONS FOR CLIMBING BY DRY ADHESION SHOWN BY A TOE OF THE LIZARD, *Gecko.*

dead and the body is in a vacuum. However, the lizards adhere with difficulty to materials having low surface tension (e.g., Teflon), and if the setae become dirty or mussed, climbing ability is impaired until after the skin is shed and replaced by new setae. To break the contacts, the lizard peels its toes off the substrate by rolling them up "backwards," starting from the tips.

The unrelated anoline lizards have shorter unbranched setae, but they function the same way. This is an amazing example of convergent evolution.

23

Swimming
and Diving

Vertebrates that live in water are said to be **aquatic.** Unfortunately, there is no term for the experts among them that swim with particular skill, speed, or endurance. All fishes are **primary swimmers**—their ancestors also swam. Other swimming vertebrates are **secondary swimmers**—their ancestors passed through a terrestrial state, and consequently they have structural and physiological handicaps which have prevented most of them from becoming entirely aquatic again.

Nearly all vertebrates can swim somewhat, and there are many expert swimmers in every class; there can be no sharp division between swimmers and nonswimmers. However, most that will receive attention here seek food and refuge in the water and can swim below the surface.

565

Relatively much has been published about swimming, yet detailed analysis is difficult, and much remains to be learned.

Advantages of Swimming and Diving

It is academic to ask about the survival value of an aquatic life to primary swimmers: It is a successful way of life, and nature has provided no alternative to most of them. As we saw in Part II of this book, the change to terrestrial life was so profound that it took about 100 million years to complete. The reverse trend back to water has been "easier" and has occurred many times. Secondary swimmers and divers may have any of the following advantages over nonswimmers:

(1) They gain access to a wide variety of aquatic foods including fishes, plankton, and larger invertebrates and plants.

(2) They escape terrestrial predators. They may also subject themselves to aquatic predators, of course, but the process of evolution has in specific instances moved in the direction of greater safety.

(3) The oceans and major inland waterways are favorable avenues for dispersal and migration. Fur seals commonly migrate 12,000 km/yr, and the gray whale's annual round trip migration is about 19,000 km.

Vertebrates That Swim and Dive

FISHES. Many fishes are slow and relatively inactive, but the performances of others are impressive. Trout can accelerate at 50 m/sec^2, achieving maximum speed of 9 body lengths/sec in 0.15 sec. Salmon swam 1000 km up one river with an expenditure of energy equivalent to an average swimming speed in quiet water of 4.2 km/hr (2.6 mile/hr). The mackerel can reach 35 km/hr and the barracuda 43 km/hr. By an ingenious device it was accurately learned that the tuna can sprint 74.6 km/hr and the wahoo 77 km/hr (48 mile/hr or 41.6 knots). These are rates of 18–21 body lengths/sec. Equal speeds have been claimed for bonito and albacore, and considerably greater speeds for marlin, sailfish, and swordfish, but the records need confirmation. For comparison, the human record is 100 m at 6.9 km/hr (4.3 mile/hr), and unlike fishes, humans cannot swim significantly faster for very short periods of time.

AMPHIBIANS and REPTILES. All members of one family of caecilians (Typhlonectidae) are exclusively swimmers. Most frogs and salamanders are aquatic or semiaquatic, but they are not sustained swimmers and they are not highly modified for swimming. The same is true of the semiaquatic iguanid and agamid lizards. Alligators, crocodiles, and turtles are more active swimmers; the sea turtles remain at sea except to lay eggs. A turtle of the Nile River dives to 10 m. Various snakes are com-

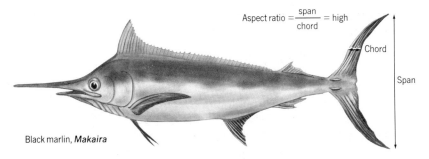

$$\text{Aspect ratio} = \frac{\text{span}}{\text{chord}} = \text{high}$$

Chord

Span

Black marlin, *Makaira*

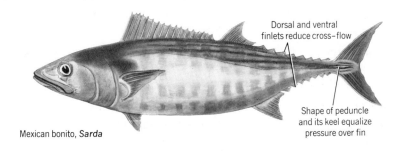

Dorsal and ventral finlets reduce cross–flow

Shape of peduncle and its keel equalize pressure over fin

Mexican bonito, *Sarda*

Body spindle-shaped

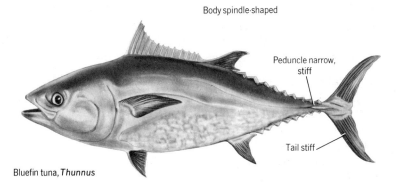

Peduncle narrow, stiff

Tail stiff

Bluefin tuna, *Thunnus*

FIGURE 23-1

EXAMPLES OF FISHES OF THE FAMILY SCOMBRIDAE showing some adaptations for fast swimming.

petent swimmers, and of the fifteen genera of sea snakes, most never leave the water. At least one species ventures hundreds of miles from shore and can remain submerged for 2 hr. About 80 million years ago there lived a group of sea lizards called mosasaurs. They had paddle-like limbs and elongate bodies. Plesiosaurs, which formed an order in the extinct reptilian subclass Euryapsida, were mostly marine. Some were huge, slow swimmers with bulky bodies and long necks; others were more streamlined. Another order of euryapsids, the highly adapted ichthyosaurs, were remarkably convergent with dolphins (Figure 1-6).

BIRDS. Even when wading birds are excluded, there are many aquatic birds. Geese, swans, and some gulls, shearwaters, petrels, and ducks swim well on the surface but do not dive. Pelicans, diving petrels, tropicbirds, boobies, and terns dive from the wing. Penguins, which swim up to 36 km/hr, and the recently exterminated great auk dive and swim underwater using their paddle-like flightless wings. Other auks, murres, puffins, and diving petrels both swim and fly with their narrow wings. Cormorants, loons, grebes, and some diving ducks are skilled divers that swim underwater with their feet. Both feet and wings are used in swimming by some of these birds. Loons may dive 55 m deep and stay underwater for 15 min.

MONOTREMES AND MARSUPIALS. The web-footed platypus seeks food on stream bottoms. The water opossum swims, dives and feeds in the water.

INSECTIVORES and RODENTS. Among insectivores, several genera of shrews, one tenrec, two genera of otter shrews, and two desmans are aquatic. Many rodents swim. The common names of some are poorly established; the list includes the mountain beaver, beaver, capybara, and nutria (all in separate families), various water rats and mice, marsh rats, and the muskrat (family Cricetidae), and aquatic rats (family Muridae). Some of these dive for periods of 6–10 min.

SIRENIANS and the HIPPOPOTAMUS. The manatee, dugong, and recently exterminated sea cow of the order Sirenia are large and sluggish (though graceful) vegetarians which are among the most modified of aquatic mammals. They can remain submerged for at least 16 min. Underwater moving pictures of the most aquatic artiodactyl, the hippopotamus, reveal surprising proficiency, yet this creature is not in the same league as the master swimmers.

CARNIVORA. The bush dog, otter-civet, and water mongoose all swim and dive. River otters are among the most agile and graceful of mammalian freshwater swimmers; the even more aquatic sea otter dives for shellfish and remains underwater for about 5 min. The suborder Pinnipedia includes seals, sea lions, and walruses which vie with dolphins in their skill as swimmers. True seals rarely reach speeds of about 15 km/hr. Sea lions and the related fur seal can sprint at about 22 km/hr. The sea lion can dive to at least 146 m, the harp seal to 273 m, and the Weddell seal to 600 m. Various seals commonly remain submerged 5–10 min and can double that period. The Wedell seal, which forages under the ice as far as several kilometers from its blow hole, can remain submerged for 1 hr.

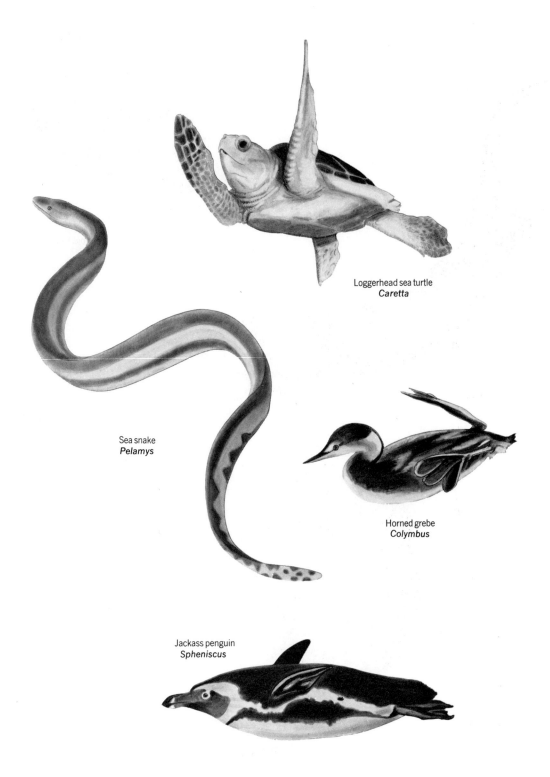

Loggerhead sea turtle
Caretta

Sea snake
Pelamys

Horned grebe
Colymbus

Jackass penguin
Spheniscus

FIGURE 23-2
EXAMPLES OF UNDERWATER SWIMMERS AMONG REPTILES AND BIRDS.

Nutria
Myocastor

Otter
Lutra

Manatee
Trichechus

CETACEANS. Whales, dolphins, and porpoises are the most highly adapted surviving secondary swimmers (Figures 1-3 and 4–11). The gray whale cruises at about 5½ km/hr (3½ mile/hr) on its long journeys. The bottlenose dolphin can swim 30 km/hr for about 7 sec and 22 km/hr for about 50 sec. The spotted dolphin can achieve its top speed of 40 km/hr (25 mile/hr) in only 2 sec, and it is said that entire schools of these speedy mariners travel at about 30 km/hr for as long as 25 min. The killer whale is considered faster, and the finback whale and roqual are said to be able to sprint to 65 km/hr (40 mile/hr). The great blue whale, the largest whale of all, is reported to be able to swim 37 km/hr (23 mile/hr) for 10 min and 27 km/hr for 2 hr!

Dolphins can remain submerged for 10-20 min. Other records are 50 min for the blue whale, 90 min for the sperm whale, and 120 min for the bottlenose whale. A dolphin was trained to dive repeatedly to 300 m . A fin whale dove 350 m. Sperm whales have gone as deep as 900 m, where the pressure is 90 atm (1323 lb/in.^{2}).

General Requirements of Swimmers and Divers

All proficient swimmers and divers must (1) reduce the resistance that water offers to motions of the moving body, (2) propel themselves in a relatively dense medium, (3) control vertical position in the water, and (4) maintain orientation and steer the body. Secondary swimmers must also (5) exclude water from their respiratory passages and ears, (6) avoid harm from crushing of gas-filled spaces, (7) alter ears and eyes to function (again) underwater, and (8) modify their respiratory and circulatory physiology to permit suspension of breathing and avoidance of the bubbling of gas in the blood (bends) on returning to the surface after a dive. Some aquatic birds and mammals must also (9) control body temperature in a medium having high thermal conductivity, and (10) adapt their reproductive biology to life in the water.

How have the various swimmers and divers met these many requirements?

Drag

Origins and Nature of Drag. The resistance that a medium (here water) offers to the motion of an object is called drag. There are several sources, or kinds, of drag. They are interdependent, but can best be presented one at a time. First is **frictional drag.** Imagine a fairly smooth, more or less spindle-shaped (i.e., fish-like) rigid object which is moving underwater. Immediately adjacent to its surface there is a very thin layer of water that adheres to the object and moves with it. A short dis-

FIGURE 23-3

EXAMPLES OF AQUATIC MAMMALS of three orders: Rodentia (above), Carnivora (center), and Sirenia (below).

tance away, the water does not move with the object at all. Between the object and the still water is the **boundary layer** where successive layers of water slide past one another; those nearest the object move nearly as fast as it does, and those more and more distant move slower and slower. The shearing forces thus produced tend to slow the moving object and are a source of drag. The boundary layer gets thicker toward the posterior end of the moving object.

If a smooth spindle-shaped object moves slowly through the water, successive layers or lamina of the boundary layer slip past one another without any eddies. Flow is said to be **laminar.** However, if the object moves fast, then where the boundary layer reaches a certain thickness, or where there are even slight roughnesses on the surface of the object, the water curls into complex eddies. The energy that moves the water in these eddies comes from the moving object and greatly increases drag. Such flow is said to be **turbulent.** Turbulent flow produces a thicker boundary layer and much more drag than does laminar flow. To swim very fast, vertebrates must reduce turbulence (though an exception is explained below).

To be realistic, let us examine the variables in some detail before seeking a simplification. Frictional drag is directly proportional to the product of the square of the speed, times the area of the object, times a value called the drag coefficient. The drag coefficient must be calculated in each instance and varies with the shape and surface texture of the object (hence with the laminar or turbulent nature of flow) and with a value

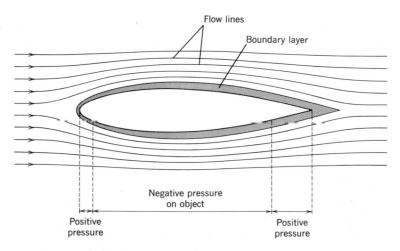

FIGURE 23-4

FLOW LINES, BOUNDARY LAYER, AND PRESSURE DISTRIBUTION WHEN FLOW AROUND A STREAM-LINED OBJECT IS LAMINAR.

called the Reynolds number. This number, in turn, is the ratio of the inertia of the medium to its viscosity, and is calculated using the speed of the object, its length, and the density and viscosity of the medium. It is a dimensionless number having high magnitude (about $10^{5.5}$–$10^{8.5}$) for large fast-swimming vertebrates.

This seems very involved, but let us unscramble it. Speed is squared in the basic formula and occurs again in the calculation of the Reynolds number. It follows that slow-swimming vertebrates have negligible drag no matter how the other variables change: Witness the unstreamlined bodies of the sluggish sea horse and trunkfish. Conversely, drag on rapid swimmers increases very fast with each increment of speed. Metabolic rate must be about doubled every time speed is increased by 1 body length/sec. It appears that the fastest swimmers closely approach the biological limits. Further, in order to swim fast, the experts must reduce as much as possible all factors other than speed that increase drag.

Swimmers cannot control the density or viscosity of the water, so these elements of the formula can be disregarded (though an exception has been suggested for tunas and their relatives). Drag increases with body size, but so does the output of the animal's power plant, and these factors nearly cancel one another. The consequence of some rather complicated physiological considerations seems to be that moderately large swimmers have some advantage. The fastest swimmers are large fishes and small whales. There remain the important variables of body shape and the nature of the surface of the body. Before we see how these are adapted for speed and efficiency, there are other sources of drag to consider.

As a spindle-shaped object moves through the water it displaces a volume of water equal to its own volume plus a volume equivalent to about $1/3$ that of the boundary layer. Also, there is a backflow of water behind the object as it moves along. Under some conditions of speed, size, and shape of object, the backfill is incomplete and separation of the boundary layer occurs which creates suction behind the object and causes water to follow in its wake. (An observer at the stern of a ship easily sees that water in the wake falls back less rapidly than water to the sides of the wake.) Also, there is positive pressure against the anterior and posterior parts of the object and negative pressure at intermediate levels. The energy needed to cause these motions of water and pressure changes is taken from the object. The resultant drag is **pressure drag**. Pressure drag is negligible when the Reynolds number is low, but important when it is high. This kind of drag is complicated to calculate.

Finally, if the object moves on the surface of the water, like a ship or duck, or close enough to the surface to cause surface

FIGURE 23-5
FLOW LINES SHOWING THE TURBULENCE CREATED BY A SWIM-
MING FISH as seen from above. Various other patterns might also be
created, depending on the variables.

waves, then energy is extracted from the object to create the
waves, and **wave drag** occurs. Resistance is greatest when the
object moves just below the surface. The variables associated
with wave drag are not well-known.

These considerations of the drag on a rigid object make it
evident that fast-swimming vertebrates require adaptations of
body form and of the nature of the surface of the body. Since
swimmers are not rigid, drag can also be reduced by certain
behavioral adaptations.

Reduction of Drag by Adaptations of Form. Pressure drag is
low when the body is long and slender, like that of a snake or
eel. Frictional drag is minimal, however, when the body is short
and plump. The best compromise is a spindle that is circular in
cross-section and thickest near the center of its length where
its diameter is $1/4$–$1/5$ of its length. The bodies of tunas, sword-
fishes, and dolphins closely approach this shape. Absence of a
functional neck (primary swimmers, cetaceans, sirenians), symmetry of the head, molding of thorax and body musculature,
and the distribution of fat and blubber all may contribute to
streamlining.

Any projections from the basic spindle usually cause turbu-
lance and increase drag. Accordingly, expert swimmers reduce
or eliminate projections not needed for propulsion and
steering: Swimmers other than mammals have no external ears
or external genitalia in their ancestry. Aquatic mammals secondarily lose their external ears and move the testes back into the
abdomen. Nipples or teats and the penis may be withdrawn
within the body contour when not functioning. Fast primary
swimmers have no limb segments between their fins and
bodies. Fast secondary swimmers have very short proximal limb
segments to again bring the feet or flippers close to the body.
The humerus of cetaceans may be only about as long as it is
wide; the femur of pinnipeds may be less than twice as long as
wide; the femur of diving birds is short and most of the leg
musculature is contained within the contour of the body. Cetaceans and sirenians have reduced the pelvic appendages to in-

ternal vestiges, and some other swimmers position the hind limbs in such a way that they do not protrude but instead extend the contour of the spindle-shaped body. The knee joints of pinnipeds, beavers, and many diving birds are constructed to allow the necessary reorientation of the limb.

Salamanders, crocodilians, and aquatic lizards hold their limbs against the body as they swim with tail and trunk. Lateral fins and flippers that propel the body, on the other hand, must protrude and present a flat surface to the water on the power stroke. Sea turtles and penguins commonly achieve a power stroke on both the downswing and upswing of the flipper by rotating it around its own long axis as the direction of the stroke changes. However, most swimmers with lateral paddles have a power stroke and a recovery stroke. Drag is reduced in various ways during the latter. Rotation of the entire appendage at its base may cause the appendage to cut the water edge-on (flipper of sea lion). The median lobes on the toes of grebes are similarly rotated on the recovery stroke, and the lateral lobes passively fold. Such flippers and lobes, and also median fins and paired appendages that are used primarily for steering, are streamlined in cross-section so that the flow of water over them is nearly laminar when they are presented to the water edge-on. Pectoral flippers of pinnipeds, wings of most diving birds, and paired fins of fishes are pressed against the body when the animal glides. Bony fishes also can reduce the area of their fins by folding. Dorsal and anal fins (including the "sail" of the sailfish) may be retracted into grooves on the body surface during fast swimming. Web-footed tetrapods flex their limbs and adduct and curl their toes on the recovery stroke.

Reduction of Drag by Adaptations of Body Surface and Behavior. If it is perfectly smooth, a rigid spindle-shaped test object achieves laminal flow over the leading 40–50% of its surface. Vertebrates can never be perfectly smooth—eyes, nostrils, gill slits, mouth, appendages, scales, feathers, hair, and scars all tend to create eddies—but fast swimmers approach a smooth body surface. Swift fishes have small smooth scales, or none at all, and are covered with slime. The small scale-like feathers of penguins and the hair of seals and otters form remarkably smooth coverings. Cetaceans and sirenians (and perhaps ichthyosaurs and plesiosaurs) are (or were) secondarily naked and slick-skinned.

(There is an interesting exception to the evolution of smooth skin among swimmers: Posterior to the widest parts of their bodies, tunas and bonitos have a corselet of thick rough scales. This does create turbulence, of course, and hence drag. The turbulence prevents separation of the boundary layer, however, which would create much more drag.)

Dolphins have shown engineers that a specialized integument can prevent or reduce turbulence in other ways, though they have not revealed all their secrets. As eddies start to form, pressure gradients pass in ripples along the body. If the integument is deformed by these ripples, it absorbs energy from the water and tends to damp out the turbulence. The very thin outer membrane of the dolphin's skin transmits pressure differences to an underlying elastically deformable layer which consists of microtubules oriented perpendicular to the skin surface. These contain an oily fluid and are interconnected by narrow channels. The fluid is apparently squeezed from regions of increased pressure to regions of decreased pressure, thus providing for small compressions and expansions of the skin surface. Photographs of fast-swimming dolphins also show large ripples in the skin. These are probably functioning in a similar way in response to pressure waves of greater amplitude, but they are as yet not well-understood. Imitating dolphin skin with rubber and silicone, engineers have reduced drag on test objects, though the original is more effective than the copy and the results have been disputed.

Importantly, fast swimmers outperform submarines and torpedos in ways that involve behavior. The drag on a swimming salmon is about half the drag on a dead salmon that is towed. The factors are complex and are understood only in general terms. It was noted that separation of the boundary layer creates suction which causes water to follow after a swimmer. However, in propelling the body, the tail fin pushes water back, thus tending to cancel this source of drag. The opercula of the fastest bony fishes (suborder Scombroidei) open alternately in synchrony with the undulations of the body. The result is that water is ejected from the gills only into the relatively low-energy water on the convex side of the body where it reduces the thickness of the boundary layer, and hence drag.

Oscillation also tends to create resistance which must be countered for maximum performance. Series of stiff finlets behind the dorsal and anal fins of fast scombroid fishes extend into the boundary layer, fencing it against cross-flow as the tail sweeps right and left (Figure 23-1). Keels on the caudal peduncle streamline it so as to reduce its drag during lateral oscillation. Also, the shape of the peduncle deflects water in such a way as to counter the tendency for water pressure to be greater at the tips of the tail fins.

When animals swim just below the surface, wave drag increases total resistance by as much as five times the minimum. It is unlikely that many vertebrates attempt to swim rapidly in that position. The "playful" leaps of dolphins may be made in part to avoid swimming at the surface while breathing.

The various animal adaptations for reducing drag can be very effective. Some cetaceans evidently attain laminar flow over nearly all of the body, an achievement that far surpasses human technology.

Nature and Principles of Propulsion in Water. Vertebrate swimmers propel themselves only with oscillating mechanisms; they have no analogs of sails, screw propellars, or jet engines (except for small forces acting at the gills). Two kinds of oscillating mechanisms are used. In one, an effective power stroke is followed by a dissimilar and less effective or ineffective recovery stroke. Examples are paired appendages such as the flippers of plesiosaurs and sea lions, the webbed feet of ducks and the platypus, and the wing-paddles of puffins. In the other kind of mechanism, each stroke is effective so the oscillations are symmetrical (or nearly so) and there is no recovery stroke. Examples are the undulating bodies of eels and water snakes, the undulating longitudinal fins of ribbon fishes, and the sweeping tail fins of trout and dolphins.

The more powerful swimmers all propel themselves by oscillations of the tail, and this mechanism will serve to illustrate the general principles of propulsion in water. Either a vertical tail fin sweeps back and forth from side to side (fishes, ichthyosaurs) or horizontal tail flukes sweep up and down (cetaceans, sirenians). Seals trail the hind feet behind the body and use them much as though they were a single vertical fin. The mechanism is the same regardless of direction of motion, so to simplify terminology in relation to space, I shall describe the action of the vertical tail of a fish.

The following factors are basic to the function of the caudal fin: (*1*) It is broad and flat so water cannot easily flow around it but instead will resist its lateral motion. In other words, it has high drag when presented broadside to the water. (*2*) Pivoting at its base, the fin membrane angles back and forth from right to left (Figure 23-6A). (*3*) The base of the tail, or supporting peduncle, moves from side to side carrying the caudal membrane with it (part B). (*4*) These two motions are timed so that the fin angles back to the left as it sweeps from left to right, and back to the right as it sweeps from right to left (part C).

As a first step in the analysis of the function of the tail we note that a consequence of the above is that except at the limits of its travel the caudal fin is constantly thrusting obliquely against the water (force F_t in Figure 23-7). The inertia of the water causes it to push with equal force in the opposite direction (F_w in the figure). Using the method presented in Chapter 19, F_w can be divided into a forward component (F_f) and a lateral component (F_l). Because of the streamlined shape of the fish, the water offers little resistance to F_f, so the body

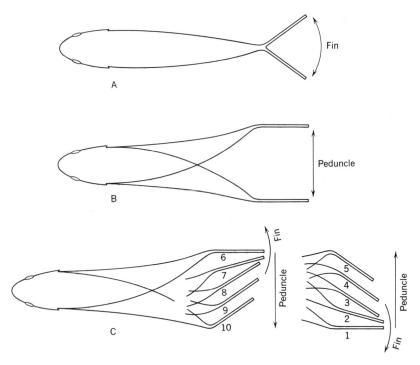

FIGURE 23-6
MOTION OF THE CAUDAL FIN OF A SWIMMING FISH.

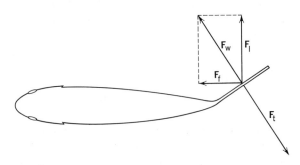

FIGURE 23-7
FORCES ACTING AT THE CAUDAL FIN OF A SWIMMING FISH.

glides forward. Force F_1 tends to cause the fish to pivot around its center of mass; the posterior part of the fish moves right (in the illustration) and the anterior part moves left (the in-torque equalling the out-torque). However, the water offers considerable resistance to sideways motion of the body. The entire body does undulate from side to side, but lateral motion of the heavier, stiffer, flatter, anterior part of the body is much less than the lateral motion of the peduncle of the tail.

In fact, the function of the tail is somewhat more complicated. Pressure fields and "lift" (as explained in Chapter 24) are created around the fin, and when the forward and undulating motions of the body are combined with the motions described above for the caudal fin and its peduncle, the result is the graceful glide shown in Figure 23-8. It is seen that the tail does not thrash back and forth creating great turbulence, but instead largely follows the path of the body, avoiding marked cross-flow of water and separation of the boundary layer.

Various aquatic vertebrates swim by undulating the body. The body is then usually long, with gradually increasing flexibility toward the tail. Examples are lampreys, eels, and water snakes. A long robust tail may functionally extend the body and continue its undulations as for salamanders, aquatic lizards, crocodilians, and an extinct suborder of whales. River and sea otters, desmans, and various rodents swim in this way but also propel themselves with their feet. Long dorsal and ventral fins commonly serve to increase the effectiveness of the body as a waterfoil (lampreys, eels). Similarly, the tail may be bilaterally compressed (salamanders, crocodilians, sea snakes, desmans, giant water shrew, muskrat). Some relatively slowswimming fishes (skates, some ribbon fishes, pipe fish) propel themselves by keeping the body straight and undulating only their fins.

Propulsion by undulation follows basically the same principles as propulsion by tail fin. Indeed, the two "methods" are not sharply demarcated. Any given segment of an undulating body or fin sweeps back and forth thrusting obliquely against the water much like an oscillating caudal fin. Compare Figure 23-9 with Figure 23-6C. Undulations move along the body as traveling waves from anterior to posterior, with the amplitude of the waves increasing to the rear. Many of these swimmers can reverse the direction of the waves and swim backward.

The action of paired flippers and wing-paddles differs in that their broadsides are presented more nearly flat-on to the water.

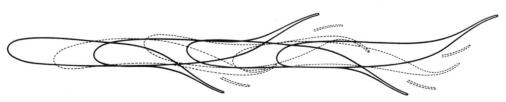

FIGURE 23-8
PATH OF A SWIMMING TROUT as seen from above.

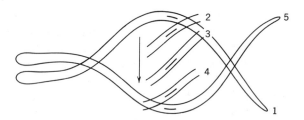

FIGURE 23-9
MOTION OF A SWIMMING EEL DURING
HALF OF A CYCLE.

(They cannot thrust in quite the same direction throughout the power stroke because they pivot at the shoulder or knee and hence describe arcs.) Accordingly, these paddles thrust in a direction more nearly opposite to the forward motion of the body (F_t in Figure 23-10), and the resistance of the water (F_w) has little lateral component (none in the figure). However, since F_w is applied beside, not behind, the body, a torque ($F_w l$) is established which tends to turn the head toward the opposite side. This torque must be countered by the stroke of a paddle on the other side of the body. The better swimmers among these animals (sea turtles, diving birds that swim with the wings, sea lion) have the paddles of the two sides of the body placed directly opposite to one another and they usually use them simultaneously. Those that swim with the hind feet may kick them together (frogs) or alternately (ducks, muskrats). If all four limbs are used in swimming (pond turtles, capybara, polar bear), diagonally opposite feet may be kicked in synchrony, though other sequences also serve. Actions of these various paddles to reduce resistance on the recovery stroke were noted above under the heading of drag.

Another kind of propulsion is used on occasion by some swimmers. If the animal can find water that has either motion (a velocity field) or a suitable pressure gradient (pressure field), it can then freeload, at least in part. One fish or whale may station itself beside and a little behind another, often larger, fish or whale, and thus benefit from the pressure drag created by the lead animal. Fishes have even been seen to follow in the wake of a submarine, apparently for this purpose. It is probable that waves caused by the wind are sometimes briefly used in similar fashion.

The most spectacular example of aquatic freeloading is the wave-riding of dolphins. Groups of dolphins may move along for many kilometers in the bow wave of a ship, seemingly without exertion. The pressure field in the front slope of the wave is parallel to the surface of the water, not to the horizontal, and

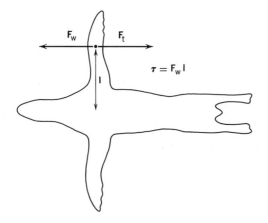

FIGURE 23-10
TORQUE RESULTING FROM THE THRUST OF THE FLIPPER OF A SEA LION.

hence has a forward component. Upwelling water thrusting against the obliquely oriented tail flukes also provides a forward impetus. The tendency to pitch forward that is created by this pressure on the tail may be compensated by the pectoral flippers. It is clear that the animal is remarkably sensitive to the pressure and velocity fields of its immediate environment and instantly compensates for every change.

Structure and Function of the Locomotor System of Swimmers. The span of the caudal fin from tip to tip divided by its chord, or average width in the direction of forward motion, is called the **aspect ratio** (Figure 23-1). (If it is easier to calculate, the equivalent equation, span^2/area, can be used.) The fastest swimmers have an aspect ratio of 5 or more, whereas slow swimmers have values of 1–2 (see bowfin, Figure 8-13). The caudal fin of fast swimmers is streamlined in cross-section, has low mass relative to the body, and is stiff—particularly on its leading edges. Such tails oscillate with low amplitude. Frequency, but not amplitude, increases with speed; during a burst of speed the tail virtually vibrates.

Rapid swimmers that propel themselves with the tail tend to have stiff vertebral spines: Centra may be long to reduce the number of intervertebral joints (sailfish), the centra may be large and platyan to reduce flexibility at the joints (some cetaceans), among fishes the zygapophyses may be unusually strong to brace the joints (marlin), or a longitudinal vertebral ligament may add rigidity (cetaceans).

There must be adequate flexibility, however, at the base of the tail and peduncle. Bony fishes usually have diarthroses

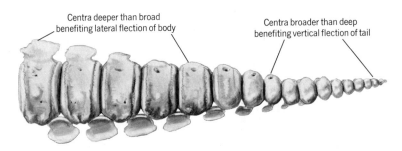

Horizontal tail flukes supported
by connective tissue, not bone

Centra deeper than broad
benefiting lateral flection of body

Centra broader than deep
benefiting vertical flection of tail

Dolphin, *Delphinus*

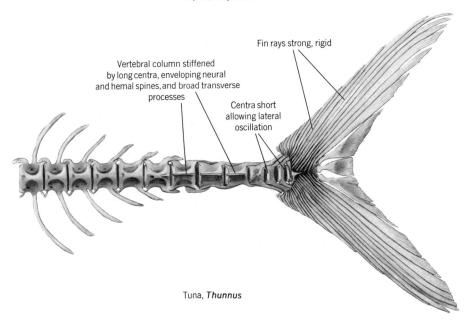

Fin rays strong, rigid

Vertebral column stiffened
by long centra, enveloping neural
and hemal spines, and broad transverse
processes

Centra short
allowing lateral
oscillation

Tuna, *Thunnus*

FIGURE 23-11
CAUDAL SKELETONS OF TWO VERY DIFFERENT FAST SWIMMERS.

where the tail fin joins the spine, and the spine of cetaceans is dorsoventrally compressed in the anal area to allow vertical flexibility at that place.

The axial musculature, which is the power plant of the tail swimmer, may weigh half as much as the entire animal. Fishes that swim constantly (tunas) or migrate long distances (salmon) have on their flanks red muscle which contracts relatively slowly, contains myoglobin, is aerobic, and has an unusually rich blood supply. The temperature of such muscle is commonly 10°C above water temperature and is not dissipated because of a countercurrent exchange mechanism (see p.

243). The white muscle which accounts for most of the bulk of the muscular system is capable of faster, anerobic contraction.

The myomeres of fishes fold to form zigzags as seen on the surface of the body, but series of interrelated cones as seen in three dimensions. The mechanics of these myomeres is complex and is still debated. The cones may extend the force of contraction of one muscle segment over several segments of the skeleton. They also ensure that muscle fibers will insert on the myosepta at oblique angles, thus providing some of the advantages of pinnate muscles. Small bones, which are a great nuisance at the dinner table, may run in the myosepta from the apex of one cone to the apex of another. The apexes of cones of the more posterior myomeres of the fastest fishes are extended into the tail as longitudinal tendons. The peduncle is therefore slender and tendinous, rather than broad and fleshy as in other fishes (Contrast Figure 23-12 with Figure 9-3).

Secondary swimmers that propel or steer themselves with paired appendages need large broad paddles. If the paddle is in some way folded on the recovery stroke to reduce drag, then it must be enlarged on the power stroke by a flexible membrane. Hence the webs between three toes of ducks, flamingos, gulls, auks, loons, and penguins; between four toes of cormorants, boobies, and pelicans; and between all five toes of frogs, pond turtles, platypus, beaver, and sea otter. Grebes, mudhens, and finfoots have lobes on the toes instead. Small aquatic mammals usually have fringes of long stiff hairs which functionally broaden the foot.

If the paddle is canted instead of folded on the recovery stroke, then there is wide range of motion at the shoulder or knee (those joints being modified accordingly), but the waterfoil itself is enlarged by rigid structures so it will stand against water pressure without muscular effort. Sea turtles and pin-

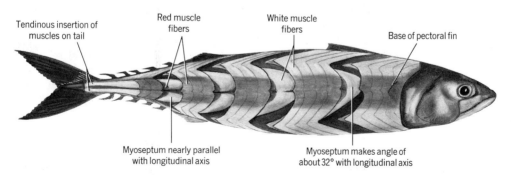

Tendinous insertion of
muscles on tail

Red muscle
fibers

White muscle
fibers

Base of pectoral fin

Myoseptum nearly parallel
with longitudinal axis

Myoseptum makes angle of
about 32° with longitudinal axis

FIGURE 23-12

AXIAL MUSCULATURE OF THE FAST-SWIMMING MACKEREL, *Scomber,* **with myomeres removed at successive levels.**

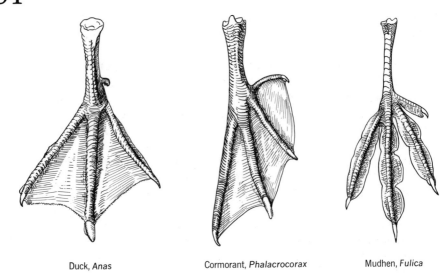

Duck, *Anas* Cormorant, *Phalacrocorax* Mudhen, *Fulica*

FIGURE 23-13
FEET OF SOME AQUATIC BIRDS.

nipeds flatten and greatly lengthen the usual compliment of metapodials and phalanges, particularly at the leading edge of the flipper. The wing skeleton of penguins, and to a lesser extent of auks, is also flattened. Ichthyosaurs evolved extra digits to broaden the paddle, one species having nine digits in all. Ichthyosaurs, plesiosaurs, and cetaceans add phalanges to one or more digits, bringing each series to from 4 to as many as 26 units. Pinnipeds extend some of the bony digits with cartilages. The integumentary membrane extends beyond the skeleton in flippers of some pinnipeds. The paddle may also be broadened by sesamoid bones, as at the elbow of penguins, or by the flattening and lateral extension of existing bones such as the pisiform of sea turtles and ulnare of penguins. Diving birds and cetaceans incorporate the forearm into the paddle; the radius and ulna become short, flat, and positioned in the same plane.

These paddles are made rigid in various ways, though some resilience remains. Spaces between the bony digits are sufficiently filled with firm tissue to brace the digits and make the surface contour of the paddle smooth. In flippers of cetaceans, synarthroses replace diarthroses, and bones are flat-ended.

Control of Vertical Position   Vertebrates that swim only on the surface are light so they will float high in the water like swans and gulls. Nondiving ducks

FIGURE 23-14

ARM SKELETONS OF SOME AQUATIC VERTEBRATES that use the pectoral appendage as a paddle. Dorsal (lateral) views of right appendage.

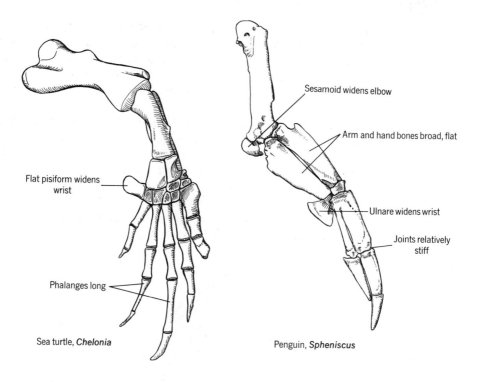

Flat pisiform widens
wrist

Phalanges long

Sea turtle, *Chelonia*

Sesamoid widens elbow

Arm and hand bones broad, flat

Ulnare widens wrist

Joints relatively
stiff

Penguin, *Spheniscus*

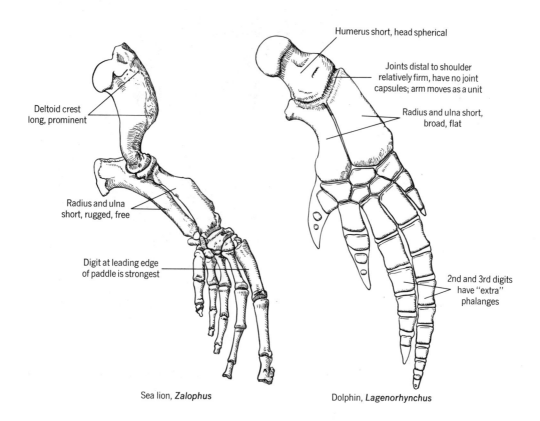

Deltoid crest
long, prominent

Radius and ulna
short, rugged, free

Digit at leading edge
of paddle is strongest

Sea lion, *Zalophus*

Humerus short, head spherical

Joints distal to shoulder
relatively firm, have no joint
capsules; arm moves as a unit

Radius and ulna short,
broad, flat

2nd and 3rd digits
have "extra"
phalanges

Dolphin, *Lagenorhynchus*

have a specific gravity of only about 0.6. A light skeleton, fat deposits, and air trapped in feathers or fur contribute to buoyancy. Vertebrates that rest on the bottom, by contrast, need to be more dense than water to maintain their position. Flat fishes and skates, which have no gas bladders, have a specific gravity of about 1.09.

Swimmers that vary their vertical position in the water maintain place in any of several ways. The hydrostatic function of the gas bladder of bony fishes was described on p. 251. Some sharks may exercise some control by selectively producing in the liver either of two metabolites which have different densities. A similar selective production of lipids occurs in a number of teleosts and in the surviving coelacanth, which has a fat-filled gas bladder. Various sharks and bony fishes that

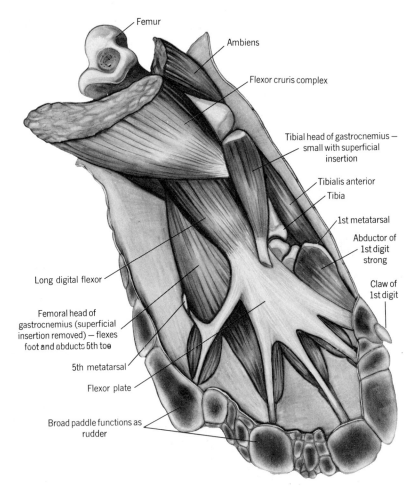

FIGURE 23-15

LEFT HIND LIMB OF THE SEA TURTLE, *Chelonia,* **seen in medial view.**

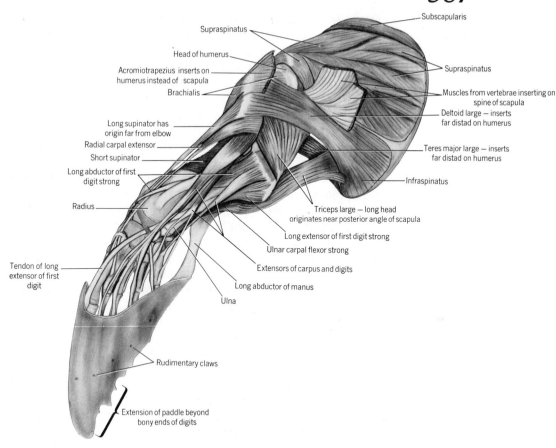

FIGURE 23-16
LEFT FORELIMB OF THE SEA LION, *Zalophus,* **seen in lateral view. (Drawn from an air-dried dissection and hence somewhat shrunken.)**

have no gas bladders and are slightly more dense than water (e.g., leopard shark, mackerel) maintain position by swimming slowly all the time, just as tetrapods breathe all the time. The source of their lift is discussed below.

Avian and mammalian divers must find ways to make their bodies more dense than those of their nondiving relatives. The bones of diving birds are less pneumatic. Their air sacs are reduced (loons) or lost (penguins). They press their feathers against the body to exclude air: Auks bubble constantly when underwater. Penguins achieve a density of 0.98.

Mammals that dive deep may hyperventilate before submerging, but they do not fill their lungs. Indeed, they may exhale during the dive. Deep-diving whales have relatively small lungs. Sirenia, which may feed while resting on the bottom or standing on their tails, have unusually heavy skeletons; their ribs are swollen and solid.

Rotation of a swimmer (or ship) around its long axis is called **roll**, rotation around its transverse axis is **pitch**, and rotation around its vertical axis is **yaw**. The body is stable if it passively tends to correct for displacements from a given position; it is unstable if a small displacement tends to increase to become a larger displacement. As for cursors (see p. 508), increased stability of swimmers reduces muscular effort but also reduces maneuverability.

Of several factors influencing stability, one is independent of forward motion. Two forces act on any submerged object: Gravity tends to make it sink, and buoyancy tends to make it rise. If the object has the same density as water, then the two forces are equal and the object neither sinks nor rises. Gravity acts on an object as though all its mass were at its **center of gravity** (CG). Buoyancy acts as though all lift were applied at its **center of buoyancy** (CB). The CB is located where the CG would lie if the object were uniformly dense throughout (like the displaced water). Vertebrates, however, are not uniformly dense: Bone, cartilage, and muscle are more dense than water, whereas fat, oil, and gas in lungs or gas bladder are less dense. The CG and CB are, therefore, usually in different places, and gravity and buoyancy act to turn the object in the water.

The diaphragm of cetaceans is oriented diagonally under long lungs placed high in the body, rather than behind short lungs placed forward in the body. This raises the CB relative to the CG and places it near, and usually a little above, the CG, thus giving the animal slight positive stability. The CG and CB of sharks are at about the same vertical level, so there is scant tendency to roll, but the CB may be a little more anterior, thus tending to lift the head. The gas bladders of bony fishes are located high in the coelomic cavity, but the heavy spine and epaxial muscles are still more dorsal, so the CG tends to lie above the CB, and the fish is unstable in regard to roll. Dead fishes float belly up.

Any fins or flippers may be extended to function as brakes, but the pectoral fins are most commonly used for this purpose. If they are lower than the CG, then the head tends to pitch downward. This tendency may be countered by canting the pectoral fins to give lift (the entire body then rising as it slows), by extending the dorsal fin, or by actions of other fins. The caudal fin may be curled into a hook as an effective supplemental brake.

Caudal fins of various shapes were named according to evolutionary relationships in Figure 8-13. Following much experimentation with the amputation of fins and with the forces exerted on models in water and in wind tunnels, tails of different shapes have also been classified according to function. Assuming the notochordal or spinal axis to be adequately rigid and

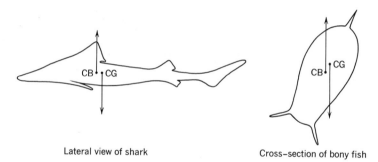

Lateral view of shark   Cross-section of bony fish

FIGURE 23-17

**INSTABILITY RESULTING FROM DIFFERENT LOCA-
TIONS OF CENTER OF GRAVITY (CG) AND CENTER
OF BUOYANCY (CB).**

the fin membranes to be passive and uniformly flexible, then symmetrical caudal fins do not cause pitch (parts A–C of Figure 23-18). Caudal fins with a down-tilted spinal axis, or with more membrane below than above the longitudinal axis of the body, or with the dorsal lobe stiffer than the ventral lobe, cause the tail to rise and the head to pitch downward (parts E–G). Tails with the reverse structure have the reverse function (parts H–J).

The traditional view has been that tails of sharks resemble part G; constant lift by the pectoral fins is then needed to counter a tendency for the head to pitch downward. However, as noted by Thompson and others, the similar tail shape of part D causes no pitch (downward thrust caused by the uptilted axis being cancelled by upward thrust caused by the more flexible ventral lobe). If an edge of the fin membrane does not neces-sarily trail passively in the water but instead can be made tem-porarily more rigid, or can be actively advanced with, or ahead of, the axis of the tail (parts K and L), then the tail might cause no pitch, or pitch either up or down as the fish chooses. Shapes D, G, and L get the tail up and out of the way as the fish swims over the bottom. Most sharks are a little heavier than water, so some lift is required much of the time.

The density of bony fishes having hydrostatic gas bladders closely approximates that of water. Hence, the tail need not be constructed to provide constant lift. Likewise, the pectoral fins need not provide constant lift, and fold against the body to reduce drag. Even though the density of these fishes is about the same as that of water, their equilibrium tends to be un-stable, so slight corrections are constantly needed. These fishes can alter the area, extension, curvature, shape, period, and amplitude of all their fins in subtle ways; their sense organs and nervous systems instantly make computations which man would find difficult and tedious to approximate.

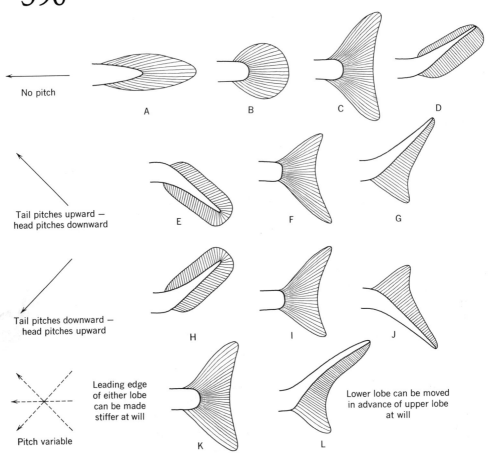

No pitch

Tail pitches upward —
head pitches downward

Tail pitches downward —
head pitches upward

Leading edge
of either lobe
can be made
stiffer at will

Pitch variable

Lower lobe can be moved
in advance of upper lobe
at will

A   B   C   D   E   F   G   H   I   J   K   L

FIGURE 23-18

FUNCTION OF CAUDAL FIN IN RELATION TO SHAPE AND OTHER VARIABLES.

Flexibility of the entire body is advantageous for maneuvering. Pinnipeds have supple necks and backs; penguins have secondarily regained much of the flexibility of the trunk which was lost in the ancestry of birds. The most maneuverable fishes live among plants and in coral reefs. Their bodies are short so they can turn (yaw) quickly, and are bilaterally compressed to allow flexibility. Fins that are anterior to the CG (including the forward part of the dorsal fin) are also used for fast turns. They trip the body around just as a dart turns in the air if thrown feather-end first.

Other Adaptations of Secondary Swimmers

Protection of Skin, Ears, and Respiratory System. Many aquatic mammals and birds trap enough air in the fur or plumage to shield the skin from wetting. The sebaceous glands of pinnipeds secrete quantities of waterproof sebum, and aquatic birds have large oil glands. The skin of cetaceans is resistant to water but not to drying. It is said that even the pouch of the water opossum is waterproof.

Cormorants and pelicans have lost the external nares. The external nares of other aquatic tetrapods, from frogs and alligators to beavers, hippopotamuses, and dolphins, are always dorsal in position, and the owner seems always to know when they are barely out of water. A ridge deflects water from the blowhole of many whales. When under water, the nares are automatically tightly closed. This is usually accomplished by sphincter muscles, but baleen whales use a large valvular plug, and toothed whales add an intricate system of pneumatic sacs so that great pressure can be resisted in each direction. Respiratory exchange may be surprisingly rapid: Whales of moderate size require only $1\frac{1}{2}$ – 2 sec to exhale and inhale 1500 liters or more of air. Tidal air is relatively great and residual air is relatively small.

Various aquatic tetrapods, from crocodilians to rodents, have modified the palate and glottis to permit chewing and swallowing under water without interference to the airway: The epiglottis is up within the nasal chamber in beavers as well as dolphins. The pelican, which plunges into water from the wing, has modified the larynx to exclude water under pressure.

The external auditory canal can be plugged or furled by pinnipeds. Cetaceans admit water to the outer part of the canal; the inner part grows shut in baleen whales. Protection of the middle ear from collapse is mentioned in the next section.

Sirenia, and Cetacea and Pinnipedia that dive to moderate depths, have a succession of 8–40 valves of smooth muscle in each bronchiole to hold air in the alveoli against pressure. Lung tissue of whales and manatees is relatively rigid due to the presence of cartilage and muscle. Their tracheas and bronchi are thick and strong. The tracheas of penguins, some petrels, sea lions, and the dugong are braced by a longitudinal partition.

Deep-diving mammals do not carry much air down. The lungs are not an oxygen store. If oxygen were exchanged, nitrogen would dissolve in the blood, only to bubble and cause great distress on the return to the surface. Alveolar collapse is probably complete at 100 m in the bottlenose porpoise. Lungs of the Weddell seal also "collapse" (or shrink markedly). The trachea is bow-shaped, not round, in cross-section to allow closure without injury. Whales have a short sternum and few fixed ribs. The remaining ribs have a single head. Thus, the thorax can also "collapse" without damage.

Adaptations of Sense Organs. The sense of taste appears to be "normal" in most aquatic vertebrates, but is rudimentary in Cetacea. Olfaction is considered to be poor in Pinnipedia and very poor in Cetacea. The structure of the nervous system clearly indicates the regression of these senses.

Hearing is acute in the more highly adapted aquatic tetrapods. Many whales have an amazing repertoire of sounds ranging from low resonant honks to high flute-like tones. In favorable circumstances some whales probably can communicate over distances of 160 km (100 mile) or more, though little is known of their language. At least most cetaceans and pinnipeds are capable of remarkably discriminating echoranging. They make sounds up to 200,000 hz emitted in pulses of from 16–400/sec. There is indirect evidence that penguins use the cavitation clicks produced by the turbulence of their own swimming as the sound source for echo-ranging (they can quickly locate fish in absolute darkness).

Underwater hearing is not dependent on a superficial tympanum or external auditory canal, provided that a gas-filled space functions as a resonator (see Chapter 16). Nevertheless, it is claimed (and also questioned) that the tough, fibrous auditory canal of cetaceans conducts sound better (even if partly occluded) than surrounding tissues. The oil-filled cavities of the heads of certain cetaceans may beam sounds and presumably return sound to the ear by specific pathways.

The tympanum and ossicles of cetaceans function in the usual way, yet are of distinctive structure: The ligamentous tympanum is mounted by a horny cone, and the ossicles are large, heavy, and firm. It is essential that the tympanum vibrate against an air space in the middle ear and, of course, that this space not collapse during dives. The middle ear of pinnipeds (and parts of the external canal?), and air sinuses communicating with the middle ear of cetaceans, are lined by highly vascular tissue which engorges during dives. The volume lost by compression of air is thus replaced by blood. Further, in cetaceans the sinuses and parts of the middle ear not adjacent to the drum are filled with a foam consisting of small air bubbles in an oil-mucus emulsion. Experiments show that these bubbles do not collapse, even under a pressure of 100 atm. Finally, the tympanic bulla of whales is strengthened by some of the thickest, most dense bone known. The bullas are loosely attached to the remainder of the skull and are cushioned in foam and blood sinuses. This permits them to function independently of each other and of the body. Directional hearing of aquatic mammals is excellent and probably is based largely on intensity discrimination.

Sirenians have poor vision, as would be expected from their sluggish habits, stationary (plant) food, and often murky environment. Baleen whales have moderately good eyesight, but have a limited field of vision and cannot see out of water. Their food is passive, and some of them dive below the level of light penetration. The food of toothed whales and pinnipeds is active, and they have excellent vision, both below and above water.

Eyes of aquatic tetrapods with good vision have secondarily acquired characteristics of the eyes of their remote ancestors among primary swimmers: The eye is large, the eyeball short along its optical axis, the lens large and spherical, and the cornea flattish or elliptical for streamlining. Lacrimal glands are reduced (pinnipeds) or absent (sirenians, cetaceans). To protect the eye from saltwater, the cornea is cornified and is bathed by the secretion of large glands in the lids. In order to withstand wave pressure, the sclera of cetaceans is very thick and tough; one can scarcely cut it with a knife. Pinnipeds can change the shape of the lens more than is usual to permit vision in and out of water.

Thermoregulation; and Response of the Circulatory System. Since the thermal conductivity of water is about 20 times greater than that of air, endothermic aquatic vertebrates must protect themselves from heat loss, particularly when inactive and when in cold seas. Air trapped in plumage or dry underfur (as of the sea otter, beaver, or fur seal) is an effective insulator. Large mammals have relatively little heat loss because of their low surface-to-volume ratio. Blubber is an effective insulator for them, coming, in extreme instances, to $1/4$ of the body weight. Flippers of cetaceans have slow circulation and countercurrent exchange, so warm outgoing blood gives its heat to the cold incoming blood. It is probable that some whales require moderate activity to maintain body temperature in arctic waters.

Conversely, swimmers must be able to dissipate heat during periods of activity when heat production may rise tenfold. The countercurrent exchange mechanism can be bypassed, and the large flat flippers (which are devoid of blubber) then serve as radiators. Also, vascular papillary ridges in the epidermis of whales dissipate heat when needed.

The circulatory physiology of air-breathing vertebrates during dives adjusts so as to supply oxygen to the brain and heart, and otherwise to avoid stress from lack of oxygen or buildup of carbon dioxide and lactic acid. Bradycardia, or slowing of the heart, is universal and occurs on submergence. The rate commonly is reduced to $1/10$ or $1/15$ of normal, and the slowing is in part preventive; its onset is faster when a deep dive is anticipated. The aorta dilates near the heart, but all arterioles constrict except those of the brain and heart. Excretion stops. The veins of pinnipeds and cetaceans (like those of fishes) have no valves. Blood volume of these swimmers, and of some diving turtles and birds, reaches two times that of comparable terrestrial vertebrates. The hepatic portal system is large, and venous sinuses may be present in the thorax and abdomen. The result is that quantities of blood stagnate in the body cavities.

Deep divers are usually not very active. Their hearts tend to be small (though that of a blue whale may still weigh 600 kg!). The metabolic rate falls off a little (pinnipeds, cetaceans, alligators, ducks). The blood of fast-swimming and deep-diving dolphins is able to carry up to three times as much oxygen as that of their opposites. The myoglobin content of divers' muscles is high. They tolerate twice as much carbon dioxide in the blood as does man. Lactic acid is stored in the muscles until breathing resumes.

The circulatory systems of swimmers show convergence in other ways that remain enigmatic. Why is the postcava doubled (a turtle, pinnipeds, cetaceans, and sirenians, but also the nonaquatic edentates and slow loris)? Why do pinnipeds have a sphincter in the postcava at the level of the diaphragm? Why are the intervertebral vessels enlarged and the jugular veins reduced (pinnipeds, cetaceans)? Why is there a venous plexus in the drainage of the kidney, or a rete to damp the flow of blood to the brain (cetaceans)?

Reproductive Biology.   Most sea snakes are viviparous and give birth at sea, as did the ichthyosaurs. Other reptiles and all birds lay their eggs or give birth on land. Among aquatic mammals, the walrus and hippopotamus sometimes give birth in the water, and Cetacea and Sirenia always do so. A single, large, precocious young is born at a time. Cetacea deliver rapidly; the calf emerges tail first. The mother whirls in the water, thus snapping the relatively short umbilical cord at a predetermined point of weakness. The newborn swims to the surface to breathe, sometimes with maternal assistance. The tail flukes are soft and curled at birth, but harden in about two days, by which time the calf can keep up with the herd.

Whale milk is thick and rich in fat. It collects in sinuses and is forced out in mouthfuls during underwater nursing. Growth of young pinnipeds and cetaceans is rapid. The calf of the blue whale, which is 7 m long at birth, gains about 90 kg/day on its mother's milk!

24

Flying
and Gliding

Some lightweight climbers can retard a fall and travel horizontally as they move downward. If control of the path followed is minimal, and the line from takeoff to landing is steeper than about 45° to the horizontal, then the animal is said to **parachute.** If some maneuvering in the air is possible and the line from takeoff to landing is less than 45° to the horizontal, then the animal is said to **glide.** If the animal is capable of sustaining itself in the air, we say it can **fly** and is **volant.**

The general principles of animal flight are well-understood, yet much remains to be learned. Aerodynamics is a complex field with some aspects which are still empirical. Even when vertebrates fly under relatively uniform conditions, they must frequently make slight adjustments to compensate for alterations of external variables. When a gull maneuvers in a changeable wind, major adjustments, many of kinds that are not possible for man-made aircraft, must be constant and nearly instantaneous. Finally, it is technically very difficult to study living, fast-flying animals, and models and stuffed animals in wind tunnels give spurious results.

Origin and Advantages of Flying and Gliding

Although they were much more primitive than modern birds, the first known birds (suborder Archaeornithes) were apparently already moderately good gliders. It is probable that birds evolved from small, bipedal, arboreal archosaurs which hopped from branch to branch, steadying themselves with outstretched forelimbs. As feathers enlarged on the margins of the forelimbs, the hops were likely extended to longer and longer glides until the power of flight was established.

An alternative theory of the origin of avian flight postulates a cursorial bipedal ancestor that held the potential wings out as it ran. However, known cursorial bipeds have small forelimbs which they do not use for balance. Also, cursors tend to have a compact foot with few toes all oriented forward—not at all the avian condition. Further, an arboreal glider can utilize gravity to increase air speed as it planes to a lower perch.

The first known bat lived 50 million years ago, but was modern in appearance and flying ability. Presumably bats evolved from small, agile mammals which scrambled about in trees seeking insects. The flying reptiles, or pterosaurs (an order of the subclass Archosauria), survived from about 180 until 65 million years ago. The earliest known representatives were already competent flyers, so nothing is certainly known of the origin of their ability to fly.

As we shall see, birds, bats, and pterosaurs differ considerably in structure. Nevertheless, in order to fly at all, certain conditions must be met, and these three groups display much convergent evolution. Flyers are among the most specialized of vertebrates.

Parachuters may drop from tree to ground to escape from less versatile predators or to move quickly to another location. Also, an inadvertent fall is not damaging. Gliders have the same advantages to a greater degree, and by gliding can forage more quickly and over a wider area than would otherwise be possible. They have the advantage over flyers that the forelimbs, being little specialized, remain useful for climbing and the manipulation of food.

The various flyers benefit from flight in different ways. Potential advantages include the following:

(1)   Flyers gain access to food that is in the air (e.g., flying insects), that must be reached from the air (terminal flowers), or that can be located from the air (rodents, fishes).

(2)   Great mobility and maneuverability enable flyers to search rapidly and efficiently for food and shelter.

(3)   Escape is provided from nonvolant predators.

(4)   By migrating, flyers can travel, according to season, to regions where climate, food supply, and nesting sites are favorable.

(5)   Dispersal is possible over distances and geographic barriers that would otherwise be insurmountable.

Parachuters and Gliders.   All parachuters are arboreal. Various tree frogs are included. These launch themselves with a jump, hold the limbs out to the side, and control their orientation so that their flat ventral surface is presented to the airstream. One species (family Hylidae) has no other adaptations for slowing its fall, yet achieves an angle of descent of about 60°. Several other tree frogs (family Rhacophoridae) are more expert, approaching an angle of descent of 45°. Their huge feet are fully webbed, and small membranes fringe the arms and span the angle between thighs and body wall.

A genus of tree snakes is capable of controlled parachuting at fairly flat angles. The body is held horizontal, the ribs are spread to the sides, and the belly is drawn in to present a concave surface to the airstream.

Some lizards can descend at about 70°, relying mainly on behavioral adaptations. Others, having fringed tails, do a little better. Several genera of geckos (family Geckonidae) have broadly webbed toes and fringes on the head and body. They can descend at nearly 45° to the horizontal with considerable maneuvering. Various small tree squirrels also parachute to break a fall by spreading legs and tail and coming down flat to the airstream.

Several fishes, at least three genera of lizards, and representatives of three orders of mammals are gliders. All have evolved broad membranes to catch an airstream, but these can function only when adequate air speed is attained. The lizards and mammals climb a tree to gain height, and then they jump. They fall steeply down until sufficient speed is attained by gravity, then they sail off following a parabolic path. Doubling the height of the takeoff more than doubles the horizontal travel.

Vertebrates That Parachute, Glide, and Fly

The fishes attain adequate air speed in a very different way. A "flying" fish (family Exocoetidae) first swims rapidly just under the surface and then emerges until only the large lower lobe of its hypocercal tail is in the water. The pectoral fins, which in one genus are as long as the body, are spread to the sides as the tail thrusts back and forth in the water. The fish skims along in this manner for from 1 to 6 m until its air speed increases to an estimated 40–70 km/hr. It then spreads the somewhat smaller pelvic fins and glides free of the water, usually for 2–4 sec, but occasionally for 10 sec or more, sometimes traveling 100 m. On landing, the fish either submerges or immediately skims again with only the tail in the water preparatory to another glide. From two to twelve glides may be made in succession. The fishes are unable to glide if there is no wind, but they can sail several meters high over the water if there is a good breeze. The fins are fixed during most of the glide, but they may be seen and heard to vibrate at takeoff and landing. The purpose of this action is not known; fin musculature is not enlarged.

Some fishes of another family (Characinidae) leap high in the air and then spread their large pectoral fins stiffly as they fall back to the water. They sometimes jump into boats. Other members of the same family have larger pectoral fins which they flutter, using enormous ventral muscles as they make long arcing excursions from the water. The aerodynamics of this near-flight is unknown.

Gliding lizards of the genus *Draco* have extensive membranes reaching on each side from the thorax to the base of the hind leg. The membranes are supported by six pairs of much-lengthened ribs. When they are not in use for gliding or display they are folded against the body. These lizards commonly glide at 20–30° to the horizontal, but can even gain elevation if they move into an updraft. Glides of 24 m have been observed.

The colugo (mammalian order Dermoptera) is a cat-sized Asiatic glider with the largest "flight" membrane of all: It extends from the throat to the wrists to the ankles to the tip of the tail. Even the toes are webbed. One animal sailed 136 m at an angle of only 5° to the horizontal.

Five species of phalangers (order Marsupialia) in three genera are gliders. They range in weight from 14 to about 1360 g. All are vivacious forest dwellers with soft fur, long bushy tails, and membranes extending from elbow to knee. They launch themselves with a leap, and commonly glide to 100 m. One sailed 540 m in six glides.

Fifteen genera of rodents are expert gliders. They range from chipmunk-size to cat-size. The very large membranes of scaly-

FIGURE 24-1

EXAMPLES OF PARACHUTERS AND GLIDERS.

"Flying" fish
Cypelurus

Parachuting frog
Rhacophorus

Gliding lizard
Draco

Parachuting gecko
Ptychozoon

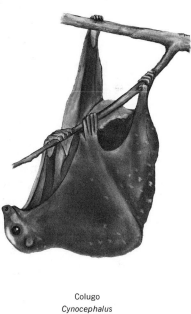

Colugo
Cynocephalus

"Flying" squirrel
Glaucomys

tailed squirrels (family Anomaluridae) are supported in part by long cartilaginous struts from the elbows. Some "flying" squirrels (family Sciuridae) have shorter struts from the wrists. Usual glides are 6–10 m in length at angles of 30–50° to the ground. However, an American species was seen to glide 50 m with a drop of only 18 m. Utilizing an updraft in a valley, a large Asiatic species glided 4500 m. Such figures really mean little—in usual circumstances all gliders seem able to glide as far as they "want" to.

Flyers. PTEROSAURS had short bodies, long necks, large bird-like heads, bat-like hind limbs, and long narrow wings supported by the arms and elongated fourth fingers. Eyes and brains were also bird-like. There were no scales and there is evidence that some had hair. It is possible that they were able to elevate the body temperature at least when flying. One group had long tails and many homodont teeth; the other group had short tails and few or no teeth. The more than 25 known genera ranged from the size of a starling to the largest of all flyers, which had a turkey-sized body but a wing span of about 7 m (23 ft)! Most pterosaurs lived along seacoasts and ate fish. It is probable that some could alight and take off into the wind from water. Perhaps some could even swim a little with their wings. They likely roosted in trees or on cliffs, scrambling about like bats when not in the air. The wing musculature of the larger species was too weak for constant flapping; they probably flew largely by soaring, depending at least in part on wind.

BATS are very successful flyers. There are about 175 living genera and more species than in any other mammalian order except Rodentia. The smallest bats weigh only 4 g. The largest

FIGURE 24-2
RESTORATION OF THE GIANT PTEROSAUR, *Pteranodon.*

weigh 900 g and have a wing span of 1.7 m (about 5 ft). Bats are variously adapted for eating insects, fruits, flowers, nectar and pollen, blood, and fish. Flight membranes of skin are supported by the arms, greatly elongated second through fifth fingers, hind limbs, and usually all or part of the tail. Some bats have fast and direct flight and others have slow and erratic flight. They commonly flit along at 10–25 km/hr (6–15 mile/hr). Maximum speeds are difficult to obtain, but speeds of 32–50 km/hr (20–31 mile/hr) have been measured with fair accuracy. Most bats are exceedingly maneuverable fliers—a necessity for creatures using echo-location, which operates at close range. The mastiff bat remains on the wing continuously for 6 hr and more. Certain bats fly at least 3000 m high. Various species forage as much as 25 km from their roosts, and some migrate 100 km one way. Numerous bats can fly carrying young weighing 50% of their own weight. One species can lift 73% of its weight!

BIRDS, the finest of all flyers, are relatively uniform in structure compared to other vertebrate classes, yet are diverse in habits and habitats. The 2 g bee hummer is the smallest. Excluding flightless birds, the extinct vulture *Teratornis* was the largest. It probably weighed about 23 kg (50 lb) and had a wing span between 3 and 4 m. Flight feathers are supported by the long arms and one robust digit (probably the second); those borne by the hand are **primary feathers** and those on the forearm are **secondary feathers.** A feathered membrane called the **patagium** spans the angle in front of the elbow.

Speed records have been obtained by following birds with a car or light plane. In spite of efforts to correct for wind and angle of flight, the figures are only approximations. Songbirds fly 16–40 km/hr (10–25 mile/hr). Ducks cruise at 50–65 km/hr, but several species can fly 100 km/hr (65 mile/hr) when pressed. A flock of sandpipers that flew diagonally in front of a plane was estimated to be flying 180 km/hr. The speed of the peregrine falcon is legendary. One bird that flew for 300 m over a field was timed at 265 km/hr (165 mile/hr). Whether this flight followed a dive was not stated. Even greater speed has been claimed for an Asiatic swift. The true capabilities of the fastest birds remain in doubt.

The sooty shearwater and sanderling migrate the 11,000–13,000 km one way between Arctic America and Patagonia. The golden plover flies 3800 km (2400 mile) nonstop from Labrador to South America. Most birds fly at altitudes below 1500 m, but migrants rarely travel as high as 6400 m (21,000 ft). Birds have been observed above 8000 m in the Himalayas. Even when at rest, mammals become unconscious at such an elevation.

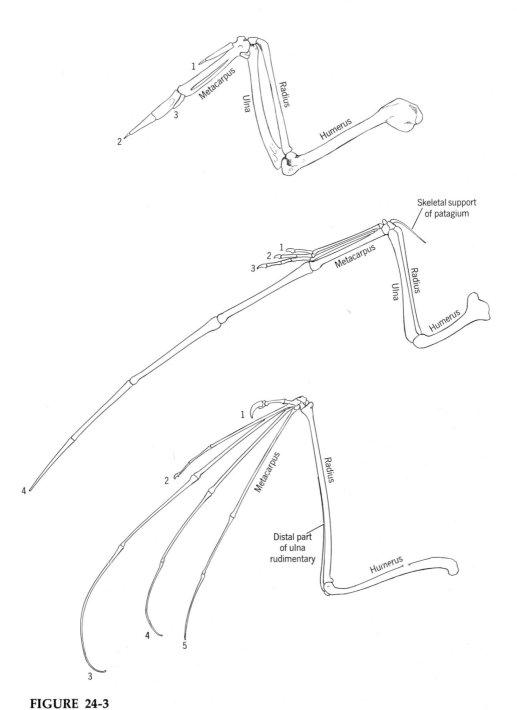

FIGURE 24-3
COMPARISON OF THE RIGHT WING SKELETONS OF A BIRD, A PTEROSAUR, AND A BAT shown by a duck, *Mergus* (above), a short-tailed pterosaur, *Pterodactylus* (center), and a fruit bat, *Pteropus* (below). The digits are numbered.

Flyers, like swimmers, move within a fluid medium. Aerodynamics and hydrodynamics are closely related fields, so the requirements of flyers parallel those of swimmers. However, air being much less dense than water, the relative importance of the variables is altered and flyers are deprived of support by flotation. All flyers must (*1*) derive sufficient upward force from their muscles or from the environment to counter the pull of gravity, (*2*) reduce drag, particularly if flights are long or fast, (*3*) propel themselves at various speeds and sometimes in restricted spaces, and (*4*) retain stability, maneuver, brake, and land according to habit. These primary requirements establish some rigid secondary requirements which focus on needs for (*5*) strength with light weight, (*6*) firmness of the trunk, and (*7*) the efficient production and utilization of power.

How do flyers meet their stringent requirements?

Since flyers are more dense than air, an upward force must act on them in order for flight to be sustained. During level flight, this force must just counter the pull of gravity, which is to say it must equal the flyer's weight. During ascending flight, flight in a downdraft, and flight while carrying young or prey, the upward force must exceed the weight of the flyer. Where does the upward force come from?

We shall first consider only level flight in still air with no flapping of the wings. These conditions are set so we can clearly distinguish upward force from backward force (drag) and forward force (propulsion).

As a first approximation, imagine a crude model of a flying bird with wings cut from thin slats of wood. If moved in an airstream in such a way that the wings meet the wind exactly edge-on, then air flows equally over and under the wings and there is no upward force (Figure 24-4A).

Now suppose that the model is improved by tilting the leading edge of the wing upward. The angle the wing makes with the airstream is called the **angle of attack,** or α. When α is small, air flows over the wing as shown in part B. It is seen that the air passing over the wing has farther to travel than the air passing under the wing. Accordingly, it must move faster and its pressure decrease. Consequently (following Bernoulli's theorem), the pressure is lower above the wing than below. The disturbed part of the airstream is thinnest near the leading edge of each wing. Air moves fastest there and creates the lowest pressure. Forces on all other parts of the wing are also in proportion to the adjacent air pressures (part C). All the forces acting on the wing that are derived from its motion can be divided into a component called **drag (D)** which by defini-tion is in line with the airstream and opposite to the direction of flight, and a component called **lift (L)** which is at right angles to **D** (part D). These forces act at the center of pressure, **X**, which is usually $\frac{1}{4}$–$\frac{1}{2}$ of the way back from the leading edge of the wing. In other words, all the actual forces together have the same effect on the wing as **D** and **L** acting at **X**.

In level flight in still air, **L** is directly upward and is the force needed to counter the pull of gravity. In ascending or descend-ing flight on fixed wings, "lift," in spite of the term, is not ver-tical but at right angles to the airstream. It does always have a vertical component (unless the flyer flies upside down or de-presses the leading edge of the wing, making α negative), but it also has a horizontal component which may be forward (descending flight) or backward (ascending flight).

As α increases from 0°, **L** also increases, and the center of pressure moves foreward to about $\frac{1}{4}$ of the way back from the leading edge of the airfoil. However, as α exceeds about 15°, the airstream above the wing suddenly ceases to flow smoothly and instead separates from the wing in strong eddies. Lift is then lost, and the flyer is said to **stall.**

There are also other variables. If an airfoil is convex on its upper surface as seen in cross-section (Figure 24-5), it is said

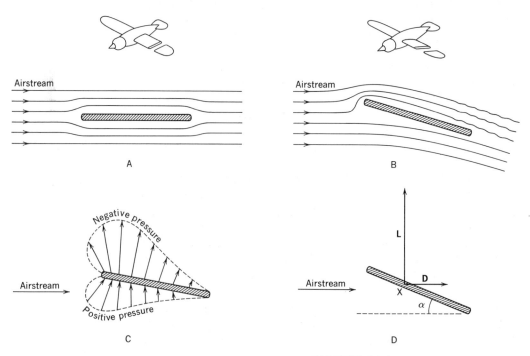

FIGURE 24-4
SOME FACTORS RELATED TO LIFT.

to have **camber.** The wings of vertebrate flyers are cambered, at least proximally. The line joining front and back edges of a cambered airfoil is called the **chord.** Further, the placement of bone and flesh in a vertebrate wing is such that it may be thicker along the leading edge than elsewhere, thus making the wing "streamlined" in cross-section. This is particularly true of bird wings. Finally, if a small second airfoil is positioned just above the leading edge of the first, a **wing slot** is created. This slot deflects air down onto the upper surface of the main wing thus energizing the region near the boundary and preventing separation. As a result, **L** is increased and α can be greater before the wing stalls. Pterosaurs and bats are unable to use slots. The first digit in the wing of a bird, together with its feathers, is called the **alula.** It ranges from $1/10$ to $3/10$ as long as the wing, and when lifted creates a wing slot. (Some birds also have slots between the tips of the primary feathers at the end of the wing. Their function is mentioned later on.)

We can now relate these variables more precisely. Lift = $1/2\rho V^2 S C_1$, where ρ = the density of air, V = air speed, S = projected area of the wing (i.e., the area of its shadow), and C_1 = the coefficient of lift. The two wings and the part of the body that joins them function as a unit in creating lift so they are best taken together in calculating S. When the wings are

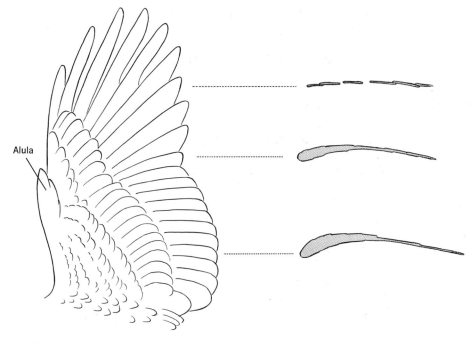

Alula

FIGURE 24-5

CAMBER AND STREAMLINING OF A BIRD WING AT SELECTED CROSS-SECTIONS.

not flapped, lift is greater toward the center of the unit than toward the extremities, or wing tips. The coefficient, C_l, is a dimensionless number which usually equals about 1.5 for birds, but can range above 2 if the wings are slotted. Its value depends on the angle of attack, the camber and streamlining of the wing, the presence and nature of wing slots, and the Reynolds number. The Reynolds number, in turn, equals the density of air, times the air speed, times the length of the average chord of the wing, divided by the viscosity of air. For vertebrate flyers, usual values are 40,000–130,000 (which is lower than corresponding numbers for swimmers).

For our purposes, this complex of variables can be simplified. Since the density and viscosity of the air are small in value and scarcely subject to control by the flyer, we can discount them when considering the generation of lift. The camber and shape of the wing influence lift largely through their relation to the angle of attack. This leaves angle of attack, wing area, and air speed as of direct importance to lift. Air speed is squared in the basic formula for lift, and occurs also in the calculation of the Reynolds number. It follows that if flyers are fast, they have enough lift even when their wings are narrow, small in area, have little camber, no slots, and function with a small angle of attack. Conversely, slow flyers need large wings with camber and maximum angle of attack in order to ascend quickly.

(There is, of course, an upward component to ascending flapping flight. This will be considered below. Also, a flyer may derive upward force from updrafts. This is discussed under soaring, gliding, and formation flying.)

Drag

Drag (the resistance the air offers to the motion of the flyer) acts horizontally backward when flight is level and in still air. It is convenient to divide the total drag (**D**) into two categories, profile drag and induced drag.

Profile drag (D_p) is all the drag acting on a hypothetical airfoil of infinite length. It is also all the drag acting on the wings and intermediate body (again taken as a unit) of a real flyer exclusive of the drag induced by motions of air around the tips of the wings. Profile drag is produced by energy lost to the environment through the friction of the air against the body, the displacement of air, the formation of pressure gradients in the air, and the creation of eddies. The value of $D_p = \frac{1}{2}\rho V^2 S C_{dp}$, where C_{dp} is the coefficient of profile drag and the other variables are the same as those in the formula for **L**. The value of C_{dp} varies with the camber, streamlining, slotting, and outline of the wing–body unit, and with the Reynolds number. (These are the same variables that influence the value of C_l, but the relationships are different here.) On the upstroke of the wings,

C_{dp} may be much reduced by birds by allowing air to pass down through the feathers. Flyers, like swimmers, can sense and control the flow of the medium over the body. In wind tunnels the energy loss by models of birds is about twice as great as that by live birds. Analysis has shown that a vulture can achieve laminar flow over much of its body.

Since air pressure below a wing that is lifting is positive relative to atmospheric pressure, and the air pressure above the wing is negative, there is a flow of air outward under the wing, around the wing tip, and inward over the wing. This flow induces a large eddy, or vortex at each wing tip and affects air flow elsewhere in a way equivalent to a slight increase in α. This takes energy from the flyer, and causes the drag called **induced drag (D_i).** The value of $D_i = \frac{1}{2}\rho V^2 S C_{di}$, where C_{di} is the coefficient of induced drag and the other variables are as before. Since D_i results from airflow around the tips of the wings, it increases as the pressure gradient around the wings increases and is smaller for narrow wings than for broad wings. Narrowness of wing is expressed by the **aspect ratio, A,** which is the span of the wings (tip of one wing to tip of the other) divided by the average width or **chord.** (Since the average chord cannot be directly measured, the equivalent formula, $A = \text{span}^2/\text{area}$, may be used instead.) The value of $C_{di} = kC_l^2/A$, where k is an empirical constant equal to about 2.

There can be neither lift nor drag unless there is forward velocity. So far we have assumed forward velocity without establishing a source for it. One way for an animal to derive or maintain forward velocity is to glide downward using gravity to convert potential energy to kinetic energy. Assume that the

Gliding, Soaring, and Formation Flying

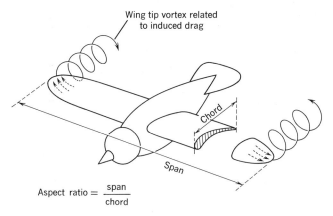

Wing tip vortex related to induced drag

Chord

Span

$$\text{Aspect ratio} = \frac{\text{span}}{\text{chord}}$$

FIGURE 24-6
THE NATURE OF THE ASPECT RATIO AND WING TIP VORTEX.

animal (whether bird, reptile, or mammal) glides with constant speed and direction in still air. The angle that its path makes with the horizontal is called the **gliding angle,** and the rate of vertical descent is called the **sinking speed.** It is as though the glider were sliding down an incline. As for a wagon that coasts down a hill, the glider's weight, **W** (see Figure 24-7), has a component, **M,** which is in the direction of motion, and a component, **N,** which is normal to the slope of the incline.

The glider's airfoils create lift and drag. Lift, drag, and gravity are the external forces that act on the glider. In order for speed and direction to remain constant, these forces must be in equilibrium. The glider therefore adjusts the angle of attack and area of its airfoils so that **L = N** and **D = M.** Similarly, **V,** the resultant of **L** and **D,** is vertical and equal to **W.** It is evident from Figure 24-7 that the greater the value of **L/D,** the flatter the angle of glide will be. A far-ranging fast-flying albatross may sink 1 unit of distance vertically while moving about 18 units

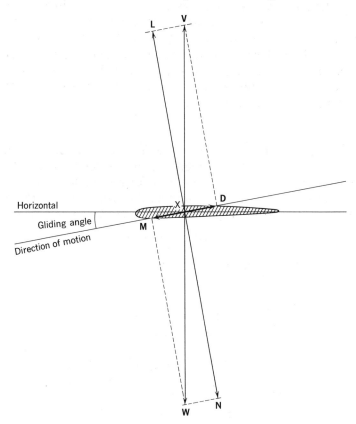

FIGURE 24-7
FORCES RELATED TO THE AIRFOIL OF A GLIDING ANIMAL moving with constant speed and direction when L/D = 5; X is the center of pressure.

horizontally. For a vulture the ratio may be 1:22. It is usually desirable for small birds to achieve minimum sinking speed. This is accomplished with a steeper angle of glide (about 1:8) but slower flight speed.

Sustained flight without flapping of the wings is soaring. There are two kinds of soaring, depending on circumstance. The first, **static soaring,** is dependent on updrafts. A flyer can stay aloft if its sinking speed relative to the surrounding air is equal to, or less than, the rate of ascent of the surrounding air relative to the ground. Like a person who walks continuously down a rising escalator, the flyer glides continuously down through rising air. Air rises when wind is deflected upward by a hill, coastline, wave, or ship. Gulls and pelicans often soar in such updrafts. Similarly, columns of air sometimes rise by convection over warm water or land. If a breeze is blowing, the columns tilt over, becoming inclined to the ground.

Vultures, hawks, and most other soaring land birds instead do their static soaring in thermals. As explained by Cone, the morning sun warms the ground, which warms a layer of air next to the ground. This air flows into bubbles which arch up and break away from the ground layer to rise like balloons through the higher cooler air, expanding as they go and drifting horizontally if there is any breeze. Unlike the gas in a balloon, however, the air in a thermal is not stagnant but circulates in a vortex shaped like a doughnut. Air constantly rises in the center of the thermal (the entire center, not just on the surface of the doughnut), radiates outward at the top of the floating bubble, descends in the margins, and turns in to rise again. The vulture soars only in the center of a thermal, wheeling constantly to remain within the hole of the doughnut. After rising with the thermal, sometimes for several thousand meters, the bird glides slowly away to the ground or into another thermal. In this way it remains on the wing for long periods and covers much distance with little expenditure of energy.

Thermals rarely rise over water. Albatrosses and some related oceanic birds are masters of **dynamic soaring,** which is dependent on the wind. This seemingly more "difficult" kind of soaring has also been analyzed in detail by Cone, yet remains controversial. *What* the bird does is more evident than *how* the bird does it. The wind (which is brisk and steady over vast areas of the oceans) has ever-decreasing velocity from about 15 m over the water down to water level. This is because of the friction between air and water. The difference in wind velocity between the top and bottom of this shear layer commonly

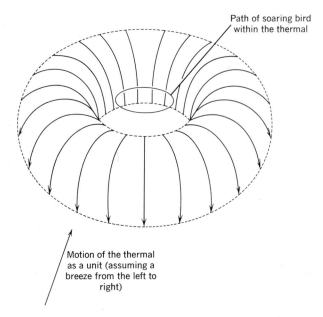

Path of soaring bird within the thermal

Motion of the thermal as a unit (assuming a breeze from the left to right)

FIGURE 24-8

SIMPLIFIED DIAGRAM OF THE CIRCULA-
TION OF AIR IN A THERMAL as seen from
above and to one side. Static soaring is done
in the air rising throughout the center of the
"doughnut."

amounts to about 60 km/hr, and the bird soars only in this zone.

Let us follow a basic flight cycle of an albatross starting with the bird flying into the wind near the top of the shear layer. Its air speed might be 70 km/hr, but its absolute speed over the water might be only 10 km/hr because of a wind speed of 60 km/hr. Now the bird turns downwind. Cone believes that as it banks in the turn the bird is able to maintain its air speed. Thus, at the completion of the turn the velocity of the wind is added to, rather than subtracted from the bird's air speed to derive its absolute speed downwind, which might be 130 km/hr. The albatross next partially flexes its wings to increase wing loading and further increases speed as it glides steeply down to the water, converting potential energy to kinetic energy as it drops. The bird now banks sharply to execute a windward turn. It still has high absolute speed when it completes this turn, because the wind has less velocity at water level. Fully extending its wings and increasing their angle of attack, the bird now completes the basic flight cycle by rising to the top of the shear layer, converting kinetic energy to potential energy again as it climbs. Absolute speed falls off, but ever-increasing wind speed maintains adequate air speed to prevent stall.

The energy seemingly extracted from the wind is sufficiently greater than the energy lost during the basic flight cycle to provide an excess that can be used for gradually maneuvering, for instance by gliding cross-wind when halfway through a turn, or by lengthening each climb to progress directly into the wind. These birds are also expert at static soaring in the updrafts over wave fronts. In these ways albatrosses travel thousands of kilometers with only infrequent flapping of the wings.

As noted in a previous section, there is an updraft of air at the wing tips. By flying in formation, ducks, geese, and some other birds utilize the upwash created by their neighbors to reduce the lift they must generate with their muscles. If the birds flew line-abreast the center birds would benefit the most. By flying in a vee formation with slightly swept back arms, all birds (including the lead bird) have about equal benefit. Nearly maximum advantage is attained when there are ten or more birds and they fly with less than half a wing span separating them horizontally. The energy saved is considerable.

Flapping Flight

Most fliers propel themselves by flapping their wings. Even expert soarers flap when landing and taking off, and when conditions needed for soaring are temporarily lost. Flapping flight is exceedingly complex. Quantitative analysis is not yet possible, though the various kinds of flapping flight are understood in general terms thanks to the research of Brown and others. The types of flight described below are representative, but they grade into one another.

Hovering. Hovering, or stationary flight in still air, represents a specialization in the evolutionary sense, yet is relatively simple aerodynamically. The small flyers that can hover make the body motionless in midair, usually to feed on the nectar of flowers which are not accessible in another way. Hummingbirds are prominent among such flyers, though some other birds and many kinds of bats can also hover.

The body is usually held nearly vertical. The wings, therefore, beat backward and forward instead of up and down. The wing tips describe a distorted figure eight (Figure 24-9). On the forward stroke, the wing, having a positive angle of attack, creates much lift, which is at right angles to the airstream but has a large vertical component. At the forward limit of the long wing stroke the wing rotates at the shoulder (the wrist and elbow are quite stiff in these birds), thus turning over and making its anatomical underside uppermost. With a positive angle of attack in the new direction the wing then sweeps

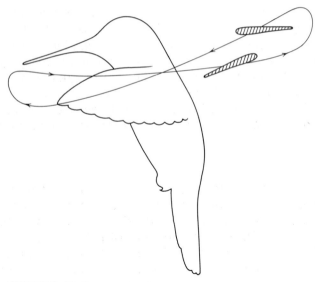

FIGURE 24-9

HOVERING FLIGHT OF A HUMMINGBIRD showing approximate path of the wing tip and angle of attack of wing on forward and backward strokes.

backward, again creating lift with a large vertical component. The horizontal components of the two strokes tend to cancel, yet by slightly altering the power of one stroke in relation to the other, the flyer can move horizontally forward or backward as needed when feeding. When the flyer is not hovering, it flies with the body horizontal.

Since the body of a hovering hummer is stationary relative to the air, the bird must move its wings very fast to establish enough air speed at the flight feathers to create lift. The frequency of the beat is 35–50/sec. The wings (and therefore the bird) are small so that their oscillations will not generate excessive inertia. The hand part of the wing is relatively long and the slower-moving arm part is short. Since the backstroke of the wings is active, the muscles that elevate the wings are relatively larger than for other birds. The entire power plant is large: The breast muscles account for 25% of the weight of the entire bird, and the little flyer may eat about twice its body weight in food per day.

Slow Ascending and Descending Flight. Most small birds of bush and forest make frequent short flights which often include steep ascents and descents. They fly slowly but flap their wings rapidly (commonly 15–25 strokes/sec) to attain the necessary air speed of the flight feathers. The body axis is held between 45° and vertical while ascending and descending. At such times the wings, therefore, move nearly backward and forward, with their tips describing flat ovals or loops. Small bats with erratic flight presumably fly much as small birds do. Their flight has been less studied, however, and the following account is based on birds.

On the forward stroke the wing is fully extended and has a positive angle of attack. Marked lift (to raise the flyer or retard its descent) and propulsion result. The backstroke, by contrast, is merely a recovery stroke having little reaction with the airstream. In order to reduce drag on the backstroke, the wing is partially flexed, thus reducing its area. The muscles that elevate the wing tend to be much smaller than those controlling the downstroke.

Wing loading is the weight of the flyer divided by the area of its wings. As was explained on p. 25, small animals have more surface area in relation to volume than do larger animals of identical proportions. Small flyers, therefore, have relatively low wing loading without having particularly large wings. Representative values for small bats are 0.07–0.17 g/cm^2 and for small birds are 0.11–0.23 g/cm^2. (There are various methods of measuring wing area, and the effect of age, sex, and season on body weight should be considered in making comparisons.)

Pigeons, gulls, ducks, hawks, owls, pheasants, and various other strong flyers of medium size can also ascend and descend at steep angles while flying slowly. Their larger size means that their wings must beat more slowly. Frequencies of 3–10 strokes/sec are usual. Further, because of the relationship between surface–volume ratios and body size, and because of the nature of the fast flight of these birds, they have moderate to high wing loadings (i.e., 0.40–1.30 g/cm^2).

When the outstretched wing sweeps downward and forward it produces much lift and a little propulsion. Since the wing loading is large and the frequency of beat is only moderate, these birds cannot afford a passive backstroke (as small birds can). Their backstroke is complicated, but three motions are particularly characteristic: (1) The wing is flexed so the tip travels close to the body, (2) the outer half of the wing (only) is turned over with what Brown calls a flick, and (3) the primary feathers rotate individually like the slats of a venetian blind so that each feather acts as a strong airfoil with positive angle of attack, yet air can pass between the feathers (see Figure 24-10). The opening of this venetian blind is the automatic consequence of the facts that (1) the shafts of the feathers are not central in their respective vanes so air pressure tends to make the vanes rotate, and (2) the flick reverses the direction of air pressure on the feathers thus enabling them to rotate in the direction not prevented by their overlap. This backstroke

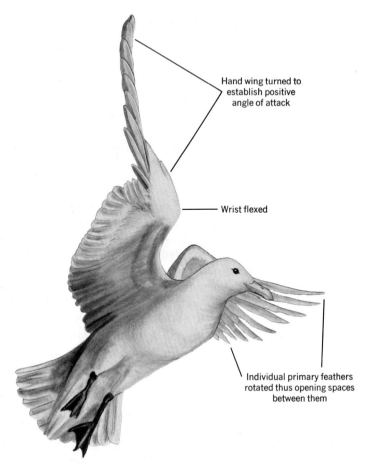

Hand wing turned to establish positive angle of attack

Wrist flexed

Individual primary feathers rotated thus opening spaces between them

FIGURE 24-10

BACKSTROKE OF WINGS IN ASCENDING FLIGHT OF A MEDIUM-SIZED BIRD shown by a gull, *Larus.*

produces even more lift than the forward stroke and also some propulsion. The slower-moving inner wing is nearly useless. Pigeons can take off and land without the inner, or secondary, feathers. The wing tip describes a figure eight in each full cycle.

Some birds of comparable size, and many larger birds, having even slower wing beats and higher wing loadings (swan, 1.7 g/cm²; heron, 2.5 g/cm²), are unable to ascend and descend steeply while flying slowly. Such birds often run to gain speed for a fast flat takeoff and land fast on water, which cushions the impact.

Fast Level Flight. Fast level flight is characteristic of the large mastiff bat and of many birds of moderate or large size such as ducks, geese, falcons, toucans, and gulls. The body is carried horizontally, and the wings beat up and down. High speed of the body provides high air speed over the wings. Consequently, the slower-flapping inner wing produces much lift on both the upstroke and downstroke. The wing beat is relatively slow, and the amplitude is small. Wing loading is usually high. Since a propulsive force is needed only to oppose drag and not to power acceleration or ascent, or to retard descent, the demand for power is relatively low. Virtually all propulsion is provided by the outer half of the wing on the downstroke. The upstroke is passive; negative air pressure over the wing may lift it without muscular effort. The outer wing does not flick over on the upstroke as for slow flight of some of the same kinds of birds. The position of the inner wing changes little on the upstroke; the outer wing is flexed somewhat at the wrist. The wing tip never moves backward in relation to the ground.

Stability and Maneuverability. As noted in previous chapters, animals must choose between stability and maneuverability. The elephant opts for stability; flyers (more than designers of aircraft) opt for maneuverability. If the body is moved slightly out of position, aerodynamic forces usually do not restore the original orientation but instead tend to quickly force the flyer farther out of position. This makes it mandatory that the sensory–nervous–motor control system correct constantly and almost instantaneously for minor displacements. The same circumstances, however, permit rapid maneuvering. Some birds occasionally do loops (hawks) or rolls (ravens), and such displays are surpassed by the aerobatics of many bats and birds as they pursue flying insects or engage in courtship flight.

A flyer can correct for roll by increasing the angle of attack of the lower wing, by increasing its surface area, or by flapping it harder than its opposite, any of which would increase its lift. (Since the same actions influence drag, compensations are needed to prevent yaw.) Also, the tail, if large, might be formed into a screw shape to correct for roll. The same kinds of behavior can, of course, be used to cause roll.

Unlike the swimmer and man-made aircraft, the flyer has no vertical fins or rudders to control yaw. Instead, it increases the drag on the wing that is tending to advance faster than its opposite. This can be done by increasing its angle of attack. Greater angle of attack, however, and also greater air speed (since it is moving ahead of the other wing) increase its lift and hence tend to cause roll. To compensate, the advancing wing is flexed to reduce its area. Dropping one foot out of the streamlined contour of the body also increases drag on that side. Further, if a bird with a large stiff tail is banking to turn, it can open and twist the tail so as to form a vertical rudder.

Wings banked, have
different angle of attack

Maneuvering
Gull, *Larus*

Wings, tail, feet
create maximum drag

Braking prior to landing
Razorbill, *Alca*

Tail fanned,
elevated

Wings unequally
spread

Taking off
Chickadee, *Penthestes*

Flipping over on landing
to hang by feet
Leafnose bat, *Macrotus*

Suction above stalling
wing lifts feathers

Webbed feet cushion
impact

High–speed landing on water
Swan, *Cygnus*

To correct for downward pitch of the body (or to initiate upward pitch) a flyer moves its wings forward, thus placing its support farther forward in relation to its center of gravity. Also, if the tail is suitably constructed, it is bent upward. The converse behaviors correct for upward pitch.

Braking and Turning. Gliders, and flyers approaching an elevated perch, may brake by swooping upward into a stall, thus using gravity for deceleration. As noted, flyers can also flap their wings to create upward force which slows descent preparatory to landing. In sustained level flight it is desirable for the ratio of **L:D** to be maximum consistent with moderate to high speed. When braking, by contrast, **D** should be high and speed should be low. This is achieved by making the angle of attack, the wing area, and the camber maximum. Birds also depress their fanned tails and bats depress the tail membrane. This is the equivalent of lowering wing flaps of man-made aircraft. The feet are thrust forward (toes spread if they are webbed) not only to be in position for landing, but also to provide drag. Further, when birds brake, they lift the alula to create a wing slot. This increases the angle of attack and decreases the velocity that can be attained before stalling occurs.

Adaptations of the legs of flyers for absorbing the shock of landing have been little studied. The impact does not shear the ball-and-socket joint at the hip, but instead is transmitted to the ilium as compression by the neck and trochanter of the femoral head. Many fast-flying water birds land into the wind and plane briefly on their webbed feet as they strike the water.

Since moving bodies tend to continue to move in straight lines, when animals turn they tend to slip sideways toward the outside of the turn. The force out of the turn is centrifugal force which varies directly with the square of the velocity and the mass of the body, and inversely with the radius of the turn. This force must be opposed by an equal and opposite centripetal force into the turn. Sharply turning cursors resist centrifugal force by friction at the ground (see p. 506), and swimmers rely on drag produced by vertical fins and a deep body. Flyers must use another method. They bank into the turn causing **L** to tilt inward (Figure 24-12). The horizontal component of **L**, or **I**, must equal the outward centrifugal force, **O**. If the flyer is not to lose altitude, the vertical component of **L**, or **V**, must equal the full value of the lift that existed before the turn started. This means that lift, and also wing loading, must increase during turns: If a 60° bank is executed, lift must be doubled. It is clear from the relation of the magnitude of **O** to

FIGURE 24-11
EXAMPLES OF FLIGHT CONTROL.

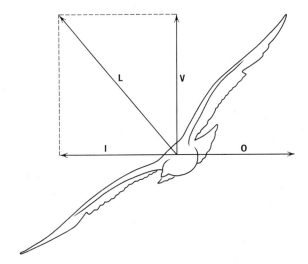

FIGURE 24-12
FORCES RELATED TO A
FLYER DURING TURNS.

mass and velocity that flyers that dodge quickly must be small, and that large flyers cannot turn sharply at high speed.

Wing Structure

The various animal flyers have strikingly different flight habits. Since form must correspond to function, wing structure is also varied. Savile has described four general types of wings, it being understood that they intergrade and that within types there is great variation of detail. The aerodynamics of flight is not yet well enough understood to fully interpret the morphology of wings.

First is the **elliptical wing.** It is characteristic of most bats and of most small and medium-sized birds of shrub and forest (e.g., sparrows, robins, crows, quails, pigeons). The specialization is for high maneuverability and precise control, often in confined spaces, and for minimum induced drag. The flyer is able to fly slowly and to ascend and descend rapidly. The wing has an elliptical outline (except where it meets the body). It is short and broad, making the aspect ratio low (commonly 3–6). Camber is moderate to marked (particularly in bats). Flapping flight is usual, though there may also be some gliding or swooping. The wing beat is moderately fast, and the amplitude is relatively great. The birds have a large alula, and in the larger and more active birds the primary feathers separate to form additional wing slots. In all, the slots may extend along the leading edge of 15–30% of the length of the wing. These slots are an antistalling device at low speeds and apparently reduce

FIGURE 24-13
THE FOUR PRINCIPAL TYPES OF WINGS.

Common murre, *Uria*

High–speed wing

Mastiff bat, *Eumops*

Long soaring wing
Albatross, *Diomedia*

Mexican fruit bat, *Artibeus*

Robin, *Turdus*

Elliptical wing

Broad soaring wing
California condor, *Gymnogyps*

turbulence at the wing tip. They may be closed in faster, level flight. The primary feathers may rotate individually when the slots open. This is the equivalent of having the entire wing twist like a propeller as it stretches out from the body, a result which is otherwise difficult to achieve with a short wing.

The **high-speed wing** is characteristic of mastiff and free-tailed bats and of swifts, swallows, falcons, shore birds, hummers, and ducks. The specialization is for high flight speed with low drag and low expenditure of energy. The wing is relatively small, so wing loading is large. The wing tapers to a slender tip, may be somewhat swept back along the leading edge, and is faired into the body behind without a sharp angle. The aspect ratio is moderately high (commonly 5–9). The hand skeleton is relatively long, and strikingly so in hummers (Figure 24-14). In cross-section the wing is thin and has little camber. Flapping is constant except perhaps for short glides when descending. The beat is rapid in relation to the size of the flyer, and the amplitude is small. There are no wing slots.

The **long soaring wing** is characteristic of albatrosses, frigate birds, gannets, terns, and gulls. Some large pterosaurs probably had similar wings. Most such birds fly over water where long wings are not a handicap. The specialization is for a high ratio of lift to drag permitting soaring at high speed with low expenditure of energy and a low gliding angle. The wing is long, slender, and pointed; the aspect ratio ranges from 9 to 18; and the span of the wings reaches more than five body lengths. The hand skeleton is relatively short in these birds (Figure 24-14), though it is not in pterosaurs. Wing loading is high. Camber is low. There are no wing slots. Landing and takeoff speed is high. Most of the birds land on water and take off into the wind—often after running to gain speed.

The **broad soaring wing** is seen in vultures, eagles, and buteo hawks. The wings of the raven and pelican are similar. The specialization is for soaring at low speed, takeoffs and landings in confined areas, high lift, and low sinking speed. The wing is

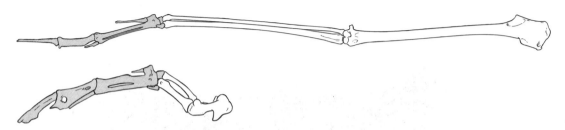

FIGURE 24-14
CONTRAST BETWEEN THE RIGHT WING SKELETONS OF A DYNAMIC SOARING BIRD AND A HOVERING BIRD shown by an albatross, *Diomedea* (above), and a hummer, *Lampornis* (below). Hand skeletons drawn to the same length.

moderately long and broad, thus giving it large area, has only moderate wing loading for such large birds, and has moderate aspect ratios (e.g., 6 or 7). Camber is marked. The alula is prominent, and terminal wing slots are conspicuous. The slots seem to correlate with broad wing tips and correspondingly large induced drag. The slots usually have U- or ⊔-shaped bases rather than the less highly evolved V-shape of some elliptical wings. Each spaced primary feather functions independently as a winglet and each is streamlined in cross-section. Muscles that elevate the wing are less developed than in birds that emphasize flapping flight: Contrast the supracoracoid muscle in Figures 24-16 and 9-9.

We have now considered the primary requirements for flight. These, however, are not enough: Even if they were provided with wings, lizards and rats (and angels) could not fly. One secondary requirement is for light weight. Flyers have light skeletons. (The skeleton of an eagle accounted for less than 7% of the total body weight—about half as much as for man.) The bones of birds and pterosaurs are hollow and air-filled. This permits the bones to have a maximum diameter and hence maximum resistance to bending forces with minimum weight. Many bones have thin inner and outer lamellae joined by spicules which tend to follow stress lines. Further, sutures tend to become obliterated. These constructions provide maximum strength with minimum hard tissue. Because bone is heavier than feathers, the bony axis of the avian tail is short. Reduction or loss of digits also saves weight for birds and pterosaurs. Modern birds and some pterosaurs lack teeth, which are heavy. (The feeding habits of most bats require that they have large ears and teeth. The fast-flying mastiff bat has enormous horizontally oriented and cambered ears. It is probable that these ears generate enough lift to help support the heavy head!)

The gonads of flyers regress when they are not active. Good flyers eat foods which, being nutritious, are lightweight in relation to the energy provided. Also, digestion is very rapid (often within an hour). Some flyers (hummers, swifts, many bats) use the legs only for perching or roosting. The legs are then very small and light. Feathers are remarkable for the loads they can bear in spite of light weight. (The feathers of an eagle accounted for 14% of the total body weight.) Feathers are stiff, yet flexible and elastic. They provide a smooth streamlined contour and are resistant to damage (see pp. 103, 104).

A second requirement is for compactness and firmness of the body so that the thrust of the wings will be transmitted to the body near its center of mass and will propel the body as a unit without deforming it or causing it to flap. Flyers have a

Further Structural Adaptations

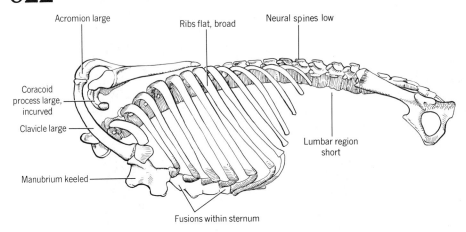

Acromion large

Ribs flat, broad

Neural spines low

Coracoid process large, incurved

Clavicle large

Lumbar region short

Manubrium keeled

Fusions within sternum

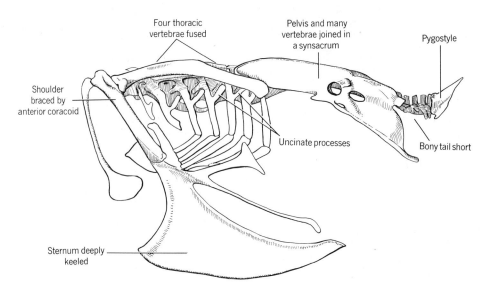

Four thoracic vertebrae fused

Pelvis and many vertebrae joined in a synsacrum

Pygostyle

Shoulder braced by anterior coracoid

Uncinate processes

Bony tail short

Sternum deeply keeled

FIGURE 24-15

SOME CHARACTERISTICS OF THE BODY SKELETON OF FLYERS shown by a fruit bat, *Pteropus* **(above), and a pheasant,** *Phasianus* **(below).**

short trunk which is stiff. Fusions between adjacent trunk vertebrae are common in bats and usual in other flyers. Pterosaurs had many sacral vertebrae and usually had fusions in the thoracic region. Birds have 12–20 vertebrae fused together in the synsacrum and may have additional fusions in the thoracic region (pheasants, falcons). Even in the absence of fusions, the shape of the joints between trunk vertebrae are such as to limit or prevent motion. Ribs tend to be broad and flat so they nearly touch edge to edge. They may fuse together proximally (some bats, many pterosaurs) or to the spine (some pterosaurs). The uncinate processes of avian ribs probably provide bracing. The

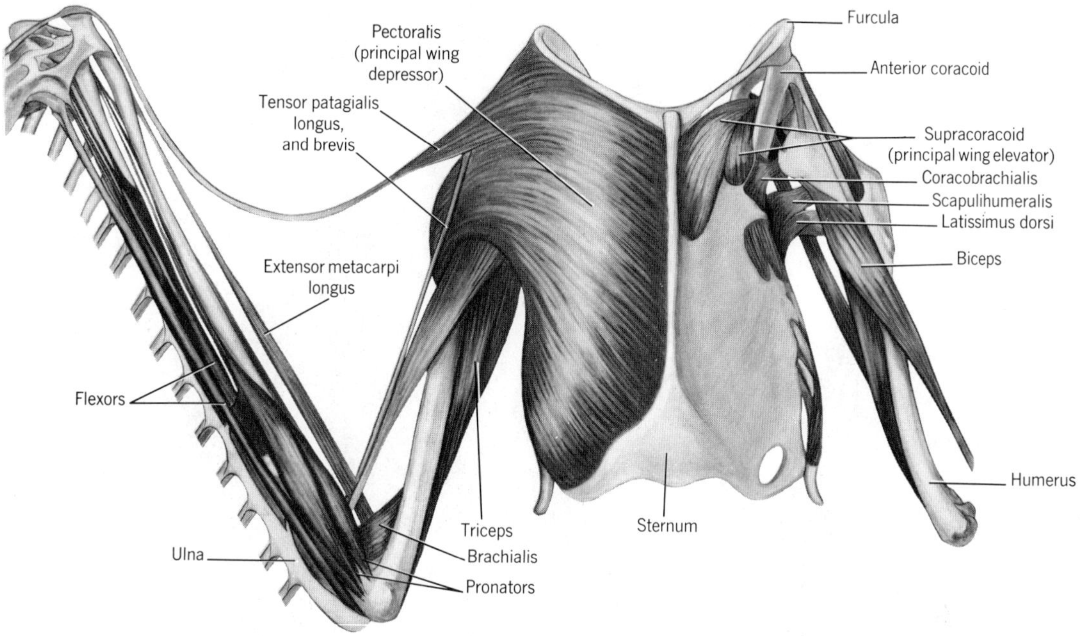

FIGURE 24-16

SOME FLIGHT MUSCLES OF A SOARING BIRD, the golden eagle, *Aquila*, seen in ventral view with the pectoralis removed on the left side. (Drawn from an air-dried dissection and hence somewhat shrunken.)

sternal ribs tend to ossify. The number of units in the sternum is reduced to three (bats) or one (birds, pterosaurs).

The body is made compact by the shortness of the bony tail (birds, some pterosaurs, and some bats) and by sharply flexing the neck during flight. The pectoral girdle is large, strong, and has unusually firm attachment to the body by the anterior coracoid (birds, pterosaurs), clavicle (bats), or even the scapula (some pterosaurs). The entire girdle is virtually immovable in birds and pterosaurs, though not in bats. The concentration of flight muscles (including elevators of the wings) on the breast of birds positions their great weight near, but a little below, the point of support of the body by the wings. The same is true to a lesser extent of bats.

A third requirement is for automatic mechanisms that save muscular effort and, therefore, both weight and energy. The joints of the wings of both birds and bats are constructed so as to limit motion to one plane. There is no supination or pronation of the forearm; all rotation is at the shoulder. The ulna (which would be required for rotation of the forearm) adheres to the radius in pterosaurs, and its distal portion is nearly lost in bats. It is retained in birds because it has a new function: The radius and ulna join the humerus and carpus in such a way

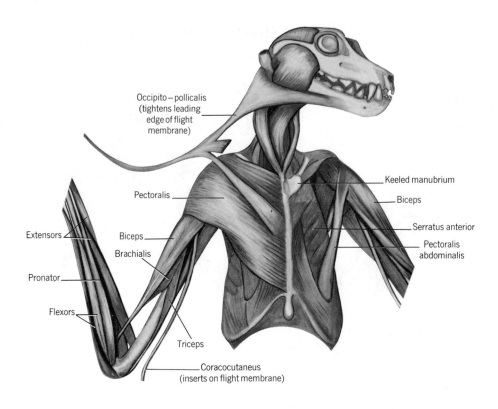

Occipito–pollicalis
(tightens leading
edge of flight
membrane)

Keeled manubrium

Biceps

Pectoralis

Serratus anterior

Extensors

Pectoralis
abdominalis

Biceps

Brachialis

Pronator

Flexors

Triceps

Coracocutaneus
(inserts on flight membrane)

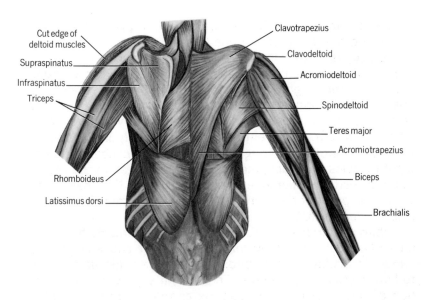

Cut edge of
deltoid muscles

Clavotrapezius

Supraspinatus

Clavodeltoid

Infraspinatus

Acromiodeltoid

Triceps

Spinodeltoid

Teres major

Acromiotrapezius

Biceps

Rhomboideus

Latissimus dorsi

Brachialis

FIGURE 24-17

SOME FLIGHT MUSCLES OF A FRUIT BAT, *Pteropus,* **seen in ventral view with the pectoralis and occipito–pollicalis removed on the left side (above), and in dorsal view with the trapezius and deltoids removed on the left side (below). (Drawn from a freeze-dried dissection and hence slightly shrunken.)**

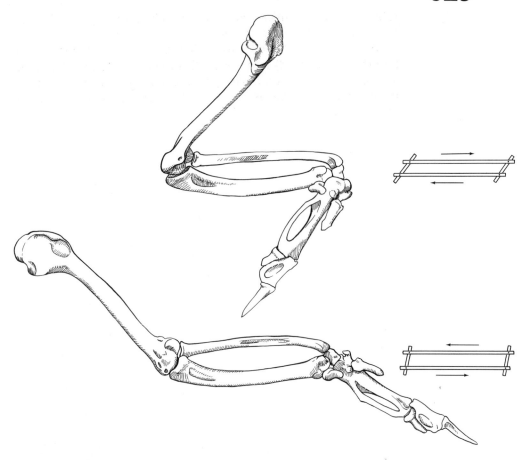

FIGURE 24-18

USE OF A PARALLELOGRAM TO EXTEND AUTOMATICALLY THE WRIST WITH THE ELBOW, shown by a pheasant, *Phasianus.*

as to form a parallelogram. When the elbow opens out, the wrist opens automatically. Other mechanisms having related functions in bats have been described by Vaughan and Norberg.

Various bats have locking devices to assist in holding the shoulder, elbow, and wrist joints at the correct angles. It is said that a bone at the elbow in the albatross can be placed in the joint to hold it open. Only a small part of the wing membrane of bats is attached to the small second digit, yet many bats use this digit and its ligamentous extension as a bow to keep the leading edge of the wing stiff.

Another requirement of flyers is for an efficient power plant. Birds have a relatively high body temperature, and it is likely that even pterosaurs could elevate the body temperature, at least when flying. Birds and bats have a high metabolic rate and efficient respiratory systems. Ventilation of the lungs is apparently synchronized with the wing beats. The heart is large, blood pressure high, and circulation rapid. Even making allowance for great efficiency, it seems certain that the energy release in avian muscles is higher than in nonflying vertebrates. Fat is the main source of energy.

For reasons explained earlier, flyers have a large cerebellum and (bats excepted) large eyes. Bats and pterosaurs roost upside down and hence have specializations of the feet to form hooks and of the pelvis and knee to allow suitable orientation of the legs.

25

Feeding

Adaptations for feeding are particularly diverse. This is, therefore, a good place to emphasize that each animal adapts not to a random combination of independent activities, but instead to a coordinated life style. Manner of locomotion is nearly always related to feeding habits, and reproductive, defensive, and other behavior are usually correlated with manner of feeding and locomotion. Parts of preceding chapters on the teeth, digestive system, sense organs, and locomotor adaptations relate directly to feeding; other parts relate indirectly.

Before studying the varied ways that vertebrates locate, secure, and process their food it is desirable to learn about some of the principle kinds of jaw mechanics.

Protrusible Jaws of Fishes. The upper jaw of fishes is said to be protrusible if it not only can be opened and closed, but also moved forward and backward in relation to the remainder of the head. The mechanism has been analyzed in England, the Netherlands, the United States, and elsewhere by R. McN. Alexander, J. W. M. Osse, K. F. Liem, B. Schaeffer, and others.

Acanthodians and the most primitive of actinopterygian fishes had large mouths. The hinge of the jaws was under the back of the braincase. Premaxilla and maxilla were not movable, and the upper jaw was not protrusible (Figure 25-1A). These fishes apparently opened the mouth wide and snapped at prey; the rapidly moving lower jaw completed closure partly by inertia. The bite was not strong.

A subsequent evolutionary stage is exemplified by the bowfin and salmon. The hinge of the jaw is nearly as far back as before, so the mouth remains large. However, the adductor of the jaw is stronger and inserts farther forward on the mandible, giving the fish a strong bite. The premaxilla is still fixed, but the maxilla is now free. Its posterior end is joined to the mandible by a ligament, and its anterior end pivots on the rostrum. Consequently, when the mandible opens, the lower end of the maxilla rocks downward and forward, thus preventing food from escaping at the corners of the mouth (part B). The upper jaw still protrudes very little or not at all.

At the next stage in the specialization of fish jaws, the hinge of the mandible moves forward under the orbit, so the mouth is smaller. The premaxilla, like the maxilla, pivots forward as the mandible is opened (part C). Further, the premaxilla is L-shaped; one arm forms part of the margin of the mouth, and the other, shorter arm forms an ascending process running backward toward the ethmoid region. The premaxilla can slide forward more or less along the axis of this ascending process, thus making the upper jaw truly protrusible. The lengthening of the head amounts to 10–20%. A complicated system of ligaments controls the motions of the bones, but the mechanism is not the same in all fishes. The position of the maxilla is commonly altered by motion of the mandible or by action of the adductor mandibulae, which may have one or more heads inserting on the upper jaw. Motion of the maxilla, in turn, moves the premaxilla. The ascending process usually is guided by a median ridge of the vomer or ethmoid as the premaxilla slides down and forward. A small cam-like bone in the ethmoid region contributes to the control of the action in some fishes. The upper jaw usually can be closed on the lower jaw both when it is protruded and when it is retracted.

Protrusible jaws have several advantages, according to species. Importantly, they permit the mouth to be closed during or immediately after sucking food into the mouth,

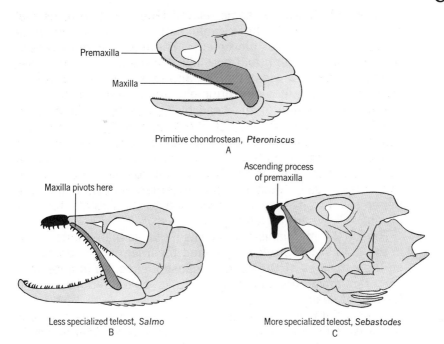

Premaxilla —

Maxilla —

Primitive chondrostean, *Pteroniscus*
A

Maxilla pivots here

Ascending process
of premaxilla

Less specialized teleost, *Salmo*
B

More specialized teleost, *Sebastodes*
C

FIGURE 25-1

STAGES IN THE EVOLUTION OF THE JAWS OF RAY-FINNED FISHES.

without blowing the food back out again (which would occur if the lower jaw were raised). Sometimes the protruded mouth points more downward than the unprotruded mouth — an advantage when taking food, including algae, from the substrate. The jaws probably can be closed a little faster when protruded because the mandible does not have as far to travel. Finally, protrusion places the mouth a little closer to food, though swimming would do the same. Perhaps half of all teleosts have protrusible jaws. They can be recognized by a transverse fold of skin just behind the margin of the upper jaw. This fold is pulled tight as the premaxillas move forward.

Cranial Kinesis. Relative motion of two or three parts of the roof of the skull is called cranial kinesis. Kinetic skulls are divided into several movable units which are linked together and usually function together to elevate and depress the upper jaw. Crossopterygians, some labyrinthodonts, caecilians, some lizards, snakes, some extinct reptiles, and birds have, or had, kinetic skulls. The mechanisms are often complex. They have been analyzed for various species, but largely on the basis of two-dimensional models.

The hinged braincase of crossopterygians is unique to that group of fishes and is associated with a distinctive musculature

(Figure 7-9). There was some related lateral motion of dermal bones of the skull; the hyomandibula served to integrate various of the complex movements. The mechanism increased the gape of the mouth and possibly also the power of the bite.

Most lizards have kinetic skulls, and for them the mechansim is relatively simple. Four principal units (some paired) are joined on each side of the head by four principal joints (Figure 25-2A). (Some struts, not shown in the figure, articulate with the mechanism.) Motion of any one unit requires motion of at least two other units and of all four joints. Movement on each side of the head is nearly restricted to one plane, and the two sides usually must function together. Kinesis allows the upper jaw to bite down on food and to manipulate food by moving relative to the lower jaw.

The kinetic mechanism of snakes includes many units (eight in the boa) which are loosely joined together and independent on the two sides of the head. In consequence, an infinite variety of relative motions is possible—a characteristic of the feeding of snakes.

The cranial kinesis of birds has four principal units and four principal joints on each side of the head as for lizards, yet is quite different and more complicated. Some birds can move the quadrate units on the two sides of the head somewhat independently, and some can slide as well as pivot the mandible

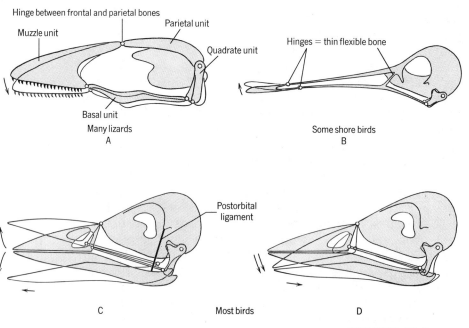

FIGURE 25-2
CRANIAL KINESIS.

on the quadrate. Motion of upper and lower bills is not necessarily confined to one plane. Further, the mandible is joined to
the braincase by the stout postorbital ligament. This appears to
link the motions of upper and lower bills in at least some birds,
though a degree of independent voluntary control seems to be
present. (The ligament may also keep the jaw closed passively).
The single dorsal hinge of the mechanism is always farther
forward than any joint of lizards, but may be at the base of the
upper bill or near the tip (compare parts B and C of Figure
25-2). Kinesis in birds allows the bill to open wide, and in so
doing to maintain its longitudinal axis (part C). It provides for
skilled manipulation of food by sliding one bill lengthwise in
relation to the other and by creating a wedge between the bills
that opens backward into the mouth, thus preventing food
from escaping (part D). Kinesis permits upper and lower bills to
close faster than otherwise because each is active. It probably
also cushions the shock of pecking for some birds.

Jaws of Mammals. The jaws of ancient fishes, most amphibians, many reptiles, and mammals are neither protrusible nor
kinetic. The jaws of carnivorous and herbivorous mammals are
further characterized by adaptations for a strong bite that
shears, crushes, or grinds, rather than for a weaker but faster
snap that catches prey which is swallowed whole. Many researchers in several countries have studied the jaw mechanics
of flesh- and plant-eating mammals, but the mechanisms are
complex and much remains to be done. What follows, therefore, is partly speculative.

Elephants apply force more or less evenly all along each
tooth row. Most ungulates do likewise for the cheek teeth, but
also can apply a lesser force at an anterior cropping mechanism. Rabbits and rodents apply force either at their anterior
gnawing incisors or at their posterior grinding cheek teeth.
Likewise, Carnivora apply greatest force with their canines at
the front of the mouth or with their **carnassials** (shearing teeth)
at the back of the mouth. The two mechanisms do not
function simultaneously. When a rabbit chews, its incisors do
not touch; when a beaver gnaws, the mandible is moved forward and the cheek teeth do not occlude.

Carnivora bite with a chopping motion. Upper and lower
tooth rows come together without sliding past one another
either front to back or side to side. The joint between mandible
and skull consists of a transversely oriented cylindrical condyle
on the mandible which rotates hinge-like in a fossa on the temporal bone. However, when the carnassial shearing mechanism
on one side of the mouth is in use, the carnassials on the other
side are not aligned and cannot function. To bring one set of
shears into play, the animal pulls the mandible to that side and

the cylindrical condyle slides a little sideways in its fossa. Further, the symphysis of the lower jaw usually is not ossified. In a bat and in dogs (which have been studied) and probably in most other Carnivora, the symphysis allows the two mandibles to alter their relative positions as one set of carnassials is positioned to shear.

Most mammalian herbivores (plant eaters) chew by sliding the lower tooth rows over the upper tooth rows. Motion may be back to front or side to side. Grinding occurs during one phase of the cycle—usually either as the mandible moves forward or from one side toward the center line. During the other, slower phase of the cycle, the jaw opens a little as it is returned to the starting position; the tongue meanwhile places more food between the tooth rows. If motion is back to front, upper and lower tooth rows are usually about the same distance apart and all cheek teeth function together. If motion is side to side, the lower tooth rows are commonly closer together than the upper rows, though the reverse is sometimes true. Either way, the animal must chew on one side of the mouth at a time. Whatever the relative motion of upper and lower tooth rows, it is nearly always crosswise or oblique (not parallel) to the orientation of the cutting ridges on the teeth.

Seen in side view, a tooth row may be straight or arched; seen in cross-section, a row may be horizontal or may slant inward or outward. The rows of one jaw (upper or lower) may be either parallel or diverging toward the back of the mouth (though usually less so than for carnivores). These factors may guide and restrict the relative motions of upper and lower tooth rows. Differential hardness of the different materials of the cheek tooth of an herbivore causes cutting ridges to form (of which more later in this chapter). For the same reason, wear also produces larger **grinding ridges** in rabbits and many artiodactyls (Figure 25-4). These serve to position and guide the tooth rows as they slide over one another. It is clear that the chewing motions of herbivores are exceedingly diverse. One should be cautious about inferring details without studying function in live animals.

The structure of the articulation between mandible and skull of herbivores is correspondingly varied. The condyle may be transverse as for carnivores. When it moves instead in a less restricting fossa (wombat, hyrax, horse), it may be longitudinally oriented; when it rocks and slips in a longitudinal groove (rabbit), it may be bluntly rounded (beaver, paca), or relatively flat (many artiodactyls). The reader might find it useful to infer from the structure and functions of the tooth rows (preceding two paragraphs) how many different kinds of motions may be required at this joint.

The symphysis between the mandibles is firmly ossified in many herbivores (sloth, hyrax, perissodactyls, camels), yet is

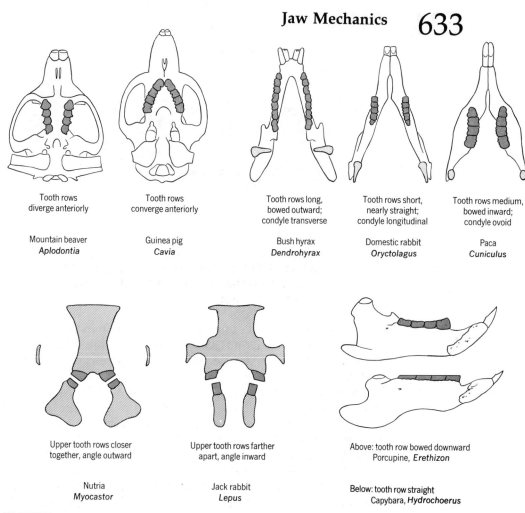

Tooth rows diverge anteriorly	Tooth rows converge anteriorly	Tooth rows long, bowed outward; condyle transverse	Tooth rows short, nearly straight; condyle longitudinal	Tooth rows medium, bowed inward; condyle ovoid
Mountain beaver *Aplodontia*	Guinea pig *Cavia*	Bush hyrax *Dendrohyrax*	Domestic rabbit *Oryctolagus*	Paca *Cuniculus*

Upper tooth rows closer together, angle outward	Upper tooth rows farther apart, angle inward	Above: tooth row bowed downward Porcupine, *Erethizon*
Nutria *Myocastor*	Jack rabbit *Lepus*	Below: tooth row straight Capybara, *Hydrochoerus*

FIGURE 25-3

SOME VARIATION IN TOOTH ROWS AND MANDIBULAR CONDYLES AMONG THREE ORDERS OF MAMMALS. Above left, skulls in ventral view; above right, mandibles in dorsal view; below left, cross-sections of skulls and mandibles at level of orbits; below right, medial views of left mandibles.

unfused and movable in rodents and most artiodactyls. The significance of this joint in herbivores invites study.

There are three adductors of the mandible. The **masseter** takes origin from the zygomatic arch and often also from the orbit or maxillary region. It inserts on the outside of the ramus and angle of the mandible. The **temporal** muscle originates on the braincase and sagittal crest, if present, and inserts on both the outside and inside of the coronoid process of the mandible. The **pterygoid** muscle originates at the base of the skull beside the palate and behind the orbit, and inserts on the inside of the ramus and angle of the mandible. Each muscle is commonly divided into several parts. These may be distinct, but they often merge with each other and (particularly for the masseter) with a complex of internal tendinous sheets. Muscle fibers are

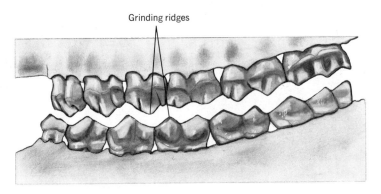

Grinding ridges

FIGURE 25-4
CHEEK TEETH OF A CARIBOU, *Rangifer,* **SHOWING GRINDING RIDGES THAT MAINTAIN ALIGNMENT DURING SIDE-TO-SIDE MOTION OF MANDIBLE.** Lateral view of left teeth.

rarely parallel even in one division of one muscle. Unfortunately, the force diagrams one sees of this three-dimensional complex of muscles are necessarily two-dimensional approximations.

The temporal muscle comprises more than half the total adductor mass of carnivores. The coronoid process provides the lever arm of this muscle, and in these animals is large. The masseter is somewhat smaller and serves in part to stabilize the articulation of the jaw. The pterygoid muscle positions the carnassials, but adds little to the force of the bite and is small.

The masseter commonly comprises about two-thirds of the adductor mass of herbivores. The pterygoid muscle is next in size and joins the masseter in producing lateral grinding motions of the mandible. The temporal muscle is relatively small, and the coronoid process, though variable, tends to be small and may be absent (rabbit, capybara).

The articulation of the jaw is about in line with the tooth rows in carnivores. Its position is various in rodents but usually is somewhat above the tooth rows. It is much above the tooth rows in most other herbivores (Figure 25-5). The high position increases the lever arm of the masseter muscle. It also causes the lower teeth to approach the upper teeth obliquely rather than perpendicularly. This might tend to roll or slice plant food as it was crushed.

Locating and Securing Food

Finding Food and Making It Accessible. The food of vertebrates ranges from microscopic (diatoms, algae) to very large, from passive (plants) or nearly so (mollusks) to very active, from nutritious (insects, worms) to low in food value (stems), from defenseless to protected and from extensively available

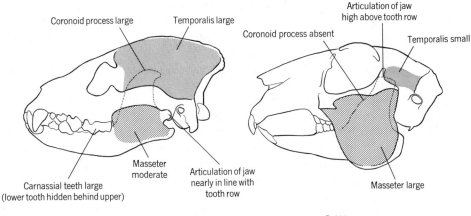

Coronoid process large   Temporalis large

Articulation of jaw
high above tooth row

Coronoid process absent   Temporalis small

Masseter
moderate   Articulation of jaw
nearly in line with
tooth row

Carnassial teeth large
(lower tooth hidden behind upper)

Masseter large

Hyena, *Crocuta*   Rabbit, *Lepus*

FIGURE 25-5

CONTRAST BETWEEN THE JAW MECHANICS OF A CARNIVORE (left) AND AN HERBIVORE (right).

to highly restricted (e.g., the nectar of only certain flowers). Food may be seasonal or constantly accessible, abundant or rare, and available with or without competition. Small wonder, therefore, that the ways that vertebrates locate and process their food are exceedingly varied. Only a summary can be offered here.

The hungry animal usually searches actively for a meal, but instead may lie in wait for prey to come within range of a sudden dash or strike (groupers, vipers). Ordinarily, the feeder perceives the food object before contact is made. This is probably true in a general sense even of the vertebrates that strain minute food particles from quantities of water. However, some larger prey is not identified before capture: Nighthawks sweep the air for small insects. Apparently responding to discontinuities on the water surface, certain bats fly low over water with the large curved claws of their feet trailing in the water ready to gaff any small fish encountered. Similarly, the bird called the skimmer flies over water with its huge mandible cutting the surface ready to scoop up any small fish in its path.

Regardless of the method of search—active or passive, directed or random—the food must be recognized when it becomes available. Any of the senses may be used, or several may be used in combination. Some vertebrates in each class are largely dependent on vision, and many rely on hearing, including echo-ranging and form-discrimination using vocal signals (many volant, aquatic, and some other mammals) or electric signals (some fishes). The chemical senses are rarely used by birds, but otherwise are commonly employed. The receptors may be outside the nose and mouth on barbles (catfish), rostrum (sturgeon), surfaces of the skin (amphis-baenians), or tip of the beak (some shore birds). Touch may also be important, the sensitive organ being the lips (manatee), snout or proboscis (elephant, pig, mole), or tongue (anteater). The use of the pit organ by certain snakes was mentioned on p. 391.

Many animals move some object to uncover food. For example, the turnstone (a bird) rolls stones with its beak in search of invertebrate food, the walrus digs clams from mud with its tusks, the towhee kicks away litter to reveal insects, the bear rips logs to find animal food, the warthog uproots tubers with its tusks, and the aye-aye (a primate) cuts bark with its strong curved incisors to get at insect larvae. Some woodpeckers drill into solid wood to secure insects: The horny beak is extra long, strong, and rapid-growing, and the bony beak is reinforced. Force is transmitted by bone and muscle to the thick and well-ossified interorbital septum of the skull.

Numerous vertebrates make food accessible by using a long reach. The arms of apes and monkeys, and the trunk of the elephant serve this purpose. The browsing giraffe reaches with a neck so long that a plexus of arteries has evolved at the base of the brain to even the blood pressure as the head swings up and down. The strikingly long neck and legs of the gerenuk (an antelope) similarly increase its reach. Several vertebrates have structures for reaching into restricted spaces. Examples are the bills of hummingbirds and creepers, and the tongues of anteaters. The third finger of the aye-aye and the fourth finger of two kinds of phalanger show remarkable convergence in that each is slender and longer than the other fingers to serve as a probe for removing insects from crevices.

FIGURE 25-6
A BROWSING
GERENUK, *Litocranius.*

Tools other than parts of the body are used by several vertebrates: The chimpanzee may use a stick as a probe and leaves as a sponge, a Galapagos finch uses cactus spines to poke into holes, an Egyptian vulture and the banded mongoose drop or throw rocks to break large eggs, and as it swims on its back, the sea otter pounds shellfish on a rock anvil placed on its chest for that purpose.

Jays hold acorns against chopping blocks as they hammer at the hulls, gulls drop crabs and mollusks onto rocks from the air to crack them open, and the beaver cuts down trees to get at the bark. Various birds follow ungulates to secure the flies and ticks attracted by those beasts. Some carnivores watch vultures to locate carrion. Certain chameleons have scent glands the secretions of which, when smeared on a branch, attract insects, which are then eaten. Alligator snapping turtles have a lure at the tip of the tongue which attracts fishes. Angler fishes attract prey by dangling in front of their jaws a lure derived from a dorsal fin spine. Those living deep in the ocean even

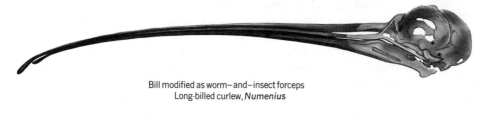

Bill modified as worm–and–insect forceps
Long-billed curlew, *Numenius*

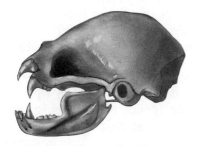

Canines and upper incisor modified as scalpels
Vampire bat, *Desmodus*

Third finger modified as insect probe
Aye-aye, *Daubentonia*

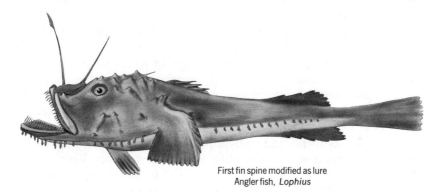

First fin spine modified as lure
Angler fish, *Lophius*

FIGURE 25-7
VARIED DEVICES FOR MAKING FOOD ACCESSIBLE.

have luminescent lures. The vampire bat makes a shallow wound on its prey using a pair of sharp, curved, upper incisors, and prevents the blood meal from clotting by the action of its saliva. The squirrel gnaws away the bracts of pine cones to get at the seeds, whereas the crossbill pries the bracts apart with its special beak. The archer fish looks up from under the water, locates insects as much as a meter away on overhanging vegetation, and then, with unerring aim, strikes them down with a forceful jet of water squirted from the mouth. The gill chamber is the force pump, and a groove on the roof of the mouth is the barrel of this water pistol.

Nature has been so inventive that many, many other devices for securing food could be listed.

Capturing and Killing Active Prey. Some fishes and snakes, and many mammals, stalk their prey, using stealth, conceal-ment, surprise, camouflage, and appropriate postures and mo-tions. Some spring onto the prospective meal. Most predators make a final short dash or longer chase, whether it be in water, on land, or in trees. Raccoons, some primates, and several rodents use the hands for capturing, or at least for controlling prey. Most hawks and owls dive on rodents from the air, se-curing the booty with their talons. Similarly, the osprey and fish owl plunge feet first into water to capture fish. The under-surface of the toes of these birds is rough, to prevent fish from slipping, and the scutes of the legs overlap so as to enter the water the "smooth way."

Several falcons dive, or "stoop" at flying birds, delivering a glancing blow with the feet and raking the victim with their strong hind talons. Various birds (pelicans, boobies, gannets, tropicbirds, kingfishers, etc.) dive head first into the water to catch fish. It is thought that the air sacs of the pelican cushion the impact. The eye is probably covered by the lower lid, and the nostrils and throat are modified to exclude water.

Various long-necked predators use the head as a catapult for capturing prey. The strike of the coiled snake is a familiar ex-ample, though high-speed photography is needed to analyze the action. Some snakes can even correct slightly the direction of the strike while it is in progress. The snake may hold the prey between the jaws or may pierce it with large sharp fangs and then withdraw temporarily. The long-necked, aquatic ple-siosaurs may have lashed out with their small heads to strike fish, with the body wheeling about but not progressing rapidly. A turtle has similar habits. Herons and egrets, which wade, and cormorants and loons, which swim under water, strike at fish by suddenly straightening the long curved neck. Dorsal and ven-tral neck muscles are strong and somewhat consolidated. Sev-eral cervical vertebrae are elongated and modified to provide a pulley for the dorsal extensors. Most of these birds use their long straight beaks as forceps to grasp the prey, but the trop-ical anhinga spears fish; fine barbs at the edge of the bill prevent loss of the food.

Many devices have evolved for capturing insects. The use of the tongue for this purpose is mentioned under a subsequent subheading. Birds that take insects on the wing have large broad mouths fringed by stiff bristles. The night-feeding frog-mouths, poorwills, and nighthawks have small horny bills but a prodigious gape. Another group of night feeders, the insec-tivorous bats, locate individual insects by echo-ranging so they do not need such large mouths. Several species use the flight membranes in catching insects. One small bat captured 175 mosquitos in 15 min. Such birds as warblers, wrens, vireos,

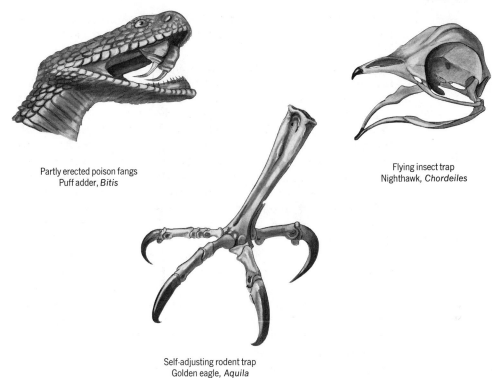

Partly erected poison fangs
Puff adder, *Bitis*

Flying insect trap
Nighthawk, *Chordeiles*

Self-adjusting rodent trap
Golden eagle, *Aquila*

FIGURE 25-8
VARIED DEVICES FOR CAPTURING ACTIVE PREY.

creepers, and thrashers that take insects from vegetation or the ground, have narrow bills that are long or short, straight or curved, according to the bird's specific habits.

Some active prey is swallowed whole and killed by smothering and the action of digestive juices—a common fate of small fishes. Somewhat larger prey must be immobilized before it can be swallowed: The secretary bird stamps reptiles to death with its long heavy legs, some fishes paralyze prey with electric shocks (see p. 214), some snakes kill with a poison bite and others by constriction, and birds that eat flesh or fish may kill with their claws or beaks. Some flesh eaters bleed or strangle their hapless victims, whereas others (African hunting dogs, several sharks, piranhas) cripple or surround the prey and then eat it alive—death comes from shock and bleeding.

Severing and Ingesting Plant Food. Various fishes and many larval anurans eat algae which may be scraped from the substrate with file-like teeth or (for the tadpoles) horny tooth analogs. Many vertebrates are said to be **herbivorous** because they eat such plant foods as leaves or stems. Relatively few adult fishes and amphibians are primarily herbivorous, but among reptiles, the giant dinosaurs, duck-billed dinosaurs, horned dinosaurs, and many turtles are, or were herbivorous. The ostrich, hoatzin, geese, and some parrots, grouse, and finches are plant eaters. Kangaroos, wallabies, wombats, langurs, sloths, rabbits, many rodents, elephants, hyraxes, sea cows, and ungulates are examples of herbivorous mammals.

The animal may bite the plants directly (duck-billed dinosaurs, iguanas, tortoises, geese). Plants may be gathered into the mouth by sensitive, mobile lips (sirenians, perissodactyls), or a protrusible tongue (giraffe, cattle), or may be fed into the mouth with hands (langurs, some rodents) or trunk (elephant). The food may be broken from the bush or tree before it reaches the mouth, but usually the leaf or stem is severed from the plant by the mouth, and this requires a cutting or cropping mechanism.

The horny margins of the mouth function as shears for tortoises and parrots. Serrate lateral teeth serve as cutting edges for the iguana. Geese have numerous lamellae at the margins of the bill which act as cutters. A specialized pair of upper and lower front teeth form a strong shearing apparatus in wombats, some sloths, hyraxes, rabbits, and rodents. These teeth have open roots and are ever-growing. Other herbivores use a cropping mechanism: Horses pinch grass between the upper and lower rows of contiguous incisors and then break it off with a sideways jerk of the head. Most artiodactyls instead break off the grass after pinching it between their lower incisors and a horny plate on the upper jaw. The analogous mechanism of kangaroos consists of a row of upper incisors which crop down against a pair of broad, procumbent lower incisors.

Filtering and Straining. Any aquatic vertebrate that feeds on floating or suspended food particles that are much smaller than itself (i.e., on plankton) must filter or strain the particles from the water. Ammocoetes, the larva of the lamprey, is the only primitive filter-feeder among vertebrates. It traps minute food particles in the sticky mucus on its complex pharynx and passes the mucus gradually back into a short intestine. All other vertebrate strainers have secondarily become specialized in various ways for their mode of feeding.

Strainers among fishes include the paddle fish, shad, and herring, and also the huge basking and whale sharks. The menhaden can strain 26 liter of water per minute. The gill rakers of these fishes are modified to form filters. They are long, slender, closely set along the gill bars, and have secondary or even tertiary branches. The larvae of various frogs are remarkably effective filter-feeders. The tadpole of *Xenopus* can filter to 0.1 μ, which challenges human technology.

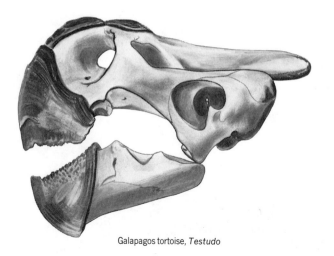

Galapagos tortoise, *Testudo*

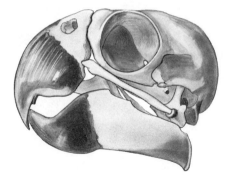

Cockatoo, *Cacatua*

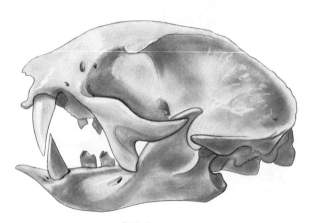

Sloth, *Choloepus*

Horse, *Equus*

FIGURE 25-9
EXAMPLES OF SHEARING AND CROPPING MECHANISMS OF PLANT EATERS.

Among ducks, the spoonbills and other dabblers strain plant or animal food from water. The bill is swung back and forth sideways, while the mandible pumps very fast. One hundred to 300 lamellae at the edge of the bill form the strainer. The flamingos are much more highly specialized and can remove diatoms and algae measuring only 0.1–0.2 mm in their longest dimension. During feeding, the large curved bill is positioned under the water "upside down," pointing toward the bird's feet. Upper and lower bills have up to twenty parallel rows of horny platelets which intermesh to form a fine screen. The platelets are less than 1 mm long, 0.1–0.2 mm apart, and frayed into

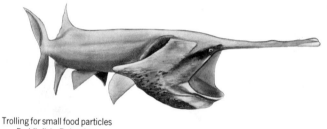

Trolling for small food particles
Paddlefish, *Polyodon*

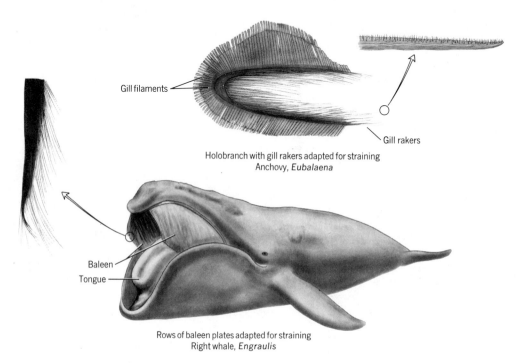

Holobranch with gill rakers adapted for straining
Anchovy, *Eubalaena*

Rows of baleen plates adapted for straining
Right whale, *Engraulis*

FIGURE 25-10
SOME ADAPTATIONS FOR STRAINING SMALL FOOD PARTICLES FROM WATER.

dozens of hair-like projections on one edge (but with variations according to species). The muscular tongue is the pump. Horny hooks on the tongue comb the food from the platelets.

The baleen whales feed mostly on krill, which consists of swarming crustaceans measuring 75 mm and less in length. The whale has 230–400 plates of baleen along each side of the upper jaw. Each plate is ribbon-like, being flat and from ¼ to 3 m long, according to species. A plate is made up like a sandwich consisting of a horny lamella on each surface and a core of horny hollow tubes running lengthwise within. This makes it light and flexible but very strong. The lamellae gradually wear off of the tip and inside margin of each plate releasing the hundreds of tubes which tangle like stiff hairs to make the filter. The plates are ever-growing at their bases to compensate for wear. Some whales swim slowly along with the mouth open and water spilling out at the corners through the baleen. Others gulp in water and then elevate the enormous tongue (heavier than an elephant in blue whales) to send water gushing out through the filter.

Gape and Suck Feeding.   Many aquatic vertebrates suck food into the mouth in a stream of water. Cephalaspids and pteraspids may have fed this way. Gape and suck feeding is very common among bony fishes; examples are sturgeons, suckers, carp, and perch. Salamander larvae, some frogs, and several turtles do the same.

The strong suction pump is provided, according to species, by expansion of the pharynx through action of opercula and gill chambers, or by expansion of the mouth and pharynx through lowering of the base of the hyoid arch, hyoid apparatus, and related structures. The volume of water sucked in at one time commonly is about 40% of the volume of the head. Fishes may supplement the sucking action by swimming closer to the food, but even if they do not do so, food usually can be sucked in from a distance equal to about ¼ the length of the head. The smaller the mouth, the greater the velocity of the incoming water: Sea horses and pipe fishes feed on active small crustaceans by whisking them into a tiny mouth on the end of a long pipette-like rostrum. Fishes that use gape and suck feeding commonly have mouths that are round or ovoid and usually are protrusible. Small teeth may be present in the mouth (they may shred food that rushes past them), but often the only teeth are in the pharynx.

(Another animal draws food in with air instead of water, and although it sucks it does not gape: The sloth bear protrudes its mobile lips into a cylinder and sucks up swarming insects and grubs with loud smacking noises. The inner upper incisors are absent and the palate is arched to provide an airway.)

Feeding with the Tongue.   The tongue of fishes may move up and down or forward and backward with the hyoid arch, to which it is firmly attached. In bony fishes the tongue may be provided with teeth, in which case it helps to hold and swallow food. The tongues of tetrapods tend to be larger, more fleshy (except in birds), and more moveable. In the different taxa it contributes to feeding in a wide variety of ways. Some examples follow.

Using high-speed photography, Gans has clarified the feeding mechanisms used by two interesting animals. The bullfrog jumps at its prey, which may be a flying insect, opens its mouth wide and (with eyes shut) captures the creature with a flip of a very sticky tongue. The tongue is anchored at the front of the mouth over a pair of special little bones. Muscles draw the mandibles together, thus depressing the small bones. This, and the hardening of another muscle that acts as a cam, catapults the tongue out in 0.05 sec.

The chameleon is a bizarre lizard that catches insects on the end of a sticky tongue which it can protrude a distance equal to its own body length in about 0.04 sec. The hyoid apparatus has a long horn which tapers as it extends forward in the midline. The horn fits within a lubricated sleeve of connective tissue which is surrounded by a specialized accelerator muscle of the tongue. The lizard opens its mouth slowly, advances the hyoid into position, aims by orienting its head just right, and then suddenly contracts the muscle which squeezes itself off of the tapering hyoid horn like a tiny man pinching himself away from a giant watermelon seed.

Having exposed an insect with its drill-like bill, the woodpecker captures it with a tongue that is sharp, cylindrical, barbed, and sticky. A highly modified hyoid apparatus enables the woodpecker to extrude its tongue as much as five times the length of its long bill.

Nectar and pollen form all or most of the diets of hummingirds, honeyeaters, sunbirds, honeycreepers, and some other birds, and also of six genera of bats, and the little honey possum. The birds all have long slender bills, variously curved according to the structure of the preferred flowers. The mammals have much longer rostrums than related species of other habits. Teeth are small and weak (bats) or reduced in number (honey possum), and jaw musculature has regressed. The tongue is invariably quite long, slender, and protrusible. It usually terminates in a brush consisting of rows or tufts of hair-like projections which slope toward the throat. This device effectively transports nectar to the mouth. The tongues of several nectar feeders have in addition, or instead, one or two narrow tubes through which nectar can be sucked into the mouth; muscles of the throat act as a suction pump.

Vampire bats feed exclusively on blood. Teeth not used for inflicting the pumpkin seed-sized wound on the host are reduced or absent. When blood oozes from the wound it is lapped up; when it flows freely, the edges of the tongue are curled up to make a tube, and the blood is sucked into the mouth. The tongue is pumped in and out to establish the suction, but the details of the mechanism have not yet been learned.

The larger mammals that feed on swarming insects are highly specialized. These include the spiny echidna, rat-sized marsupial anteater, aardvark, pangolins, and the giant and smaller "true" anteaters. The rostrum is usually long. The tongue is

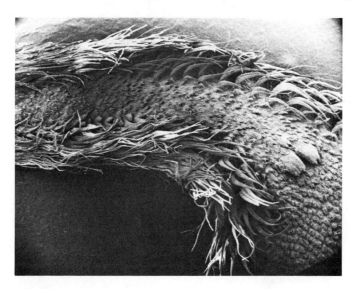

FIGURE 25-11
TONGUE OF A NECTAR-FEEDING AND POLLEN-FEEDING BAT, *Leptonycteris,* **under 19× magnification. Dorsal surface with tip of tongue out of picture to left. Photograph taken with a scanning electron microscope by John Mais.**

cylindrical, sticky, mobile, and exceedingly long (reaching three times the length of the head). In several anteaters it is anchored at the posterior end of the breastbone rather than in the throat. Vascular spaces in the base of the tongue of echidnas function as erectile tissue, and a central, thick-walled artery in the tongue of pangolins has the same purpose. These animals are protected in several ways from ant bites: Pangolins (and others?) can close the nostrils. Most anteaters have a very thick, tough skin, and the eyelids and even the cornea may be thickened.

Swallowing Food Whole: Small, Medium, and Large.   Many mammals and most nonmammals swallow much or all of their food whole with little or no prior slicing, grinding, or other processing in the mouth. The nature and degree of adaptation required relates to the size of the food objects relative to the size of the feeder.

Liquid and soft foods, and food objects that are of small size relative to the size of the mouth can be swallowed with few special provisions. This includes most of the seeds, berries, insects, and fishes eaten. Indeed, it includes most of the food taken by filterers, strainers, gape and suck feeders, and tongue feeders, and most food of all kinds eaten by nonpredatory birds. About 30% of all fishes feed on larvae or adult insects.

Processing Food in Mouth and Pharynx

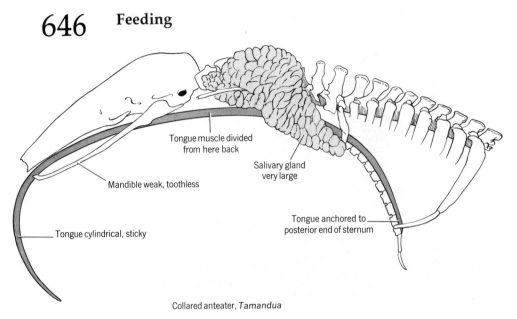

Tongue muscle divided
from here back

Salivary gland
very large

Mandible weak, toothless

Tongue anchored to
posterior end of sternum

Tongue cylindrical, sticky

Collared anteater, *Tamandua*

FIGURE 25-12

SOME ADAPTATIONS OF A MAMMALIAN ANTEATER.

If all the food eaten by a kind of animal is liquid, soft, or small (e.g., blood, honey, diatoms, ants, termites), and teeth are not needed for purposes other than feeding (defense, grooming), then the teeth tend to become small and simple (blood- and nectar-feeding bats, marsupial anteater), then peg-like, wanting in enamel, single-rooted or open-rooted (aardwolf, aardvark, armadillo), and finally lost (minnows, echidna, pangolin, anteaters). Many such animals have protrusible jaws or long rostrum, small mouths, and weak mandibles.

If part or most of the food eaten is of larger size relative to the size of the feeder, then, of course, mouths tend to be larger and jaws stronger. If the food eaten is active or slippery, then teeth tend to be small and of simple structure, yet sharp and numerous to prevent the escape of prey before it can be swallowed. Examples are the teeth of many salamanders, lizards, crocodilians, ichthyosaurs, plesiosaurs, seals, sea lions, and toothed whales.

Some fishes (moray, toadfish) have teeth that are hinged so that they tilt back toward the throat as prey enters the mouth, but stand erect when pressure is in the other direction. Sea turtles have numerous horny, backward-pointing spines in the gullet. Some pterosaurs had teeth, and others were toothless. The merganser (a bird) has horny tooth analogs at the edge of the bill to prevent the escape of fishes. If the food is plant material which is to be swallowed in large pieces, then the animal is commonly toothless (various dinosaurs, turtles).

Coming to the other end of the scale, some flesh, fish, and egg eaters swallow exceedingly large food objects whole. One

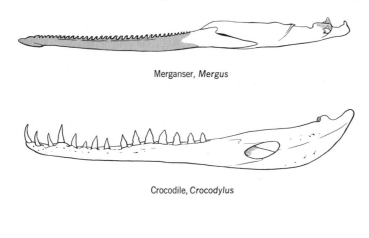

Merganser, *Mergus*

Crocodile, *Crocodylus*

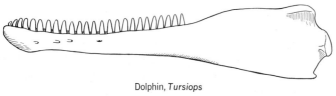

Dolphin, *Tursiops*

FIGURE 25-13
LEFT MANDIBLES OF VERTEBRATES THAT SWALLOW ANIMAL FOOD OF MODERATE SIZE REQUIRING HOLDING BUT LITTLE SHEARING.

fish with a distensible belly gulps down prey as large as itself. Penguins, loons, pelicans, cormorants, boobies, auks, roadrunners, and secretary birds swallow very large prey. Most snakes swallow large food, and several kinds that are adapted for eating eggs can manage eggs having a diameter three and more times the diameter of the head.

First, the object must be oriented end-on to the throat, with any large scales pointing the "smooth way." Pelicans (and, to a lesser extent, some other birds) have a distensible gular, or throat pouch to hold large fishes until they can be positioned for swallowing. Birds may toss a fish in the air and catch it with the desired orientation; snakes gradually work the jaws around to the correct position. Birds usually hold the bill up so gravity will assist in swallowing. They may also suddenly extend the neck, thus thrusting against the inertia of the food to aid in working the food back. Frogs draw the eyes down toward the mouth to help push on a large mouthful. They may use the hands to stuff in a particularly large insect.

Some fishes can protract and retract the upper jaw mechanism, with the small backward-slanting marginal teeth alternately pulling on the prey, releasing and reaching forward, and pulling again. When the prey reaches the pharynx, pharyngeal

Egg-eating snake, *Dasypeltis*

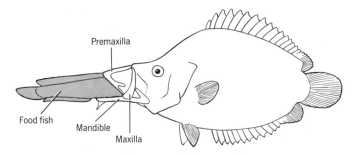

Nandid fish, *Monocirrhus*

FIGURE 25-14
EXAMPLES OF VERTEBRATES THAT SWALLOW LARGE FOOD OBJECTS WHOLE.

teeth function in a similar way. The remarkably adapted head of the snake is even more specialized. The teeth not only curve toward the throat, but the bones bearing them can be rotated so the teeth angle inward or outward at will. (Egg-eating snakes, however, tend to reduce or lose the teeth.) The lower jaw has no symphysis, so the mandibles can separate widely. The angles of the jaw pivot laterally from the skull, and the upper and lower jaws of some species can work forward and backward one side at a time.

Swallowers of large prey have extra large and distensible throats and huge stomachs. Snakes have no sternum so that large food items can distend the body by spreading the ribs. The airway to the lungs is kept open during the slow swallowing process by protruding the epiglottis out of the mouth like a snorkel below the prey being ingested. Several egg-eating snakes have evolved a remarkable method of breaking an egg after it is swallowed. Sharp, midventral projections from about a dozen vertebrae actually penetrate the esophagus. When the egg is in position, powerful muscles (attachments are augmented for the purpose) squeeze the egg against the projections, thus breaking the shell. Some species pass the crumpled shell back through the gut, whereas others hold the egg contents in place with special valves and regurgitate the shell.

Road-runners, secretary birds, and some large frogs may swallow half of a large snake or lizard and wait for it to digest so there will be room to swallow the remainder.

Cutting, Shearing, and Tearing of Animal Food. The subdividing of animal food by shearing is in some instances functionally similar to the ingestion of plant food by severing. Predators that eat mostly flesh are said to be **carnivorous.** Examples are crocodilians, various extinct reptiles, raptorial

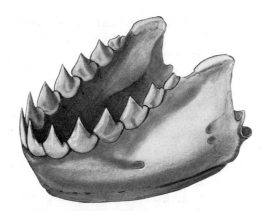

FIGURE 25-15

MANDIBLE AND SHEARING TEETH
OF THE PIRANHA FISH)

birds (hawks and owls), the Carnivora, and one family of Marsupialia. Predators that eat mostly fishes are termed **piscivorous.** About 30% of all fishes are piscivorous, but most of them swallow their prey whole without first subdividing the hapless victim. Many sharks do bite and shear their prey, however, and to a lesser extent the long, sharp teeth of the barracuda, pike, gar, and pickerel may prepare the prey for swallowing. Animals that eat carrion are called **scavengers.** Albatrosses, fulmars, and their relatives are scavengers at sea; vultures, caracaras, adjutant storks, jackals, and hyenas are scavengers on land.

Animal food requires little or no chewing to make it digestible, but many carnivores (including some piscivores) and most scavengers must cut, shear, or tear at least part of their food to prepare pieces that are small enough to swallow. Animals that eat insects are said to be **insectivorous.** If the feeder is small relative to the insect, then it too must subdivide the prey before swallowing it. What are the adaptations of such feeders?

In general the teeth tend to be conical or blade-like, sharp, strong, thecodont (except in fishes), and relatively few in number (again with exceptions among fishes). Many sharks, piranhas, and various extinct carnivorous reptiles have (or had) teeth shaped like pointed blades with straight or serrate margins (see teeth of Selachii, Figure 6-5, and teeth of piranha, Figure 25-15). In some instances upper and lower teeth shear past one another. Synapsid reptiles were moderately heterodont, and mammalian carnivores are markedly so. (The very long canines have a greater role in fighting and in capturing prey than in eating it.) Incisors and anterior premolars are most commonly adapted for cutting. Carnivora have one pair of teeth on each side of the mouth (upper fourth premolar and lower

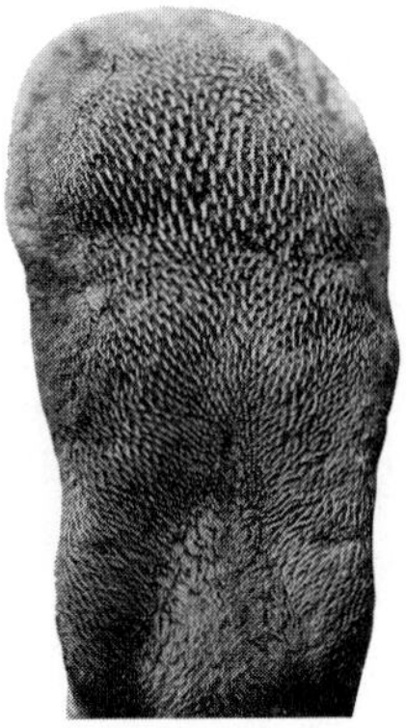

**FIGURE 25-16
TONGUE OF A
MOUNTAIN LION**
showing rasping surface.

first molar in living forms), called **carnassial teeth,** which are enlarged, lengthened, and positioned so as to form a powerful shearing mechanism for slicing tendons and skin (Figure 6-7). The carnassials of hyenas are as effective as large tin-snips (Figure 25-5).

Some piscivorous fishes have teeth on the bony tongue. Reptiles and birds may have analogous horny spines on the tongue. Close-set horny projections on the tongues of some carnivores—notably the big cats—transform the organ into a rasp with which it licks the flesh from bones. The horny teeth of hagfishes can also cut flesh.

Raptorial and scavenging birds, having no teeth in their ancestry, must divide their food another way. Hawks and owls have the horny upper bill curved down at the tip to make a formidable meat hook. Pushing down with its talons and pulling up with its beak, the bird tears its prey apart. Some extinct reptiles had similar beaks. The smaller shrike and kestrel, which eat grasshoppers and beetles, also have strong hooked bills and may use claws or thorns in dismembering an insect.

Finally, various flesh eaters, including certain sharks and crocodilians, may grasp the prey in the powerful jaws and then violently twist, thrash, and roll to dismember the food object.

Grinding of Plant Food. Once it is free in the mouth, plant food is swallowed with little or no chewing by some tortoises and birds, but is very thoroughly chewed by most herbivores. This means that feeding must be slow. Most artiodactyls swallow their food a first time with little chewing, but regurgitate it, one bolus at a time, to be chewed as cud at the animal's leisure. As much as 7–10 h/day may be spent chewing. (Some marsupials regurgitate and chew an occasional bolus, but spend little time at it.) Large quantities of saliva must be secreted, particularly if the food is dry.

In order to triturate course vegetation, a grinding mill is needed. As explained on p. 25, the requirements are most stringent for large animals, so the adaptations of ungulates and elephants will be mentioned first. Because there is one kind of job to do, the cheek teeth are similar to one another; the functional premolars become molariform in nature. The occlusal surfaces of the teeth are flat and large in area: A single mammoth tooth may present a grinding surface of more than 250 cm^2. Such large and heavy teeth must be supported by multiple roots.

Further, provision is made for long wear because the teeth are used so much of the time and because the material ground is itself coarse and often has an admixture of grit. To provide for long wear, the roots are deep in the jaw of the young animal and the crown (the part of the tooth above the roots, not just

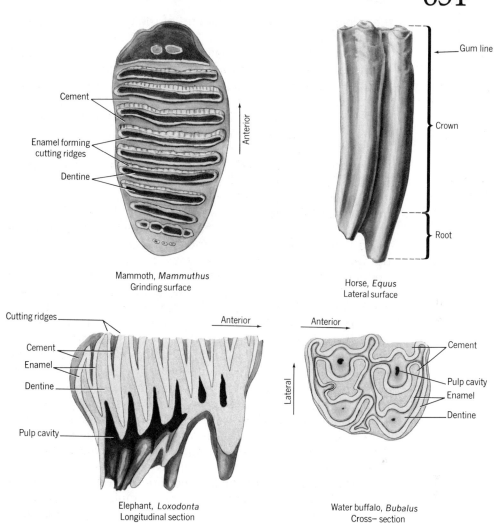

Cement

Enamel forming cutting ridges

Dentine

Anterior

Mammoth, *Mammuthus*
Grinding surface

Gum line

Crown

Root

Horse, *Equus*
Lateral surface

Cutting ridges

Cement

Enamel

Dentine

Pulp cavity

Anterior

Lateral

Elephant, *Loxodonta*
Longitudinal section

Anterior

Cement

Pulp cavity

Enamel

Dentine

Water buffalo, *Bubalus*
Cross– section

FIGURE 25-17
CHEEK TEETH ADAPTED FOR GRINDING.

above the gums) is high. Such teeth are called **hypsodont** (= high + tooth). As the exposed part of the tooth wears down, the roots slowly rise higher in the jaw, and bone fills in under them. This exposes more of the crown. When most of the crown is gone and the roots are just under the gums, the animal, probably now very old, may suffer from malnutrition.

Elephants have evolved a different mechanism which assures complete wear of all of the crown. The teeth lie in a groove in the jaw rather than in sockets. Each groove is deep at the back of the jaw and slopes upward toward the front. Six teeth form in succession at the back of each groove and migrate slowly forward. Each erupts first at its forward corner and starts to wear there long before the posterior part of the tooth is in service. As forward migration continues, a tooth rises in the sloping groove to compensate for wear. Finally, the crown of each part of each tooth wears off completely as it reaches the anterior end of the tooth row. The short roots are pushed up out of the gums and fall away. Only two of these great teeth function at one time in each jaw. As the last tooth moves along, the groove is filled in from behind by spongy bone. The series of six teeth lasts the elephant the 70 or more years that it will live.

Hypsodont teeth have another adaptation. Their occlusal surfaces are broad and flat, like millstones. But in order to grind effectively, millstones must not be smooth. The enamel of these teeth folds up and down, in and out of the substance of the tooth, thus forming a succession of hills and valleys. The pattern of folding is precise for each species, but varies widely between species. The valleys may remain open (many artiodactyls) or may be filled with cement shortly before the tooth erupts (horses, elephants). In the latter case, as the tooth wears there will be a horizontal succession of enamel–dentine–enamel–cement. Since enamel is harder than dentine or cement, it is a bit more resistant to wear and projects a little above the other materials to form **cutting ridges.** This provides and maintains the needed roughness.

The cheek teeth of small herbivores are less specialized in that they are smaller and usually have less complicated patterns of infolded enamel, yet they may be more specialized in being open-rooted and ever-growing (rabbits, some rodents).Individual teeth of herbivorous reptiles never achieve the complexity of those of the large mammalian grazers and browsers, but nature provided the duck-billed dinosaurs with an effective analog. Several hundred relatively simple teeth became pressed together to form a large rough grinding plate. Some

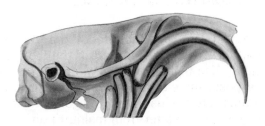

FIGURE 25-18
DISSECTION OF THE SKULL OF A GOPHER, *Thomomys,* **SHOWING OPEN-ROOTED, EVER-GROWING TEETH.**

vegetarian fishes have crowded thousands of small teeth together in a comparable way. Others have no teeth in the mouth, but shred plant food with pharyngeal teeth.

Crushing and Cracking Nonedible Material. Some animals must crush or crack shells, hulls, woody seeds, or other hard materials to make digestible food available. These animals are termed **durophagus.** Examples of fishes that crush shells to secure the soft meat within are the stingray, chimaeras, dipnoans (Figure 6-5), some cichlid fishes, and the porcupine fish. The parrot fish crushes bits of coral to get at the polyps. These fishes have powerful jaws which are usually autostylic. Most have few teeth which are pavement-like and strong. Stingrays have numerous teeth which fit together like bathroom tiles to form crushing plates (Figure 25-19).

As their name implies, the extinct euryapsid reptiles called placodonts had large flat teeth for crushing mollusks. Several lizards and several turtles crush snails. The molars of the sea otter and walrus are flattened for the same purpose. The platypus has horny plates for crushing snails and other food.

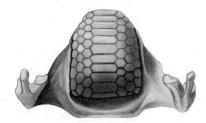

Stingray, *Holorhinus*

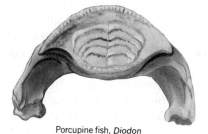

Porcupine fish, *Diodon*

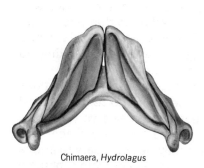

Chimaera, *Hydrolagus*

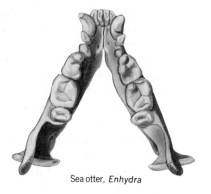

Sea otter, *Enhydra*

FIGURE 25-19
MANDIBLES AND LOWER TEETH OF DUROPHAGUS VERTEBRATES.

Grain- and nut-eating birds have particularly muscular gizzards for grinding their food. Grit is eaten to make the mill more effective. Similarly, sturgeons, the gizzard shad, and the mullet, which are detritus-eaters, have muscular stomachs with tough linings for crushing miscellaneous invertebrates.

Some other vertebrates crack rather than crush their hard foods. The grackle (a bird) cuts acorns open against a keel on the inside of the upper bill. The hawfinch cracks cherry and olive pits. The beak is very stout and deep at the base, and the jaw adductors are remarkably extensive. It has been claimed that this bluebird-sized finch can bite a pit with a force of more than 45 kg. The dusky shark has a biting pressure of 3000 kg/cm^2. Hyenas crack open the bones of large ungulates to secure the marrow. Their adaptations include extra heavy jaws and teeth, enormous jaw muscles and saggital crest, and early closure of cranial sutures. Since the food is touched only by the points of the carnassials and several other teeth, great pressure is generated.

Storage and Digestion

The morphology of the digestive tract correlates sufficiently well with function so that the approximate eating habits and diet of a vertebrate usually can be determined from its digestive system, though the structure of the system in some fishes and marine mammals remains puzzling.

If the vertebrate feeds on nutritious food (which digests quickly), if the food is ingested as small particles, and if feeding is slow but frequent or nearly continual, then there need be no temporary storage of food, and a short gut is adequate. Many filterers and strainers, parasites that rasp the flesh of their hosts, many gape and suck feeders, and some feeders on algae, nectar, mollusks, and insects are in this category. They have relatively small stomachs or no stomach at all (cyclostomes, holocephalians, dipnoans), and the gut tends to be short and relatively straight (though it is unexpectedly long in baleen whales). Nectar-feeding birds have the entrance and exit to the muscular stomach close together so that nectar can pass directly into the intestine. The blood eaten by vampire bats passes directly from esophagus to intestine, backing up into the stomach only when the foregut is full.

Carnivores, scavengers, and fish eaters also eat nutritious foods, so their guts also tend to be short (except as noted below), but these animals often eat large meals very quickly. Large prey may be swallowed whole, food may be available only after a successful hunt or wait, or it may be necessary to gulp a meal to prevent a fellow carnivore or scavenger from robbing a meal. Consequently, capacious temporary storage is essential.

The esophagus of most such animals is remarkably distensible. The fishes tend to have long straight stomachs. Carnivorous reptiles and mammals have a simple but distensible stomach. Most of the birds have crops that are pleated when empty and enormous when full. Their gizzards are relatively thin-walled and nonmuscular. The stomach of pinnipeds is simple, but that of whales is inexplicably complicated. The small intestine of most carnivores and scavengers is relatively uniform and short, but in piscivorous vertebrates other than fishes it may be curiously long (penguins, pinnipeds, some toothed whales), thick-walled, and small in internal diameter. The hind gut of flesh eaters is short, and caeca are small or absent. See the viscera of the carnivores shown in Figures 10-7 and 10-9.

Scales, bones, feathers, and hair are avoided by some of these animals but are swallowed by many. These may be digested (small bones), regurgitated (by raptorial birds), or passed through the gut.

Somewhat similar to the requirements of carnivores are those of feeders on swarming insects. Anteaters have a roomy stomach which in some species is heavily cornified and muscular. Their salivary glands are very large (Figure 25-12). Their guts are moderately short.

Leaves and stems are relatively low in food value. Accordingly, they must be eaten in large quantity, and storage is needed. Further, the cellulose wall of plant cells cannot be broken down chemically by enzymes of vertebrates, so thorough mechanical grinding in the mouth or stomach is required. Finally, if the cellulose itself is to be utilized as food, bacterial fermentation is essential and this, in turn, means that the food must remain many hours in a spacious fermentation chamber.

Plant-eating birds have a very large crop which is positioned relatively far posterior and ventral in the body so that its mass, when full, will not interfere with the bird's stability. Further, there is a muscular gizzard to substitute for teeth in grinding the food. The crop of the hoatzin is also muscular, has a horny lining, and shreds the leaves that the bird eats.

Because they have no crop, herbivorous reptiles and mammals have a large stomach. The stomach is huge in artiodactyls, sirenians, kangaroos, langurs, and sloths, and is complex in that it is divided into several compartments having somewhat different functions.

The gut of herbivores is relatively long, particularly if abundant roughage is eaten, or if the animal is large. The intestine is much coiled, often in set patterns, and the long hind gut is of large diameter. The caecum (or caeca) is always large but is particularly capacious if the stomach is relatively small as in perissodactyls, swine, elephants, and various rodents. The

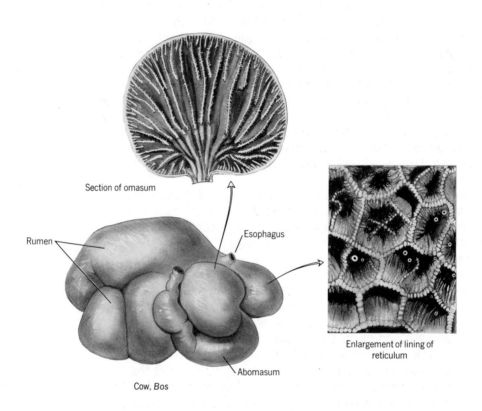

Section of omasum

Rumen

Esophagus

Enlargement of lining of
reticulum

Abomasum

Cow, *Bos*

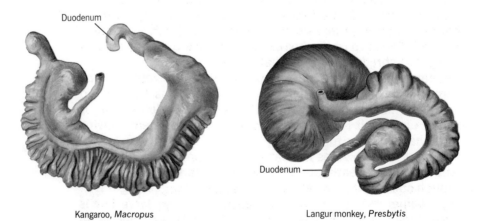

Duodenum

Kangaroo, *Macropus*

Duodenum

Langur monkey, *Presbytis*

FIGURE 25-20
COMPLEX STOMACHS OF HERBIVOROUS MAMMALS IN THREE ORDERS.

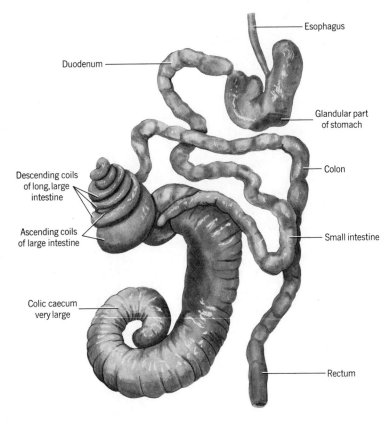

Muskrat, *Ondatra*

FIGURE 25-21
GUT OF AN HERBIVOROUS RODENT.

folds, villi, and microvilli of the digestive tube increase its surface at least 600-fold. The epithelial lining of the gut is subjected to considerable wear and is constantly replaced; the
turnover time is 2–3 days.

Importantly, most herbivores have commensal populations of
bacteria which feed on the cellulose fraction of the food. The
byproducts are fatty acids, amino acids, and vitamins which
are then digested by the herbivores. Most such fermentation
takes place in the stomach if that organ is complex, and otherwise in the intestine and caecum.

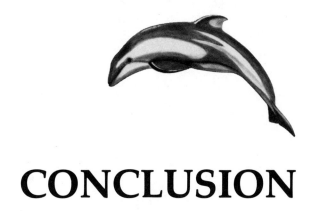

CONCLUSION

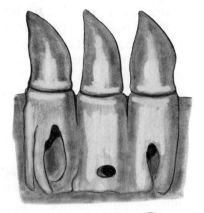

Vertebrate Morphology: Diversity, Unity, and Challenge

Vertebrate morphology is too vast a subject to be "covered" by one book, one course, one student, or one professor. This book has merely summarized descriptive anatomy, and even in areas selected for emphasis it has omitted much. Some additional adaptations which could be subjects for supplemental reading or special reports are the following.

Provisions for defense and escape include a diversity of structures and functions, among which are armor, teeth, claws, horns, antlers, tusks, beaks, spines, quills, camouflages, and mechanisms for stinging, poisoning, crushing, inflating the body, dropping the tail, and producing slime, irritants, or noxious odors.

There are structural as well as physiological and behavioral adaptations for living at high elevations, deep in the sea, and in caves. Some vertebrates are modified for walking on snow, mud, or floating vegetation. Adaptations for prenatal development and parturition include placental morphology, length and structure of the umbilical cord, and shape of the uterine cavity in relation to the shape, flexibility, and densities of different parts of the fetus. The larynx, syrinx, or other sound-producing mechanism exhibits marked specializations, as does the ear.

Although adaptations for resisting heat and drought are mostly physiological and behavioral, there are structural mechanisms that radiate heat to the environment (e.g., ears of elephants and jack rabbits, wings of bats, flippers of sea lions and dolphins), lower the temperature and hence the humidity of expired air (in kangaroo rats), and selectively cool arterial blood flowing to the brain (some artiodactyls). Adaptations for resisting cold include thick insulating layers of blubber, feathers or fur, and countercurrent exchange mechanisms that allow the appendages of arctic birds and mammals to remain at much lower temperatures than the body without draining heat from the body.

Man has received scant attention in this book, but he also has distinctive locomotor skills. Even if other animals could have the incentive to try and the patience to learn such "artificial" activities as gymnastics, diving, skiing, figure skating, and pole vaulting, none could match his ability. Also, man is not only uniquely expert but also impressively skilled at throwing. The mechanics of throwing is complex and difficult to analyze. A combination of structural adaptations favor the achievement of high velocity at the hand, just as they favor high velocity at the feet of cursors. Behavioral modification (i.e., training) assures that velocity of the projectile (ball, javelin, discus, shot) is increased by rotating several joints in the same direction at the same time, by the application of force to the object for as long as possible, by conserving the linear momentum of a runup, and by conserving angular momentum as successively lighter parts of the body and arm are made to rotate in sequence.

It is less useful to stress how much of the accumulated knowledge about vertebrate structure has been omitted from our course of study (since it has, nonetheless, included much) than to emphasize the unity of what has been presented and the extent of the engaging material that remains to be learned.

The subject matter of Parts II and III of this book are not as distinct as may at first seem apparent. Structure that characterizes the major vertebrate taxa, and the relatively general changes in structure that relate major taxa in time were stressed in Part II, whereas structure patterns that adapt vertebrates to specific modes of locomotion and feeding were presented in Part III. This distinction was made because (*1*) the emphasis of much current research justifies greater stress on function than has been usual in textbooks of morphology, and (*2*) in my judgement, when the material here presented in Part III is mixed in with that of Part II, it may seem to constitute digressions and exceptions which cloud the story of the broad outline of evolution.

Nonetheless, this division is merely a convenience in presenting and analyzing the subject. The material of the two parts of the book is not, after all, independent and significantly different in kind. Much function *was* included in Part II; the adaptations described in Part III are *also* the product of evolution. Distinctions between structure that characterizes major taxa and structure that adapts to specific behaviors break down if carried far: Osteichthyes is surely a major taxon, yet nearly all its members share one general mode of locomotion. One could neither study the phylogeny of fishes without reference to adaptation for swimming, nor analyze the structure of fins without reference to evolutionary relationships. A double-circuit heart and complex cerebellum (described in Part II) characterize the class Aves, yet are just as necessary for masterful flight as are airfoils and flight muscles (described in Part III).

Further, and importantly, essentially the same evolutionary processes that produced such generally beneficial structures as jaws, gills, lungs, and paired appendages also produced structures of more limited benefit such as keeled sternums, suspensory ligaments, and adhesive mechanisms. Evolution provides that individual animals tend to survive, that each species tends to multiply, and that the environment tends to be used to the utmost. Specialization permits specific kinds of animals to survive without serious competition in closer proximity to other kinds of animals than would be possible if all shared the same requirements.

Unity of our subject also emerges from the recognition of concepts and relationships that cut across chapter headings. In coming to the end of the book it would be helpful, therefore, to turn again to the principles and considerations discussed in Chapter 1 to be sure that they have not been lost sight of. It was necessary at the outset to state the ideas in general terms because the vocabulary and specifics of our subject were not yet at hand. Now the student might augment that introductory statement by gleaning additional examples of each concept from the body of the book.

664 Vertebrate Morphology

I enjoy vertebrate morphology because it shows me that the animals can do so many things so well. A runner can dash faster than I drive my car on freeways. A springer can leap 14 times higher than its body length. A digger can thrust with 32 times its body weight. A climber can walk upside down on polished glass. A swimmer can sprint twice as fast in water as a man can run on land. A flyer can dart backwards as well as forwards. A feeder can swallow objects three times larger than its head, and another can strain from water food particles having 1/5000 the diameter of a pencil lead.

There are living direction finders, strain gauges, force multipliers, flow equalizers, suction cups, bifocal lenses, locking devices, pressure sensors, self-lubricating bearings, electric generators, gas analyzers, echo-rangers, depth gauges, recoil mechanisms, countercurrent exchangers, and compact computers. The list could go on and on, yet we have not recognized nearly all of nature's methods, let alone understood them. It will be long before we know "very much" of what there is to learn about vertebrate structure and behavior. In the meantime, the continued observation and interpretation of form and function promise to increase man's pleasure and wonder in the complexity, diversity, and perfection of the vertebrate body.

References

Bock, W. J. 1963. Evolution and phylogeny in morphologically uniform groups. Am. Naturalist 97: 265–285.

Bolk, L., E. Göppert, E. Kallius, and W. Lubosch (eds.). 1931–1939. Handbuch der vergleichenden Anatomie der Wirbeltiere. Urban und Schwartzenberg, Berlin and Vienna. 6 v. Many specialists have contributed to this comprehensive work.

Clark, W. E. L. 1965. The tissues of the body; an introduction to the study of anatomy. Clarendon Press, Oxford. 423 p. Excellent for supportive, vascular, and nervous tissues and skin.

Colbert, E. H. 1969. Evolution of the vertebrates. 2nd ed. John Wiley & Sons, New York. 535 p. Good survey of vertebrate paleontology.

Cole, F. J. 1949. A history of comparative anatomy. Macmillan, London. 524 p.

**Parts I & II
General
References**

665

Darlington, P. J. 1957. Zoogeography; the geographical distribution of animals. John Wiley & Sons, New York. 675 p.

DeBeer, G. R. 1958. Embryos and ancestors. 3rd ed. Oxford University Press, London. 197 p.

Dellman, H. D. 1971. Veterinary histology; an outline atlas. Lea and Febiger, Philadelphia. 305 p.

Goodrich, E. S. 1930. Studies on the structure and development of vertebrates. Macmillan, London. 837 p. (Republished by Dover, New York, 1958.)

Gordon, M. S., G. A. Barthalomew, A. D. Grinnell, C. B. Jørgensen, and F. N. White. 1972. Animal physiology; principles and adaptations. 2nd ed. Macmillan, New York. 592 p.

Grassé, P. P. (ed.). 1950–1970. Traite de zoologie; anatomie, systématique, biologie. Masson et Cie, Paris. Vertebrates, v. 13–17. A monumental work.

Hildebrand, M. 1968. Anatomical preparations. University of California Press, Berkeley. 100 p.

Huxley, J. S. 1932. Problems of relative growth. Methuen, London. 276 p. A classic.

Irvine, W. 1955. Apes, angels, and victorians; the story of Darwin, Huxley, and evolution. McGraw-Hill, New York. 399 p.

Jaeger, E. C. 1955. A source-book of biological names and terms. 3rd ed. Thomas, Springfield, Ill. 319 p.

Mayr, E. 1963. Animal species and evolution. Harvard University Press, Cambridge, Mass. 797 p.

Mayr, E. 1969. Principles of systematic zoology. McGraw-Hill, New York. 428 p.

Patt, D. I., and G. R. Patt. 1969. Comparative vertebrate histology. Harper & Row, New York. 438 p.

Piveteau, J. (ed.). 1952–1958. Traité de paléontologie. Masson et Cie, Paris. Vertebrates, v. 4–7.

Raup, D. M., and S. M. Stanley. 1971. Principles of paleontology. Freeman, San Francisco. 388 p.

Rensch, B. 1960. Evolution above the species level. Columbia University Press, New York. 419 p. (Originally in German, 1954.)

Romer, A. S. 1959. The vertebrate story. 4th ed. University of Chicago Press, Chicago. 437 p. Structure of vertebrates, living and extinct.

Romer, A. S. 1966. Vertebrate paleontology. 3rd ed. University of Chicago Press, Chicago. 468 p. An advanced textbook with extensive summary of classification and list of references.

Simpson, G. G. 1949. The meaning of evolution. Yale University Press, Newhaven, Conn. 364 p. Excellent general reference, readable and nontechnical. An abridged paperback edition available from The New American Library.

Simpson, G. G. 1953. The major features of evolution. Columbia University Press, New York. 434 p.

Simpson, G. G., A. Roe, and R. C. Lewontin. 1960. Quantitative zoology. Rev. ed. Harcourt, Brace, Jovanovich, New York. 440 p.

Torrey, T. W. 1967. Morphogenesis of the vertebrates. 2nd ed. John Wiley & Sons, New York. 448 p. Combines anatomy and embryology.

Willier, B. H., P. A. Weiss, and V. Hamburger. 1955. Analysis of development. Saunders. Philadelphia. 735 p. A comprehensive reference work.

Young, J. Z. 1962. The life of vertebrates. 2nd ed. Oxford University Press, New York and Oxford. 820 p. An analytical account organized by systematic groups rather than by organ systems.

**Chapters 3 & 4
Vertebrate
Animals**

Berg, L. S. 1947. Classification of fishes both recent and fossil. Edwards, Ann Arbor, Mich. 517 p. In Russian and English. Lists anatomical characters of each group.

Brodal, A., and R. Fänge. 1963. The biology of Myxine. Universitetsforlaget, Oslo. 588 p.

Brown, M. E. (ed.). 1957. The physiology of fishes. Academic Press, New York. 2 v.

Gans, C. A., d'A. Bellairs, and T. S. Parsons. 1969–. Biology of the reptilia. Academic Press, New York. Several volumes published; others projected.

Greenwood, P. H., D. E. Rosen, S. H. Weitzman, and G. S. Myers. 1966. Phyletic studies of teleostean fishes, with a provisional classification of living forms. Am. Mus. Nat. Hist., Bull. 131, Article 4: 341–455.

Hardesty, M. W., and I. C. Potter. 1971–1972. The biology of lampreys. Academic Press, New York. 2 v.

Heilmann, G. 1926. The origin of birds. Witherby, London. 208 p.

Marshall, A. J. (ed.). 1960–1961. The biology and comparative physiology of birds. Academic Press, New York. 2 v. Excellent reference for anatomy and physiology.

Marshall, N. B. 1965. The life of fishes. Weidenfeld and Nicolson, London. 402 p.

Noble, G. K. 1931. The biology of the amphibia. McGraw-Hill, New York. 577 p. (Republished by Dover, New York. 1954.)

Parsons, T. S., and E. E. Williams. 1963. The relationships of the modern amphibia: a re-examination. Quart. Rev. Biol. 38: 26–53.

Simpson, G. G. 1945. The principles of classification and a classification of mammals. Am. Mus. Nat. Hist., Bull, 85, 350 p. Still the most authoritative classification of living and extinct mammals.

Simpson, G. G. 1961. Principles of animal taxonomy. Columbia University Press, New York. 247 p.

Stoll, N. R. (Chairman of Editorial Committee). 1961. International code of zoological nomenclature adopted by the XV international congress of zoology. International Trust for Zoological Nomenclature, London. 176 p.

Thompson, K. S. 1969. The biology of lobe-finned fishes. Biol. Rev. 44: 91–154. Evaluation of theories on evolution and function.

Thomson, K. S. 1971. The adaptation and evolution of early fishes. Quart. Rev. Biol. 46: 139–166.

Van Tyne, J., and A. J. Berger. 1959. Fundamentals of ornithology. John Wiley & Sons, New York. 624 p. Includes a chapter on anatomy and an account of each family.

Watson, D. M. S. 1954. A consideration of ostracoderms. Roy. Soc. London, Philosophical Trans., Ser. B, Biol. Sci. 238: 1–25.

Young, J. Z. 1957. The life of mammals. Oxford University Press, New York and Oxford. 820 p. Broad interpretation of mammalian morphology.

References

Goodrich, E. S. 1907. On the scales of fish, living and extinct, and their importance in classification. Zool. Soc. London, Proc., 1907: 751–774. Comprehensive account, still basically sound.

Hörstadius, S. 1950. The neural crest. Oxford University Press, London. 111 p.

Maderson, P. F. A. 1965. The structure and development of the squamate epidermis, p. 129–153. In A. G. Lyne and B. F. Short (eds.), The biology of the skin and hair growth. American Elsevier, New York.

Maderson, P. F. A. 1967. A comment on the evolutionary origin of vertebrate appendages. Amer. Naturalist 101: 71–78.

Maderson, P. F. A. (ed.). 1972. The vertebrate integument. Symposium in Am. Zool. 12: 12–171. Excellent collection of articles.

Montagna, W. 1956. The structure and function of skin. Academic Press, New York. 356 p. Only human skin is described.

Moss, M. L. 1968. The origin of vertebrate calcified tissues, p. 360–371. In T. Ørvig (ed.), Current problems of lower vertebrate phylogeny. Wiley-Interscience, New York.

Moss, M. L. 1968. Bone, dentin, and enamel and the evolution of vertebrates, p. 37–65. In P. Person (ed.), Biology of the mouth. Am. Assoc. for the Advancement of Sci., Washington.

Ørvig, T. 1951. Histologic studies of placoderms and fossil elasmobranchs. I. The endoskeleton, with remarks on the hard tissues of lower vertebrates in general. Arch. för Zool., Ser. 2, 2: 321–454. An important work on the classification of hard tissues with a theory on the origin of dermal armor.

Parakkal, P. F., and N. J. Alexander. 1972. Keratinization; a survey of vertebrate epithelia. Academic Press, New York. 59 p. Excellent illustrations.

Stensiö, E. 1962. Origine et nature des ecailles placoïdes et des dents, p. 75–85. In Collogues Internationaux du Centre National de la Recherche Scientifique, No. 104, Problèmes actuels de paléontologie (évolution des vértebrés), Paris, 29 mai-3 juin, 1961.

Butler, P. M. 1956. The ontogeny of molar pattern. Biol. Rev. Cambridge Philosophical Soc. 31: 30–70.

Edmund, A. G. 1960. Tooth replacement phenomena in the lower vertebrates. Roy. Ontario Museum, Toronto. Life Sciences Division, Contribution 52. 190 p.

Gregory, W. K. 1934. A half century of trituberculy. The Cope–Osborn theory of dental evolution, with a revised summary of molar evolution from fish to man. Am. Philosophical Soc., Philadelphia, Proc. 73: 169–317.

Kurten, B. 1954. Observations on allometry in mammalian dentitions; its interpretation and evolutionary significance. Acta Zool. Fennica 85: 1–13.

Moss, M. L. 1970. Enamel and bone in shark teeth; with a note on fibrous enamel in fishes. Acta. Anat. 77: 161–187.

Peyer, B. (translated and edited by R. Zangerl). 1968. Comparative odontology. University of Chicago Press, Chicago. 347 p. Includes discussion of the nature of hard tissues.

Crompton, A. W. 1963. The evolution of the mammalian jaw. Evolution 17: 431–439.

DeBeer, G. R. 1937. The development of the vertebrate skull. Oxford University Press, London and New York. 552 p.

Frazzetta, T. H. 1968. Adaptive problems and possibilities in the temporal fenestration of tetrapod skulls. J. Morphol. 125: 145–158.

Jollie, M. T. 1957. The head skeleton of the chicken and remarks on the anatomy of this region in other birds. J. Morphol. 100: 389–436.

Jollie, M. T. 1960. The head skeleton of the lizard. Acta Zool. 41: 1–64.

Miles, R. S. 1968. Jaw articulation and suspension in *Acanthodes* and their significance, p. 109–127. *In* T. Ørvig, (ed.), Current problems of lower vertebrate phylogeny. Wiley-Interscience, New York.

Parrington, F. R. 1949. A theory of the relations of lateral lines to dermal bones. Zool. Soc. London, Proc. 119: 65–78.

Parrington, F. R. 1956. The patterns of dermal bones in primitive vertebrates. Zool. Soc. London, Proc. 127: 389–411. Interpretive, not descriptive.

Parrington, F. R., and T. S. Westoll. 1941. On the evolution of the mammalian palate. Roy. Soc. London, Philosophical Trans. Ser. B 230: 305–355.

Romer, A. S. 1956. Osteology of the reptiles. University of Chicago Press, Chicago. 772 p.

Watson, D. M. S. 1953. The evolution of the mammalian ear. Evolution 7: 159–177.

Chapter 7
Head Skeleton

Burt W. H. 1960. Bacula of North American mammals. University of Michigan Museum Zool., Misc. Publ. 113: 1–76.

Eaton, T. H. 1951. Origin of tetrapod limbs. Am. Midland Naturalist 46: 245–251.

Enlow, D. H. 1962. A study of the post-natal growth and remodeling of bone. Am. J. Anat. 110: 79–102.

Enlow, D. H., and D. B. Harris. 1964. A study of the postnatal growth of the human mandible. Am. J. Orthodontics 50: 25–50. Analysis of remodeling during growth.

Evans, F. G. 1939. The morphology and functional evolution of the atlas-axis complex from fish to mammals. New York Acad. Sci., Ann. 39: 29–104. Out of date as to interpretation of proatlas and occipital condyles.

Gregory, W. K. 1935. The pelvis from fish to man; a study in paleomorphology. Am. Naturalist 69: 193–210. Emphasis on primates.

Gunter, G. 1956. Origin of the tetrapod limb. Science 123: 495–496.

Guthrie, D. M., and J. R. Banks. 1970. Observations on the function and physiological properties of a fast paramyosin muscle — the notochord of amphioxus (*Branchiostoma lanceolatum*). J. Exptl. Biol. 52: 125–138.

Haines, R. W. 1942. The evolution of epiphyses and of endochondral bone. Biology Rev. 17: 267–292.

Lacroix, P. 1951. The organization of bones. Blakiston, Philadelphia. 235 p. Excellent account of histology, development, and growth of long bones.

Montagna, W. 1945. A reinvestigation of the wing of the fowl. J. Morphol. 76: 87–113.

Chapter 8
Body Skeleton

Mookerjee, H. K. 1936. The development of the vertebral column and its bearing on the study of organic evolution. Indian Sci. Cong., Proc. 23: 307–343. Review article by a major contributor to the field.

Ridewood, W. G. 1921. On the calcification of the vertebral centra in sharks and rays. Roy. Soc. London, Philosophical Trans., Ser. B, 210: 311–407.

Schaeffer, B. 1941. The morphological and functional evolution of the tarsus in amphibians and reptiles. Am. Mus. of Nat. Hist., Bull. 78: 395–472.

Wake, D. B. 1970. Aspects of vertebral evolution in the modern Amphibia. Forma et Functio 3: 33–60.

Westoll, T. S. 1958. The lateral fin-fold theory and the pectoral fins of ostracoderms and early fishes, p. 180–211. In T. S. Westoll (ed.), Studies on fossil vertebrates. University of London Press, London.

Williams, E. E. 1959. Gadow's arcualia and the development of tetrapod vertebrae. Quart. Rev. Biol. 34: 1–32. Important reevaluation of a long-held theory.

Chapter 9
Muscles and
Electric Organs

Cheng, Cze-Ching. 1955. The development of the shoulder region of the opossum Didelphis virginiana, with special reference to the musculature. J. Morphol. 97: 415–472. Illustrates use of embryology for establishing homologies of muscles.

Grundfest, H. 1960. Electric fishes. Sci. Am. 203(4); 115–124, Oct.

Haines, R. W. 1934. The homologies of the flexor and adductor muscles of the thigh. J. Morphol. 56: 21–49.

Howell, A. B. 1937. Morphogenesis of the shoulder architecture: Part VI, Therian mammalia. Quart. Rev. Biol. 12: 440–463.

Huxley, H. E. 1958. The contraction of muscle. Sci. Am. 199 (5): 67–82, Nov.

Lissmann, H. W. 1958. On the function and evolution of electric organs in fish. J. Exp. Biol. 35: 156–191.

Romer, A. S. 1924. Pectoral limb musculature and shoulder-girdle structure in fish and tetrapods. Anat. Record 27: 119–143.

Straus, W. L., Jr. 1946. The concept of nerve-muscle specificity. Biol. Rev. 21: 75–91.

Sullivan, G. E. 1962. Anatomy and embryology of the wing musculature of the domestic fowl (Gallus). Australian J. Zool. 10: 458–518.

Chapter 10
Digestive System
and Coelom

Butler, G. W. 1889. On the subdivision of the body-cavity in lizards, crocodiles, and birds. Zool. Soc. London, Proc. 1889: 452–474.

Clark, R. B. 1964. Dynamics in metazoan evolution; the origin of the coelom and its segments. Clarendon Press, Oxford. 313 p.

Elias, H. 1955. Liver morphology. Biol. Rev. 30: 263–310. Review article by a leading student of the liver.

Gorham, F. W., and A. C. Ivy. 1938. General function of the gall bladder from the evolutionary standpoint. Field Museum Nat. Hist., Chicago, Zool. Ser. 22: 159–213.

Hill, W. C. O. 1926. A comparative study of the pancreas. Zool. Soc. London, Proc. 1926: 581–631.

Mitchell, P. C. 1901. On the intestinal tract of birds. Linnaean Soc. London, Trans. (Zool.), Ser. 2, 8: 173–275.

Mitchell, P. C. 1916. Further observations on the intestinal tract of mammals. Zool. Soc. London, Proc. 1916: 183–251. Broadly comparative study of gross structure.

Pernkopf, E. 1930. Beiträge zur vergleichende Anatomie des vertebraten Magens. Z. Anat. 91: 329–390. Reference on comparative structure of the stomach.

Slijper, E. J. 1946. Die physiologische Anatomie der Verdauungsorgane bei den Vertebraten. Tabulae Biologicae 21, Pars 1: 1–81. Tabulated data based on 280 references.

Tucker, R. 1958. Taxonomy of vertebrate salivary glands. Systematic Zool. 7: 74–83.

Warner, E. D. 1958. The organogenesis and early histogenesis of the bovine stomach. Am. J. Anat. 102: 33–63. General comments on ontogenetic and phylogenetic origins of the parts of mammalian stomachs.

**Chapter 11
Respiratory
Systems and
Gas Bladder**

Atz, J. W. 1952. Narial breathing in fishes and the evolution of internal nares. Quart. Rev. Biol. 27: 366–367.

Gans, C. 1970. Strategy and sequence in the evolution of the external gas exchangers of ectothermal vertebrates. Forma et Functio 3: 61–104.

Gray, J. E. 1954. Comparative study of the gill area of marine fishes. Biol. Rev. 107: 219–225.

Hughes, G. M. 1963. The comparative physiology of vertebrate respiration. Harvard University Press, Cambridge, Mass. 145 p.

Jones, F. R. H., and N. B. Marshall. 1953. The structure and functions of the teleostean swim bladder. Biol. Rev. 28: 16–83. Comprehensive.

McLaughlin, R. F., W. S. Tyler, and R. O. Canada. 1961. A study of the subgross pulmonary anatomy in various mammals. Am. J. Anat. 108: 149–158.

Salt, G. W., and E. Zeuthen. 1960. The respiratory system, p. 363–409. In A. J. Marshall (ed.), Biology and comparative physiology of birds. Academic Press, New York.

Schmidt-Nielsen, K. 1971. How birds breathe. Sci. Am. 225(6): 72–79, Dec.

Staub, N. C. 1963. The interdependence of pulmonary structure and function. Anesthesiology 24: 831–854.

Tenney, S. M., and J. E. Remmers. 1963. Comparative quantitative morphology of the mammalian lung: diffusing area. Nature 197: 54–56, Jan.

Wolf, S. 1933. Zur Kentnis von Bau und Funktion der Reptilienlunge. Zool. Jahrb. Abt. für Anat. und Ontogenie der Tiere 57: 139–190.

Woskoboinikoff, M. 1932. Der Apparat der Kiemenatmung bei den Fischen. Zool. Jahrb., Abt. für Anat. und Ontogenie der Tiere 55: 315–488.

**Chapter 12
Circulatory
System**

Barnett, C. H., R. J. Harrison, and J. D. W. Tomlinson. 1958. Variations in the venous systems of mammals. Biol. Rev. 33: 442–487. A review, with extensive bibliography.

Foxon, G. E. H. 1955. Problems of the double circulation in vertebrates. Biol. Rev. 30: 196–228. Review of heart structure in relation to function.

Huntington, G. S. 1911. The development of the lymphatic system in the reptiles. Anat. Record 5: 261–276.

Jordan, H. E. 1933. The evolution of blood-forming tissues. Quart. Rev. Biol. 8: 58–76.

Kampmeier, O. F. 1969. Evolution and comparative morphology of the lymphatic system. Thomas, Springfield, Ill. 620 p.

Mossman, H. W. 1948. Circulatory cycles in the vertebrates. Biol. Rev. 23: 237–255. Analysis of general structure and function of elements in the system.

O'Donoghue, C. H. 1920. The blood vascular system of the tuatara, *Sphenodon punctatus*. Roy. Soc. London, Philosophical Trans., Ser. B, 210: 175–252. Monograph on a seemingly primitive reptilian circulatory system.

Quiring, D. P. 1949. Collateral circulation. Lea & Febiger, Philadelphia. 142 p. Describes functional importance of an overlooked aspect of the circulatory system.

Rennick, B. R., and H. Gandia. 1954. Pharmacology of smooth muscle valve in renal portal circulation of birds. Soc. Exptl. Biol. and Med., Proc. 85: 234–236.

Robertson, J. I. 1913. The development of the heart and vascular system of Lepidosiren paradoxa. Quart. J. Microscop. Sci. 59: 53–132. A species of evolutionary significance.

Satchell, G. H. 1971. Circulation in fishes. Cambridge University Press. 131 p. Summary of current knowledge of anatomy and physiology.

Shearer, E. M. 1930. Studies on the embryology of circulation in fishes. Parts I and II. Am. J. Anat. 46: 393–459.

Yoffey, J. M., and F. C. Courtice. 1956. Lymphatics, lymph and lymphoid tissue. Edward Arnold, London. 510 p.

Chapter 13
Urogenital
System

Asdell, S. A. 1964. Patterns of mammalian reproduction. 2nd ed. Cornell University Press, Ithaca, N. Y. 670 p. Reference work of wide coverage.

Cowles, R. B. 1958. The evolutionary significance of the scrotum. Evolution 12: 417–418.

Fraser, E. A. 1950. The development of the vertebrate excretory system. Bio. Rev. 25: 159–187.

Marshall, E. K., Jr. 1934. The comparative physiology of the kidney in relation to theories of renal secretion. Physiol. Rev. 14: 133–159.

Sadleir, R. M. F. S. 1973. The reproduction of vertebrates. Academic Press, New York. 227 p.

Scheibler, T. H. 1959. Morphologie der Nieren. Handbuch der Zool. Band 8, Lieferung 21: 4(7): 1–84. Detailed review of gross and fine structure of mammalian kidneys with extensive bibliography.

Smith, H. W. 1932. Water regulation and its evolution in the fishes. Quart. Rev. Biol. 7: 1–26. Explains physiological basis of kidney structure and function.

Smith, H. W. 1953. From fish to philosopher. Little, Brown, Boston. 264 p. Entertaining narrative of vertebrate evolution with emphasis on excretion.

Sperber, I. 1944. Studies on the mammalian kidney. Zool. Bidrag från Uppsala 22: 249–431. Excellent interpretation of variation in gross and subgross structure.

Witschi, E. 1948. Migration of the germ cells of human embryos from the yolk sac to the primitive gonadal folds. Contrib. to Embriol. 32: 67–80.

**Chapters 14 & 15
Nervous System**

Aronson, L. R., and H. Kaplan. 1968. Function of the teleostean fore-brain, p. 107–125. *In* Ingle (ed.), The central nervous system and fish behavior. University of Chicago Press, Chicago.

Eccles, J. C. 1969. The development of the cerebellum of vertebrates in relation to the control of movement. Naturwissenschaften 56: 525–534.

Everett, N. B. 1971. Functional neuroanatomy. 6th ed. Lea & Febiger, Philadelphia. 357 p. Human neuroanatomy, concise.

Gardner, E. 1963. Fundamentals of neurology. Saunders, Philadelphia. 349 p. Short and simple text on human neuroanatomy.

Hassler, R., and H. Stephan (eds.). 1967. Evolution of the forebrain. Plenum, New York. 464 p. Forty-two technical papers from an international symposium. Useful for advanced students in the absence of any recent summaries on comparative neurology.

House, E. L., and B. Pansky. 1967. A functional approach to neuroanatomy. McGraw-Hill, New York. 550 p. Terse, well-illustrated, fairly detailed treatment of human neuroanatomy.

Kappers, C. U. A., G. C. Huber, and E. C. Cosby. 1936. The comparative anatomy of the nervous system of vertebrates, including man. (Reprinted, 1960, Hafner, New York.) 3 vols. Technical and detailed, but still valuable, especially on vertebrates other than man. Each section concludes with a resume.

Luria, A. R. 1970. The functional organization of the brain. Sci. Am. 222(3): 66–78, March.

Nieuwenhuys, R. 1962. Trends in the evolution of the actinopterygian forebrain. J. Morphol. 111: 69–88.

Petras, J. M., and C. R. Noback (consulting eds.). 1969. Comparative and evolutionary aspects of the vertebrate central nervous system. New York Acad. Sci., Ann. 167, Art. 1. 513 p. Thirty-three technical papers presented at a conference; useful for advanced students.

Truex, R. C., and M. B. Carpenter. 1969. Human neuroanatomy. 6th ed. Williams & Wilkins, Baltimore. 673 p. Thorough; good illustrations.

Vastola, E. F. 1968. Localization of visual functions in the mammalian brain; a review. Brain, Behavior & Evolution 1: 420–471.

Woodburne, L. S. 1967. The neural basis of behavior. Merrill, Columbus, Ohio. 378 p. Provides a good background in human neuroanatomy.

**Chapter 16
Sense Organs**

Allison, A. C. 1953. The morphology of the olfactory system in vertebrates. Biol. Rev. 28: 195–244. Excellent review article.

Am. Soc. Zool., symposium. 1966. The vertebrate ear. Am. Zool. 6: 368–466.

Cahn, P. H. (ed.). 1967. Lateral line detectors. Indiana University Press, Bloomington. 512 p. Research reports by many authors.

Cold Spring Harbor symposium. 1966. Sensory Receptors. Cold Spring Harbor Laboratory of Quantitative Biology, New York. Symposia on quantitative biology No. 30. 649 p. Contains 56 technical papers on a wide range of subjects.

Detwiler, S. R. 1956. The eye and its structural adaptations. Am. Scientist 44: 45–72.

Dijkgraaf, S. 1963. The functioning and significance of the lateral-line organs. Biol. Rev. 38: 51–105.

Eakin, R. M. 1970. A third eye. Am. Scientist 58: 73–80.

Kelly, D. E. 1962. Pineal organs: photoreception, secretion, and development. Am. Scientist 50: 597–625.

Walls, G. L. 1967. The vertebrate eye and its adaptive radiation. Haffner, New York. 785 p. (Originally published 1942.) Still much the best reference on the comparative morphology of the eye.

Chapter 17 Endocrine Glands

Bern, H. A. 1967. Hormones and endocrine glands of fishes. Science 158: 455–462.

Gorbman, A. (ed.). 1967. Recent developments in endocrinology. Am. Zool. 7: 81–169.

Gorbman, A., and H. A. Bern. 1962. A textbook of comparative endocrinology. John Wiley & Sons, New York. 468 p.

Matty, A. J. 1966. Endocrine glands in lower vertebrates, v. 2: 44–138. *In* Intern. Rev. Gen. Exptl. Zool., Academic Press, New York.

Quay, W. B. 1969. The significance of the pineal. Soc. Endocrinology, Mem. 18: 423–445.

Turner, C. D., and J. T. Bagnara. 1971. General endocrinology. 5th ed. Saunders, Philadelphia. 659 p.

Part III General References

Alexander, R. McN. 1967. Functional design in fishes. Hutchinson, London. 160 p. Excellent; discusses swimming, buoyancy, respiration, feeding, and sense organs.

Alexander, R. McN. 1968. Animal mechanics. University of Washington Press, Seattle. 346 p. Far-ranging application of mechanical principles to animal functions.

Gray, J. 1953. How animals move. Cambridge University Press, Cambridge. 114 p. Shorter and more elementary than next reference.

Gray, J. 1968. Animal locomotion. Norton, New York. 479 p. Application of mechanics to locomotion. Excellent source on fishes, amphibians, snakes, and birds. Extensive bibliography.

Dyson, G. H. G. 1964. Mechanics of athletics. University of London Press, London. 203 p.

Hertel, H. 1963. Structure, form, and movement. Otto Krausskopf-Verlag, Germany. English edition, 1966, Reinhold, New York. 251 p. An engineer's analysis of body mechanics. Emphasis on swimming and flying.

Smith, J. M., and R. J. C. Savage. 1956. Some locomotory adaptations in mammals. J. Linnean Soc., Zool. 42: 603–622. Study of torque and gait in analysis of structure.

Thompson, D'A. W. 1961. On growth and form. Cambridge University Press, Cambridge. 346 p. Updated and shorter edition of a classic first published in 1917.

Walker, E. P. 1968. Mammals of the world. 2nd ed. John Hopkins, Baltimore. 2 v. All genera are described and illustrated by photographs. Invaluable for visualizing body form.

**Chapter 18
Structural
Elements
of the Body**

Bassett, C. A. 1965. Electrical effects in bone. Sci. Am. 213(4); 18–25, Oct.

Bock, W. J., and B. Kummer. 1968. The avian mandible as a structural girder. J. Biomechanics 1: 89–96.

Currey, J. D. 1962. Stress concentrations in bone. Quart. J. Microscopical Sci. 103: 111–133. Analysis of the affect of lacunae, pits, and projections on the strength of bone.

Elliott, D. H. 1965. Structure and function of mammalian tendon. Biol. Rev. 40: 392–421. Review article with extensive bibliography.

Enlow, D. H. 1968. The human face; an account of the postnatal growth and development of the craniofacial skeleton. Harper & Row, New York. 303 p. Includes excellent accounts of the nature of cartilage and bone and skeletal morphogenesis.

Evans, F. G. 1957. Stress and strain in bones. Thomas, Springfield, Ill. 243 p.

Frost, H. M. 1967. An introduction to biomechanics. Thomas, Springfield, Ill. 151 p. Good summary of properties of structural tissues, but does not include bone-muscle systems.

Gardner, E. 1950. Physiology of movable joints. Physiol. Rev. 30: 127–176. Review article with extensive bibliography.

Hall, M. C. 1966. The architecture of bone. Thomas, Springfield, Ill. 346 p. Illustrates bone sections showing internal structure.

Kummer, B. 1966. Photolastic studies on the functional structure of bone. Folia Biotheoretica 6: 31–40.

Vis, J. H. 1957. Histological investigations into the attachment of tendons and ligaments to the mammalian skeleton. Koninkl. Ned. Akad. van Wetenschappen, Proc. Ser. C, 60: 147–157.

Whiting, H. P. 1961. The pelvic girdle in amphibian locomotion. Zool. Soc. London, Symposium No. 5: 43–57. Analysis of ilio-sacral joint of anurans.

**Chapter 19
Mechanics
of Support
and Movement**

Blackwood, O. H., W. C. Kelly, and R. M. Bell. 1963. General physics. 3rd ed. John Wiley & Sons, New York. 685 p. First 151 pages provide a concise introduction to the mechanics relevant to Chapter 19 of this book.

Gans, C., and W. J. Bock. 1965. The functional significance of muscle architecture—a theoretical analysis. Ergebnisse der Anatomie und Entwicklungsgeschichte 38, IV: 116–142.

Gregory, W. K. 1912. Notes on the principles of quadrupedal locomotion and on the mechanism of the limbs of hoofed animals. New York Acad. Sci., Ann. 22: 267–294. Excellent for graviportal adaptations.

Hill, A. V. 1950. The dimensions of animals and their muscular dynamics. Sci. Prog. 38: 209–230. Force, velocity, power, and efficiency are related to load and size.

Kummer, B. 1959. Bauprinzipien des Säugerskeletes. Georg Thieme Verlag, Stuttgart. 235 p. Mechanics of support of the static skeleton with emphasis on the spine and femur.

Osborn, H. F. 1929. The titanotheres of ancient Wyoming, Dakota, and Nebraska. U.S. Geological Survey, Monograph 55, v. 2: 703–945. See Chapter IX for an analysis of graviportal adaptations. Many illustrations.

Slijper, E. J. 1946. Comparative biologic-anatomical investigations on the vertebral column and spinal musculature of mammals. Akad. van Wetenschappen, Afd. Natuurkunde, Tweede sectië, 42(5), 128 p.

Tricker, R. A. R., and B. J. K. Tricker. 1967. The science of movement. American Elsevier, New York. 284 p. Simple presentation of principles of mechanics as applied to athletics.

Tucker, R. 1954. Studies in functional and analytical craniology. Australian J. Zool. 2: 381–430. Structure of skull related to stresses produced in chewing.

Williams, M., and H. R. Lissner. 1962. Biomechanics of human motion. Saunders, Philadelphia. 147 p. Good application of mechanical principles to physical therapy.

**Chapter 20
Running and
Jumping**

Camp, C. L., and N. Smith. 1942. Phylogeny and function of the digital ligaments of the horse. University of California, Mem. 13: 69–124.

Eaton, T. H., Jr. 1944. Modifications of the shoulder girdle related to reach and stride in mammals. J. Morphol. 75: 167–171.

Hall-Craggs, E. C. B. 1965. An analysis of the jump of the lesser galago (*Galago senegalensis*). Zool. Soc. London, Proc. 147: 20–29.

Hatt, R. T. 1932. The vertebral columns of ricochetal rodents. Am. Mus. Nat. Hist., Bull. 63, Article 6: 599–738. Extensive survey with functional interpretation.

Hildebrand, M. 1965. Symmetrical gaits of horses. Science 150(3697): 701–708.

Hildebrand, M. 1966. Analysis of symmetrical gaits of tetrapods. Folia Biotheoretica 6: 9–22.

Howell, A. B. 1965. Speed in animals. Their specializations for running and leaping. Hafner, New York. 270 p. (Originally published in 1944.)

Manter, J. T. 1938. The dynamics of quadrupedal walking. J. Expl. Biol. 15: 522–540. Exemplary experimental analysis of the walking of a cat.

Snyder, R. C. 1962. Adaptations for bipedal locomotion of lizards. Am. Zool. 2: 191–203.

**Chapter 21
Digging; and
Locomotion
without
Appendages**

Agrawal, V. C. 1967. Skull adaptations in fossorial rodents. Mammalia 31: 300–312.

Campbell, B. 1939. The shoulder anatomy of the moles; a study in phylogeny and adaptation. Am. J. Anat. 64: 1–39.

Chapman, R. N. 1919. A study of the correlation of the pelvic structure and the habits of certain burrowing mammals. Am. J. Anat. 25: 185–219.

Ellerman, J. R. 1959. The subterranean mammals of the world. Roy. Soc. of South Africa., Trans. 35: 11–20.

Gans, C. 1968. Relative success of divergent pathways in amphisbaenian specialization. Am. Naturalist 102: 345–362.

Gans, C. 1970. How snakes move. Sci. Am. 222(6): 82–86, 88, 93–96, June.

Hisaw, F. L. 1923. Observations on the burrowing habits of moles (Scapanus aquaticus machrinoides). J. Mammalogy 4: 79–88.

Lehmann, W. H. 1963. The forelimb architecture of some fossorial rodents. J. Morphol. 113: 59–76.

Reed, C. A. 1951. Locomotion and appendicular anatomy in three soricoid insectivores. Am. Midland Naturalist 45: 513–671.

Reed, C. A., and W. D. Turnbull. 1965. The mammalian genera Arctoryctes and Cryptoryctes. Fieldiana: Geology 15: 99–170. Description of the forelimb of extinct mole-like mammals.

Shimer, H. W. 1903. Adaptations to aquatic, arboreal, fossorial and cursorial habits in mammals. III. Fossorial adaptations. Am. Naturalist 37: 819–825.

Yalden, D. W. 1966. The anatomy of mole locomotion. J. Zool., London, 149: 55–64.

Erikson, G. E. 1963. Brachiation in New World monkeys and in anthropoid apes. Zool. Soc. London, Symposium No. 10: 135–164. Reviews morphological adaptations.

Hiller, U. 1968. Untersuchungen zum Feinbau und zur Funktion der Haftborsten von Reptilien. Z. Morphol. Tiere 62: 307–362. Some publications in English have equally fine illustrations, but this study put dry adhesion climbing of reptiles on a sound physical basis.

Jones, F. W. 1953. Some readaptations of the mammalian pes in response to arboreal habits. Zool. Soc. London, Proc. 123: 33–41. Adaptations in porcupines and anteaters.

Miller, R. A. 1943. Functional and morphological adaptations in the forelimbs of the slow lemurs. Am. J. Anat. 73: 153–183.

Napier, J. R., and A. C. Walker. 1967. Vertical clinging and leaping—a newly recognized category of locomotor behavior of primates. Folia Primatologica 6: 204–219.

Noble, G. K., and M. E. Jaeckle. 1928. The digital pads of the tree frogs; a study of the phylogenesis of an adaptive structure. J. Morphol. Physiol. 45: 259–292.

Oxnard, C. E. 1968. The architecture of the shoulder in some mammals. J. Morphol. 126: 249–290. Canonical analysis distinguishes the scapula in climbers from that in nonclimbers.

Richardson, F. 1942. Adaptive modifications for tree-trunk foraging in birds. University of California Publ. Zool. 46: 317–368.

Spring, L. W. 1965. Climbing and pecking adaptations in some North American woodpeckers. Condor 67: 457–488.

**Chapter 22
Climbing**

Affleck, R. J. 1950. Some points in the function, development and evolution of the tail in fishes. Zool. Soc. London, Proc. 120: 349–368.

Anderson, H. T. 1966. Physiological adaptations in diving vertebrates. Physiol. Rev. 46: 212–243.

Brett, J. R. 1965. The swimming energetics of salmon. Sci. Am. 213(2): 80–86, Aug.

**Chapter 23
Swimming
and Diving**

Harrison, R. J., and J. D. W. Tomlinson. 1963. Anatomical and physiological adaptations in diving mammals, p. 115–162. *In* J. D. Carthy and C. L. Duddington (eds.), Viewpoints in biology 2. Butterworths, London.

Howell, A. B. 1929. Contributions to the comparative anatomy of the eared and earless seals (genera *Zalophus* and *Phoca*). U.S. Nat. Museum, Proc. 73: 1–142. Descriptive and interpretive.

Howell, A. B. 1970. Aquatic mammals. Dover, New York. 338 p. (Originally published in 1930.) Broad and basic; more interpretive than most anatomical monographs of the period.

Kramer, M. O. 1965. Hydrodynamics of the dolphin. Advances in Hydroscience 2: 111–130. Dolphin performance is analyzed and compared to products of human technology.

Lang, T. G. 1966. Hydrodynamic analysis of cetacean performance, p. 410–432. *In* K. S. Norris (ed.), Whales, dolphins, and porpoises. University of California Press, Berkeley.

Lawrence, B., and W. E. Schevill. 1956. The functional anatomy of the delphinid nose. Harvard University, Museum Comp. Zool., Bull. 114: 103–151.

Nursall, J. R. 1956. The lateral musculature and the swimming of fish. Zool. Soc. London, Proc. 126: 127–143.

Storer, R. W. 1960. Evolution in the diving birds. Proc. XIIth Intern. Ornithological Congr., Helsinki 1958: 694–707.

Thomson, K. S. 1971. The adaptation and evolution of early fishes. Quart. Rev. Biol. 46: 139–166. Includes functional interpretations of the shape of body and tail.

Walker, W. F., Jr. 1971. Swimming in sea turtles of the family Cheloniidae. Copeia 1971: 229–233.

Walters, V. 1962. Body form and swimming performance in the scombroid fishes. Am. Zool. 2: 143–149.

Willemse, J. J. 1966. Functional anatomy of the myosepta in fishes. Proc. Akad. Wetenschappen, Ser. C 29: 58–63.

Chapter 24
Flying and
Gliding

Brown, R. H. J. 1953. The flight of birds, II. Wing Function in relation to flight speed. J. Exptl. Biol. 30: 90–103.

Brown, R. H. J. 1963. The flight of birds. Biol. Rev. 38: 460–489. A later paper by a leading student of bird flight.

Cone, C. D., Jr. 1962. Thermal soaring of birds. Amer. Sci. 50: 180–209.

Cone, C. D., Jr. 1964. A mathematical analysis of the dynamic soaring of the albatross, with ecological interpretations. Virginia Inst. of Marine Sci., Gloucester Point, Virginia, Special Sci. Rept. 50. 104 p.

Cottam, C., C. S. Williams, and C. A. Sooter. 1942. Flight and running speeds of birds. Wilson Bull. 54: 121–131.

Leen, N., and A. Novik. 1969. The world of bats. Holt, Rinehart, and Winston, New York. 171 p. Outstanding photographs.

Lissaman, P. B. S., and C. A. Shollenberger. 1970. Formation flight of birds. Science 168: 1003–1005. May 22.

Norberg, U. M. 1970. Functional osteology and myology of the wing of *Plecotus auritus* Linnaeus (Chiroptera). Arkiv Zool. (Stockholm), Ser. 2, 22: 483–543.

Norberg, U. M. 1972. Bat wing structures important for aerodynamics and rigidity (Mammalia, Chiroptera). Z. Morphol. Tiere 73: 45–61.

Raspet, A. 1960. Biophysics of bird flight. Smithsonian Inst., Ann. Rept. 1960: 405–425. An analysis of static soaring by a vulture.

Savile, D. B. O. 1957. Adaptive evolution of the avian wing. Evolution 11: 212–224.

Schmidt, H. 1960. Der Flug der Tiere. Verlag Woldemar Kramer, Frankfurt. 164 p. Good survey of parachuting, gliding, and flying in all animals with many illustrations; nontechnical.

Urry, D., and K. Urry. 1969. Flying birds. Harper & Row, Scranton, Pa. 192 p. Excellent photographs.

Vaughan, T. A. 1959. Functional morphology of three bats: Eumops, Myotis, Macrotus. University of Kansas Publ. Nat. Hist. 12: 1–153.

Welty, J. C. 1962. The life of birds. Saunders, Philadelphia 546 p. Includes anatomical material and a chapter on flight.

Chapter 25
Feeding

Alexander, R. McN. 1966. The functions and mechanisms of the protrusible upper jaws of two species of cyprinid fish. J. Zool., London 149: 288–296.

Am. Soc. Zool., symposium. 1961. Evolution and dynamics of vertebrate feeding mechanisms. Am Zool. 1: 177–234. Six articles.

Becht, G. 1953. Comparative biologic-anatomical researches on mastication in some mammals. Koninkl. Nederl. Akad. van Wetenschappen, Proc. 56: 508–527.

Böker, H. 1937. Einführung in die vergleichende biologische Anatomie der Wirbeltiere. v. 2, 258 p. Fischer, Jena. An important early contribution to functional morphology in spite of outmoded concepts of evolution; profusely illustrated.

Frazzetta, T. H. 1962. A functional consideration of cranial kinesis in lizards. J. Morphol. 111: 287–319.

Frazzetta, T. H. 1966. Studies on the morphology and function of the skull in the Boidae (Serpentes). Part II. Morphology and function of the jaw apparatus in *Python sebae* and *Python molurus*. J. Morphol. 118: 217–296. Fine functional analysis.

Gans, C. 1952. The functional morphology of the egg-eating adaptations in the snake genus *Dasypeltis*. Zoologica 37: 209–244.

Gans, C. 1961. A bullfrog and its prey. Nat. Hist. 70: 26–37.

Gans, C. 1967. The chameleon. Nat. Hist. 76: 53–59. Analyses the unique feeding mechanism.

Jenkin, P. M. 1957. The filter-feeding and food of flamingoes (Phoenicopteri). Roy. Soc. London, Philosophical Trans. Ser. B., No. 674, Vol. 240: 401–493.

Liem, K. F. 1970. Comparative functional anatomy of the Nandidae (Pisces: Teleostei). Fieldiana: Zool. 56: 1–166. Analyses mechanism for swallowing entire large prey.

Luling, K. H. 1963. The archer fish. Sci. Am. 209(1): 100–104, 106, 108, July.

Smith, J. M., and R. J. G. Savage. 1959. The mechanics of mammalian jaws. School Sci. Rev. 141: 289–301.

Zusi, R. 1967. The role of the depressor mandibulae muscle in kinesis of the avian skull. U.S. Natl. Museum, Proc. 123: 1–28.

Index

Numbers in boldface indicate pages where the entry is illustrated.

681

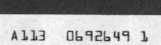